FROSTING AND ICING FOR EFFICIENT ENERGY USE IN ENGINEERING APPLICATIONS

FROSTING AND ICING FOR EFFICIENT ENERGY USE IN ENGINEERING APPLICATIONS

Edited by

MENGJIE SONG

Department of Energy and Power Engineering, School of Mechanical Engineering, Beijing Institute of Technology, Beijing, P.R. China

LONG ZHANG

Department of Energy and Power Engineering, School of Mechanical Engineering, Beijing Institute of Technology, Beijing, P.R. China

ELSEVIER

Elsevier
Radarweg 29, PO Box 211, 1000 AE Amsterdam, Netherlands
125 London Wall, London EC2Y 5AS, United Kingdom
50 Hampshire Street, 5th Floor, Cambridge, MA 02139, United States

ISBN: 978-0-443-15495-9

For Information on all Elsevier publications
visit our website at https://www.elsevier.com/books-and-journals

Publisher: Megan Ball
Acquisitions Editor: Fran Kennedy-Ellis
Editorial Project Manager: Joshua Mearns
Production Project Manager: Sruthi Satheesh
Cover Designer: Miles Hitchen

Typeset by MPS Limited, Chennai, India

Contents

6. Simulation of the impact and icing processes of water droplets on spheres

Xuan Zhang, Xin Liu, Jingchun Min, Yiqing Guo, Xiaomin Wu and Yunda Fei

7. Growth distribution characteristics of micro-scale trapped air bubbles in ice and their impact on the freezing process

Keke Shao, Mengjie Song, Long Zhang, Xuan Zhang, Zekang Zhen, Sirui Yu, Shiqiang Zhou and Runmiao Gao

8. Research on mechanism and measurement technology of frost formation of micro water frosting

Xiaoqin Zhi, Yinghe Qi, Wenhan Shu and Limin Qiu

9. Frost formation and growth on cold surfaces: a review on experimental studies

Yuan Wang, Xiaofei Yue and Duo Zhang

10. A sustainable defrosting technology for air source heat pump systems

Long Zhang, Longping Wu, Jiatai Jiang, Mengjie Song, Xuan Zhang, Keke Shao, Han Shi and Yiqiang Jiang

11. High-efficiency frost-free air source heat pumps

Pengfei Zhang, Zhihua Wang, Longxia Ma and Fenghao Wang

12. Atomization equipment for simulating icing environment

Huanyu Deng, Mengjie Song, Xiaohong Liu and Xiaofeng Li

13. The thermodynamic evolution of landfast ice in the polar regions

Jiechen Zhao

14. Ice prediction technology and applications for typical stationary and moving surfaces

Sirui Yu, Mengjie Song, Long Zhang, Xuan Zhang, Zekang Zhen, Keke Shao, Mohasan Muhammad and Lizhen Huang

15. Ultrasonic application on ice detection and removal

Jifeng Zhang, Yuan Tian, Zhonghua Shi, Wen Luo, Huanyu Zhao and Yudong Guo

16. Experimental investigation of thermal deicing characteristics and energy efficiency

Jiankai Dong, Yiqiang Jiang, Jiqin Li and Teng Xie

17. Dynamic ice-melting characteristics of carbon dioxide laser irradiation under natural convection

Zekang Zhen, Mengjie Song, Long Zhang, Xuan Zhang, Keke Shao, Sirui Yu, Han Shi and Runmiao Gao

18. Conclusions and future work

Mengjie Song, Xuan Zhang and Long Zhang

List of contributors

Haoting Cai Shanghai Key Laboratory of Multiphase Flow and Heat Transfer in Power Engineering, School of Energy and Power Engineering, University of Shanghai for Science and Technology, Shanghai, P.R. China

Ping Cheng Shanghai Key Laboratory of Multiphase Flow and Heat Transfer in Power Engineering, School of Energy and Power Engineering, University of Shanghai for Science and Technology, Shanghai, P.R. China; MOE Key Laboratory of Power Machinery and Engineering, School of Mechanical Engineering, Shanghai Jiao Tong University, Shanghai, P.R. China

Huanyu Deng Beijing Institute of Radio Measurement, Beijing, P.R. China

Jiankai Dong School of Architecture and Design, Harbin Institute of Technology, Harbin, Heilongjiang, P.R. China

Yunda Fei Department of Energy and Power Engineering, School of Mechanical Engineering, Beijing Institute of Technology, Beijing, P.R. China

Runmiao Gao Department of Energy and Power Engineering, School of Mechanical Engineering, Beijing Institute of Technology, Beijing, P.R. China

Yubo Gao Department of Energy and Power Engineering, School of Mechanical Engineering, Beijing Institute of Technology, Beijing, P.R. China

Yudong Guo Wuhan Aviation Instrument Co., Ltd., Wuhan, P.R. China

Yiqing Guo Key Laboratory for Thermal Science and Power Engineering of Ministry of Education, Department of Energy and Power Engineering, Tsinghua University, Beijing, P.R. China

Lizhen Huang Department of Energy and Power Engineering, School of Mechanical Engineering, Beijing Institute of Technology, Beijing, P.R. China

Jiatai Jiang Department of Energy and Power Engineering, School of Mechanical Engineering, Beijing Institute of Technology, Beijing, P.R. China

Yiqiang Jiang School of Architecture and Design, Harbin Institute of Technology, Harbin, Heilongjiang, P.R. China; Department of Building and Thermal Engineering, School of Architecture, Harbin Institute of Technology, Harbin, P.R. China

Jiqin Li School of Architecture and Design, Harbin Institute of Technology, Harbin, Heilongjiang, P.R. China

Kailiang Li Department of Energy and Power Engineering, School of Mechanical Engineering, Beijing Institute of Technology, Beijing, P.R. China

Xiaofeng Li Beijing Institute of Radio Measurement, Beijing, P.R. China

Xiaohong Liu Beijing Institute of Radio Measurement, Beijing, P.R. China

Xiaoliang Liu Department of Energy and Power Engineering, School of Mechanical Engineering, Beijing Institute of Technology, Beijing, P.R. China

Xin Liu Chinese Aeronautical Establishment, Beijing, P.R. China

Zhongxin Liu Department of Energy and Power Engineering, School of Mechanical Engineering, Beijing Institute of Technology, Beijing, P.R. China

Wen Luo College of Aerospace and Civil Engineering, Harbin Engineering University, Harbin, P.R. China

Longxia Ma School of Human Settlement and Civil Engineering, Xi'an Jiaotong University, Xi'an, Shaanxi, P.R. China

Jingchun Min Key Laboratory for Thermal Science and Power Engineering of Ministry of Education, Department of Engineering Mechanics, Tsinghua University, Beijing, P.R. China

Mohasan Muhammad Department of Energy and Power Engineering, School of Mechanical Engineering, Beijing Institute of Technology, Beijing, P.R. China

Yinghe Qi Institute of Refrigeration and Low Temperature, Zhejiang University, Hangzhou, P.R. China

Limin Qiu Institute of Refrigeration and Low Temperature, Zhejiang University, Hangzhou, P.R. China

Keke Shao Department of Energy and Power Engineering, School of Mechanical Engineering, Beijing Institute of Technology, Beijing, P.R. China; School of Mechanical Engineering, Hanyang University, Seoul, South Korea

Han Shi Department of Energy and Power Engineering, School of Mechanical Engineering, Beijing Institute of Technology, Beijing, P.R. China

Zhonghua Shi Xi'an Electronic Engineering Research Institute, Xi'an, P.R. China

Wenhan Shu Institute of Refrigeration and Low Temperature, Zhejiang University, Hangzhou, P.R. China

Mengjie Song Department of Energy and Power Engineering, School of Mechanical Engineering, Beijing Institute of Technology, Beijing, P.R. China; School of Mechanical Engineering, Hanyang University, Seoul, South Korea

Yuan Tian College of Aerospace and Civil Engineering, Harbin Engineering University, Harbin, P.R. China

Fenghao Wang School of Human Settlement and Civil Engineering, Xi'an Jiaotong University, Xi'an, Shaanxi, P.R. China

Yuan Wang College of Aerospace Science and Engineering, National University of Defense Technology, Changsha, Hunan, P.R. China

Zhihua Wang School of Human Settlement and Civil Engineering, Xi'an Jiaotong University, Xi'an, Shaanxi, P.R. China

Longping Wu Department of Energy and Power Engineering, School of Mechanical Engineering, Beijing Institute of Technology, Beijing, P.R. China

Xiaomin Wu Key Laboratory for Thermal Science and Power Engineering of Ministry of Education, Department of Energy and Power Engineering, Tsinghua University, Beijing, P.R. China

Teng Xie School of Architecture and Design, Harbin Institute of Technology, Harbin, Heilongjiang, P.R. China

Chun Yang School of Mechanical and Aerospace Engineering, Nanyang Technological University, Singapore

Sirui Yu Department of Energy and Power Engineering, School of Mechanical Engineering, Beijing Institute of Technology, Beijing, P.R. China

Xiaofei Yue College of Aerospace Science and Engineering, National University of Defense Technology, Changsha, Hunan, P.R. China

Chaoyang Zhang Paris Elite Institute of Technology, Shanghai Jiao Tong University, Shanghai, P.R. China

Duo Zhang College of Aerospace Science and Engineering, National University of Defense Technology, Changsha, Hunan, P.R. China

Hui Zhang School of Mechanical and Aerospace Engineering, Nanyang Technological University, Singapore

Jifeng Zhang College of Aerospace and Civil Engineering, Harbin Engineering University, Harbin, P.R. China

Long Zhang Department of Energy and Power Engineering, School of Mechanical Engineering, Beijing Institute of Technology, Beijing, P.R. China

Pengfei Zhang School of Human Settlement and Civil Engineering, Xi'an Jiaotong University, Xi'an, Shaanxi, P.R. China

Xuan Zhang Department of Energy and Power Engineering, School of Mechanical Engineering, Beijing Institute of Technology, Beijing, P.R. China

Huanyu Zhao Liaoning Provincial Key Laboratory of Aircraft Ice Protection, AVIC Aerodynamics Research Institute, Shenyang, P.R. China

Jiechen Zhao Qingdao Innovation and Development Base (Centre) of Harbin Engineering University, Qingdao, Shandong Province, P.R. China; Key Laboratory for Polar Acoustics and Application of Ministry of Education (Harbin Engineering University), Harbin, Heilongjiang Province, P.R. China

Yugang Zhao Shanghai Key Laboratory of Multiphase Flow and Heat Transfer in Power Engineering, School of Energy and Power Engineering, University of Shanghai for Science and Technology, Shanghai, P.R. China

Zekang Zhen Department of Energy and Power Engineering, School of Mechanical Engineering, Beijing Institute of Technology, Beijing, P.R. China

Xiaoqin Zhi Institute of Refrigeration and Low Temperature, Zhejiang University, Hangzhou, P.R. China

Shiqiang Zhou Department of Energy and Power Engineering, School of Mechanical Engineering, Beijing Institute of Technology, Beijing, P.R. China

CHAPTER

1

Introduction of frosting and icing

Mengjie Song

Department of Energy and Power Engineering, School of Mechanical Engineering, Beijing Institute of Technology, Beijing, P.R. China

1.1 Frosting and icing phenomenon

Frosting on the surfaces of plant leaves is a well-known phenomenon in nature. This phenomenon was recorded in *The Reed*, which is included in the collection of ancient Chinese poetry, *Classic of Poetry*, wrote in about 2700 years ago. As described in the second verse of this love poem, *"Dew and frost gleam,"* the dew on the surface of reed leaves changed into frost, with the short nucleation process of a droplet clearly described. As mentioned in the fourth verse, *"Beyond the stream,"* the reed was located on the riverside, where the air had a high relative humidity (RH). The poem may be set on an autumn day at dusk when the air temperature was low. Clearly, the following three conditions, (1) cold surface, (2) high RH, and (3) low air temperature, are prerequisites for the frost phenomenon. Therefore, we may easily understand that the frosting phenomenon is commonly observed in refrigeration and air-conditioning fields, and it occurs because of their operating conditions [1].

The Reed

Green, green the reed, dew and frost gleam.
Where's she I miss? Beyond the steam.
Upstream I go, the way is long.
Downstream I go, she's there among.

Frost crystals deposited on these heat transfer surfaces act as fins [2], which may improve the heat transfer between the surfaces and the air during the early initial period of frost formation. However, after significant frost accumulation and after a continuous frost layer has been formed, this frost layer will become an insulation and will increase the heat transfer resistance owing to its porous characteristics; therefore, the heat transfer is severely degenerated. As previously mentioned, if the frost accumulation is not removed on time or without effective control, then as the frost layer thickness increases, the frost that has formed on the

Frosting and Icing for Efficient Energy Use in Engineering Applications
DOI: https://doi.org/10.1016/B978-0-443-15495-9.00018-6

fins of the heat exchangers blocks the air flow passage; this leads to the decrease in the coefficient of performance (COP) and may even cause system malfunction [3]. It has been reported that there is as much as a 50%–75% decrease in heat transfer and a substantial increase in the pressure drop, which resulted from the frost deposition on a compact heat exchanger in an air source heat pump (ASHP) unit [4]. Frost on the surface of a spark plug of a hydrogen engine will cause ignition failure and cannot start.

Harmful frosting phenomena mainly occur in various industry processes that occur under ordinary low-temperature conditions [5,6]. However, owing to the increased practical and engineering applications, frost formation on surfaces of very low temperatures has been receiving increasing attention from both industries and scientific researchers. For example, in the liquefied natural gas (LNG) field, passive-evaporation technology uses air as a heat source to heat the LNG; thus, frost may be deposited on surfaces of LNG evaporators that have very low temperatures [7,8]. When a cryogenic tank is exposed to humid atmospheric air, frost formation may have positive effects because the frost that would form on the cold tank surface would act as insulation, thus mitigating heat transfer into the tank [9]. In the field of aerospace, regarding the airplanes that fly during the night, frost deposits may occur on their wings owing to the low air temperature [10]. The −112°C surfaces of liquid-oxygen feed-line brackets were studied by Ferrick et al. [11], on which frost deposition may appear during the prelaunch period following the fuel loading of space shuttles. On the surface of the precooler in a gas turbine, due to the cold source of cryogenic fluids before the air enters the compressors, frost deposition and block of airflow passage may occur [10]. Similar frosting problems may occur and deposit on the launching rocket's surface, thus causing communication satellites to fail to enter a correct synchronous orbit [9].

To clearly understand the frosting mechanism, frost formation phenomena on a flat plate surface are investigated and discussed widely [12]. There are two types of frosting due to the temperature of the cold surface, condensation frosting and desublimation frosting. In the open literatures, the entire condensation frosting process, on an ordinary low-temperature surface, may be divided into four periods, namely (1) the droplet condensation period, (2) the solidified liquid tip-growth period, (3) the frost layer growth period, and (4) the frost layer full growth period. This classification renders the frosting process more easily understood. Clearly, only the first droplet condensation period is at the liquid status, and the later three periods are at the solid status. In the second period, the thickness of the frost layer increases a lot, and the later two periods increase its density. Between the (3) and (4) periods, there are periodic reverse melting and regrowing of ice crystals, which directly reflects the increase of frost density. An entire desublimation frosting process, on an ultra-low temperature surface, does not contain the liquid status of water. Frost is formed from water vapor directly.

As we know, frost is just a low-density type of ice. Ice is very common in nature and industry and often plays a negative role. For example, the ice blockage in the throttling expansion channel of small cryogenic chillers due to a small amount of water vapor solidification will cause safety hazards and function failures. The icing of wind turbine blades will reduce power generation, cause mechanical failure, and endanger human and equipment safety. Icing on transmission lines can lead to major accidents such as short circuits, insulator flashover, and short line toppling. Ice on aircraft surfaces and runways is likely to cause accidents. In the field of transportation, icy roads/bridge decks will cause car accidents due to tire sliding. Freezing of bogies and pantographs of trains in cold weather seriously threatens the safety of high-speed rail. In the polar navigation field, ice in the channel will hinder the passage of ships. Sea ice will endanger the safety of ships. Icing on the surface of the ship will increase the weight of the ship, form surface obstacles on the ship, cause electrical interference, threaten the safety of navigation and operation personnel, and even delay the aircraft. In the field of coal mines, coal mine icing will cause material adhesion, and icing at the mine inlet will create a safety hazard. In electronic communications, frost, ice, and snow on radar and radome surfaces can interfere with radio reception. In the field of food refrigeration and biomedicine, the freezing of raw ingredients and organ tissues will lose nutrients or lose activity due to the destruction of cell walls during the freeze-thaw process.

But frost and ice are not all hazards, and they can play positive effects in some fields. For example, natural ice can be used to store food materials or relieve heat. Artificial ice and snow sports venues, artificial ice and snow landscapes, and edible ice can enrich daily cultural and sports life. Ice storage in the air conditioning system can reduce peak and fill valleys in power consumption. In the field of climate and environment, the content and proportion of gases in microbubbles in polar ice cores can be used to decode the evolution history of the earth's climate and environment [13], and the frost layer in the cryogenic liquid working medium in the aerospace field can also optimize the vaporization and transportation of cryogenic working medium [14].

1.2 Frost and ice investigations

In view of the extensive influence of frost and ice in the above fields, to fundamentally understand frost and ice and then accurately predict and control frost and ice growth, the research on frost and ice mechanism, as a typical interdisciplinary problem, has gradually become a hot problem that needs to be solved in the field of refrigeration and cryogenic science and even engineering thermophysics in recent years. Fig. 1.1A and B, respectively,

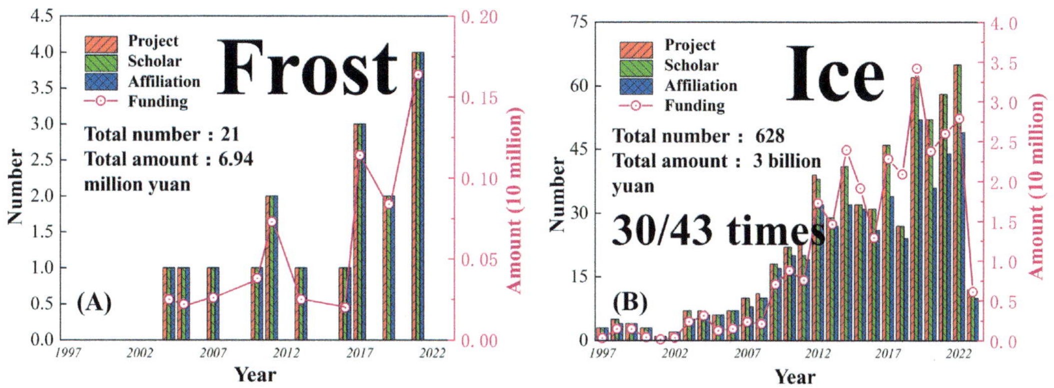

FIGURE 1.1 Frost and ice projects from the NSFC (1997–2022, 26 years in total). (A) Frost projects; (B) Ice projects.

sum up the number of projects, the number of scholars, the number of units, and the amount of funding for the "frost" and "ice" projects initiated by the National Natural Science Foundation of China (NSFC) from 1997 to 2022. The following conclusions can be found. (1) The number of research projects on frost and ice is increasing. (2) The volume of frost research is much smaller than that of ice research, with the total number of projects being 21 and 628, respectively, a difference of 30 times. The total amount of the project is 6.94 million yuan and 295 million yuan, respectively, a difference of 43 times. (3) Although the total amount of the frost and ice project is more than 300 million yuan, it is far less than the amount of funding for research topics such as batteries, hydrogen energy, and working media. (4) Frost-related projects are mainly within the discipline of refrigeration and cryogenic engineering, but ice-related projects involve aerospace, ship traffic, meteorological physics, and other fields, including "antiice, deicing, antiice, ice control, ice accumulation, ice covering" and "ice, ice body, ice conditions, ice, ice core, ice crystal" and other engineering ice and natural ice keywords.

Although it is difficult to find records of earlier frosting and icing in foreign countries, as shown in Fig. 1.2, from the identification of refrigeration industry organizations in many countries around the world with hexagonal ice crystal snowflake elements, it can be seen that frosting and icing are at least one of the representatives of human initial understanding of refrigeration. Second, the Japan Refrigeration and Air Conditioning Society, established in 1922, has a long-standing "Frosting and Defrosting Technical Committee." It was listed as one of the nine major technical committees along with "Food, Working Fluid, Compressor, Heat Exchanger, New Refrigeration System, Desiccant Adsorption and

FIGURE 1.2 Frosting and icing logos of many refrigeration organizations worldwide.

Absorption, Natural Working Fluid Air Conditioning System, and Solid Liquid Phase Change Heat Storage." Third, Professor Richard Errett Smalley, a Nobel laureate from Rice University, included frost and ice challenges among his list of humanity's top 10 problems for the next 50 years. The top two are energy and water, and the issue of frost and ice is closely related to both. With the common occurrence of extremely low-temperature weather in recent years and the continuous increase in Arctic shipping capacity, basic research on frost and ice is of great significance to humans from ancient and modern times, both at home and abroad.

1.3 Objectives, scopes, and outline

Frost and ice accumulated on the cold surface directly affect its thermal, mechanical, and optical properties, etc. For example, the frost layer could act as thermal resistance for the heat exchanger, and the ice layer on the surface of airplane wing deteriorates its flight capability. It is easy to understand that frost and ice are unevenly distributed on the surface at different densities. Therefore, to accurately predict and control the frosting and icing process, a series of investigations have been carried out, and the study results will be given in this book. The objectives of the research work and the scopes presented in this book are as follows.

Chapter 1 introduces the background, the process, period, and characteristics of frosting and icing, states the motivation to develop new methods or technologies to solve the frosting and icing problem, and elaborates the objectives and scopes of the research work presented, as well as the outline of this book.

Chapter 2 develops a theoretical model of the icing processes of sessile water droplets on the plate surface and spherical surface, as well as the icing process of pendant droplets on the plate surface, with the effect of gravity considered. Based on the model, the icing characteristics are studied under different droplet volumes, contact angles, and surface temperatures.

Chapter 3 introduces the Lattice Boltzmann simulation of pure and nanofluid droplet icing process with dynamical interface capturing to elucidate the underlying principles of heat and mass transfer during droplet freezing. The phenomena of freezing front fragmentation, changes in droplet interface, and motion of immersed particles can be captured in these models.

Chapter 4 reports the freezing characteristics and mechanism of a multicomponent droplet, containing components such as dissolved gases, ions, other liquids, large organic molecules, and solid particles. The multicomponents cause a nonnegligible change of the fluid properties, altering icing dynamics and even engendering different shapes of the frozen form.

Chapter 5 investigates the mechanism of impact and icing of water droplets on flat surfaces. The spreading and height factors of impact droplets at room temperature will be obtained from experiments, with the influence of Weber number and contact angle studied. Based on the inherent morphology of water droplets at the onset of impact, a theoretical model for the maximum spreading factor of ellipsoidal water droplets under various initial aspect ratios will be established.

Chapter 6 introduces the simulation results of the impact and icing processes of water droplets on spheres. On the basis of the Volume of Fluid method applied to capture the multiphase-interface flow, the Dynamic Contact Angle model and the Solidification/

Melting model are coupled to solve the droplet impact and icing characteristics on a curved surface with high precision.

Chapter 7 shows the growth distribution characteristics of micro-scale trapped air bubbles in ice and their impact on the freezing process to accurately predict and control the icing process and then to develop and optimize various antiicing technologies. The nucleation mechanism, growth process, distribution characteristics, and stability of trapped air bubbles will be reviewed from the micro and macro scales.

Chapter 8 introduces the current research on the frost formation of microwaters from the frost formation mechanism and measurement methods. The nucleation mechanism of the microwater, the growth characteristics of the frost layer, and the deposition and sublimation characteristics of the microwater are analyzed. The principles, application scenarios, visual measurement technology, optical measurement method, and low-temperature micromass measurement technology are described.

Chapter 9 shows the review results of experimental investigations on frost formation and growth on cold surfaces. The main influencing parameters of frost accumulation are discussed, where air RH, cold surface temperature, and surface structure are of relatively significant dominance. The effects of topographical texture on frost formation and growth are particularly concentrated.

Chapter 10 introduces a sustainable defrosting technology for air source heat pump units, which has few negative impacts on the heating performances of an ASHP under the normal heating mode. Phase change material is used, and there are no predefrosting and postdefrosting stages in the sustainable defrosting modes because they do not need the reversing operation of a four-way valve and shutdown.

Chapter 11 proposes a novel frost-free air source heat pump system, which is coupled to an extra heat exchanger coated by a solid desiccant with an energy storage device. Furthermore, a control strategy optimization is proposed, and a technical and economic analysis of the proposed system in comparison with the conventional ones is conducted.

Chapter 12 introduces a study of atomization equipment for simulating the icing environment. The arrangement of 10 nozzles in the atomizer was adjusted, and a model of the atomization equipment was independently developed, with the ice shape and the maximum ice thickness on the test rod as validation parameters.

Chapter 13 shows the thermodynamic evolution of landfast ice in the polar regions, based on the annual observations of first-year and second-year landfast ice carried out in the vicinity of Zhongshan Station in East Antarctica between December 2011 and December 2012. The first melting ponds emerged in early December 2011. Following a powerful hurricane in late January 2012, landfast ice partially cracked.

Chapter 14 reviews the ice prediction technology and applications for typical stationary and moving surfaces, roads, power transmission lines, and wind turbine blades and aircraft, respectively. It utilizes both hardware and software to acquire input parameters such as current meteorological data and uses these pieces of information to predict future parameters like ice thickness.

Chapter 15 introduces the ultrasonic application on ice detection and removal. Based on the frequency-domain finite element method, the propagation process of certain ultrasonic guided wave mode-frequency pairs in the icing plate is simulated, the mode conversion of ultrasonic guided waves is quantitatively analyzed, and the ice parameters are accurately estimated.

Chapter 16 reports the experimental investigation of thermal deicing characteristics and energy efficiency. To reduce deicing time and improve efficiency, the impact factors and energy efficiencies of two thermal deicing methods, hot air, and infrared ray, are investigated, with the ideal air temperature and velocity for hot air deicing given.

Chapter 17 presents the results of dynamic ice-melting characteristics of carbon dioxide laser irradiation under natural convection. To accurately evaluate and control the laser-induced ice melting process, the experiments on melting characteristics of clear ice and bubble ice under horizontal and vertical carbon dioxide laser irradiation are carried out with laser powers less than 60 W.

Chapter 18 draws conclusions from the research work presented in this book and recommends possible future work on the development of frosting/icing mechanism study and anti/defrosting/icing technologies.

References

[1] Z.L. Liu, Y.W. Dong, Y.X. Li, An experimental study of frost formation on cryogenic surfaces under natural convection conditions, International Journal of Heat and Mass Transfer 97 (2016) 569–577.

[2] L.Y. Huang, Z.L. Liu, Y.M. Liu, Y.J. Gou, J.T. Wang, Experimental study on frost release on fin-and-tube heat exchangers by use of a novel anti-frost paint, Experimental Thermal and Fluid Science 33 (2009) 1049–1054.

[3] Z.L. Liu, H.Y. Wang, X.H. Zhang, S. Meng, C.F. Ma, An experimental study on minimizing frost deposition on a cold surface under natural convection conditions by use of a novel anti-frosting paint. Part I. Anti-frosting performance and comparison with the uncoated metallic surface, International Journal of Refrigeration 29 (2) (2006) 229–236.

[4] A.F. Emery, B.L. Siegel, Experimental measurements of the effects of frost formation on heat exchanger performance, Proc AIAA/ASME Thermo Phys Heat Transfer Conf 139 (1990) 1–7.

[5] Y. Yao, Y.Q. Jiang, S.M. Deng, Z.L. Ma, A study on the performance of the airside heat exchanger under frosting in an air source heat pump water heater/chiller unit, International Journal of Heat and Mass Transfer 47 (2004) 3745–3756.

[6] V.S.J. Nascimento, F.R. Loyola, C.J.L. Hermes, A study of frost build-up on parallel plate channels, Experimental Thermal and Fluid Science 60 (2015) 328–336.

[7] S.P. Chen, J.L. Lai, J.S. Yin, Experiment on frost characteristic of airwarmed cryogenic finned-tube vaporizer, Journal of Refrigeration 31 (4) (2010) 26–30.

[8] Y.W. Kuang, C.C. Yi, W. Wang, Numerical simulation of frosting behavior and its effect on a direct-contact ambient air vaporizer, Journal of Natural Gas Science and Engineering 27 (2015) 55–63.

[9] M.A. Dietenberger, A frost formation model and its validation under various experimental conditions. NASA Contract Report 3595, NASA Scientific and Technical Information Branch (1982).

[10] M.B. Bragg, D.C. Heinrich, W.O. Valarezo, Effect of under-wing frost on a transport aircraft airfoil at flight Reynolds number, Journal of Aircraft 31 (6) (1994) 1372–1379.

[11] M.G. Ferrick, N.D. Mulherin, R.B. Haehnel, B.A. Coutermarsh, G.D. Durell, T.J. Tantillo, et al., Evaluation of ice release coatings at cryogenic temperature for the space shuttle, Cold Regions Science and Technology 52 (2) (2008) 224–243.

[12] Y. Hayashi, A. Aoki, S. Adachi, K. Hori, Study of frost properties correlating with frost formation types, Journal of Heat Transfer 99SerC (2) (1977) 239–245.

[13] P.D. Neff, A review of the brittle ice zone in polar ice cores, Annals of Glaciology 55 (68) (2014) 72–82.

[14] M.J. Song, C.B. Dang, Review on frost characteristics measurement and calculation, International Journal of Heat and Mass Transfer 124 (2018) 586–614.

CHAPTER

2

A general model for the icing characteristics of water droplets on cold surfaces

Yubo Gao[1], *Xuan Zhang*[1], *Xin Liu*[2], *Mengjie Song*[1], *Long Zhang*[1], *Xiaoliang Liu*[1], *Kailiang Li*[1] *and Jingchun Min*[3]

[1]Department of Energy and Power Engineering, School of Mechanical Engineering, Beijing Institute of Technology, Beijing, P.R. China [2]Chinese Aeronautical Establishment, Beijing, P.R. China [3]Key Laboratory for Thermal Science and Power Engineering of Ministry of Education, Department of Engineering Mechanics, Tsinghua University, Beijing, P.R. China

Nomenclatures

Symbol	Description	Unit
a	Semi-major axis of the ellipse model	m
b	Semi-minor axis of the ellipse model	m
c	Specific heat	kJ/(kg ·°C)
e	Eccentricity of the ellipsoid model	/
g	Gravity acceleration	m/s^2
h	Enthalpy	J
k	Thermal conductivity	W/(m ·°C)
r	Radius	m
t	Time	s
v	Freezing rate	m/s
v_n	Moving velocity of the freezing front	m/s

Frosting and Icing for Efficient Energy Use in Engineering Applications
DOI: https://doi.org/10.1016/B978-0-443-15495-9.00016-2

v_{slip}	Slipping velocity along the freezing front	m/s
x	Radial coordinate	m
y	Height coordinate	m
z	Axial coordinate	m
A	Base area	m^2
Bo	Bond number	/
$D0$	Equivalent diameter of the droplet	m
H	Freezing front height	m
L	Latent heat of solidification	kJ/kg
R	The coordinates of the droplet profile underneath the freezing front	m
RS	Sphere radius	m
T	Temperature	°C
V	Volume	μL
Greek symbol	**Description**	**Unit**
$\Delta\rho$	The difference in density between water and the surrounding air	kg/m^3
σ	Water surface tension	N/m
β	Ice mass fraction	%
ρ	Density	kg/m^3
θ	Contact angle	rad
θ_d	Dynamic growth angle	rad
θ_R	Receding angle	rad
η	Characteristic slipping velocity	m/s
ε	Time deviation	%
φ	Formal parameter	rad
χ	Formal parameter	rad
Subscript	**Description**	**Unit**
0	Initial	/
i	Ice	/
t	Time	/
m	Water-ice mixture	/
w	Water	/
Exp	Experiment	/
F	Freezing	/

Model	Model	/
N	Nucleation	/
Ref	Reference	/
S	Sperical	/
P	Pedant	/

2.1 Introduction

Freezing and frosting are widely present in various fields. In the field of transportation, freezing of roads may greatly reduce the friction between the wheels and the road surface, causing vehicles to slip or brake failure and ultimately leading to traffic accidents [1,2]. In the field of aerospace, frosting on wings and propeller blades may alter the aerodynamic and flight characteristics, increase flight resistance, and threaten flight safety [3,4]. In the field of electricity, freezing may increase the pressure on power lines, and in severe cases, it may cause power outages by crushing the wires [5,6]. The freezing of individual droplets is an important component of the freezing and frosting process [7]. Therefore, the study of the freezing process of droplets on cold plates not only helps us better understand the mechanism and characteristics of freezing and frosting but also provides the scientific basis and technical support for preventing and controlling freezing and frosting.

The icing process of droplets on a cold surface can be divided into five stages [8–10], as shown in Fig. 2.1. At the supercooling stage, the droplet temperature continuously decreases until the nucleation temperature is reached. The nucleation occurs when water droplets reach a certain degree of supercooling, which is the beginning of recalescence. At the recalescence stage, the water droplet changes from liquid to an opaque ice–water mixture and releases latent heat, thus causing the temperature to rise to the freezing point. The time for nucleation and recalescence is very short, which is in recalescence [11]. At the freezing stage, the droplet temperature remains constant at the freezing point, and the freezing front moves from bottom to top, ultimately forming a protruding tip [12,13]. At the solid cooling stage, the temperature of frozen droplets gradually decreases to the temperature of the cold surface. The icing processes are affected by many factors, such as droplet volume [14], supercooling degree [15], additional field [16], surface characteristics [17], and flow environment [11].

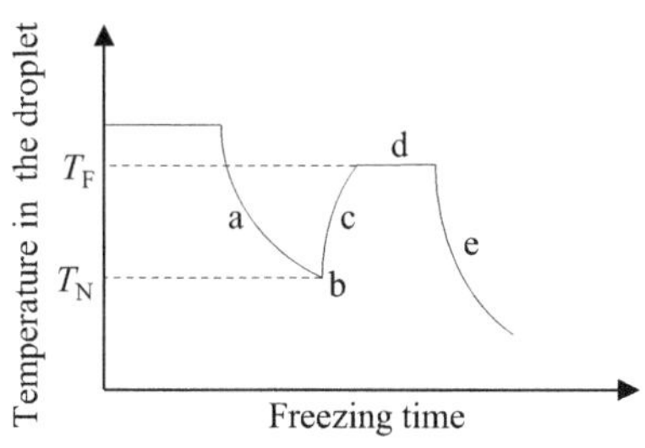

FIGURE 2.1 Temperature transition at five stages during droplet freezing: (a) supercooling, (b) nucleation, (c) recalescence, and (d) freezing. (e) solid cooling.

For the icing process of liquid droplets on a horizontal surface, Aderson et al. [18] used dynamic growth angle coupled boundary slip to establish an icing model of liquid droplets on plate surfaces considering gravity. The results indicate that the frozen profile is consistent with the experiment, while the freezing time differs significantly. Zhang et al. [19] established an icing model considering both undercooling and gravity which reduces the deviation of freezing time between the model and experiment from 40% to 10%. Zhang et al. [20] further assumed that the frozen front is spherical and established a freezing model. The results showed that the calculated results were in good agreement with the experimental results, and the freezing time was within 2% of the measured values.

For the icing process of liquid droplets on an inclined surface, Dang et al. [21] comprehensively considered deformation and diffusion and established a theoretical model for droplet icing on an inclined plane based on the enthalpy method. The results indicate that droplets on surfaces with larger inclined angles are more prone to icing. Strarostin et al. [22] conducted experimental studies on frozen tips on inclined and inverted planes. The results showed that the angle of the frozen tip was not affected by any factors. Vinh et al. [23] established a two-dimensional numerical model of droplet solidification on a vertical wall by a front-tracking/finite difference method. The results indicate that the conical tip of the frozen droplet still exists, but it moves downwards, resulting in an asymmetric droplet finally. The deformation and tip displacement of droplets increases with the increase of the Bond number. Zeng et al. [24] studied the icing process of pendant and sessile droplets experimentally. The results showed that gravity significantly affects the freezing process by changing the initial shape of droplets.

For the icing process of liquid droplets on a spherical surface, Ju et al. [25–27] studied the impacting and freezing processes of water droplets on different cold spherical surfaces experimentally. The results indicate that symmetric water droplets on spherical and cylindrical surfaces also exhibit a protruding tip after freezing, and the shape of the tip is universal. Zhang et al. [27] found that changes in temperature and concave radius during the freezing process do not lead to significant changes in the shape of frozen droplets. Troung et al. [28] conducted numerical research on the freezing process of liquid droplets on curved surfaces using the axisymmetric three-phase front tracking method. The results show that a more convex surface completes the solidification process earlier, while a more concave surface prolongs the entire solidification time.

At present, the freezing process on cold surfaces is mainly concentrated on the horizontal surface, and most of them are experiments and numerical simulations. Additionally, there is little research considering the inclined, inverted, and curved surfaces. However, the above surfaces are very common in industrial activities [29,30]. In this chapter, a theoretical model of the freezing process on horizontal, inverted, and spherical surfaces is established, and based on this, icing characteristics such as freezing time, freezing height, freezing profile, as well as the effects of surface temperature, contact angle, and droplet volume on the above icing characteristics are explored. Fig. 2.2 presents the outline of this chapter.

2.2 Icing characteristics of sessile water droplets on cold plate surfaces

Freezing of sessile water droplets on plate surfaces is very common in daily production and life, such as freezing and frosting on roads and refrigeration equipment. In this

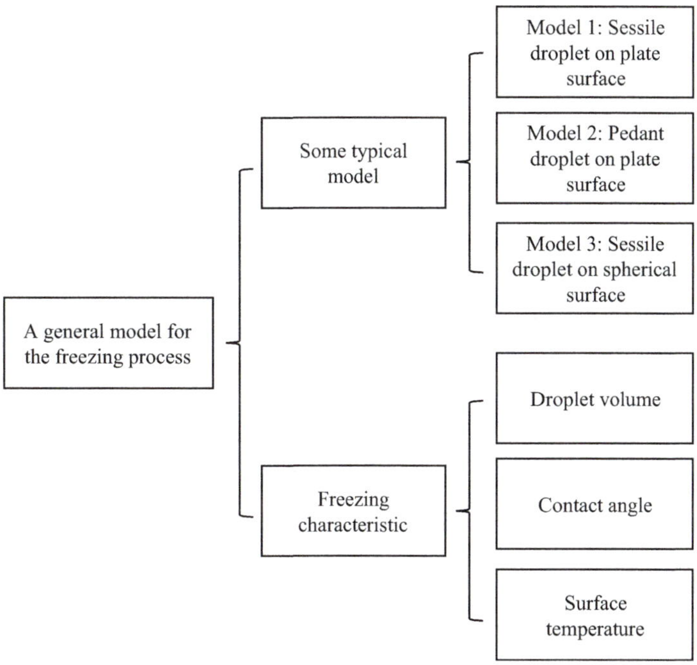

FIGURE 2.2 Outline of the general model for the freezing process.

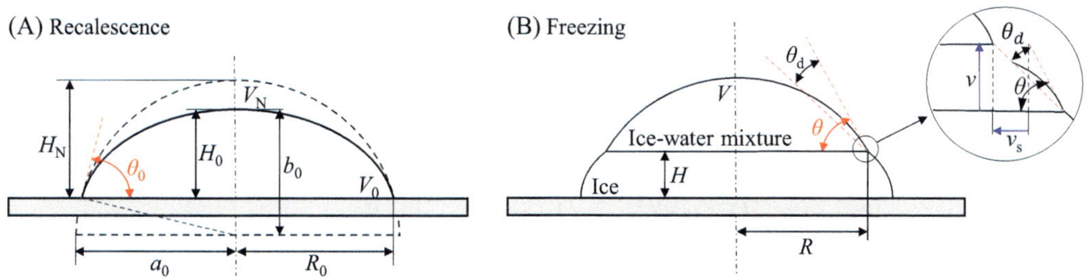

FIGURE 2.3 Stages of the icing process of a sessile water droplet on a cold plate surface: (A) Recalescence and (B) Freezing stages.

chapter, a theoretical model for sessile droplets on plate surfaces is established, and based on this, studies on the effect of droplet volume, contact angle, and surface temperature on icing characteristics are conducted.

2.2.1 Theoretical model

As shown in Fig. 2.3, a two-dimensional axisymmetrical model is developed. This model includes the initial, nucleation-recalescence, and freezing stages. In order to simplify the model, the following assumptions are made:

1. The heat transfer between droplets and the surrounding environment can be ignored [31,32].
2. Due to the low temperature and short freezing time, the droplet evaporation can be ignored.
3. During the freezing process, the droplet always maintains axial symmetry, and its axis is perpendicular to the cold plate [19,24,32].
4. During the nucleation and recalescence stage, the contact angle remains constant [31,32].
5. Under the influence of gravity, the ice–water mixture in the upper part of the freezing front can be simplified into an ellipsoid whose eccentricity is affected by the volume [14,31].
6. The temperature inside the ice layer is linear, while the temperature of the ice–water mixture remains at 0°C.
7. Using the physical properties at average temperature instead of those that vary with temperature [19,24,31,32].
8. The influence of small bubbles in ice on the thermal conductivity and density is ignored [32–34].
9. The frozen front has a surface parallel to the cold plate [21,31].

2.2.1.1 *Initial stage*

The initial (solid line) and recalescence (dotted line) profiles of the sessile droplets on the plate surface are shown in Fig. 2.3A. Under the combined action of gravity and surface tension, the droplet profile presents as an ellipsoid rather than a spherical cap. Neglecting the density of air, the Bond number can be described as $Bo = \Delta\rho_w g D_{eq}^2/\sigma$, where ρ_w is the density of water, g is the gravity acceleration, σ is the surface tension of water, D_{eq} is the equivalent droplet diameter calculated by the droplet volume, that is, $D_0 = (6V_0/\pi)^{1/3}$.

Using the Young-Laplace equation [14,31,35], the initial profile of a droplet can be defined as follows:

$$\frac{y''}{(1+y')^{3/2}} + \frac{y'}{x(1+y')^{1/2}} = \frac{2}{r} - \frac{\Delta\rho g y}{\sigma} \tag{2.1}$$

where x and y are the horizontal and vertical distances from the vertex of the droplet, r is the radius of curvature at the vertex of the droplet, $\Delta\rho$ is the difference in density between water and the surrounding air, and σ is the surface tension of water. Due to the difficulties in the iteration when calculating the freezing process, an ellipsoid was used to fit the droplet profile, from which the semi-major axis, a_0, semi-minor axis, b_0, of the ellipse, and initial base radius, R_0, can be obtained.

2.2.1.2 *Recalescence stage*

During the nucleation and recalescence stage, the volume and physical properties of droplets change dramatically, which may have a significant impact on the subsequent freezing stage. The volume and other properties after the nucleation/recalescence stage are determined by the ice mass fraction, which can be calculated from the energy balance based on the fact that the droplet temperature is restored to the equilibrium freezing point of 0°C due to the latent heat release.

The water enthalpy can be represented by the following:

$$h_w = \begin{cases} h_{w,Ref} + c_{w,T_F}(T - T_{Ref}), T_{Ref} < T \le T_F \\ h_{w,Ref} + c_{w,T_F}(T_F - T_{Ref}) + \int_{T_F}^{T} c_w(T)dT, T > T_F \end{cases} \tag{2.2}$$

where T_{Ref} is the reference temperature ($T_{Ref} < T_c$), T_F is the freezing temperature, $h_{w,\ Ref}$ is the water enthalpy evaluated at T_{Ref}, c_{w,T_F} and c_w (T) are the specific heats of water at T_F and T.

The energy conservation during the nucleation and recalescence stages can be expressed as follows:

$$h_{w,T_N} = h_{i,T_F}\beta_i + h_{w,T_F}(1 - \beta_i) \tag{2.3}$$

or

$$\beta_i = \frac{h_{w,T_F} - h_{w,T_N}}{h_{w,T_F} - h_{i,T_F}} \tag{2.4}$$

where β_i is the ice mass fraction at the end of the nucleation and recalescence stage, h_{w,T_F} and h_{w,T_N} is the water enthalpies evaluated at T_F (freezing temperature) and T_N (nucleation temperature), h_{i,T_F} is the ice enthalpy at T_F.

According to Eq. (2.3), the enthalpy variation due to the temperature change can be expressed as follows:

$$h_{w,T_F} - h_{w,T_N} = c_{w,T_F}(T_F - T_N) \tag{2.5}$$

The latent heat of the solidification can be expressed as follows:

$$h_{w,T_F} - h_{i,T_F} = L \tag{2.6}$$

Therefore, the mass fraction of ice can be expressed as follows:

$$\beta_i = \frac{c_{w,T_F}(T_F - T_N)}{L} = \frac{c_{w,T_F}}{c_i} St \tag{2.7}$$

where $T_F - T_N$ is the supercooling degree, and St is the Stefan number, given by $St = c_i\ (T_F - T_N)/L$, which is the ratio of the sensible to latent heat.

The corresponding physical parameters at the end of the nucleation and recalescence stage can be represented by β_i. For example, solidification latent heat and density can be expressed as follows:

$$L_m = (1 - \beta_i)L \tag{2.8}$$

$$\rho_m = (1 - \beta_i)\rho_w + \beta_i\rho_i \tag{2.9}$$

Thus, the droplet volume at the end of the nucleation and recalescence stage can be expressed as follows:

$$V_N = \frac{\rho_w V_0}{\rho_m} \tag{2.10}$$

During the nucleation and recalescence process, the base radius and contact angle of the droplet remain constant. Using the ellipsoidal equation, V_N can also be expressed as follows:

$$V_N = \begin{cases} \pi a_N^2 b_N \left(\frac{2}{3} - \sin \varphi_N + \frac{\sin^3 \varphi_N}{3} \right), \theta_0 < \frac{\pi}{2} \\ \pi a_N^2 b_N \left(\frac{2}{3} - \sin \varphi_N - \frac{\sin^3 \varphi_N}{3} \right), \theta_0 \geq \frac{\pi}{2} \end{cases} \tag{2.11}$$

and

$$\tan\theta_0 = \begin{cases} \frac{b_N}{a_N \tan\varphi_N}, \theta_0 < \frac{\pi}{2} \\ -\frac{b_N}{a_N \tan\varphi_N}, \theta_0 \geq \frac{\pi}{2} \end{cases} \tag{2.12}$$

where $\varphi_N = \arccos(R_N/a_N)$, R_N is the base radius of droplets during the recalescence stage, and a_N and b_N are the semi-major and semi-minor axes after the recalescence stage. Solutions of Eqs. (2.11) and (2.12) yield a_N and b_N. Thus, the droplet profile eccentricity, e_N, can be expressed as follows:

$$e_N = \frac{\sqrt{a_N^2 - b_N^2}}{a_N} \tag{2.13}$$

2.2.1.3 *Freezing stage*

At the freezing stage, the freezing front moves from bottom to top, and the ice–water mixture gradually solidifies into ice. The mass balance between the ice–water mixture and ice can be expressed as follows:

$$\rho_m \frac{dV}{dt} = -\pi \rho_i R^2 \frac{dH}{dt} \tag{2.14}$$

where V is the volume of the ice–water mixture, R and H are the radius and height of the freezing front shown in **0**, ρ_m, and ρ_i are the density of the ice–water mixture and ice.

The energy balance at the freezing front can be expressed as a Stefan problem

$$q_i - q_m = \rho_i L_m \frac{dH}{dt} \tag{2.15}$$

Assuming that the internal temperature of the ice layer follows a linear pattern. However, the temperature of the ice–water mixture remains at 0°C. Assuming that the temperature inside the frozen part of the droplet follows a linear distribution while the temperature of the ice–water mixture remains at 0°C, Eq. (2.15) can also be described as follows:

$$\frac{k_i \Delta T}{H} = \rho_i L_m \frac{dH}{dt} \tag{2.16}$$

By integrating Eq. (2.13), the relationship between the freezing front height and freezing time can be written as follows:

$$H = \sqrt{2\frac{k_i \Delta T}{\rho_i L_m} t} \tag{2.17}$$

At any given moment, the volume of the ice–water mixture still satisfies the ellipsoidal model and can therefore be expressed as follows:

$$V = \begin{cases} \pi a^2 b\left(\frac{2}{3} - \sin\varphi + \frac{\sin^3\varphi}{3}\right), \theta < \frac{\pi}{2} \\ \pi a^2 b\left(\frac{2}{3} - \sin\varphi + \frac{\sin^3\varphi}{3}\right), \theta \geq \frac{\pi}{2} \end{cases} \tag{2.18}$$

where $\varphi_1 = \arccos(R/a)$, a and b are the long and short semi-axes of the ellipsoid. Also, the relation between the contact angle and the slope at the tri-junction satisfies Eq. (2.12), that is,

$$\tan\theta = \begin{cases} \frac{b}{a\tan\varphi}, \theta < \frac{\pi}{2} \\ \frac{b}{a\tan\varphi}, \theta \geq \frac{\pi}{2} \end{cases} \tag{2.19}$$

Due to the varying contact angle during the freezing stage, Eqs. (2.18) and (2.19) alone are not sufficient to determine the droplet profile. Therefore, Eq. (2.20) as an additional boundary equation is necessary.

$$e = \frac{\sqrt{a^2 - b^2}}{a} = \frac{V}{V_N} e_N \tag{2.20}$$

At the late freezing stage, slipping phenomena may occur at the tri-junction [36]. As shown in Fig. 2.3B, there is an angle between the movement direction of the tri-junction and the local tangent of the gas–liquid interface, which is the dynamic growth angle θ_d. The decreasing rate of the freezing front is related to the freezing rate (v) and the dynamic contact angle, which can be described as follows:

$$\frac{dR}{dt} = -\frac{1}{\tan(\theta - \theta_d)}\frac{dH}{dt} = -\frac{v}{\tan(\theta - \theta_d)} \tag{2.21}$$

The horizontal slipping velocity (v_S) at the tri-junction can be described as follows:

$$v_s = v\left(\frac{1}{\tan(\theta - \theta_d)} - \frac{1}{\tan\theta}\right) \tag{2.22}$$

Besides, the slipping velocity is also related to the contact angle, which can be described as follows:

$$v_{slip} = \begin{cases} \frac{\eta(\theta_R - \theta)}{\theta}, \theta < \theta_R \\ 0, \theta \geq \theta_R \end{cases} \tag{2.23}$$

where θ_R is the receding angle and η is the characteristic slipping velocity, and both of them need to be obtained from experiments. As for the sessile water droplet on a cold plate surface, the characteristic slipping velocity, η, is selected as 0.8 mm/s, and the receding angle, θ_R, is chosen as 0.08π.

2.2.2 Model validation

To verify the accuracy of the freezing model, an experimental setup was designed, as shown in Fig. 2.4 and experiments on the icing process of sessile and pendant droplets were conducted.

Meanwhile, the experimental droplet profiles (Fig. 2.5) and freezing time (Table 2.1) at different surface temperatures were compared with the results of the theoretical model. For the droplet profile, the initial and final droplet profiles generated by the experiment and model fit well under different *Bo*. For the freezing time, the average deviation is 10.15%, which is within an acceptable range.

2.2.3 Results and discussion

Based on the above theoretical model, calculations listed in Table 2.2 are conducted, and their impact on icing characteristics is explored.

2.2.3.1 *Effect of droplet volume*

When the droplet volume is 10 μL, $Bo \approx 1$. Therefore, the volumes of 1, 5, 10, 20, and 50 μL are selected to study the effect of droplet volume on the icing characteristics.

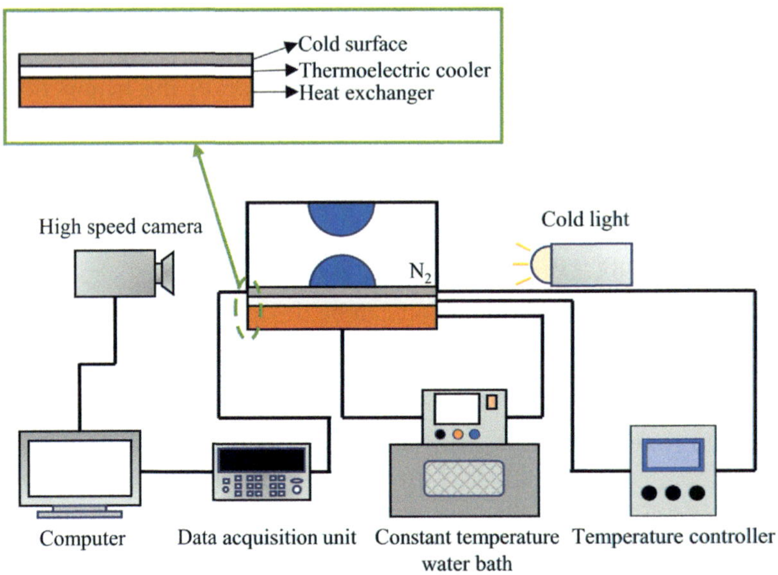

FIGURE 2.4 Schematic diagrams of experimental setup.

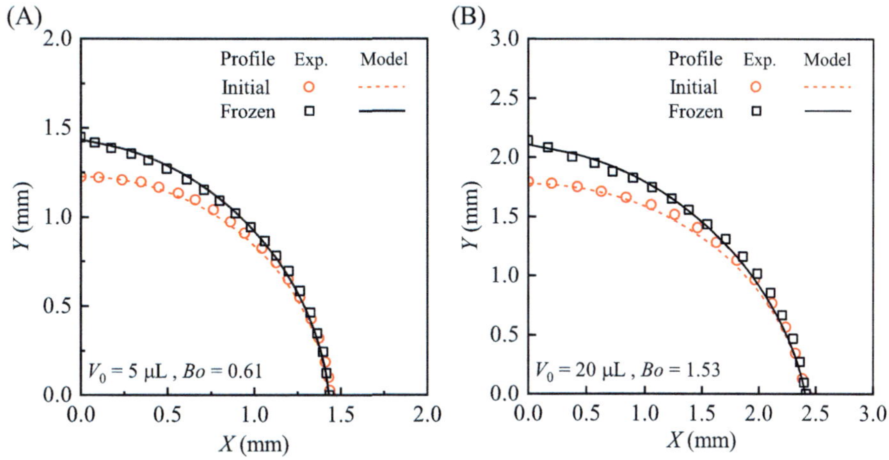

FIGURE 2.5 Comparison between experimental and theoretical profiles under different Bond number: (A) $V_0 = 5\ \mu L$ ($Bo = 0.61$) and (B) $V_0 = 20\ \mu L$ ($Bo = 1.53$).

TABLE 2.1 Model parameters and freezing times.

V_0 (μL)	η (mm/s)	θ_R (%)	$t_{F,Exp}$ (s)	$t_{F,Model}$ (s)	ε (%)
5	0.8	14.4	9.08	8.21	9.58
20	0.8	14.4	15.96	17.67	10.71

TABLE 2.2 Calculation conditions for the theoretical model.

Droplet volume, V_0 (μL)	Contact angle, θ (°)	Surface temperature, T_0 (°C)
1, 5, 10, 20, 50	30, 60, 90, 120, 150	−5, −10, −15, −20, −25

Fig. 2.6A and B shows the initial and final profiles at different droplet volumes. It can be seen that the larger the droplet volume, the larger the base area, and the greater the deviation of the initial profile from the arc. After the freezing of the droplet, the base radius is basically unchanged, but the height and volume increase significantly. This is because the density of ice is smaller than that of water; ignoring the evaporation during the icing process, the volume of ice is larger than that of water under the same mass. Fig. 2.6C and D shows the freezing time and freezing height at different droplet volumes. It can be seen that both the freezing time and freezing height increase with the increase of droplet volume. The freezing time is related to the base area, while the freezing height is related to the droplet profile. On the one hand, a larger droplet volume leads to a larger substrate area, thereby strengthening the heat transfer between the droplets and the cold plate. On the other hand, a larger droplet volume means that more heat is needed to complete the freezing process.

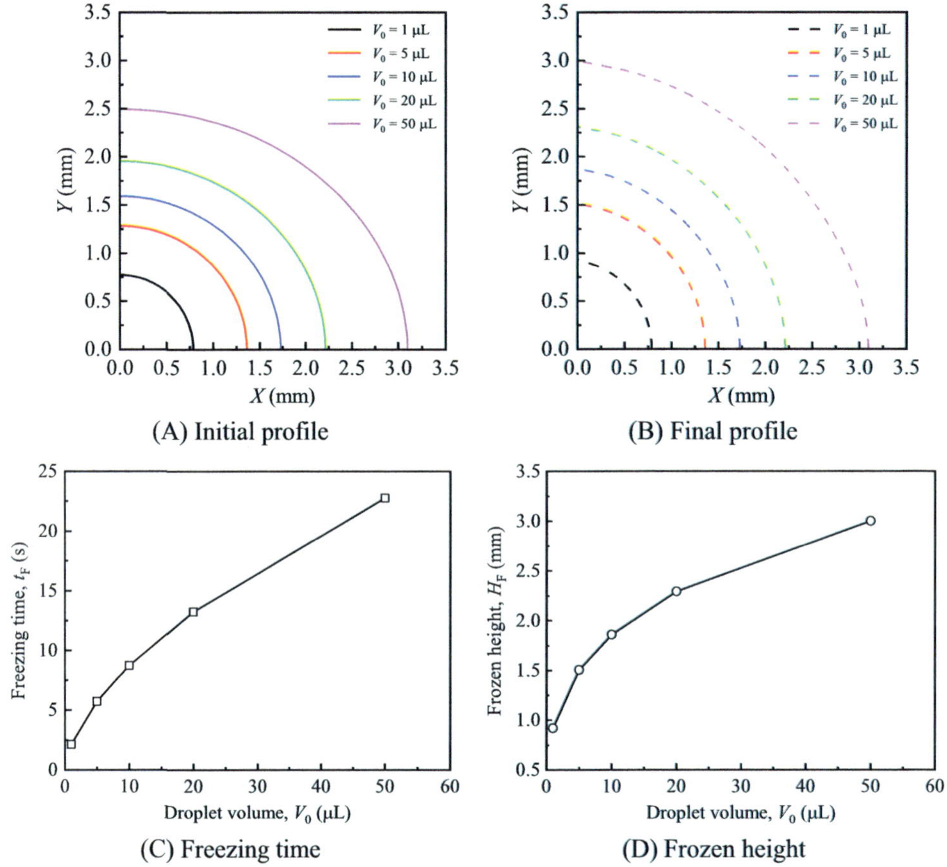

FIGURE 2.6 Icing characteristics of sessile droplets at different droplet volumes: (A) Inital profile, (B) final profile, (C) freezing time, and (D) frozen height.

2.2.3.2 *Effect of contact angle*

Considering the influence of contact angle on icing characteristics, simulations were conducted on the icing process of 5 μL and 50 μL droplets at contact angles of 30°, 60°, 90°, 120°, and 150°, respectively.

Fig. 2.7A and B shows the initial and final profiles at different contact angles. It can be seen that the base area decreases with the increase of contact angle. Fig. 2.7C and D shows the freezing time and freezing height at different contact angles. It can be seen that at the same droplet volume, the freezing time and freezing height both increase with the increase of contact angle. The freezing time is related to the base area. The increasing contact angle leads to a smaller base area, and the heat transfer between the cold surface and the droplet is weakened, resulting in an extended freezing time. However, the freezing height is related to the profile. Under the same volume, the smaller the contact angle, the higher the height, which can be clearly observed in Fig. 2.6A and B. Similarly, droplets with larger

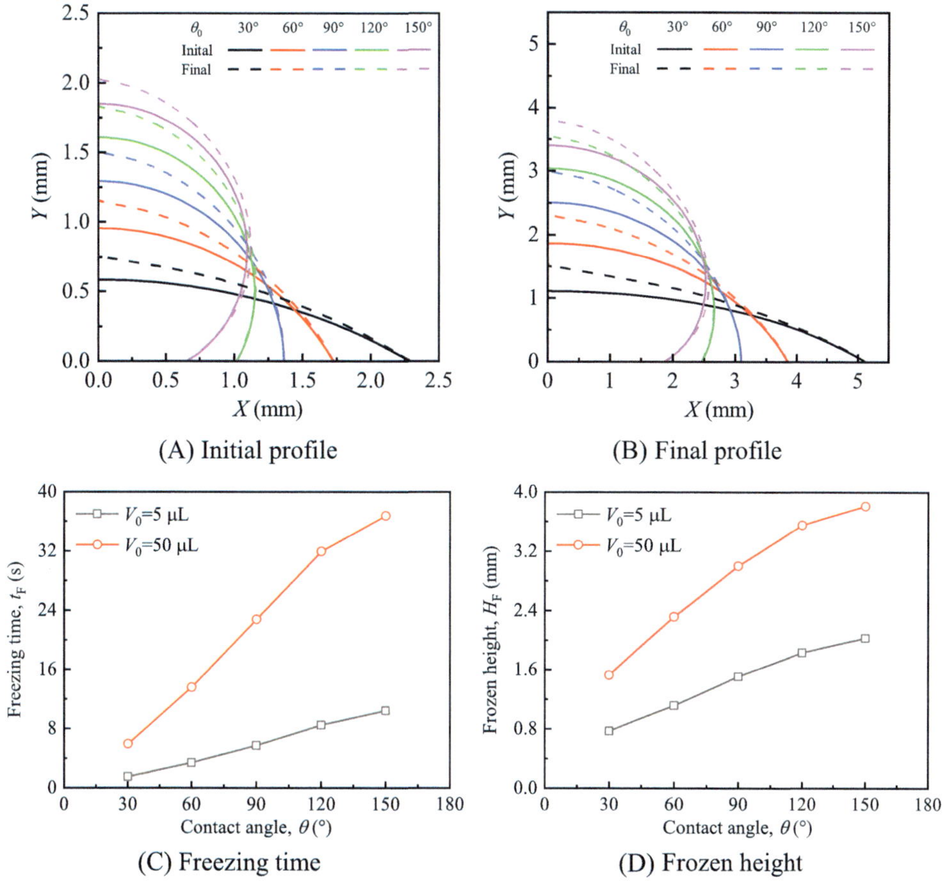

FIGURE 2.7 Icing characteristics of sessile droplets at different contact angles: (A) Inital profile, (B) final profile, (C) freezing time, and (D) frozen height.

volumes have longer freezing times and higher freezing heights and are more sensitive to changes in contact angles, which is consistent with what we observed in Section 2.2.2.1.

2.2.3.3 *Effect of surface temperature*

Considering the influence of surface temperature on freezing characteristics, simulations were conducted on the freezing process of 5 and 50 μL droplets at surface temperatures of −5, −10, −15, −20, and −25 °C, respectively.

Fig. 2.8A and B shows the initial and final profiles under different surface temperatures, and Fig. 2.8C and D shows the freezing time and frozen height at different contact angles. It can be seen that the droplet profile is almost unaffected by surface temperature, so the freezing height is also independent of surface temperature. The freezing time decreases with the decrease in surface temperature, which is due to the lower surface temperature leading to greater undercooling and promoting heat transfer between the droplets and the cold plate.

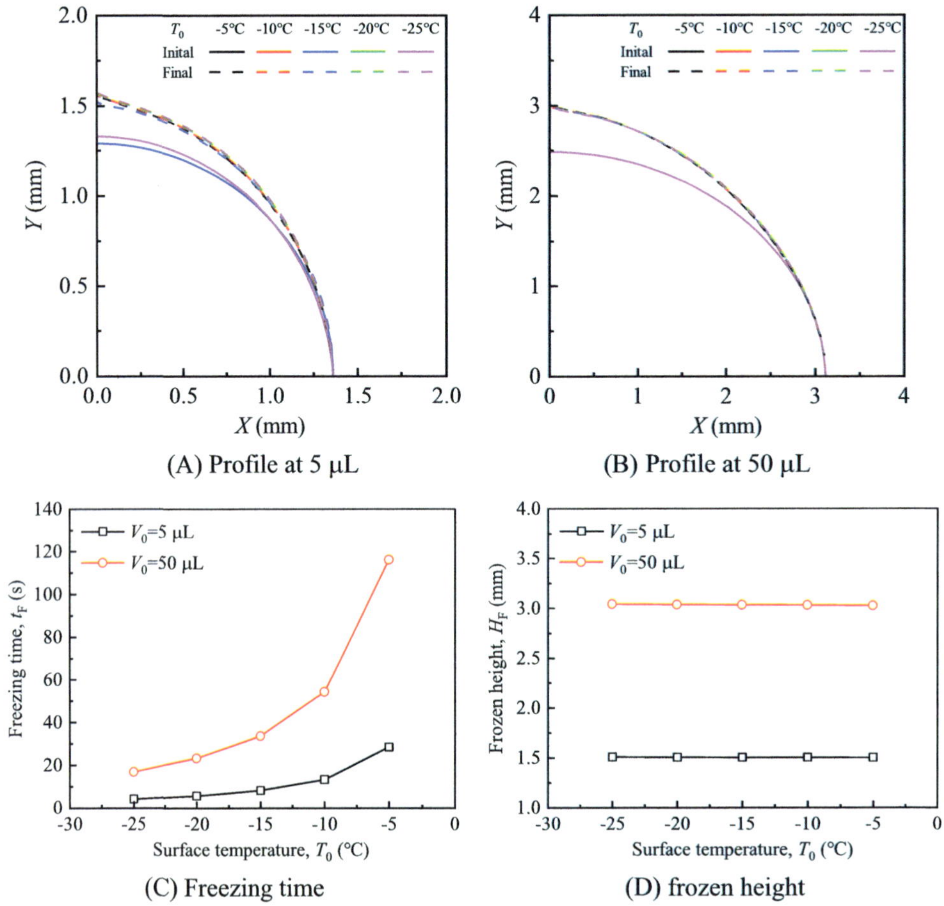

FIGURE 2.8 Icing characteristics of sessile droplets at different surface temperatures: (A) Profile at 5 μL, (B) profile at 50 μL, (C) freezing time, and (D) frozen height.

2.3 Icing characteristics of pendant water droplets on cold plate surfaces

Freezing of pendant droplets on plate surfaces is also very common in daily life, such as freezing on condensate collection systems and so on. In this chapter, a theoretical model for pendant droplets on plate surfaces is established, and based on this, studies on the effect of droplet volume, contact angle, and surface temperature on icing characteristics are conducted.

2.3.1 Theoretical model

The model of a pendant droplet on a plate surface is shown in Fig. 2.9. The theoretical model of the pendant droplet is roughly the same as that of the sessile droplet, with the following two main differences.

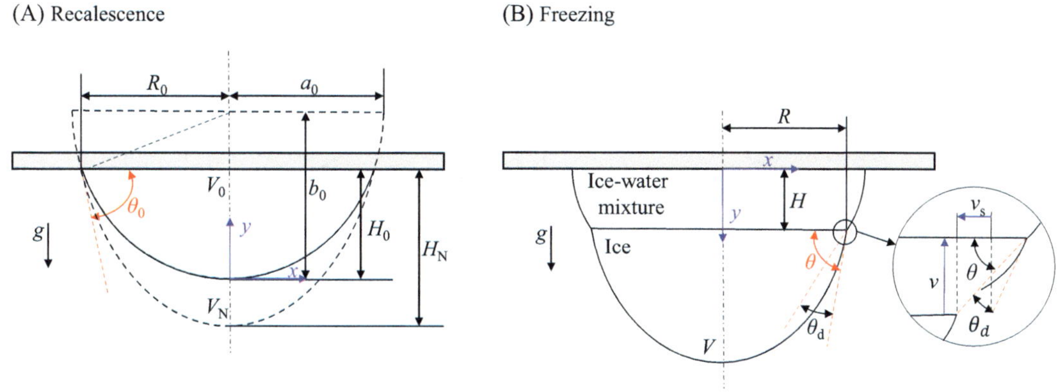

FIGURE 2.9 Stages of the icing process of a pendant water droplet on a cold plate surface: (A) Recalescence and (B) freezing stages.

First, the Young Laplace equation used to describe the initial profile is different. The initial profile of a droplet is determined by both gravity and surface tension. The additional pressure on the curved liquid surface always points towards the center of curvature of the liquid surface. As for the sessile droplet, the additional pressure is in the same direction as gravity, while for the pendant droplet, the additional pressure is in the opposite direction of gravity. Therefore, the Young Laplace equation for pendant droplets can be written as follows:

$$\frac{y''}{(1+y')^{3/2}} + \frac{y'}{x(1+y')^{1/2}} = \frac{2}{r} - \frac{\Delta \rho g y}{\sigma} \tag{2.24}$$

Second, the receding angle is different. As for the characteristic slipping velocity, η, is selected as 0.8 mm/s, which is the same as a sessile droplet. While for the receding angle, θ_R, is chosen as 0.13π, which means the difference in the influence of gravity on the freezing tip and different tip shapes.

2.3.2 Model validation

Experiments on the icing process of pendant droplets on cold plate surfaces were carried out using the experimental setup introduced in Section 2.2.2. Meanwhile, the experimental droplet profiles (Fig. 2.10) and freezing time (Table 2.3) at different surface temperatures were compared with the results of the theoretical model. For the droplet profile, the initial and final droplet profiles generated by the experiment and theoretical model fit well under different *Bo*. For freezing time, the average deviation is 10.94%, which is greater than that of sessile droplets but still within an acceptable range.

2.3.3 Results and discussion

Based on the above theoretical model, calculations listed in Table 2.2 are conducted, and their impact on icing characteristics is explored.

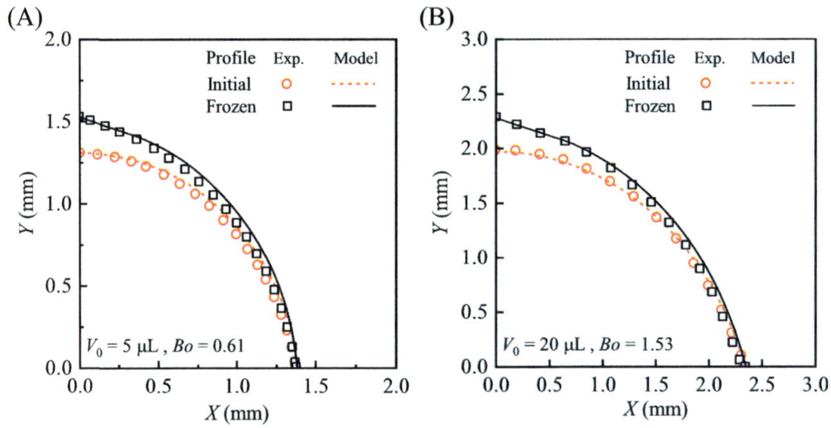

FIGURE 2.10 Comparison between experimental and theoretical profiles under different Bond number: (A) $V_0 = 5$ μL ($Bo = 0.61$) and (B) $V_0 = 20$ μL ($Bo = 1.53$) .

TABLE 2.3 Model parameters and freezing times.

Droplet volume, V_0 (μL)	Characteristic slipping velocity, η (mm/s)	Receding angle, θ_R (°)	Freezing time by experiment, $t_{F, Exp}$ (s)	Freezing time by model, $t_{F, Model}$ (s)	Deviation, ε (%)
5	0.8	23.4	10.44	9.50	9.00
20	0.8	23.4	18.32	20.68	12.88

2.3.3.1 *Effect of droplet volume*

Fig. 2.11A and B shows the initial and final profiles of sessile and pendant droplets under different droplet volumes. It can be seen that as the droplet volume increases, the deviation between the initial profile and the arc becomes larger and larger, and this phenomenon is particularly evident when $Bo > 1$. Compared with the initial profile, the difference between the final profile of sessile and pendant droplets is reduced, which can be attributed to the reduction of the volume of the ice water mixture caused by the advancement of the freezing front. From Fig. 2.11C and D, as the droplet volume becomes bigger, the freezing time and frozen height of both sessile and pendant droplets increase, and the difference between the two is related to droplet volume. When $Bo < 1$, the freezing time and freezing height of sessile and pendant droplets have a high degree of overlap, and the placement of liquid droplets doesn't affect the freezing characteristics. When $Bo > 1$, there is a significant difference between sessile and pendant droplets in terms of freezing time and frozen height, and the difference is positively correlated with Bo, which is related to the droplet profile.

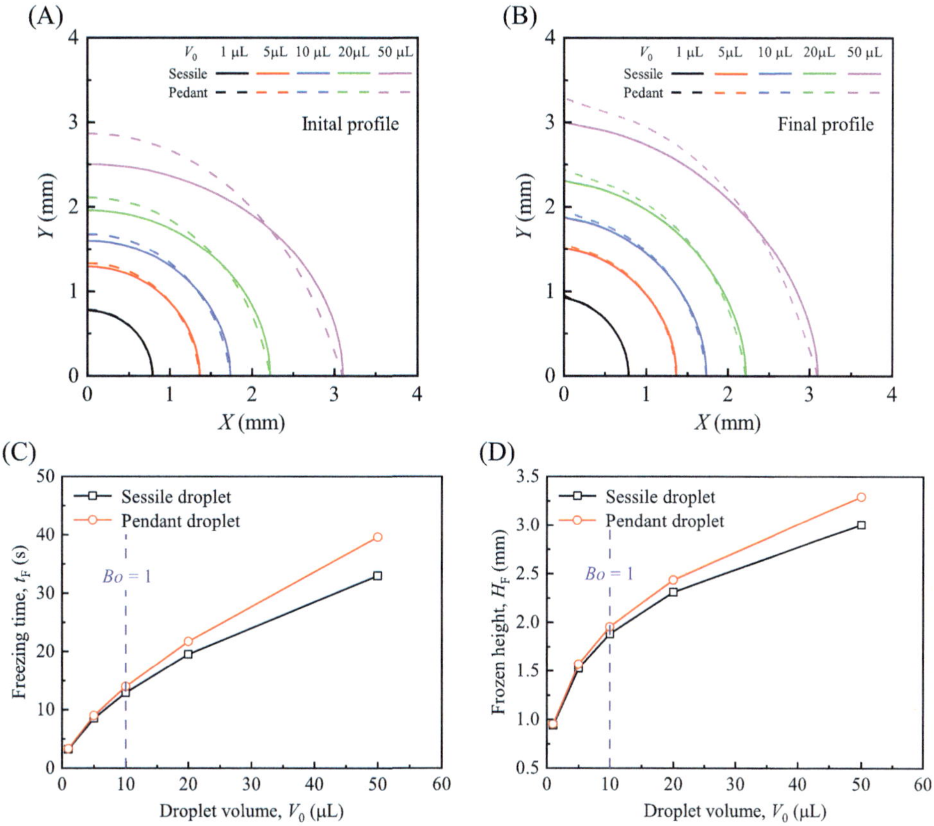

FIGURE 2.11 Icing characteristics of sessile and pendant droplets at different droplet volumes: (A) Inital profile, (B) final profile, (C) freezing time, and (D) frozen height.

2.3.3.2 *Effect of contact angle*

Due to the influence of gravity, a larger contact angle is bound to occur for pendant droplets. Nevertheless, the difference is not so notable. Thus, the contact angles used in this chapter are all that of sessile droplets.

The profiles of sessile and pendant droplets under different contact angles are shown in Fig. 2.12. Both for sessile and pendant droplets, a larger contact angle results in a smaller base area. On the assumption that the base area is the same, the pendant droplet has a smaller contact angle than the sessile droplet, which means a higher profile. For $Bo < 1$, there is little difference in the profiles between sessile and pendant droplets. While for $Bo > 1$, the difference significantly increases and is positively correlated with the contact angle, especially at 150°. The freezing time and frozen height are shown in Fig. 2.13. The freezing time and frozen height of both sessile and pendant droplets are positively correlated with the contact angle, and the difference becomes apparent under a bigger Bo and contact angle. For $Bo > 1$, the difference between the two

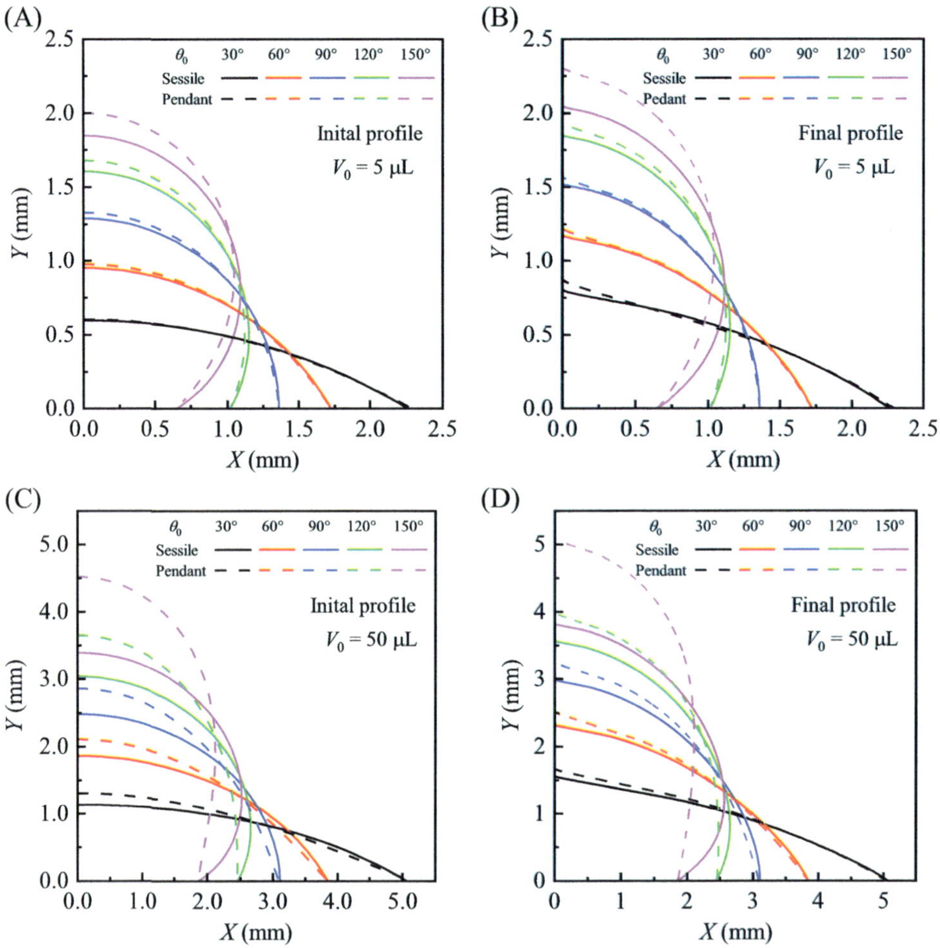

FIGURE 2.12 Initial and final profiles at different contact angles: (A) Inital profile at $V_0 = 5$ μL, (B) final profile at $V_0 = 5$ μL, (C) inital profile at $V_0 = 50$ μL, and (D) final profile at $V_0 = 50$ μL.

becomes significant, and the slope of the pendant droplet curve is slightly larger, indicating that the freezing process of the pendant droplet is more affected by the contact angle.

2.3.3.3 *Effect of surface temperature*

As shown in Fig. 2.14, the final profiles of pendant and sessile droplets at different surface temperatures are roughly the same under the same volume. For $Bo \leq 1$, the profile of the sessile droplet is basically consistent with that of the pendant droplet. For $Bo > 1$, different profiles are presented for sessile and pendant droplets, with a smaller contact angle and a higher frozen height for the pendant droplet. The above results suggest that the surface temperature has almost no effect on the droplet profile. From Fig. 2.15, it can be seen

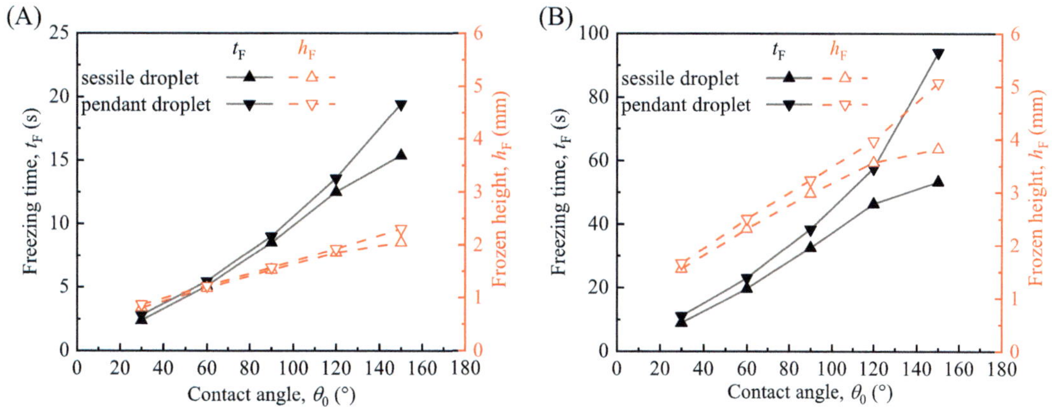

FIGURE 2.13 (A) Freezing time and (B)frozen height at different contact angles.

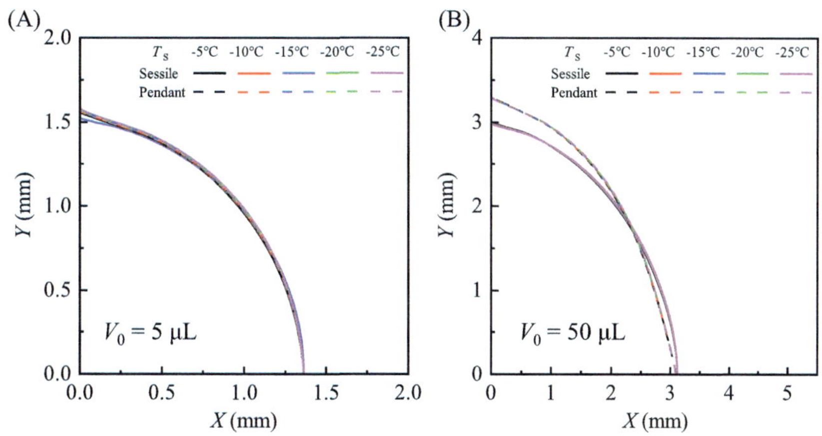

FIGURE 2.14 Final profiles under the droplet volume of (A) 5 μL and (B) 50 μL at different surface temperatures.

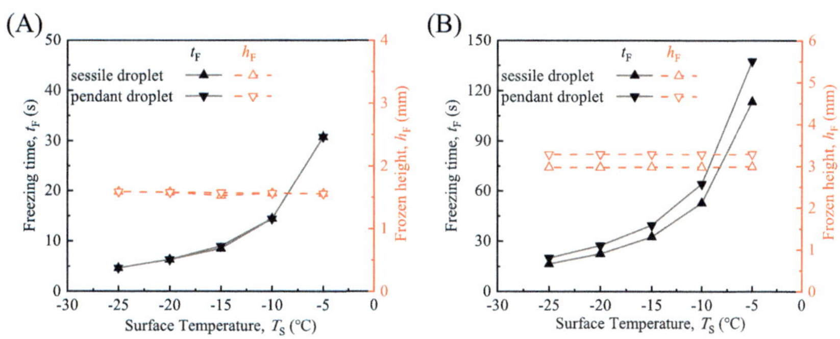

FIGURE 2.15 (A) Freezing time and (B) frozen height at different surface temperatures.

that surface temperature mainly affects the freezing time and has a relatively small impact on the frozen height. The higher surface temperature weakens the heat transfer between the cold plate and the droplets, resulting in an increase in freezing time. Meanwhile, when $Bo < 1$, the difference in freezing time and freezing height between sessile and pendant droplets is small, while for $Bo > 1$, the difference increases.

2.4 Icing characteristics of sessile water droplets on cold spherical surfaces

Freezing of sessile droplets on spherical surfaces is also very common in daily life, such as the freezing of wires and fruit surfaces. In this chapter, a theoretical model for sessile droplets on spherical surfaces is established, and based on this, studies on the effect of droplet volume, contact angle, and surface temperature on icing characteristics are conducted.

2.4.1 Theoretical model

The model of the pendant droplet on the plate surface is shown in Fig. 2.16. During the initial and recalescence stages, the spherical and plate models are roughly the same. However, during the freezing stage, experimental results show that the freezing front can be simplified into a series of arcs concentric with the sphere rather than a plane, which is the biggest difference from the plate surface.

The radius of the freezing front can be described as follows:

$$R_{S,t} = R_{S,0} + H \tag{2.25}$$

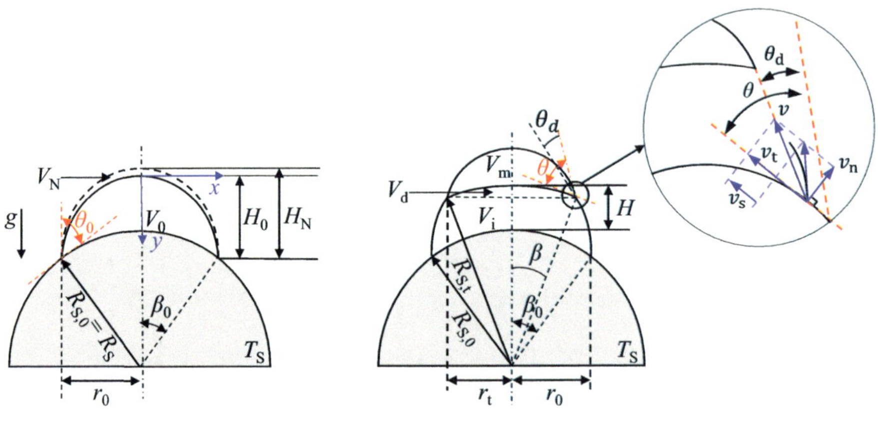

FIGURE 2.16 Stages of the icing process of a sessile water droplet on a cold spherical surface: (A) Recalescence and (B) freezing stages.

where R_S is the spherical radius, $R_{S,t}$ is the circumferential radius of the freezing front, and h_t is the height of the freezing front on the axis of symmetry. The movement of the freezing front on the axis can be simplified as a one-dimensional Stephen problem, that is,

$$H = \sqrt{2\frac{k_i(T_F - T_S)}{\rho_i L_m}t} \tag{2.26}$$

The conservation of mass during the freezing stage can be expressed as follows:

$$\rho_m \frac{dV_m}{dt} = -\rho_i \frac{dV_i}{dt} \tag{2.27}$$

in which the volume of ice can be expressed as follows:

$$V_i = \int_{R_s}^{R_{s,t}} 2\pi R^2 (1 - \cos\chi) dR \tag{2.28}$$

where $\chi = \arcsin(r/R)$, r and R express the coordinates of the droplet profile underneath the freezing front, which could be acquired from the temporal radii of the tri-junction point (r_t) and the freezing front ($R_{S,t}$).

Using the ellipsoid model, V_m can be expressed as follows:

$$V_m = \pi a^3 b \int_{\varphi}^{\frac{\pi}{2}} \cos^3\xi d\xi - V_d \tag{2.29}$$

$$V_d = \frac{\pi r_t^3}{3}\left(\frac{2 - 3\cos\beta + \cos^3\beta}{\sin^3\beta}\right) \tag{2.30}$$

where V_d is the volume of the spherical crown between the plane of the triple-phase line and the freezing front, a and b are the long and short semi-axes of the ellipsoid, $\beta = \arcsin(r_t/R_{S,t})$, φ is a parameter. Also, the relation between the contact angle and the slope at the tri-junction can be described as follows:

$$\frac{b}{a} = \tan(\theta + \beta)\tan\varphi \tag{2.31}$$

The ice–water mixture during the freezing stage can be approximated as an ellipsoid with an eccentricity of the following:

$$e = \frac{\sqrt{a^2 - b^2}}{a} = \frac{V_m}{V_N} e_N \tag{2.32}$$

where e_N is the eccentricity of the ellipsoid after the nucleation and recalescence stage.

The moving velocity of the freezing front is defined as follows:

$$v_n = \frac{dR_{S,t}}{dt} \tag{2.33}$$

The slip velocity of the freezing front is related to the front movement velocity, contact angle, and dynamic growth angle, that is,

$$v_{\text{slip}} = v_{\text{n}} \left[\frac{1}{\tan(\theta - \theta_{\text{dg}})} - \frac{1}{\tan\theta} \right] \tag{2.34}$$

$$\tan(\theta - \theta_{\text{dg}}) = \frac{v_{\text{n}}}{v_{\text{t}}} = -\frac{dR_{\text{S,t}}/dt}{R_{\text{S,t}} d\beta/dt} \tag{2.35}$$

Similar to the plate surface, the slip velocity can also be expressed as follows [36]:

$$v_{\text{slip}} = \begin{cases} \dfrac{\eta(\theta_{\text{R}} - \theta)}{\theta}, \theta < \theta_{\text{R}} \\ 0, \theta \geq \theta_{\text{R}} \end{cases} \tag{2.36}$$

where θ_{R} is the receding angle and η is the characteristic slipping velocity, and both of them need to be obtained from experiments. As for the sessile water droplet on a cold spherical surface, the characteristic slipping velocity, η, is selected as 0.8 mm/s, and the receding angle, θ_{R}, is chosen as 0.15π.

The relationship between the receding angle and diameter ratio can be expressed as follows:

$$\theta_{\text{R}} = \left[\left(\frac{D_0}{D_{\text{S}}} \right)^{1.35} + 0.08 \right] \pi \tag{2.37}$$

where D_0 is the equivalent diameter of the droplet, and D_{S} is the sphere diameter.

2.4.2 Model validation

To verify the accuracy of the theoretical model, the initial and final profiles (Fig. 2.17) and freezing times (Table 2.4) under different surface temperatures in reference [25] were

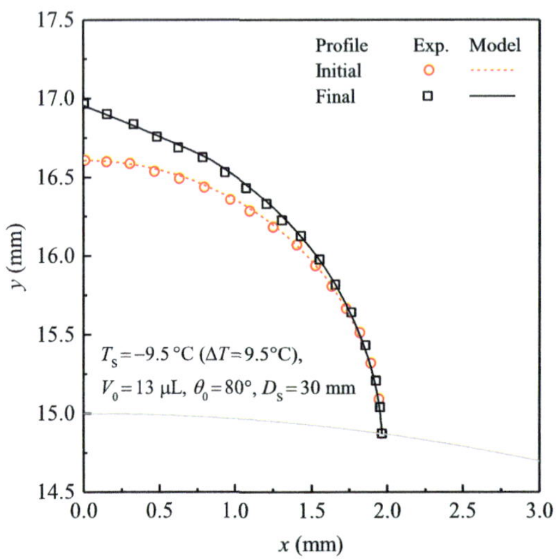

FIGURE 2.17 The initial and final profiles in reference and theoretical model.

TABLE 2.4 Model parameters and freezing times.

Surface temperature, T_0 (°C)	Droplet volume, V_0 (μL)	Contact angle, θ_0 (°)	Sphere diameter, D_S (mm)	Freezing time by experiment, $t_{F,\ exp}$ (s)	Freezing time by model, $t_{F,\ model}$ (s)	Deviation, ε (%)
−5.0				51.5	49.36	−4.16
−9.5	13	80	30	24.8	23.94	−3.47
−14.0				16.0	15.44	−3.50

TABLE 2.5 Calculation conditions for the theoretical model.

Droplet volume, V_0 (μL)	Contact angle, θ (°)	Surface temperature, T_0 (°C)	Sphere diameter, D_S (mm)
1, 10, 50	60, 90, 120	−10, −15, −20	8, 10, 20, 50, 100

compared with the results of the theoretical model. For the droplet profile, the initial and final droplet profiles generated by experimental and theoretical models match well when $T_0 = -9.5$°C. For the freezing time, when $T_0 = -5$, 9.5, and 14.0°C, the experimental and theoretical model's freezing time deviations were 4.16%, 3.47%, and 3.50%, respectively, with an average error of 3.71%, which is within an acceptable range.

2.4.3 Results and discussion

Based on the above theoretical model, calculations listed in Table 2.5 are conducted, and their impact on icing characteristics is explored.

From Fig. 2.19, it can be seen that the freezing characteristics of droplets on spherical surfaces exhibit similar variation patterns to those on the upright and inverted planes.

The base area is affected by the droplet volume and surface temperature but is independent of surface temperature. A larger droplet volume and smaller contact angle will result in a larger substrate area, and the freezing time will also be prolonged. In addition, reducing the surface temperature will also result in a longer freezing time. The frozen height is related to the freezing profile and increases with the increase of droplet volume and contact angle, independent of surface temperature. The spherical radius also affects the above indicators. A larger sphere radius causes an increase in the base area and a shortened freezing time. Meanwhile, the radius of the sphere will change the initial profile, thereby changing the final profile, and the frozen height will also be changed and decrease with the increase of the sphere radius (Fig. 2.18).

2.5 Correlations of freezing time

The influence of droplet volume, contact angle, and surface temperature on droplet freezing characteristics has been qualitatively discussed in the previous section. Among

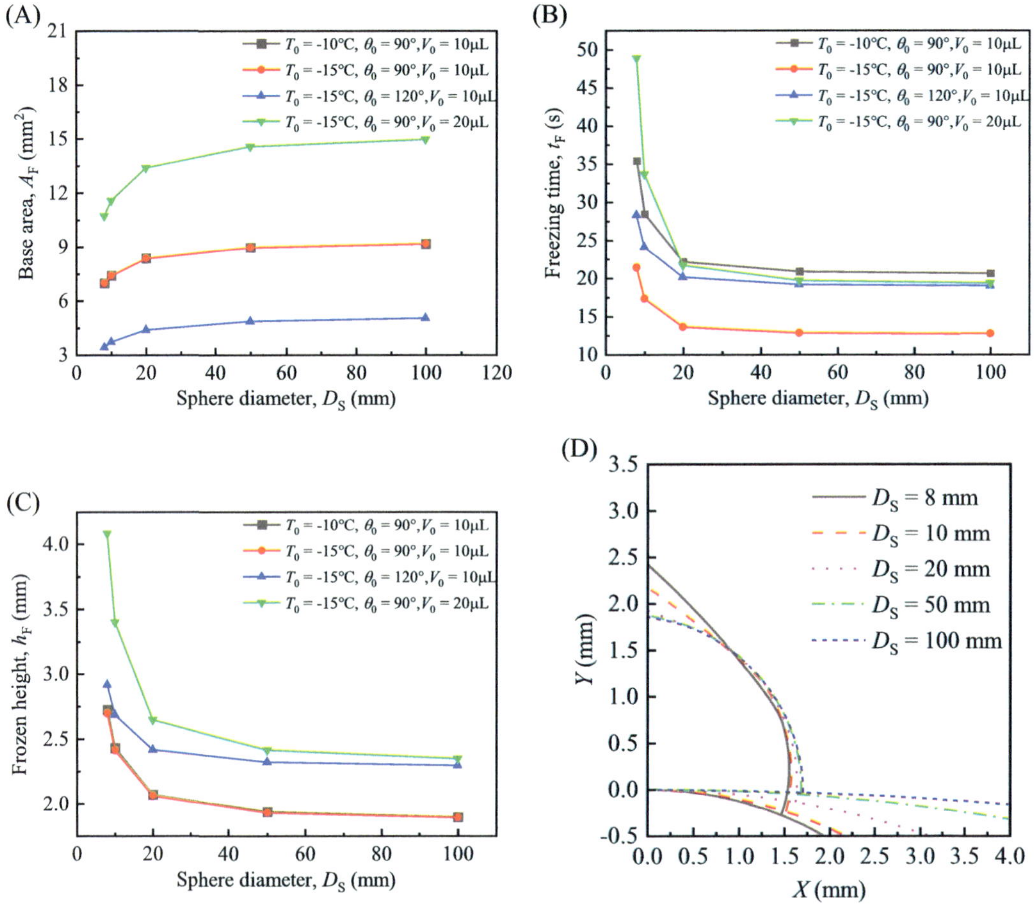

FIGURE 2.18 Icing characteristics of water droplets at different sphere diameters: (A) Base area, (B) freezing time, (C) frozen height, and (D) final profile.

the freezing characteristics, freezing time is undoubtedly one of the most important indicators. Therefore, it is particularly important to establish a quantitative relationship between freezing time and droplet volume, contact angle, and surface temperature.

2.5.1 Sessile and pendant droplets on cold plate surfaces

For sessile and pendant droplets on cold plate surfaces, the freezing time can be calculated using Eq. (2.17), that is,

$$t_F = \frac{\rho_i L_m}{2k_i(T_F - T_N)} H_F^2 \tag{2.38}$$

where H_F is the frozen height, and the initial height of the sessile droplet can be described as follows:

$$H_{0,S} \sim \sqrt[3]{\frac{3V_0}{\pi}\frac{1-\cos\theta_0}{2+\cos\theta_0}} \tag{2.39}$$

The frozen height can be expressed as follows:

$$H_{F,S} = 1.20 H_{0,S} \tag{2.40}$$

Therefore,

$$t_F \sim \frac{\rho_i L_m V_0^{2/3}}{k_i(T_F - T_N)}\left(\frac{1-\cos\theta_0}{2+\cos\theta_0}\right)^{2/3} \tag{2.41}$$

In practical applications, the profile is also affected by droplet volume (Bo) and contact angle. Thus, the freezing time can be written as follows:

$$t_F = f(Bo,\theta_0)\frac{\rho_i L_m V_0^{2/3}}{k_i(T_F - T_N)}\left(\frac{1-\cos\theta_0}{2+\cos\theta_0}\right)^{2/3} \tag{2.42}$$

where $f(Bo,\theta_0)$ is a function related to Bo and contact angle. By using the experimental data for fitting, the expression of $f(Bo,\theta_0)$ can be obtained as follows:

$$f(Bo,\theta_0) = \frac{Bo^n}{2}\sin\left(\frac{\theta_0}{2}\right), n = \begin{cases} -0.14, Bo \le 1 \\ -0.14, Bo > 1, \text{sessile droplet} \\ -0.11, Bo > 1, \text{pendant droplet} \end{cases} \tag{2.43}$$

So that the freezing time of the sessile droplet can be expressed as follows:

$$t_F = \frac{Bo^{-0.14}}{2}\sin\left(\frac{\theta_0}{2}\right)\frac{\rho_i L_m V_0^{2/3}}{k_i(T_F - T_N)}\left(\frac{1-\cos\theta_0}{2+\cos\theta_0}\right)^{2/3} \tag{2.44}$$

Dimensionless treatment was applied to the freezing time of pendant droplets, resulting in the following:

$$\frac{t_{F,P}}{t_{F,S}} = \begin{cases} 1 & , Bo \le 1 \\ Bo^{1/4} & , Bo > 1 \end{cases} \tag{2.45}$$

To verify the accuracy of the above formula, the model data, experimental data, and relevant data in the literature were compared, and the results are shown in Fig. 2.19 It can be seen that the deviation of all data points is within ± 25%.

2.5.2 Sessile droplets on cold spherical surfaces

The freezing time of sessile droplets on cold spherical surfaces is related to the radius ratio of the droplets to the sphere, as shown in Fig. 2.20 Using experimental data, the

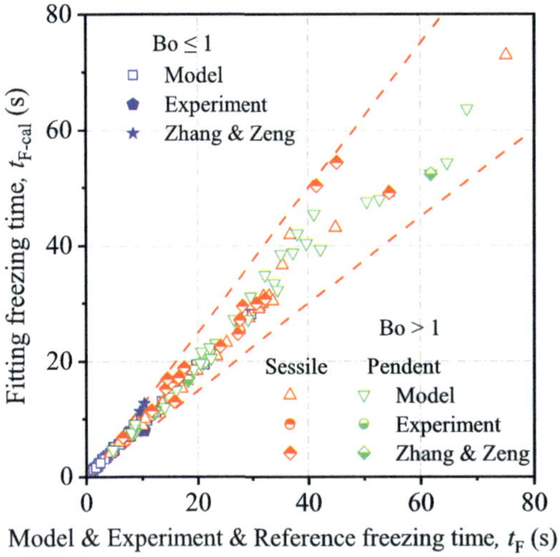

FIGURE 2.19 Comparisons of the freezing heights between model/experiments/reference.

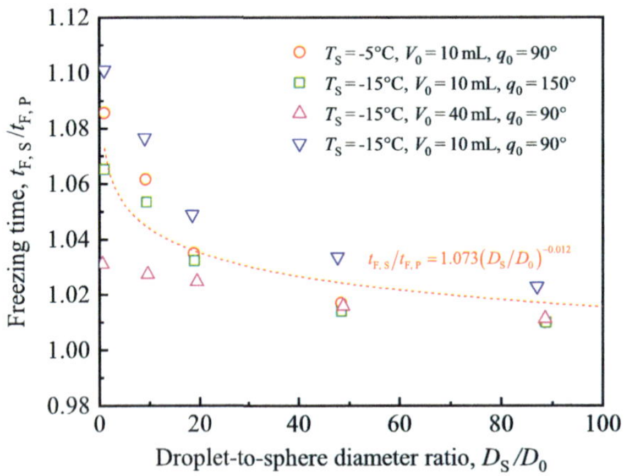

FIGURE 2.20 Comparisons of the freezing heights between model/experiments/reference.

relationship between the freezing time of sessile droplets on the sphere and plane can be fitted by the following:

$$\frac{t_{F,S2}}{t_{F,S}} = 1.073\left(\frac{D_S}{D_0}\right)^{-0.012} \tag{2.46}$$

where D_S is the sphere diameter and $D_0 = \sqrt[3]{6V_0/\pi}$ is the equivalent diameter of the droplet.

2.6 Summary

In this chapter, the theoretical models of the icing processes of the sessile and pendant water droplets on cold plate surfaces and the sessile water droplets on cold spherical surfaces are summarized and analyzed for comparison. The effects of droplet volume, contact angle, and surface temperature on icing characteristics (including freezing time, freezing height, and droplet profile) are studied. The main conclusions are as follows:

1. The theoretical models of the icing processes of the sessile and pendant water droplets on cold plate surfaces and the sessile water droplets on cold spherical surfaces are roughly the same. Compared with the sessile droplets on cold plate surfaces, the pendant droplets change in their Young Laplace equation due to the change in the direction of the additional pressure. Since the freezing front of the sessile water droplets on cold spherical surfaces is a curved surface, there is a difference in the calculation of the volume of ice and ice–water mixture compared to the plate surface. At the same time, the relationship between the major and minor axes of the ellipse at the triple junction has also changed. In addition, the receding angles selected for the three surfaces are not the same.
2. For the sessile and pendant water droplets on cold plate surfaces, the base area of the droplet increases with the increase of droplet volume and the decrease of contact angle, independent of surface temperature. The freezing time increases with the increase of droplet volume, contact angle, and surface temperature. The contact angle reduces the heat transfer area while the surface temperature slows down the heat transfer between the cold surface and the droplet by reducing the temperature difference, thereby prolonging the freezing time. The frozen profile is influenced by the droplet volume and contact angle but is independent of the surface temperature. However, under the same volume, contact angle, and surface temperature, the frozen profile height of the pendant droplet is higher, and the freezing time is longer.
3. The spherical radius also has an impact on the icing characteristics. The larger the spherical radius, the larger the base area, and thus the shorter the freezing time. This phenomenon is more pronounced under small spherical radii.

Acknowledgments

This work is supported by the Beijing Municipal Natural Science Foundation (Nos. 3244046 and 3232032), National Natural Science Foundation of China (Nos. 52076013 and 52306003), Key Laboratory of Icing and Anti/De-icing (Nos. IADL20230113 and IADL20220304), China Postdoctoral Science Foundation (No. 2022M720423), and Young Elite Scientist Sponsorship Program by BAST (Nos. BYESS2023352).

References

[1] Z.K. Zhen, R.M. Gao, M.J. Song, X. Zhang, Y.J. Xu, T.Z. Zhan, An experimental study on the effect of CO_2 laser powers on melting characteristics of clear ice - Part II Vertical irradiation, Applied Thermal Engineering 246 (2024) 122946.

[2] X. Zhang, Y.B. Gao, X. Liu, J.C. Min, K.L. Li, M.J. Song, Modelling of the icing processes of sessile supercooled water droplets on cold spheres, Applied Thermal Engineering 243 (2024) 122596.

[3] S.N. Chang, W.D. Yu, M.J. Song, M.Y. Leng, Q.Y. Shi, Investigation on wavy characteristics of shear-driven water film using the planar laser induced fluorescence method, International Journal of Multiphase Flow 118 (2019) 242–253.

[4] D.Y. Ma, S.N. Chang, K. Wu, M.J. Song, Experimental study on the water film thickness under spray impingement based on planar LIF, International Journal of Multiphase Flow 130 (2020) 103329.

[5] S.R. Yu, M.J. Song, R.M. Gao, J. Bae, X. Zhang, S.Q. Zhou, A review of icing prediction techniques for four typical surfaces in low-temperature natural environments, Applied Thermal Engineering 241 (2024) 122418.

[6] R.M. Gao, M.J. Song, C.Y.H. Chao, S.L. Lin, L. Zhang, X. Zhang, Review on condensation frosting and defrosting experiments for superhydrophobic surfaces, Applied Thermal Engineering 236 (2024) 121691.
[7] Y.L. Li, M.X. Li, C.B. Dang, X.T. Liu, Effects of dissolved gas on the nucleation and growth of ice crystals in freezing droplets, International Journal of Heat and Mass Transfer 184 (2022) 122334.
[8] M.J. Song, C.B. Dang, T. Higashi, E. Hihara, Review of experimental data associated with the solidification characteristics of water droplets on a cold plate surface at the early frosting stage, Energy Building 223 (2020) 110103.
[9] X. Zhang, X.M. Wu, J.C. Min, Freezing and melting of a sessile water droplet on a horizontal cold plate, Experimental Thermal and Fluid Science 88 (2017) 1–7.
[10] C. Reichl, C. Sandstrom, F. Hochwallner, F. Linhardt, M. Popovac, J. Emhofer, Frosting in heat pump evaporators part A Experimental investigation, Applied Thermal Engineering 199 (2021) 117487.
[11] S. Jung, M.K. Tiwari, N.V. Doan, D. Poulikakos, Mechanism of supercooled droplet freezing on surfaces, Nature Communications 3 (1) (2012) 615.
[12] Y. Wang, Z.G. Wang, Observation of sessile droplet freezing on textured micropillar surfaces via visualization and thermography, Journal of Coatings Technology and Research 16 (2019) 869–879.
[13] A.G. Marín, O.R. Enríquez, P. Brunet, P. Colinet, J.H. Snoeijer, Universality of tip singularity formation in freezing water drops, Physical Review Letters 113 (5) (2014) 54301.
[14] G. Chaudhary, R. Li, Freezing of water droplets on solid surfaces an experimental and numerical study, Experimental Thermal and Fluid Science 57 (2014) 86–93.
[15] A. Virozub, I.G. Rasin, S. Brandon, Revisiting the constant growth angle estimation and verification via rigorous thermal modeling, Journal of Crystal Growth 310 (24) (2008) 5416–5422.
[16] M. Prasad, N.J. English, S.N. Chakraborty, Effect of static electric fields on liquid water, its structure, dynamics, and hydrogen bond asymmetry a molecular dynamics simulation study of TIP4P/2005 water model, The Journal of Chemical Physics 159 (5) (2023) 054504.
[17] F.J. Montes Ruiz-Cabello, S. Bermudez-Romero, P.F. Ibanez-Ibanez, M.A. Cabrerizo-Vilchez, M.A. Rodriguez-Valverde, Freezing delay of sessile drops Probing the impact of contact angle, surface roughness and thermal conductivity, Applied Surface Science 537 (2021) 147964.
[18] D.M. Anderson, M.G. Worster, S.H. Davis, The case for a dynamic contact angle in containerless solidification, Journal of Crystal Growth 163 (3) (1996) 329–338.
[19] X. Zhang, X.M. Wu, J.C. Min, X. Liu, Modelling of sessile water droplet shape evolution during freezing with consideration of supercooling effect, Applied Thermal Engineering 125 (2017) 644–651.
[20] X. Zhang, X. Liu, J.C. Min, X.M. Wu, Shape variation and unique tip formation of a sessile water droplet during freezing, Applied Thermal Engineering 147 (2019) 927–934.
[21] Q. Dang, M.J. Song, X. Zhang, L. Pekař, S.H. Hosseini, Modelling study on freezing process of water droplet on inclined cold plate surface with droplet dynamic behavior considered, International Journal of Heat and Mass Transfer 197 (2022) 123327.
[22] A. Starostin, V. Strelnikov, L.A. Dombrovsky, S. Shoval, E. Bormashenko, On the universality of shapes of the freezing water droplets, Colloid and Interface Science Communications 47 (2022) 100590.
[23] V.N. Duy, T.V. Vu, A numerical study of a liquid drop solidifying on a vertical cold wall, International Journal of Heat and Mass Transfer 127 (2018) 302–312.
[24] H. Zeng, S. Lyu, D. Legendre, C. Sun, Influence of gravity on the freezing dynamics of drops on a solid surface, Physical Review Fluids 7 (2022) 103605.
[25] J. Ju, Z. Jin, H. Zhang, Z. Yang, J. Zhang, The impact and freezing processes of a water droplet on different cold spherical surfaces, Experimental Thermal and Fluid Science 96 (2018) 430–440.
[26] Z. Jin, H. Zhang, Z. Yang, The impact and freezing processes of a water droplet on different cold cylindrical surfaces, International Journal of Heat and Mass Transfer 113 (2017) 318–323.
[27] H. Zhang, Z. Jin, M. Jiao, Z. Yang, Experimental investigation of the impact and freezing processes of a water droplet on different cold concave surfaces, International Journal of Thermal Sciences 132 (2018) 498–508.
[28] T.V. Vu, Three-phase solidification of a liquid compound droplet on a curved surface, International Journal of Heat and Mass Transfer 181 (2021) 121881.
[29] J. Kim, S. Byun, J. Lee, K.S. Lee, Frost growth behavior according to the cold surface inclination angle, International Journal of Heat and Mass Transfer 146 (2020) 118841.
[30] S.P. Chen, F.S. Xie, S.T. Yao, H.Y. Han, Z.X. Chang, Experiment study on unsteady frosting of cryogenic vertical plate surface under natural convection conditions, Advanced Materials Research 516–517 (2012) 24–29.

[31] X. Zhang, X. Liu, X.M. Wu, J.C. Min, Simulation and experiment on supercooled sessile water droplet freezing with special attention to supercooling and volume expansion effects, International Journal of Heat and Mass Transfer 127 (2018) 975–985.

[32] X. Zhang, X. Liu, J.C. Min, M.J. Song, K.K. Shao, B. Huang, A comparative study of property-constant and property-variable icing models, International Journal of Green Energy (2023) 1–13.

[33] N. Azuma, T. Miyakoshi, S. Yokoyama, M. Takata, Impeding effect of air bubbles on normal grain growth of ice, Structural Geology 42 (2012) 184–193.

[34] W. Huang, Z. Li, X. Liu, H. Zhao, S. Guo, Q. Jia, Effective thermal conductivity of reservoir freshwater ice with attention to high temperature, Annals of Glaciology 54 (62) (2013) 189–195.

[35] X. Zhang, K.L. Li, Z.B. Zhu, W.Z. Fang, F.Q. Zhu, C. Yang, Droplet impact and freezing dynamics on ultracold surfaces: a scaling analysis of central-concave pattern, Applied Thermal Engineering 239 (2024) 122135.

[36] D.M. Anderson, S.H. Davis, Fluid flow, heat transfer, and solidification near tri-junctions, Journal of Crystal Growth 142 (1) (1994) 245–252.

CHAPTER

3

Lattice Boltzmann simulation of droplet icing process with dynamical interface capturing

Chaoyang Zhang[1], *Hui Zhang*[2], *Chun Yang*[2] *and Ping Cheng*[3]

[1]Paris Elite Institute of Technology, Shanghai Jiao Tong University, Shanghai, P.R. China [2]School of Mechanical and Aerospace Engineering, Nanyang Technological University, Singapore [3]MOE Key Laboratory of Power Machinery and Engineering, School of Mechanical Engineering, Shanghai Jiao Tong University, Shanghai, P.R. China

3.1 Introduction

The process of droplet freezing, a ubiquitous occurrence in both natural phenomena and industrial operations, represents a profound shift from the liquid to solid state. Grasping the intricate mechanics of solidification heat transfer in liquid droplets stands poised to revolutionize various sectors, including refrigeration, aerospace, and food processing [1]. Moreover, this understanding serves as a cornerstone in propelling the innovation of antiicing and de-icing technologies, essential for averting the perils and economic losses entailed by inadvertent freezing events [2].

When pristine water droplets crystallize upon subfreezing substrates, their eventual frozen form exhibits a distinctively sharp, pointed tip—a hallmark manifestation attributed to the volumetric expansion resulting from the contraction of ice density [3,4]. This pointed apex, characterized by a specific cone angle, emerges as a universal metric contingent upon the density ratio between water and ice [5,6]. Through the lens of this dynamic model, a comprehensive depiction of a droplet's morphological evolution and interfacial profiles during the freezing process unfolds. Furthermore, leveraging thermal transfer models augmented with phase-change dynamics has paved the way for diverse simulation methodologies aimed at encapsulating the entirety of the freezing process.

Meanwhile, the strategic incorporation of the solvent phase of colloidal or nanosuspension compounds into the solidification of bulk liquids has catalyzed a paradigm shift in the

Frosting and Icing for Efficient Energy Use in Engineering Applications
DOI: https://doi.org/10.1016/B978-0-443-15495-9.00008-3

fabrication realm, giving rise to bioinspired porous materials and composites through methodologies known as "freeze-casting" or "ice-templating". This innovative approach hinges on the entrapment or segregation of particles by the advancing front of solidification within the bulk fluid matrix. However, despite the burgeoning interest in freeze-casting, scant attention has been paid to the dynamics of liquid droplets harboring suspensions within. Intriguingly, Zhao et al. [7] unearthed a notable divergence in the freezing dynamics of nanofluid droplets vis-a-vis their pure water counterparts, wherein the frozen nanofluid droplets manifested a flattened plateau resembling a trapezoidal round table, eschewing the conventional pointed tip morphology. This deviation hints at a nuanced interplay where dispersed nanoparticles are repelled from the freezing front, thus compelling segregated particles to impel the liquid phase via capillary forces, ultimately yielding a flat plateau postfreezing.

Delving deeper into droplet freezing scenarios featuring dispersed nanoparticles, advanced simulation methodologies become indispensable. Foremost among these is the lattice Boltzmann (LB) method (LBM), renowned for its prowess in elucidating phase-change heat transfer phenomena spanning boiling, condensation, droplet evaporation, melting, and solidification [8]. Notably, the LBM offers a mesoscopic lens, facilitating the simulation of multiphase flows replete with intricate interfacial dynamics. Building upon the foundation laid by Shan-Chen's multicomponent LB model [9,10], a pioneering LB model tailored to simulate the freezing process of sessile water droplets, accounting for volumetric expansion effects, has been proposed [11,12]. This innovative endeavor aims to extend the model's applicability to scenarios involving suspended particles by integrating the immersed boundary-LB approach [13], thereby enabling the simulation of nanofluid droplet freezing processes complete with particle dispersion dynamics.

In essence, this chapter serves as a compendium of groundbreaking simulation methodologies, culminating in the proposition of an innovative axisymmetric LB model poised to comprehensively unravel the intricacies of droplet freezing processes, including those featuring volumetric changes and nanoparticle dispersion.

3.2 The droplet solidification lattice Boltzmann model

3.2.1 The triple solidification phase-change model

To model the intricate dynamics of droplet freezing and shape formation, we employ the Shan–Chen LBM [9]. This method utilizes the BGK (Bhatnagar–Gross–Krook) evolution equation to accurately track the liquid–gas interface. To enhance the fidelity of our model, we augmented the original equation with a mass source term. This addition allows us to accurately represent the density change effect within the system [11,12]. For both two-dimensional (2D) and three-dimensional (3D) systems, we utilize either the D2Q9 or D3Q19 LBM, respectively (as shown in Fig. 3.1).

The BGK equation governing the evolution of the distribution functions is expressed as follows:

$$f_i(\mathbf{x}+\mathbf{e}_i\delta_t, t+\delta_t) - f_i(\mathbf{x}, t) = -\frac{1}{\tau}\left(f_i(\mathbf{x}, t) - f_i^{eq}(\mathbf{x}, t)\right) + \Delta f_i(\mathbf{x}, t) + \delta_t\Phi_i \tag{3.1}$$

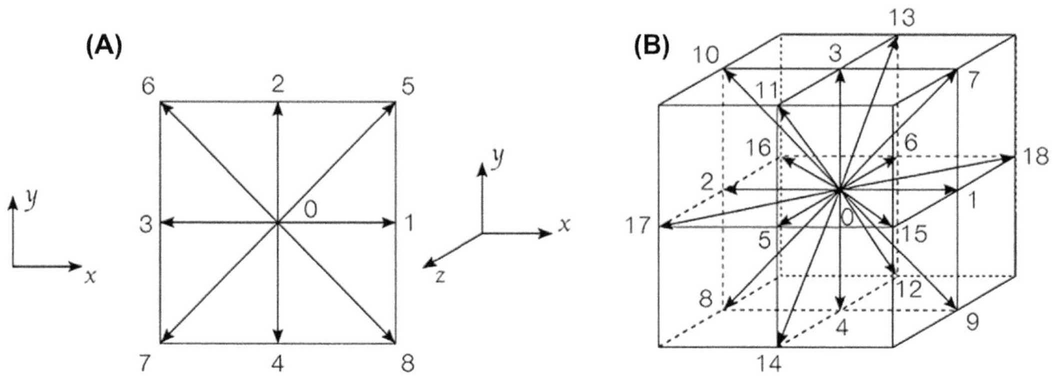

FIGURE 3.1 D2Q9 and D3Q19 lattice models. The values following D and Q indicate the number of simulation dimensions and velocities, respectively.

where f_i is the density distribution function, $\mathbf{x}$ and t, respectively, are the current position and time, $\mathbf{e}_i$ represents the velocities in different directions ($i=9$ in D2Q9, $i=19$ in D3Q19), τ is the relaxation time related to the kinematic viscosity ν, Φ_i is the mass source term, and $f_i^{eq}(\mathbf{x},t)$ is the corresponding equilibrium distribution function given by

$$f_i^{eq} = \omega_i \rho \left[1 + \frac{\mathbf{e}_i \cdot \mathbf{u}}{c_s^2} + \frac{(\mathbf{e}_i \cdot \mathbf{u})^2}{2c_s^4} - \frac{\mathbf{u}^2}{2c_s^2} \right] \tag{3.2}$$

where ω_i represents the weighting coefficients, and c_s denotes the lattice sound speed. The term $\Delta f_i(\mathbf{x}, t)$ encapsulates the body force, which is precisely defined within the exact difference method:

$$\Delta f_i(\mathbf{x}, t) = f_i^{eq}\left(\rho(\mathbf{x}, t), \mathbf{u} + \frac{\mathbf{F}\delta_t}{\rho} \right) - f_i^{eq}(\rho(\mathbf{x}, t), \mathbf{u}) \tag{3.3}$$

in which $\mathbf{F}$ is the total body force term exerting on the particles and is given by

$$\mathbf{F} = \mathbf{F}_f(x) + \mathbf{F}_s(\mathbf{x}) + \mathbf{F}_g(\mathbf{x}) \tag{3.4}$$

The interparticle interaction force ($\mathbf{F}_f$), fluid–solid interaction force ($\mathbf{F}_s$), and gravitational force ($\mathbf{F}_g$) are pivotal components of the equation. Following the methodology outlined in references, the calculation of $\mathbf{F}_f$ can be expressed as

$$\mathbf{F}_f(\mathbf{x}) = -\beta c_0 \psi(\mathbf{x}) g \nabla \psi(\mathbf{x}) - \frac{(1-\beta)}{2} c_0 g \nabla \psi^2(\mathbf{x}) - \frac{1}{2} c_s^2 c_0 g \psi(\mathbf{x}) \nabla \left(\nabla^2 \psi(\mathbf{x}) \right) \tag{3.5}$$

where β is the weighting factor, and $\psi(\mathbf{x})$ is the so-called effective mass and applied to calculate from the macroproperties, from a given equation of state (EOS) [14]:

$$\psi(\rho(\mathbf{x})) = \sqrt{\frac{2\left(p - \rho c_s^2\right)}{c_0 g}} \tag{3.6}$$

where g is the interaction strength which will be eliminated in Eq. (3.5).

From the distribution function f_i, one can derive the density and velocity within the flow field in each numerical step by applying the following relationships:

$$\rho = \sum_i f_i, \rho \mathbf{u} = \sum_i \mathbf{e}_i f_i \tag{3.7}$$

Additionally, the macrovelocity is updated in accordance with the momentum balance before and after the collision process:

$$\sim \mathbf{u} = \mathbf{u} + \frac{\mathbf{F}\delta_t}{2\rho} \tag{3.8}$$

3.2.2 Solid–liquid phase change with volume change

During freezing events, the streaming process of the distribution function is partitioned into two distinct segments: the freezing segment, which is reflected back with a fraction [15,16]B, and the remainder, which continues to stream [17]. Consequently, the evolution equation is modified as follows:

$$f_i(\mathbf{x} + \mathbf{e}_i\delta_t, t + \delta_t) - f_i(\mathbf{x}, t) = (1 - B)\left[-\frac{1}{\tau}(f_i(\mathbf{x}, t) - f_i^{eq}(\mathbf{x}, t)) + \Delta f_i(\mathbf{x}, t)\right] + B\Omega_i^s + \delta_t k_i\left(\mathbf{x} + \frac{1}{2}\mathbf{e}_i\delta_t, t + \frac{1}{2}\delta_t\right) + \delta V_i \tag{3.9}$$

$$\Omega_i^s = f\bar{i}(\mathbf{x}, t) - f_i(\mathbf{x}, t) + f_i^{eq}(\rho, \mathbf{u}_s) - f\,\bar{i}^{eq}(\rho, \mathbf{u}) \tag{3.10}$$

In the equation, $\mathbf{u}_s$ denotes the velocity of the freezing interface, and $\bar{i}$ represents the opposite direction of i. The weighting factor B is intricately linked to the volume fraction of the solid phase fs. This solid phase fraction fs is associated with its enthalpy under solid or fluid states:

$$B = \frac{\tau - 0.5}{0.5 - f_s + \tau} f_s \tag{3.11}$$

In scenarios where the density ratio $\gamma = \rho_s/\rho_l = 1.0$ during the liquid–solid phase transition, the last term δV_i in Eq. (3.1) simplifies to 1. However, for water freezing, where the density ratio $\gamma < 1.0$, volume expansion occurs during the solidification process. In our numerical treatment, we assume the density of ice to be equal to that of the liquid. Consequently, the expansion volume can be calculated as mass increases, and this effect is incorporated within the additional mass source term, which accounts for volume expansion:

$$\delta V_i = f_i^{eq}(\delta\rho, \sim \mathbf{u}) \tag{3.12}$$

$$\delta\rho = (1 - \gamma)\rho\frac{\partial f_s}{\partial t} \tag{3.13}$$

Due to the incorporation of additional mass during simulation, the apparent density of the freezing solid may appear larger than the saturated liquid density ρ_l, leading to a discrepancy between the obtained expansion volume and the expected value. To address this issue, when the density exceeds ρ_l, an approach is proposed to mitigate the compressibility of both the liquid and solid phases. This is achieved by computing the equilibrium distribution function as

$$f_i^{eq} = \begin{cases} \omega_i \rho_l \left[1 + \dfrac{\mathbf{e}_i \cdot \mathbf{u}}{c_s^2} + \dfrac{(\mathbf{e}_i \cdot \mathbf{u})^2}{2c_s^4} - \dfrac{\mathbf{u}^2}{2c_s^2} \right] & i = 0 \\ \omega_i \dfrac{\rho - \omega_0 \rho_l}{1 - \omega_0} \left[1 + \dfrac{\mathbf{e}_i \cdot \mathbf{u}}{c_s^2} + \dfrac{(\mathbf{e}_i \cdot \mathbf{u})^2}{2c_s^4} - \dfrac{\mathbf{u}^2}{2c_s^2} \right] & i \neq 0 \end{cases} \tag{3.14}$$

By employing this treatment of f_i^{eq}, we ensure mass conservation while adjusting the distribution in surrounding directions. When the density is smaller than ρ_l, Eq. (3.2) is still applicable.

To correct for the volume expansion assumption and obtain the true value of the solid phase density, thus eliminating the effect of the additional mass source, the following correction is utilized to get the real density field:

$$\tilde{\rho} = \rho(1 - f_s + \gamma f_s) \tag{3.15}$$

3.2.3 The axisymmetric multiphase model

In the context of a 2D axisymmetric system described in cylindrical coordinates $x = (r, z)$, as opposed to 2D Cartesian coordinates $x = (x, y)$, the velocity vector is transformed from $u = (u_x, u_y)$ to $u = (u_r, u_z)$. Upon comparing the Navier–Stokes equations expressed in these two coordinate systems, it becomes evident that the inclusion of an appropriate source term is necessary to restore the axisymmetric continuity equation and momentum equations [18]. These equations are articulated as follows:

$$f_i(\mathbf{x} + e_i\delta_t, t + \delta_t) - f_i(\mathbf{x}, t) = -\frac{1}{\tau}\left(f_i(\mathbf{x}, t) - f_i^{eq}(\mathbf{x}, t)\right) + \Delta f_i(\mathbf{x}, t) + \delta_t k_i\left(\mathbf{x} + \frac{1}{2}e_i\delta_t, t + \frac{1}{2}\delta_t\right) \tag{3.16}$$

The induced source term k_i is the induced source term due to the coordinate transformation, and it has the following form according to

$$k_i = \omega_i\left[-\frac{\rho u_r}{r} + \frac{1}{c_s^2}(e_{iz}K_z + e_{ir}K_r)\right] \tag{3.17}$$

$$K_r = \frac{e_{ir}}{r}\left[2\mu\left(\frac{\partial u_r}{\partial r} - \frac{u_r}{r}\right) - \rho u_r^2\right] \tag{3.18}$$

$$K_z = \frac{e_{iz}}{r}\left[\mu\left(\frac{\partial u_z}{\partial r} + \frac{\partial u_r}{\partial z}\right) - \rho u_r u_z\right] \tag{3.19}$$

In addition to modifying the source term in the evolution equation, adjustments must also be made to the gradient and Laplace operators in force calculations to accommodate cylindrical coordinates, as outlined in

$$\nabla\psi = \nabla_c\psi \tag{3.20}$$

$$\nabla^2\psi = \nabla_c^2\psi + \frac{1}{r}\frac{\partial\psi}{\partial r} \tag{3.21}$$

Hence, the interaction force between fluid particles must be adapted to:

$$F_f(x) = -\beta c_0\psi(x)g\nabla_c\psi(x) - \frac{(1-\beta)}{2}c_0 g\nabla_c\psi^2(x) + \frac{1}{2}c_s^2 c_0 g\psi(x)\nabla_c\left(\frac{1}{r}\frac{\partial\psi(x)}{\partial r}\right) \tag{3.22}$$

When addressing the gradient and Laplace terms, we employ the following second-order compact schemes:

$$\frac{\partial\psi}{\partial x} = \frac{1}{2c_s^2\delta_t}\sum\nolimits_\alpha \omega_i e_i[\psi(x+e_i) - \psi(x-e_i)] \tag{3.23}$$

$$\frac{\partial^2\psi}{\partial x^2} = \frac{1}{c_s^2\delta_t}\sum\nolimits_\alpha \omega_i e_i[\psi(x+e_i) + \psi(x-e_i) - \psi(x)] \tag{3.24}$$

3.2.4 The lattice Boltzmann—immersed boundary model

In the immersed boundary method depicted in Fig. 3.2, solid boundaries are represented by Lagrangian points superimposed onto the surrounding Eulerian lattice nodes of the fluid.

The Lagrangian points, denoted as $\mathbf{X}_B = (X_B, Y_B)$, correspond to the Eulerian lattice nodes $\mathbf{x}_{ij} = (x_i, x_j)$ described in Section 1.1. To enforce the no-slip boundary condition, body forces are applied to lattice nodes near Lagrangian points using the Dirac delta transition function. In this model, the particle number N_p is assumed to be spherical with diameter d_p ($r_p = 0.5\ d_p$ is the radius). Lagrangian points $\mathbf{X}_B$ (numbered N_B) are evenly distributed on each particle's surface, each with a corresponding velocity $\mathbf{U}_B$. Utilizing the Dirac delta function, the temporal velocity $\mathbf{u}_B$* at boundary points $\mathbf{X}_B$ is related to velocities at $\mathbf{x}_{ij}$ as described by [19]

$$u_B^*(\mathbf{X}_B, t+\delta_t) = \sum\nolimits_{i,j} \sim \mathbf{u}\left(\mathbf{x}_{ij}, t+\delta_t\right)D_{ij}\left(\mathbf{x}_{ij} - X_B\right)\delta_x\delta_y \tag{3.25}$$

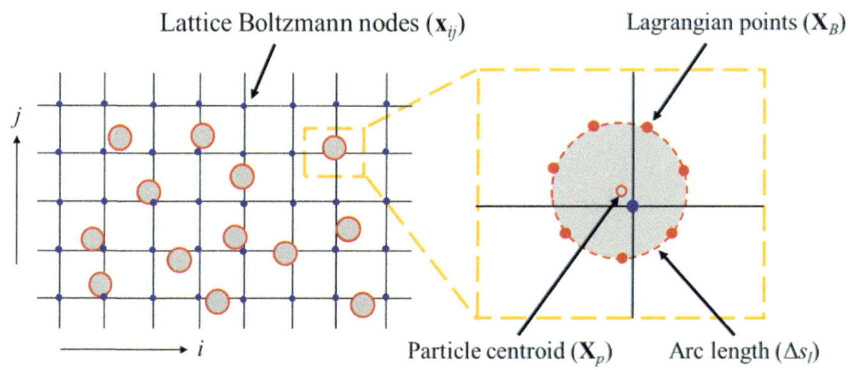

FIGURE 3.2 Schematic diagram of immersed-boundary particles represented by Lagrangian points ($\mathbf{X}_B$) in a fluid represented by Eulerian lattice nodes ($\mathbf{x}_{ij}$).

where D_{ij} is the weighting function proposed by [20]

$$D_{ij}(\mathbf{x}_{ij}-\mathbf{X}_B)=\frac{1}{h^2}\delta_h\left(\frac{x_i-X_B}{h}\right)\cdot\delta_h\left(\frac{x_j-Y_B}{h}\right) \tag{3.26}$$

$$\delta_h(r)=\begin{cases}\frac{1}{4h}\left(1+\cos\left(\frac{\pi r}{2}\right)\right), & |r|\le 2\\ 0, & |r|>2\end{cases} \tag{3.27}$$

where $h=1$ is the lattice spacing. The hydrodynamic force on the Lagrangian points is obtained from [21]

$$\mathbf{F}_B(\mathbf{X}_B,t+\delta_t)=2\rho_p\frac{\mathbf{U}_B-\mathbf{u}_B^*(X_B,t+\delta_t)}{\delta_t} \tag{3.28}$$

Then body force at the node $\mathbf{x}_{ij}$ is as follows:

$$\mathbf{F}_b(x_{ij},t+\delta_t)=\sum\nolimits_B\mathbf{F}_B(\mathbf{X}_B,t+\delta_t)D_{ij}(\mathbf{x}_{ij}-\mathbf{X}_B)\Delta s_l \tag{3.29}$$

where $\Delta s_l=\pi d_p/N_B$ represents the arc length between two adjacent points on the particle boundary. This term contributes to an additional hydrodynamic force on the lattice nodes, constituting a component of the fluid-particle interaction force $\mathbf{F}_p$.

The motion of particles can be computed from Newton's law based on the hydrodynamic force acting on the Lagrangian points. The center of the particle is denoted as $\mathbf{X}_p$, from which we obtain the particle's translational velocity $\mathbf{U}_p$ and angular velocity $\mathbf{\Omega}_p$:

$$\mathbf{F}_p(X_p)=\sum\nolimits_B\mathbf{F}_B(\mathbf{X}_B)\Delta s_l=M_p\frac{d\mathbf{U}_p}{dt} \tag{3.30}$$

$$\mathbf{T}_p=\sum\nolimits_B(\mathbf{X}_B-\mathbf{X}_p)\times\mathbf{F}_B(\mathbf{X}_B)\Delta s_l=I_p\frac{d\mathbf{\Omega}_p}{dt} \tag{3.31}$$

where M_p and I_p represent the mass and moment of the particle, respectively. Subsequently, the positions of the particles at each step can be tracked using $\frac{d\mathbf{X}_p}{dt}=\mathbf{U}_p$.

In the next time step, the Lagrangian points velocity $\mathbf{U}_B$ is calculated from $\mathbf{U}_p$ and $\mathbf{\Omega}_p$, which is

$$\mathbf{U}_B(\mathbf{X}_B)=\mathbf{U}_p+\mathbf{\Omega}_p\times(\mathbf{X}_B-\mathbf{X}_p) \tag{3.32}$$

While the discussion primarily focuses on particle–fluid interaction, in a droplet freezing system with suspended particles, interactions become more complex due to collisions between particles, walls, and phases. Here, we assume particles solely reside in the liquid phase. To model their behavior, we employ a generalized Lennard-Jones potential, acting as a repulsive force during collisions with other particles, walls, and interfaces [19]:

$$\mathbf{F}_m{}^{p-p}(\mathbf{X}_p)=\begin{cases}\varepsilon\sum_{n,n\neq m}\left[\left(\frac{2r_p}{X_R^{m,n}}\right)^{2\hat{n}}-2\left(\frac{2r_p}{X_R^{m,n}}\right)^{\hat{n}}\right]\frac{\mathbf{X}_p^m-\mathbf{X}_p^n}{(2r_p)^2}, & X_R^{m,n}\le 2r_p+\zeta\\ 0, & X_R^{m,n}>2r_p+\zeta\end{cases} \tag{3.33}$$

$$\mathbf{F}_m{}^{p-w|g|s}(\mathbf{X}_p)=\begin{cases}\varepsilon\sum_{n}\left[\left(\frac{r_p}{X_R^{m,w|g|s}}\right)^{2\hat{n}}-2\left(\frac{r_p}{X_R^{m,w|g|s}}\right)^{\hat{n}}\right]\frac{\mathbf{X}_p^m-\mathbf{X}_{w|g|s}}{r_p{}^2}, & X_R^{m,w|g|s}\le 2r_p+\zeta\\ 0, & X_R^{m,w|g|s}>2r_p+\zeta\end{cases} \tag{3.34}$$

where $\varepsilon = r_p^2$. In Eqs. (3.33) and (3.34), m donates the particle under force, n any other particle, and $w|g|s$ the wall, gas, or ice interface. $X_R^{m,n} = \chi|\mathbf{X}_p^m - \mathbf{X}_p^n|$ represents the distance between particles, while $X_R^{m,w|g|s} = \chi|\mathbf{X}_p^m - \mathbf{X}_{w|g|s}|$ denotes the distance between a particle (at $\mathbf{X}_p^m$) and the interface or wall. χ adjusts the reaction distance, $\hat{n}$ relates to collision elasticity, and ζ represents the threshold. We choose $\chi = 2.0$, $\hat{n} = 1$, and $\zeta = 2d_p$ for our simulation. Note that the liquid–gas and liquid–ice interfaces are modeled using diffusion formulation, assuming their positions $\mathbf{X}_g$ and $\mathbf{X}_s$ to lie between liquid and air/ice phases. When a particle nears both interfaces simultaneously ($X_R^{m,g} < 2d_p$ and $X_R^{m,s} < 2d_p$), it is considered trapped within the ice phase and held fixed in computation, reflecting the slower particle movement as the freezing front advances within the liquid.

In our multiphase model (Fig. 3.1), where particle size is significantly smaller than a lattice length, the particle centroid does not align with lattice points, resembling nanoparticle behavior relative to the bulk liquid. To simplify force calculations, we assume a hypothetical concentration distribution $\xi_p(\mathbf{x}_{ij})$ of a particle on Eulerian lattice nodes based on the Dirac delta function. Following Zhang et al. [22], the capillary force on the particle is given by

$$\mathbf{F}_c^p\left(X_p\right) = \sum_{ij}\left[-\gamma D_{ij}(\mathbf{x}_{ij} - \mathbf{X}_p)\sum_{x'} G_p(\mathbf{x}, \mathbf{x}')\psi(\mathbf{x}')\right] \tag{3.35}$$

where G_p represents the parameter denoting the strength of fluid–particle interaction, $\mathbf{x}'$ denotes the surrounding nodes of $\mathbf{x}$, γ is the weighting factor determined by particle size and set as $\gamma = 1$ in our simulation. $\psi(\mathbf{x}) = \sqrt{(p - \rho/3)/3}$ is the "effective mass" derived from the pressure p and density ρ.

In summary, the total force applied on the particle in Eq. (3.20) should be updated to incorporate the capillary force term given by

$$\mathbf{F}_p(\mathbf{X}_p) = \sum_B \mathbf{F}_B(\mathbf{X}_B)\Delta s_l + \mathbf{F}_m{}^{p-p}(\mathbf{X}_p) + \mathbf{F}_m{}^{p-w}(\mathbf{X}_p) + \mathbf{F}_m{}^{p-g}(\mathbf{X}_p) + \mathbf{F}_m{}^{p-s}(\mathbf{X}_p) + \mathbf{F}_c^p(\mathbf{X}_p) \tag{3.36}$$

The capillary force from particles attracts the fluid phase. Assuming the force on one particle is evenly distributed among boundary points, the reaction force on the lattice is obtained by summing forces from all particles on one lattice:

$$\mathbf{F}_{c,p}(\mathbf{x}) = \sum_p\left[\sum_B \frac{1}{N_B}\mathbf{F}_p(\mathbf{X}_p)D_{ij}(\mathbf{x}_{ij} - \mathbf{X}_B)\Delta s_l\right] \tag{3.37}$$

Combining the hydrodynamic force $\mathbf{F}_b$ and the capillary force $\mathbf{F}_{c,p}$, we observe the effects on fluid lattices caused by particles. However, it is worth noting that since the particles are small in our model, their hydrodynamic impact on the velocity field is limited to the macro/meso-scale. Therefore this force can be safely ignored to ensure smoother flow continuity.

3.2.5 The enthalpy thermal lattice Boltzmann model

Without considering viscous dissipation, the energy equation can be written as

$$\frac{\partial T}{\partial t} + \nabla\cdot(\tilde{u}T) = \nabla\left(\frac{\lambda}{\tilde{\rho}c_p}\nabla T\right) + \phi \tag{3.38}$$

where λ and c_p represent the thermal conductivity and specific heat capacity, respectively. ϕ is the source term, comprising three components in the cylindrical coordinate system. The first component arises from the thermal property disparity among the gas, liquid, and freezing solid phases, ensuring continuous heat flux across the interface:

$$\phi_1 = \frac{1}{\tilde{\rho} c_p} \nabla \cdot (\lambda \nabla T) - \nabla \cdot \left(\frac{\lambda}{\tilde{\rho} c_p} \nabla T \right) \tag{3.39}$$

Drawing from Eqs. (3.20) and (3.21), this term in the axisymmetric system can be formulated as

$$\phi_1 = \frac{1}{\tilde{\rho} c_p} \nabla_c \cdot (\lambda \nabla_c T) - \nabla_c \cdot \left(\frac{\lambda}{\tilde{\rho} c_p} \nabla_c T \right) + \frac{1}{r \tilde{\rho} c_p} \frac{\partial (\lambda \nabla_c T)}{\partial r} - \frac{1}{r} \frac{\partial \left(\frac{\lambda}{\tilde{\rho} c_p} \nabla_c T \right)}{\partial r} \tag{3.40}$$

$$\phi_2 = \frac{\Delta H_{ls}}{c_p} \frac{\partial f_s}{\partial t} \tag{3.41}$$

$$\phi_3 = -T \frac{\tilde{u}_r}{r} + \frac{\lambda}{r \tilde{\rho} c_p} \nabla T \tag{3.42}$$

For solving the energy equation, another distribution function for temperature g_i is utilized to simulate the thermal field, with the evolution equation provided as

$$g_i(\boldsymbol{x} + \boldsymbol{e}_i \delta_t, t + \delta_t) - g_i(\boldsymbol{x}, t) = -\frac{1}{\tau_T} \left(g_i(\boldsymbol{x}, t) - g_i^{eq}(\boldsymbol{x}, t) \right) + \delta_t \omega_i \phi \tag{3.43}$$

$$g_i^{eq} = \omega_i T \left[1 + \frac{\boldsymbol{e}_i \cdot \tilde{\boldsymbol{u}}}{c_s^2} + \frac{(\boldsymbol{e}_i \cdot \tilde{\boldsymbol{u}})^2}{2 c_s^4} - \frac{\tilde{\boldsymbol{u}}^2}{2 c_s^2} \right] \tag{3.44}$$

$$\tau_T = \frac{3\lambda}{\tilde{\rho} c_p} + 0.5 \tag{3.45}$$

Utilizing above evolution equations, the temperature and enthalpy can be computed using

$$T = \sum\nolimits_i g_i \tag{3.46}$$

$$H = f_s c_{p,s} T + (1 - f_s) c_{p,f} T + f_s \Delta H_{ls} \tag{3.47}$$

Considering both liquid and gas phases within the fluid phase, the gas fraction f_g is employed to adjust the enthalpy for refining the liquid–gas and solid–gas interfaces, as follows:

$$\tilde{H} = f_g (H_l + \Delta H_{gl}) + (1 - f_g) H \tag{3.48}$$

$$f_g = \frac{\rho_l - \rho}{\rho_l - \rho_g} \tag{3.49}$$

$$f_s = \begin{cases} 0 & H \geq H_l \\ \dfrac{H - H_s}{H_l - H_s} & H_s < H < H_l \\ 1 & H \leq H_s \end{cases} \tag{3.50}$$

where H_l represents the enthalpy of the liquid phase at the onset of solidification phase change, while H_s denotes the enthalpy of the solid phase when the phase change process concludes. The disparity between H_l and H_s signifies the latent heat ΔH_{ls} associated with the solid–liquid phase change, with the corresponding temperature point (the freezing point) designated as T_f.

3.3 Computation domain and boundary conditions

Fig. 3.3 illustrates the computational domain for freezing a nanofluid droplet on a subcooled plate. Initially, small particles are dispersed within the water droplet. As the freezing front, representing the liquid–ice interface, progresses upward from the bottom, the particles are carried along, aggregating as depicted in the enlarged view. We assume symmetry in the droplet due to a uniform initial distribution of particles. Hence, we employ a computational domain of size $L_x \times L_y$, with the left side serving as the symmetric boundary. The substrate height at the bottom of the domain is H_s, and a constant temperature T_w is maintained at the bottom of the substrate.

Conjugate heat transfer across the fluid–solid interface at the upper part of the substrate is managed following the approach by [23]. Wall boundary conditions for the velocity field and thermal isolation boundary conditions for the temperature field are applied to the right and upper sides of the computation domain, respectively.

On the substrate, initialization of a liquid droplet and surrounding gas phase is conducted using the multiphase LBM. This model incorporates the Peng–Robinson (P–R) equation of state, expressed as follows:

$$p = \frac{\rho R_g T_0}{1 - b\rho} - \frac{a\rho^2 \varepsilon(T_0)}{1 + 2b\rho - b^2\rho^2} \tag{3.51}$$

where ω is the acentric factor (chosen as 0.344 for water), $a = 2/49$, $b = 2/21$, and $R_g = 1$ is the gas constant. Maintaining a constant T_0 ensures a predetermined liquid–gas density ratio during freezing. We set $T_0 = 0.85\ T_c$ with liquid density $\rho_l = 6.64$ and gas density $\rho_g = 0.32$, The droplet's initial shape is a quarter circle with radius R_0, and density across its interface is determined through interpolation:

$$\rho = \frac{1}{2}(\rho_l + \rho_g) - \frac{1}{2}(\rho_l - \rho_g) \times \tanh\left(\frac{2\sqrt{(r - r_0)^2 + (z - z_0)^2} - R_0}{W}\right) \tag{3.52}$$

In the equation, (r_0, z_0) represents the droplet's center coordinates, and $W = 5$ indicates the interface thickness. The initial contact angle of the droplet on the substrate is 90 degrees. To achieve varying droplet shapes under the same volume, we adjust surface wettability to target a specific contact angle θ. Following Ding and Spelt's method [24], we calculate the effective mass of surface nodes on the plate (subscripted as 0) relative to the effective mass of upper liquid nodes (subscripted as 1 and 2). This is expressed as

$$\psi_{i,0} = \psi_{i,2} + \tan\left(\frac{\pi}{2} - \theta\right) |\psi_{i+1,1} - \psi_{i+1,1}| \tag{3.53}$$

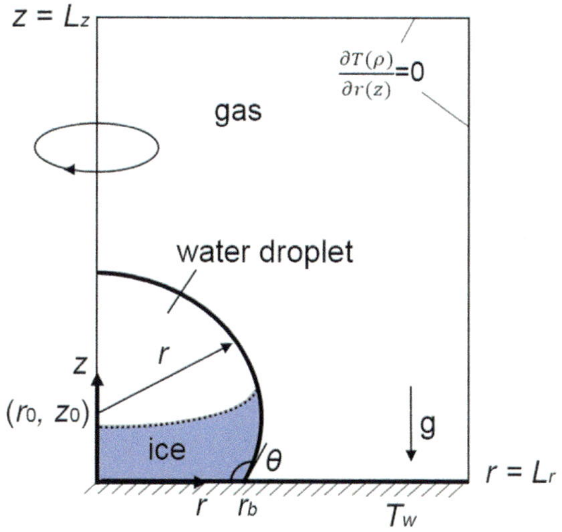

FIGURE 3.3 Schematic of the axisymmetric simulation domain with a water droplet placed on the substrate surface. The boundary conditions used for computations are indicated.

3.4 Sessile droplet freezing with volume change

In this section, simulation results of pure liquid droplet freezing will be presented. At first, model validation is conducted by comparing with experimental data for a water droplet on a subcooled surface with varying contact angles. Then the evolution of droplet shape during freezing and the eventual formation of the tip under different solid–liquid density ratios are investigated. Additionally, the cone angle at the tip and the angle between the freezing front and the interface are examined and compared with assumptions from prior research.

3.4.1 Model validation and comparison with experiment

Fig. 3.4 illustrates the progression of the droplet freezing process at various stages. The numerical simulation initializes just before the nucleation phase. Over time, the liquid-ice freezing front ascends, and the ice edge, or the triple contact line point, progressively deviates from its original profile due to volume expansion effects.

In Fig. 3.5, the computed change in droplet volume is shown, defined as $V = \iint 2\pi r dr dz$ for both the liquid water and solid ice phases. The rate of volume expansion decreases with increasing time. Upon completion of the freezing process, the volume ratio approaches the density ratio of liquid and ice.

Using consistent dimensionless parameters, Fig. 3.6 juxtaposes experimental images with numerical simulation results. Dashed lines in the figures represent the horizontal freezing front. Quantitative comparison of the experimental and simulation outcomes regarding the final droplet height reveals relatively small error. These results signify a reasonable agreement between experimental observations and simulation predictions.

Fig. 3.7 presents comparisons between experiments and simulations for the freezing front dimensionless height and radii throughout the freezing process. Both on hydrophilic

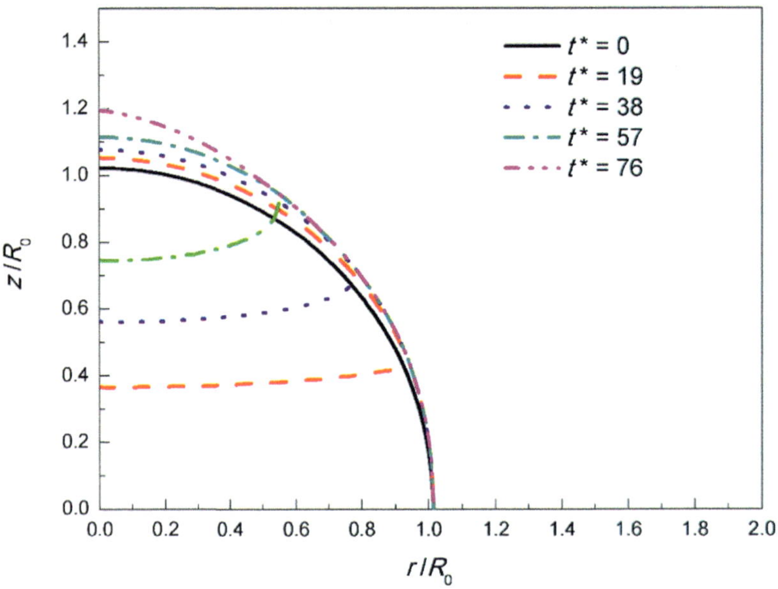

FIGURE 3.4 Time evolution of the droplet profiles at different freezing stages.

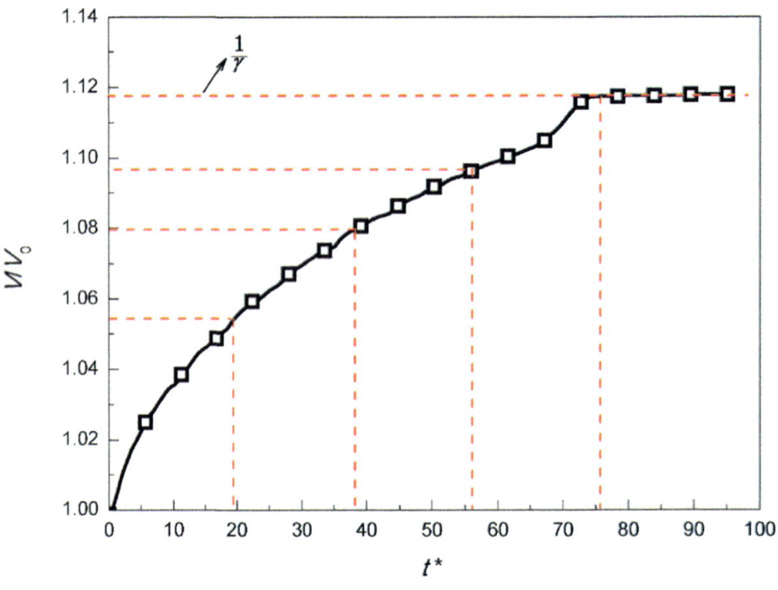

FIGURE 3.5 Droplet volume change during the entire freezing process (here V_0 is the original liquid droplet volume at $t^* = 0$).

and hydrophobic surfaces, the numerical and experimental curves closely align, except for the final stage. The freezing front radii exhibit nearly linear time-dependence, while the freezing front height increases with time at a gradually decreasing speed. During the early stage, the experimental and numerical freezing front positions almost overlap, but the difference between them gradually increases afterward. This divergence stems from the

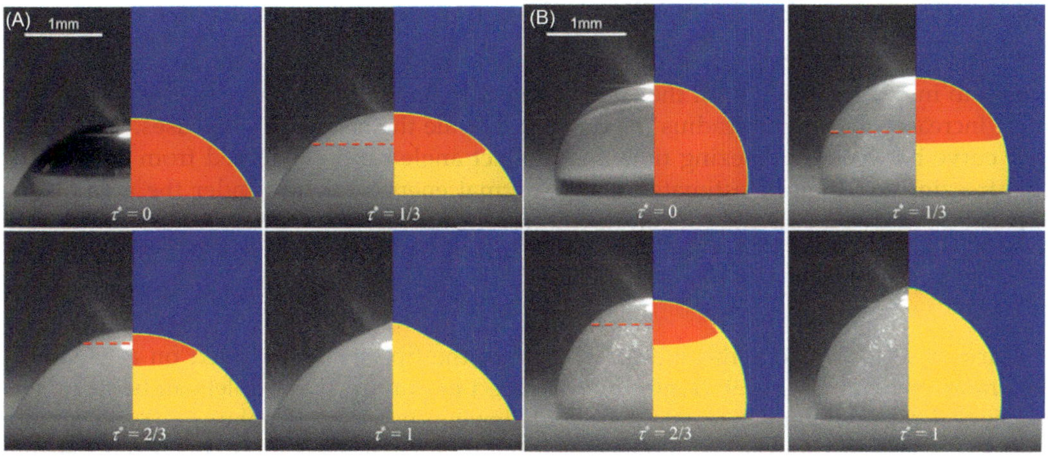

FIGURE 3.6 Comparison of water droplet freezing shapes at different times between experiment (left) and simulation (right).

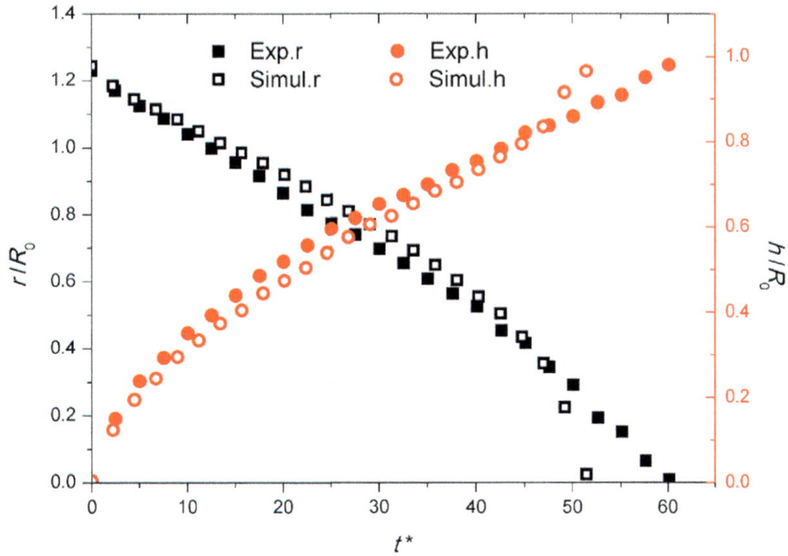

FIGURE 3.7 Comparison of the droplet freezing front height and radii between experimental and simulation results.

experimental freezing process initiating from the nucleation stage, where droplet expansion occurs rapidly. Moreover, the presence of an ice skeleton within the unfreezing water phase enhances heat transfer and reduces the water-ice latent heat released. Consequently, this leads to a smaller St number in the experiment compared to the simulation, resulting in a lower freezing velocity in the simulation for most of the freezing period.

3.4.2 Effects of surface wettability and volume expansion on droplet freezing

In Fig. 3.8, the freezing time for droplets on surfaces with different contact angles is shown to increase rapidly with the contact angle. It is evident to know that as the contact angle increases, the bottom radius (r_b) decreases while the height (h_t) increases. A relationship curve between the freezing time and contact angle can be obtained from the results. This freezing time can be estimated using thermal energy balance under the assumption of pseudo-steady and 1D conduction heat transfer in the ice:

$$t = \frac{\rho V H_{fg}}{\pi r_b^2 q''} \tag{3.54}$$

Marin et al. provided a theoretical analysis of the tip formation based on a geometric illustration shown in Fig. 3.9. Neglecting the effect of gravity, the droplet can be treated as a spherical cap, allowing us to establish a relationship between its volume (V) and bottom radius (r_b). This relationship is derived from the total latent heat released during freezing in the numerator, and the heat transferred to the cold substrate surface per unit time in the denominator:

$$V = r_b^3 f_V(\theta) = \frac{\pi r_b^3}{3}\left(\frac{2 - 3\cos\theta + \cos^3\theta}{\sin^3\theta}\right) \tag{3.55}$$

The dimensionless total freezing time t^* (with respect to $\theta = 90°$) is estimated as a function of contact angle, as displayed in Fig. 3.10. Additionally, based on numerically simulated r_b

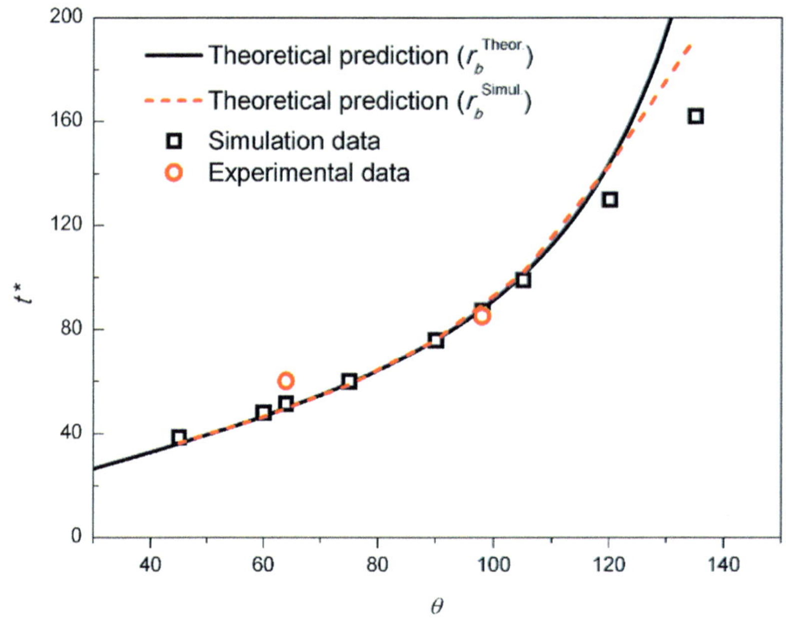

FIGURE 3.8 Total freezing time versus contact angle with a comparison among the simulations, the theoretical predictions, and the experiments.

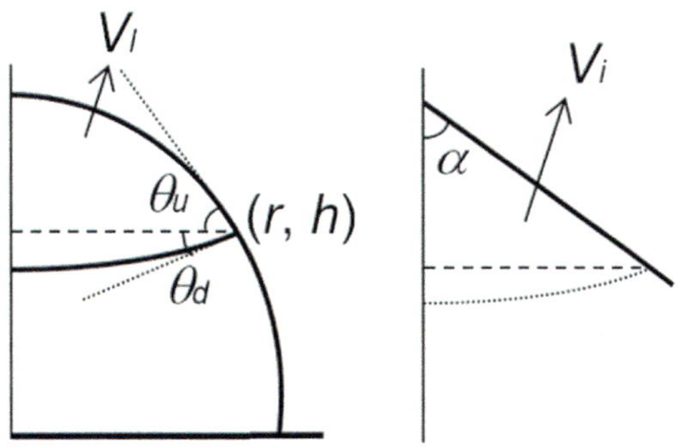

FIGURE 3.9 Sketch of the tip geometry on the droplet at the last freezing stage with the freezing front-to-interface angles.

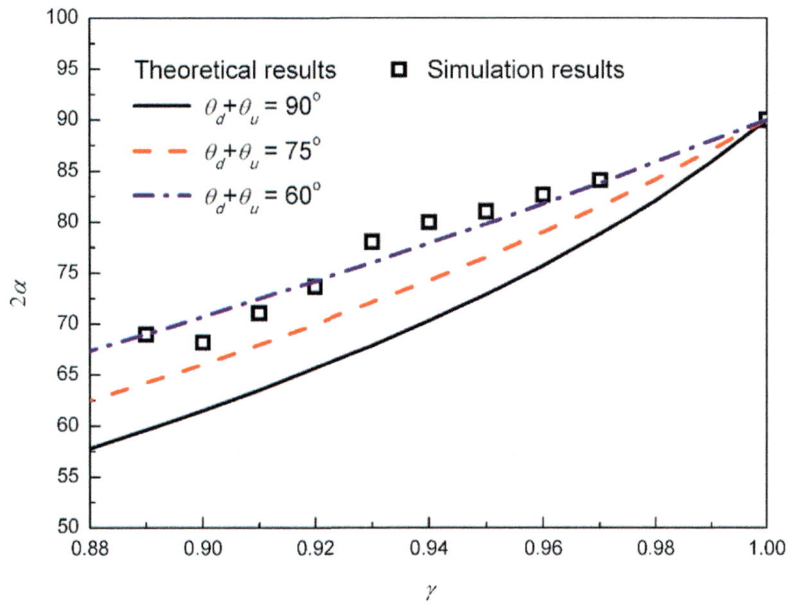

FIGURE 3.10 Cone angle versus volume expansion coefficient.

and Eq. (3.55), another prediction of t^* is plotted for comparison. Except for larger contact angles (i.e., $\theta > 130°$), both the numerical and analytical predictions align well with the experimental results. Differences arise due to higher contact angles resulting in larger droplet heights h_t and reduced contact areas with the bottom surface.

Marin et al. provided a theoretical analysis of tip formation. Close to the final freezing stage, the remaining liquid volume V_l comprises a spherical cap and a circular truncated cone with angles θ_u and θ_d. Thus V_l can be expressed as

$$V_l = r^3 f_V(\theta_u) + r^3 f_V(\theta_d) \tag{3.56}$$

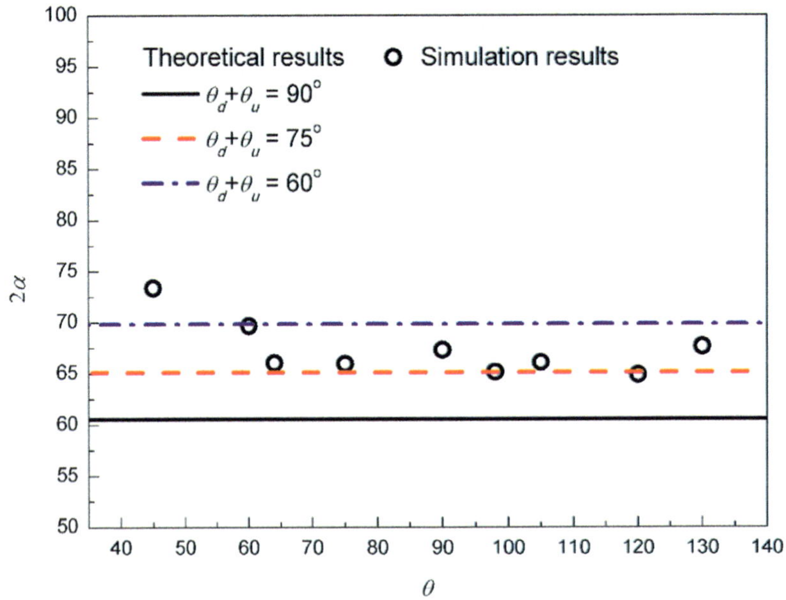

FIGURE 3.11 Cone angle versus contact angle.

After freezing is complete, the solid ice volume V_i is assumed to comprise a spherical cone and the same circular truncated cone as given by

$$V_i = \frac{\pi}{3}\frac{r^3}{\tan\alpha} + r^3 f_V(\theta_d) \tag{3.57}$$

The relationship between the volumes V_l and V_i is determined by $V_l = \gamma V_i$. Fig. 3.11 illustrates the simulated cone angle 2α versus density ratio γ and the contact angle θ, respectively. The simulated cone angle 2α aligns closer with specific lines, indicating a consistent trend. Fig. 3.12 presents the time evolution of freezing front angles, θ_u and θ_d, showing that their sum remains approximately constant throughout most of the freezing period. Therefore our simulation results suggest that the freezing front-to-interface angle $\theta_u + \theta_d$ does not remain constant even at the final stage of freezing, indicating nonuniform freezing front velocities at different radial locations.

3.5 Sessile nanofluid droplet freezing

In this section, we adopt a computation domain of size $L_x \times L_y$, with the left side acting as the symmetric boundary (as shown in Fig. 3.13). The substrate height at the bottom of the domain is set to H_s, and a constant temperature T_w is applied to the bottom of the substrate. After determining the initial droplet shape with contact angle θ, particles are introduced into the domain as follows:

1. Randomly initialize a coordinate within the computation domain as the center of the particle.
2. Check if the coordinate lies within the liquid phase (excluding the interface).

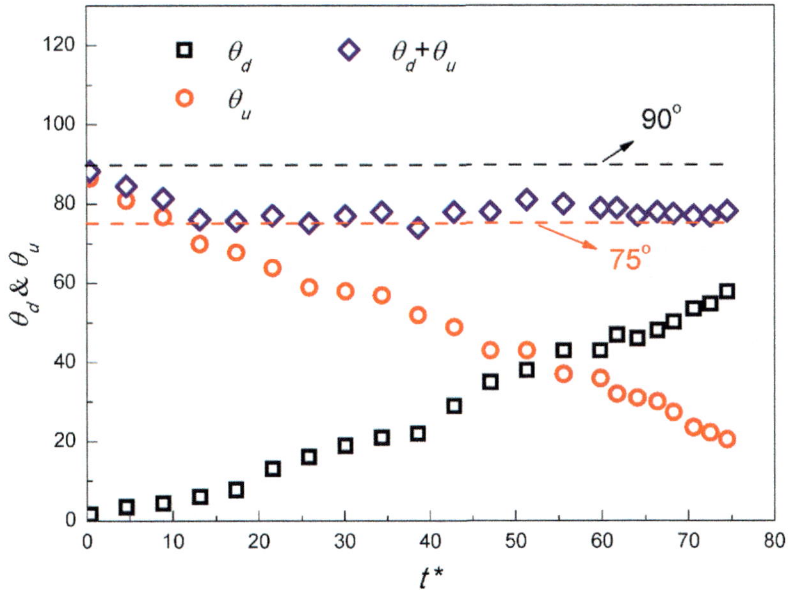

FIGURE 3.12 Time evolution of angles θ_u and θ_d for the case of $\theta = 90°$ and $\gamma = 0.9$.

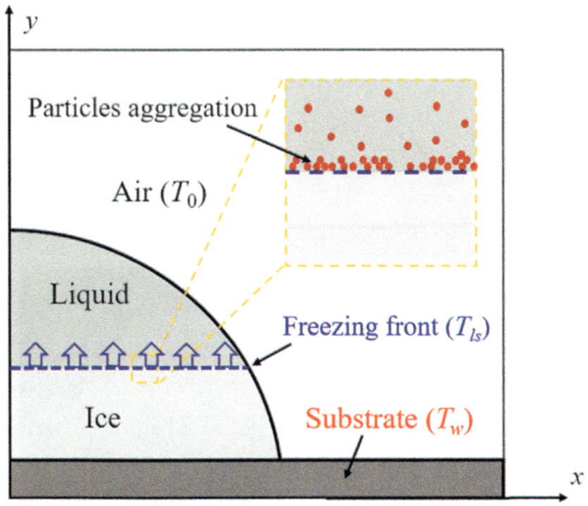

FIGURE 3.13 Schematics of the 2D simulation domain with a freezing droplet placed on a cold plate.

3. Verify if the coordinate is sufficiently distant (distance greater than $2d_p$) from other particles.
4. If conditions (2) and (3) are met, mark the coordinates as a separate particle.
5. Repeat steps 1–4 until the desired particle count N_p is reached, concluding the particle generation loop.

Following the outlined steps, we achieve a relatively uniform distribution of particles within the droplet. In our simulation, the particle diameter is $d_p = 0.1$, resulting in a droplet-to-particle size ratio of $R_0/r_p = 1000$. Assuming a droplet radius of 1.42 mm, the particle size approximates 1.42 μm, consistent with real particle sizes used in experiments.

Furthermore, we neglect gravitational effects due to the small sizes of both droplet and particles. For the solid particle phase, its density is set at 20 ρ_l, the interaction strength G_p is set at -0.6, and the Lagrangian point number N_B is 6.

3.5.1 Droplet shape formation

Fig. 3.14 illustrates the freezing process of nanofluid droplets, combining simulated profiles with experimental photographs. Initially, the droplet, with a contact angle of 60 degrees and temperature T_0, is rapidly cooled at the substrate bottom. The freezing front

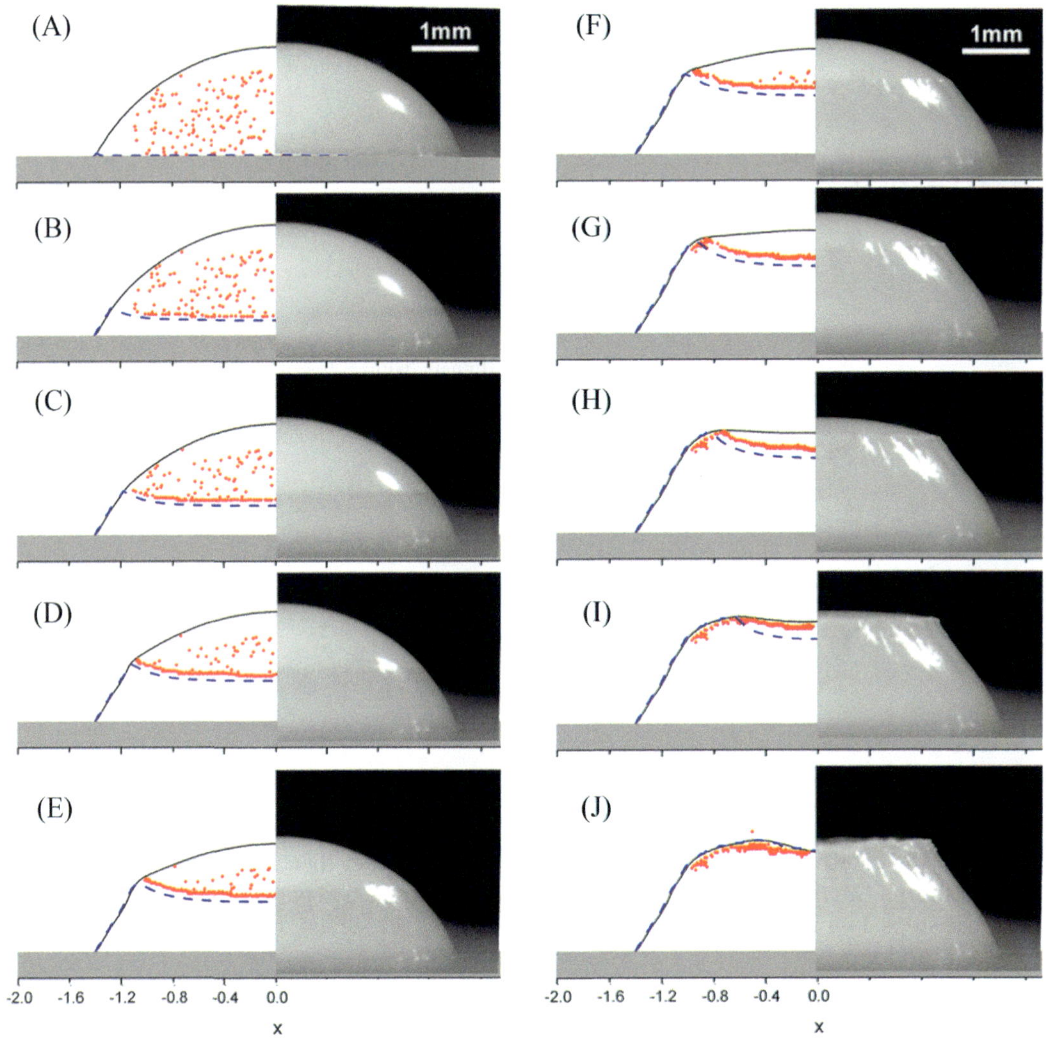

FIGURE 3.14 Simulated dynamic process of freezing droplet with particles. The right sides are snapshots from the experiment of freezing nanofluid droplet.

emerges from the droplet bottom, progressing upward with a concave shape over time. A layer of particles accumulates on the freezing front, inducing a capillary force that drives central liquid toward the tri-junction. This results in a larger curvature radius for the frozen droplet interface. Eventually, as the remaining liquid flattens, particles become trapped in the ice phase, covering the upper droplet interface. This process highlights the dynamic interplay between particle movement and droplet solidification.

Fig. 3.15 shows the freezing front height evolution (h/R_0) from the droplet bottom. Compared to pure droplet freezing, the nanofluid case exhibits slower moving speed for h/R_0. Initially, experimental velocity is higher due to enhanced heat conductivity.

Fig. 3.16 depicts volume evolution for both pure and particle droplets, with identical initial and ice droplet volumes. The volume ratio closely matches our set density ratio of liquid and ice ($\gamma = 0.916$). Under the same volume and density ratio, frozen volume evolution differs between the two cases. In the middle of freezing, ice region volume is larger for the particle case due to larger freezing front area and smaller freezing front height compared to the pure droplet. These effects result in more heat flux being transferred from droplet bottom to liquid phase.

In Fig. 3.17, velocity vectors at intermediate time ($\tau^* = 0.4$) are depicted for both pure water and nanofluid cases. Velocity scale is three times larger in nanofluid droplet than in pure water. In pure water case, bulk liquid's velocity is upward with a small value due to volume expansion effect. In the case of particles inside droplet, flood contour revealing particles' aggregating area shows horizontal flow from upside and center to triple-phase contact region, driven by capillary force between aggregating particles and fluid phase

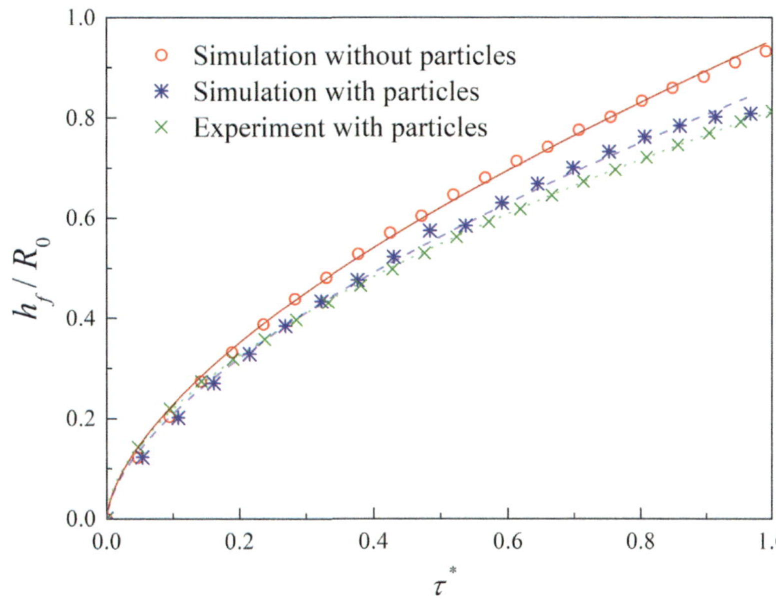

FIGURE 3.15 Evolution of the freezing front height in droplets without and with particles inside.

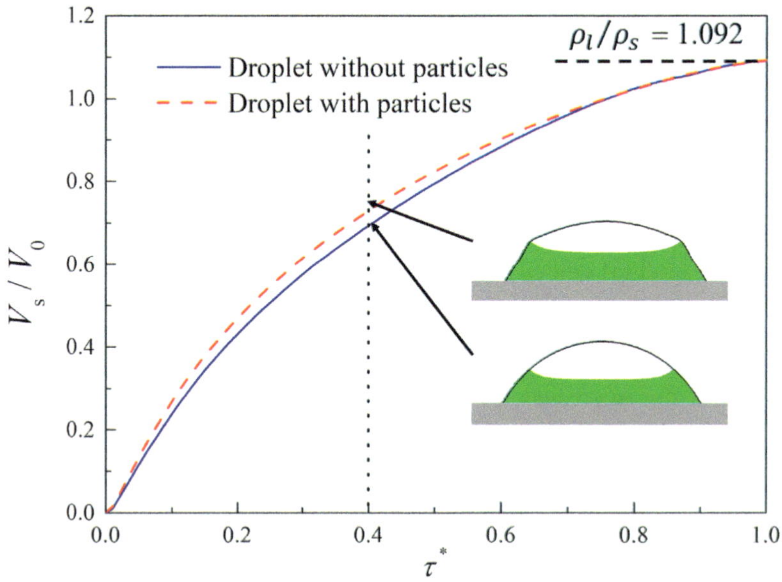

FIGURE 3.16 Ice volume during droplet freezing in droplet without and with particles inside.

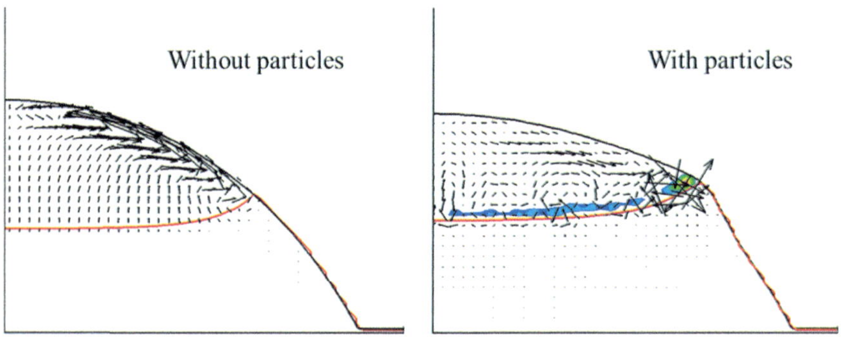

FIGURE 3.17 The velocity vectors in droplet without and with particles inside at $\tau^* = 0.4$.

(Eq. 3.27), believed to be the main reason for the formation of frozen droplet shape with particles inside.

3.5.2 Particle distributions

To examine the impact of substrate surface wettability on droplet shape, the variation was made while maintaining constant droplet volume. A parameter φ, representing accumulation of adjacent particles on lattice grids, described particle distribution in the droplet.

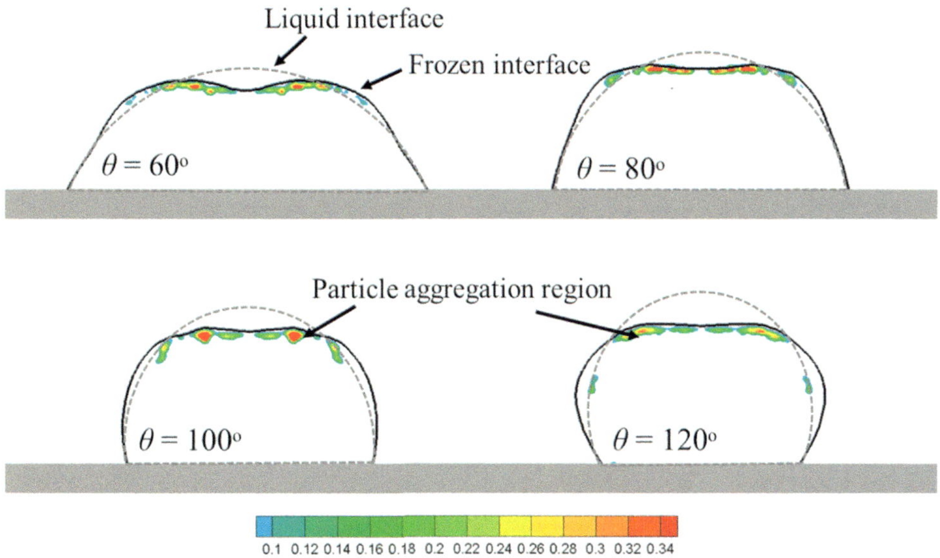

FIGURE 3.18 Frozen droplet shape with different contact angles at particle volume content $\Phi = 0.36\%$. Dash lines represent liquid droplet profiles.

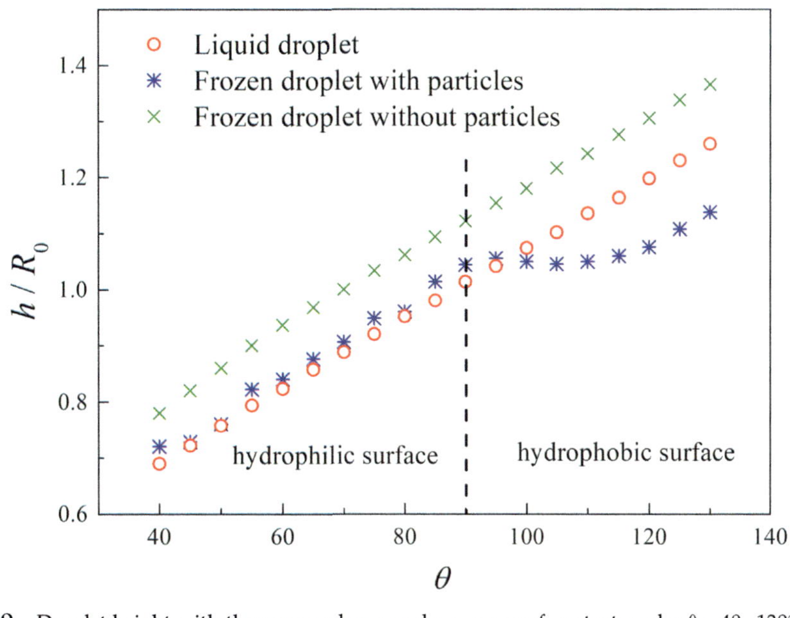

FIGURE 3.19 Droplet height with the same volume under a range of contact angles θ = 40–130°.

Fig. 3.18 illustrates frozen droplet profiles and particle distribution maps for four contact angles: $\theta = 60°$, 80°, 100°, and 120°. In these figures, particles primarily concentrate at the top of frozen droplets, with fewer particles near the interface for larger contact angles ($\theta = 120°$). On the plateaus formed at the top, particles tend to distribute symmetrically in

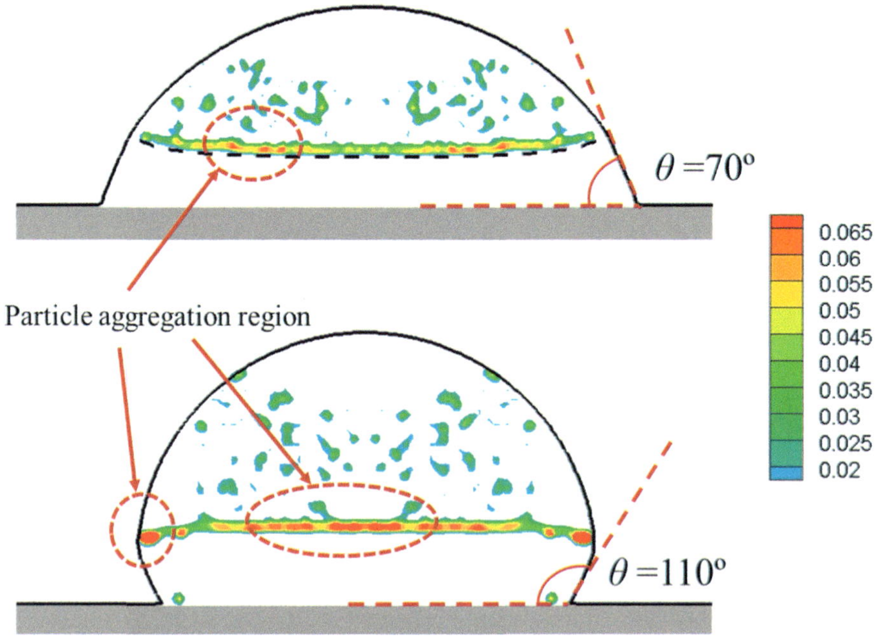

FIGURE 3.20 Particle distribution and freezing front shape at $t^* = 0.2$ with droplet on hydrophilic ($\theta = 70°$) and hydrophobic ($\theta = 110°$) surfaces.

two parts rather than centrally. Additionally, it was observed that frozen droplet height does not increase with increasing contact angle.

This particle aggregation phenomenon is confirmed in Fig. 3.19, depicting initial liquid droplet height, frozen height of pure droplet case, and nanofluid droplet case across a range of contact angles $\theta = 40–130°$. Initial liquid droplet height exhibits linear increase with contact angle, as does pure frozen height, which surpasses the initial height.

However, in the nanofluid case, the frozen height exhibits a different trend depending on whether the contact angle exceeds $\theta = 90°$. The frozen height linearly increases, similar to the pure droplet height. Conversely, when $\theta > 90°$, the frozen height remains relatively unchanged as the contact angle increases. To understand this difference, we analyzed particle distribution characteristics at an early stage for hydrophilic and hydrophobic surfaces. In Fig. 3.20, both cases are depicted at dimensionless times $t^* = 0.2$ with contact angles $\theta = 70°$ and $\theta = 110°$, respectively.

Distinct particle distributions were observed on the freezing fronts, particularly at the two sides. For the hydrophilic droplet, the freezing front exhibited a concave shape, with particles distributed relatively uniformly. Conversely, for the hydrophobic droplet, the freezing front displayed a convex shape, resulting in more particles aggregating at the tri-junction region. As the number of particles increased, the capillary force exerted by particles on both sides grew, leading to wider expansion in the horizontal direction and a smaller frozen droplet height.

3.6 Conclusions

This chapter outlines LBMs used to simulate the freezing process of sessile droplets undergoing liquid–solid phase change. The novel LB model integrates a multiphase entropy model, addressing phase change with volume expansion effects due to density differences, alongside the immersed boundary method to model numerous particles. By comparing LB simulations with experimental data on frozen droplet shape and freezing front results for water droplets, distinct freezing dynamics and droplet shapes resulting from particle segregation and attractive interaction forces with the liquid are observed. These findings offer mechanistic insights into complex multiphase and phase change systems, broadening the application of the LBM to particle-suspended multiphase flow research.

References

[1] M.-H. Kim, H. Kim, K.-S. Lee, D.R. Kim, Frosting characteristics on hydrophobic and superhydrophobic surfaces: A review, Energy Conversion and Management 138 (2017) 1–11.
[2] Y. Lin, H. Chen, G. Wang, A. Liu, Recent progress in preparation and anti-icing applications of superhydrophobic coatings, Coatings 8 (2018) 208.
[3] D. Anderson, M.G. Worster, S. Davis, The case for a dynamic contact angle in containerless solidification, Journal of Crystal Growth 163 (1996) 329–338.
[4] P. Hao, C. Lv, X. Zhang, Freezing of sessile water droplets on surfaces with various roughness and wettability, Applied Physics Letters 104 (2014).
[5] A.G. Marin, O.R. Enriquez, P. Brunet, P. Colinet, J.H. Snoeijer, Universality of tip singularity formation in freezing water drops, Physical Review Letters 113 (2014)054301.
[6] M. Nauenberg, Theory and experiments on the ice–water front propagation in droplets freezing on a subzero surface, European Journal of Physics 37 (2016).
[7] Y. Zhao, C. Yang, P. Cheng, Freezing of a nanofluid droplet: from a pointy tip to flat plateau, Applied Physics Letters 118 (2021).
[8] T. Lei, K.H. Luo, D. Wu, Generalized lattice Boltzmann model for frosting, Phys Rev E 99 (2019)053301.
[9] X.S.H. Chen, Simulation of non-ideal gases and liquid-gas phase transitiona by lattice Boltzmann equation, Physical Review E 49 (1994) 2941.
[10] X. Shan, Multicomponent lattice Boltzmann model from continuum kinetic theory, Physical Review, E, Statistical, Nonlinear, and Soft Matter Physics 81 (2010)045701.
[11] C. Zhang, H. Zhang, W. Fang, Y. Zhao, C. Yang, Axisymmetric lattice Boltzmann model for simulating the freezing process of a sessile water droplet with volume change, Physical Review E 101 (2020).
[12] C. Zhang, H. Zhang, Y. Zhao, C. Yang, An immersed boundary-lattice Boltzmann model for simulation of deposited particle patterns in an evaporating sessile droplet with dispersed particles, International Journal of Heat and Mass Transfer 181 (2021).
[13] J. Favier, A. Revell, A. Pinelli, A Lattice Boltzmann–Immersed Boundary method to simulate the fluid interaction with moving and slender flexible objects, Journal of Computational Physics 261 (2014) 145–161.
[14] A. Kupershtokh, Calculations of the action of electric forces in the lattice Boltzmann equation method using the difference of equilibrium distribution functions, Proc. 7th Int. Conf. on Modern Problems of Electrophysics and Electrohydrodynamics of Liquids, 152, St. Petersburg State University, St. Petersburg, Russia, 2003, p. 155 155.
[15] D. Chatterjee, S. Chakraborty, An enthalpy-based lattice Boltzmann model for diffusion dominated solid–liquid phase transformation, Physics Letters. A 341 (2005) 320–330.
[16] D. Chatterjee, S. Chakraborty, An enthalpy-source based lattice Boltzmann model for conduction dominated phase change of pure substances, International Journal of Thermal Sciences 47 (2008) 552–559.

[17] X. Li, J. Zhao, P. Cheng, A lattice Boltzmann model for condensation and freezing of dry saturated vapor about a cryogenic spot on an inclined hydrophobic surface, International Journal of Heat and Mass Transfer 114 (2017) 628–639.
[18] S.A. Reijers, H. Gelderblom, F. Toschi, Axisymmetric multiphase lattice Boltzmann method for generic equations of state, Journal of Computational Science 17 (2016) 309–314.
[19] X.D. Niu, C. Shu, Y.T. Chew, Y. Peng, A momentum exchange-based immersed boundary-lattice Boltzmann method for simulating incompressible viscous flows, Physics Letters. A 354 (2006) 173–182.
[20] C.S. Peskin, The immersed boundary method, Acta Numerica 11 (2002) 479–517.
[21] S.K. Kang, Y.A. Hassan, A comparative study of direct-forcing immersed boundary-lattice Boltzmann methods for stationary complex boundaries, International Journal for Numerical Methods in Fluids 66 (2011) 1132–1158.
[22] X. Zhang, X. Liu, J. Min, X. Wu, Shape variation and unique tip formation of a sessile water droplet during freezing, Applied Thermal Engineering 147 (2019) 927–934.
[23] Q. Li, Y.L. He, G.H. Tang, W.Q. Tao, Lattice Boltzmann model for axisymmetric thermal flows, Physical Review. E, Statistical, Nonlinear, and Soft Matter Physics 80 (2009)037702.
[24] H. Ding, P.D. Spelt, Wetting condition in diffuse interface simulations of contact line motion, Physical Review. E, Statistical, Nonlinear, and Soft Matter Physics 75 (2007)046708.

CHAPTER

4

Freezing of a multicomponent drop: characteristics and mechanism

Yugang Zhao[1], *Haoting Cai*[1], *Chun Yang*[2] *and Ping Cheng*[1,3]

[1]Shanghai Key Laboratory of Multiphase Flow and Heat Transfer in Power Engineering, School of Energy and Power Engineering, University of Shanghai for Science and Technology, Shanghai, P.R. China [2]School of Mechanical and Aerospace Engineering, Nanyang Technological University, Singapore [3]MOE Key Laboratory of Power Machinery and Engineering, School of Mechanical Engineering, Shanghai Jiao Tong University, Shanghai, P.R. China

4.1 Introduction

For all phase-change phenomena, icing or solidification is widely considered to be the last piece of the jigsaw, particularly in drop form where a free convective interface and a solid base play important roles, drawing increasing research attention. In this very recent decade, water freezing in droplet form has been studied extensively for a wide spectrum of purposes, ranging from exploring the limit of supercooling [1–3], obtaining different types of ice [4–6], analyzing the effects of surface properties on icing dynamics [7,8], or even assembling frozen droplets to form complex mesoscale 3D structures [9,10]. A list of significant mechanisms has been revealed: metastable quasiliquid layers which can sustain even at subcooling temperatures covering ice surfaces [11,12], kinetic behaviors such as dislodging and explosion accompany with the freezing dynamics [13,14], interplay among condensed drops with (ice bridging) [15,16] and without physical touches (freezing halo) [17–19], evolutions of morphologies for impinging droplets on subcooled surfaces [20–22], nucleation kinetics in special scenarios like nanometer-sized and suspended droplets in a power ultrasound flow [23,24], influences of gravity [25,26], system alignments [27,28] surface chemistry and topological features [29,30], and externally applied physical fields [31–33].

Frosting and Icing for Efficient Energy Use in Engineering Applications
DOI: https://doi.org/10.1016/B978-0-443-15495-9.00011-3

Occasionally, and almost inevitably in many applications, water droplets are mixed with a second or even third component, which may present in the forms of solid particles, ions, dissolved gaseous phases, another miscible fluid (e.g., alcohol in water), or immiscible fluid (e.g., oil in water), uniformly dispersed in the droplets or some region with significantly higher concentration (e.g., near the free interface). The added components can change the thermophysical properties of the liquid mixture, promote or hinder icing nucleation, alter the profiles of dynamic contact lines and freezing fronts, stimulate an interfacial flow that can yield a net mass redistribution, and modulate the morphologies of frozen droplets. Meanwhile, during the freezing process, the added components interplay with the advancing freezing interface, via freezing segregation or engulfment, and accordingly themselves are locally or globally relocated, engendering the formation of micrometer-sized bubbles, droplets, and particle clusters inside or atop the main frozen droplets [34,35]. Some of these changes can vanish at the moment of remelting, restoring a uniform liquid mixture, and other changes, however, can sustain for a relatively long term [36,37]. The freezing of such mixture droplets, with both the intriguing freezing characteristics and their underlying mechanisms, has numerous applications in freshwater harvest [38,39], biochemistry engineering [40,41], atmospheric studies [42,43], and among others.

This chapter aims to provide a comprehensive up-to-date review on the topic of freezing of a multicomponent drop. First, we list a few important features in freezing of a pure water drop, such as the limit of supercooling and preferential sites of icing nucleation, two successive freezing stages, advancing dynamics of the freezing interface, frozen drop profiles, and also freezing-induced kinetic behaviors, to denote later how these features are changed when adding new components. As shown in Fig. 4.1, second, we discuss the changes caused by the added new components in terms of physicochemical properties, nucleation kinetics, the morphology of the phase front, freezing segregation behavior, in-drop flow, and mass redistribution. Third, we place specific emphasis on six popular scenarios, including with dissolved gases, ionic components, solid nanoparticles, miscible, and immiscible fluids, to illustrate the variability of freezing outcomes. We believe that this review will improve the existing understanding on the freezing of a multicomponent drop and also benefit the understanding on phase change in other complex systems.

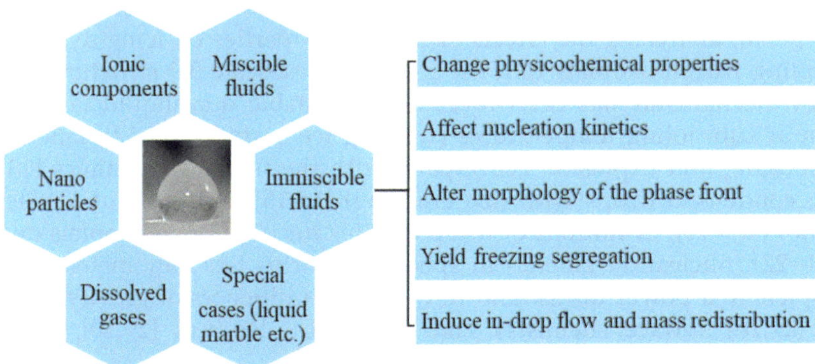

FIGURE 4.1 Different types of added components and their impacts on the freezing dynamics of a multicomponent drop.

4.2 Freezing characteristics of a pure water drop

4.2.1 Supercooling and sites of nucleation

Icing nucleation, referred to as a stochastic process forming new ice clusters from the mother liquid domain, is well-developed by Fletcher using a classical thermodynamic approach [44]. Arranging water molecules from the chaotic state to a long-range order crystalline structure with a diameter larger than the critical size of tens of nanometers generally occurs at a temperature much lower than the equilibrium freezing point, termed the degree of supercooling or subcooling.

For a pure water drop, icing nucleation preferentially occurs at the solid–water interface where the heterogeneous nucleation mode is exposed to a much smaller energy barrier. According to Fletcher's description of heterogeneous nucleation, a smaller drop reduces the possibility of icing nucleation and thus can sustain an appreciably larger degree of supercooling. Inada et al. [45] evaluated the effect of drop size varying from micrometer to submilimeter on icing nucleation. Their directional experimental results show that the degree of supercooling reduces substantially when the interface area decreases from 1×10^{-8} to 1×10^{-10} m^2. Meanwhile, icing nucleation is initiated within a narrow temperature gap of 2°C for the interface smaller than 1×10^{-10} m^2, while this temperature scatters with increasing interface area.

There is a longstanding debate whether special locations such as the triple line are preferred sites for nucleation in the context of the whole solid–water interface. Gurganus et al. [46,47] designed a microscopic high-speed imaging system and studied the incidence of surface freezing for millimeter-sized drops resting on a smooth homogeneous silicon wafer. As shown in Fig. 4.2A with direct image sequences captured, both contact nucleation (at the triple line) and immersion nucleation (at the contact interface) can be observed. After collecting more nucleation incidences, they plotted the locations of hundreds of freezing events on the horizontal plane as shown in Fig. 4.2B for two different contact angles. In contrast to prior reports in the literature, they found the nucleation sites distributed uniformly across the contact surface with no noticeable preference for the triple line.

4.2.2 Two successive freezing stages

After icing nucleation is initiated, the following in-drop freezing process occurs in two successive stages, particularly for drops with an appreciatable degree of supercooling. Jung et al. [48] studied the freezing process of a sessile water drop when homogeneous nucleation is triggered at the free interface by an unsaturated cold nitrogen shear flow. Image sequences of the first recalescence stage and the second isothermal stage are shown in Fig. 4.3A. In the rapid recalescence stage (a time step of 6 Ms between snapshots), partially frozen ice starts from the site of nucleation and covers the whole water drop, yielding a gradually cloudy appearance. Depending on the degree of supercooling upon the icing nucleation, the solid ratio varies. Most of the latent heat released in this stage is used to heat up the remaining liquid instead of dissipating to the surrounding air, restoring macroscopically the drop temperature to the equilibrium freezing temperature. In the significantly slower isothermal stage (an average time step of 4 seconds between snapshots),

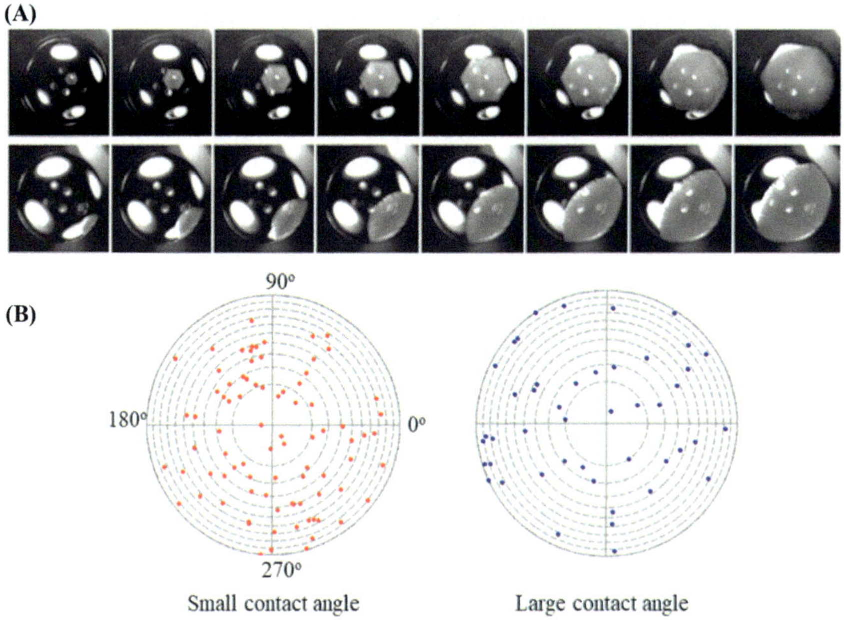

FIGURE 4.2 (A) Freezing dynamics in the form of immersion nucleation initiated at the central contact interface and contact nucleation initiated at the triple line, respectively. Time steps between images are 1 Ms. (B) Distribution of nucleation sites on a clean hydrophilic surface (left panel) and a silanized hydrophobic surface. Source: *(A) Reprinted with the permission from C. Gurganus, A.B. Kostinski, R.A. Shaw, Fast imaging of freezing drops: no preference for nucleation at the contact line, Journal of Physical Chemistry Letters, 2(12) (2011) 1449–1454. https://pubs.acs.org/doi/10.1021/jz2004528. ©2011 American Chemical Society. (B) Reprinted with the permission from C. Gurganus, A.B. Kostinski, R.A. Shaw, High-speed imaging of freezing drops: still no preference for the contact line, The Journal of Physical Chemistry C, 117(12) (2013) 6195–6200. ©2013 American Chemical Society.*

a smooth freezing front proceeds from the substrate base, filling micrometer-sized gaps among ice dendrites formed in the recalescence stage.

Released latent heat in this stage is mostly dissipated via the subcooled substrate, while the temperature change of the liquid residual is negligible.

Thermophysical properties of the drop, the substrate, and the surroundings modulate these two freezing stages, such as the duration and characteristics of each stage. Wang et al. [50] evaluated quantitatively the effects of substrate supercooling and wettability on two freezing stages. Colder substrate induces an earlier icing nucleation and a faster freezing rate, while the average recalescence duration remains unchanged. Compared to the bare plain brass, the coated hydrophobic substrate can largely suppress icing nucleation. Also, they found an appreciatable thermal hysteresis associated to the recalescence stage from the infrared thermographic results, showing the temperature rise as the result of recalescence latent heat release can last more than one order of magnitude than the recalescence duration.

Drop size also plays an important role in the two freezing stages. Buttersack and Bauerecker [49] experimentally studied the freezing of suspended subcooled drops, with their emphasize placed on the correlation among drop size, dendritic freezing velocity, and heat transfer coefficient. They found that the dendritic freezing front is almost planar,

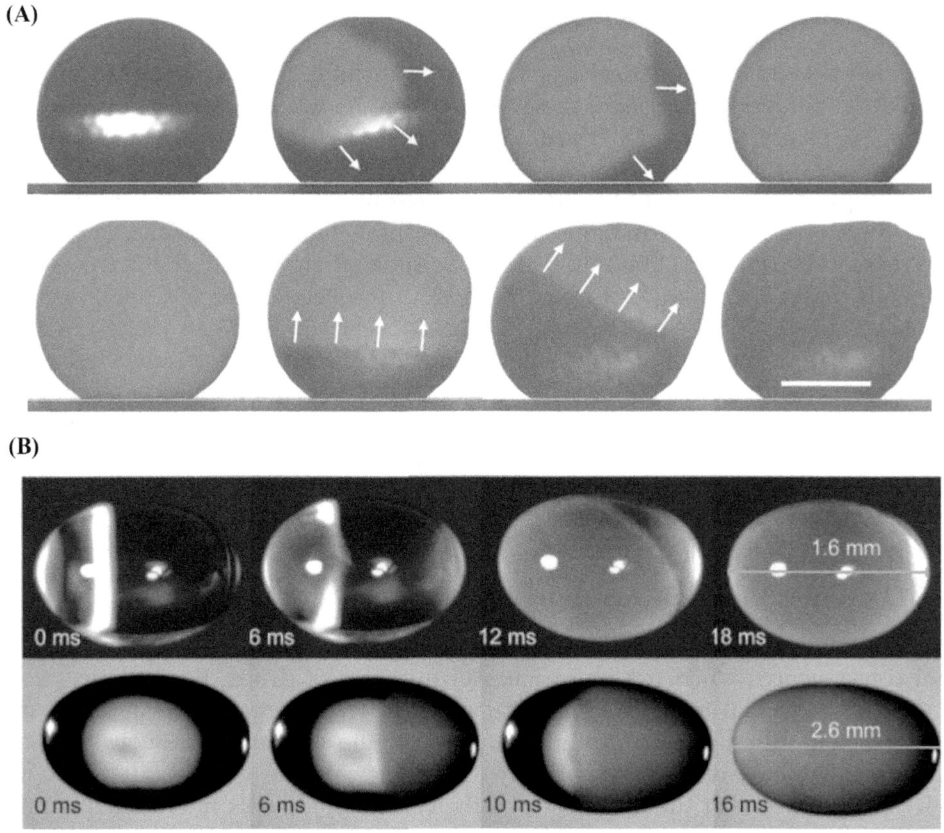

FIGURE 4.3 (A) Freezing of a sessile water drop exposed to a supercooled nitrogen flow, the recalescence stage (upper panel, time step of 6 Ms) and the isothermal stage (lower panel, average time step of 4 s). (B) Dendritic freezing of levitated drops of two different sizes [49]. Source: *(A) Reprinted with the permission from S. Jung, M.K. Tiwari, N.V. Doan, D. Poulikakos, Mechanism of supercooled droplet freezing on surfaces, Nature Communications, 3 (2012) 615. ©2012 Springer Nature. (B) Reprinted with the permission from T. Buttersack, S. Bauerecker, Critical Radius of supercooled water droplets: on the transition toward dendritic freezing, The Journal of Physical Chemistry. B, 120(3) (2016) 504–512. ©2016 American Chemical Society.*

and it propagates considerably faster for a smaller drop size as shown in Fig. 4.3B. More importantly, their finite element method simulation results show the presence of a critical drop size in the range of 0.1–1 μm below which only one-stage freezing occurs. Furthermore, Buttersack et al. [51] pointed out that conventionally two-stage freezing can transit to one-stage freezing even for millimeter-sized drops at the hypercooling temperature of around −64°C.

4.2.3 Advancement of the freezing interface

As the ice phase formed in the recalescence stage is a partially solidified spongy network, most literature describes only the freezing interface profile of the quasiequilibrium isothermal stage while neglecting protrusions or dendrites. Anderson et al. [52] found that conditions of the freezing interface can be dominative for the product of freezing drops and provided a simple geometrical description by assuming the freezing interface to be planar, that is, parallel to the solid substrate. Although such an assumption is relatively simplifying, a planar freezing interface as described is consistent with one-dimensional heat conduction.

A more explicit description of the advancing freezing interface needs more inputs from mass and heat conservation equations across the solid–liquid, liquid–air, and solid–air interfaces. Chaudhary and Li [53] found that the propagation velocity of the freezing interface depends on the drop temperature at the prerecalescence instant and the initial contact angle. They numerically solved the enthalpy-based heat conduction equation and found that profiles of the freezing interface interpreted via temperature counters are concave, particularly for smaller contact angles. Vu et al. [54] developed a numerical method allowing dynamically tracking the profile of the freezing interface and the free interface. Once adding the volume change term with the assumption of axisymmetry, they could even predict the instant of complete freezing, that is, the formation of a singular tip.

4.2.4 3D profiles for final frozen drops

One intriguing property of water is that it expands upon freezing, yielding some unusual consequences when freezing occurs in the form of drops. For instance, a sessile water drop of the initial truncated spherical cap shape ends into a conical tip at the last stage of freezing, although surface tension prevents the formation of such singularities. Marín et al. [55] measured systematically the angles of the conical tip as shown in Fig. 4.4A. They found that the magnitude of the conical angle fails into a narrow range of $140 \pm 10^{\circ}$, while this value is independent of the substrate supercooling and wettabilities. Similar behaviors were also reported for another two materials, silicon and germanium, which also expand upon freezing [58,59]. Starostin et al. [60] experimentally studied the change of conical angle when drop freezing occurs on inclined or even upside-down substrates. Their results confirm the universal mechanism of the tip formation, and the magnitude of the conical angle was not influenced by the spatial orientation of the substrate, contact angle hysteresis, and drop size.

More types of morphologies can be obtained when a water drop impacts and freezes on the supercooled substrate instead of being a sessile drop. Fang et al. [61] attributed the frozen drop morphologies to the interaction between spreading and freezing. By varying the impact velocity and substrate supercooling, they observed a list of different freezing morphologies: central cap, single ring, double rings, and basin. In practical applications such as aircraft icing and frozen rain, impact and freezing occur more frequently on inclined substrates. Zhu et al. [20] systematically investigated the freezing dynamics of drop impact and freezing on a supercooled inclined substrate (45°) and obtained four types of frozen morphologies as shown in Fig. 4.4B, termed elliptical cab, half ring + cap I, half ring + cap II, and lastly half ring + sing ring as the impact velocity increased.

Aside from the main feature of the overall frozen drop morphologies, there are also noticeable microscale features. Schiaffino and Sonin [56] and Xiong et al. [57] studied the

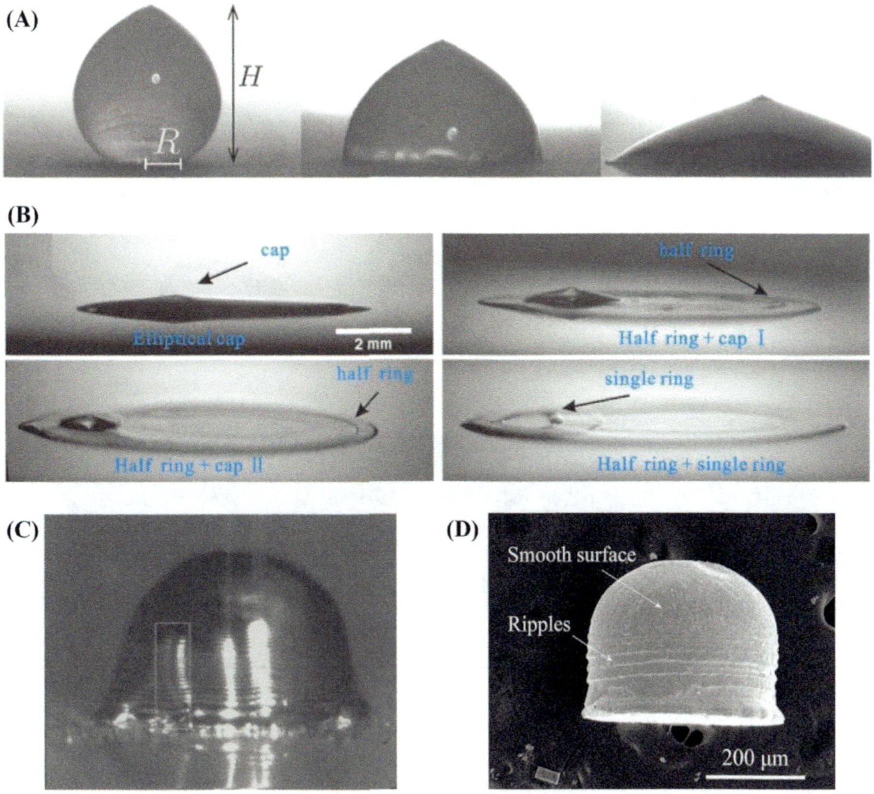

FIGURE 4.4 (A) Morphologies of frozen sessile drops. (B) Morphologies of drops from impact and freezing on an inclined surface. (C) Morphologies of frozen mercury drop. (D) Morphologies of frozen solder drop. Source: *(A) Reprinted with the permission from [55]. ©2016 American Physical Society. (B) Reprinted from [20]. ©2021 with permission from Elsevier. (C) Reprinted with the permission from [56]. ©1997 American Institute of Physics. (D) Reprinted from [57]. ©2017 with permission from Elsevier.*

freezing dynamics of impact metal melt drops and found multiple tiny striations adjacent to the contact line as shown in Fig. 4.4C and D. For viscous fluids, oscillations interface with the advancing freezing interface, yielding perturbations of local pressure and leaving striations in the final frozen phase.

4.2.5 Freezing-induced dynamic and kinetic behaviors

To complete freezing, drops have to undergo a set of successive processes, such as sustaining a considerable degree of supercooling, nucleation at preferential sites, and phase propagation in two freezing stages. In many scenarios, very intriguing dynamic and even kinetics behaviors have been reported recently. Lambley et al. [62] investigated the freezing of supercooled drops resting on microtexture-engineered surfaces as shown in Fig. 4.5A. The explosive evaporative flux simulated by the latent heat released in the recalescent

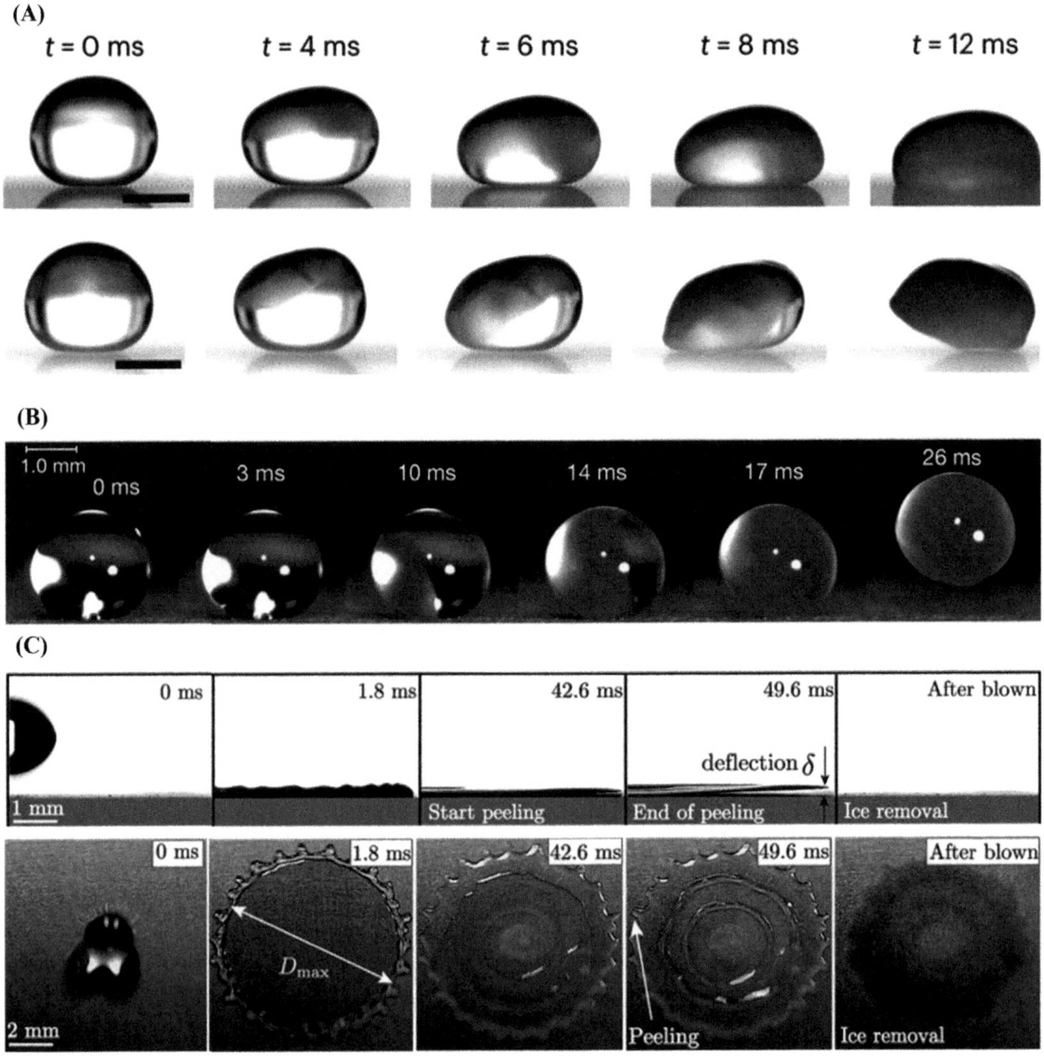

FIGURE 4.5 (A) Impalement (upper panel) and expulsion (lower panel) behaviors occur during the recalescent freezing of a supercooled drop on a microtextured surface. (B) Water drop spontaneously departs the substrate upon freezing. (C) Drop impacts, spreads, freezes, and peels off the substrate. Source: *(A) Reprinted with the permission from [62]. Copyright 2023 Springer Nature. (B) Reprinted with the permission from [63]. Copyright 2015 Springer Nature. (C) Reprinted with the permission from [64]. Copyright 2023 Springer Nature.*

freezing stage results in a force that can either push the drop to penetrate microtextures, inducing the wetting transition from Cassie-Baxter to Wenzel state, or expel the drop from the substrate. Schutzius et al. [63] observed that a water drop can spontaneously depart from the carrying substrate at the onset of the freezing as shown in Fig. 4.5B. Once the adhesion between the drop and the substrate base is small enough, that is, poor wetting on superhydrophobic surfaces, vapor flux can boost the levitation, lifting away the freezing drop and accordingly maintaining a clear surface. Fang et al. [64] reported a different drop departure phenomenon where a water drop impacts on a supercooled smooth hydrophobic surface as shown in Fig. 4.5C. A thermal–mechanical model was developed that attributes the peeling behavior to the competition among thermal contraction of the ice phase, cohesion energy among ice molecules, and adhesion energy among ice and the substrate.

4.3 How added components modulate freezing dynamics

Once a new component is added, intentionally for fundamental studies or inevitably in many applications, the drop freezing behaviors will be altered to a considerable large extent, from liquid drop shape, impact dynamics, and wetting state to the maximum degree of supercooling, nucleation kinetics, profile of the freezing front, and finally the frozen morphologies.

4.3.1 Effect on thermophysical properties of fluids

The thermal conductivity of mixtures is of considerable importance, in determining the rate of heat transfer in the drop freezing process. Hereby, we use base fluids with dispersed nanoparticles, that is, referred to as nanofluids, to demonstrate the effect of added nanoparticles on the thermal conductivity of nanofluids. Murshed et al. [65] prepared a set of nanofluids by dispersing rod-shaped and spherical-shaped TiO_2 nanoparticles into deionized water. The conductivity of these two nanofluids was measured using a transient hot-wire apparatus. Their results show that the thermal conductivity of nanofluids increases nonlinearly with particle concentration, and the particle size and shape also affect the magnitude of this enhancement. Buongiorno et al. [66] collaborated with over 30 organizations worldwide, using many different measuring approaches and experimentally measuring the thermal conductivity of varying base fluids, particle types, sizes, shapes, and concentrations. Their results agree well with Maxwell's effective medium theory, showing that anomalous enhancement in the thermal conductivity of nanofluids is unlikely to occur. Fan et al. [67] and Hossain et al. [68] proposed to use thermal enchantment of nanofluids to promote heat transfer in phase change materials for energy storage applications.

The viscosity of the multiphase mixture affects drastically the drop deformation dynamics upon impact and intra-drop flow during the freezing process [69,70]. Murshed et al. [71] conducted a combined experimental and theoretical study on the viscosity of an aqueous-based nanofluid. They found that the nanofluid viscosity increases with the particle concentration, faster than predictions from classic models. Sundar et al. [72] prepared magnetic nanofluids by mixing nanodiamond and nickel nanoparticles and measured their viscosity at varying loading of nanoparticles and different base fluids. Their results show that the ratio of viscosity enhancement varies for water and ethylene glycol with the same particle loading. Zhang et al. [73] examined the effect of pH on the stability of nanofluids, which in turn affects the viscosity. Further away from the isoelectric point, the liquid becomes more stable, yielding lower viscosity.

Surface tension determines the wetting mode and thus the effective heat transfer area when drops freeze. Both adding surface active agents or nanoparticles can drastically change the surface tension of nanofluid, engendering different drop shapes. Chen et al. [74] measured the surface tension of different nanofluids using the pendant drop method. Nanofluid drops suspended in air evaporate and get concentrated, and accordingly, their surface tension decreases. They also found a hysteresis where an evaporating drop can achieve lower surface tension than the bulk nanofluid of a small concentration. Bhuiyan et al. [75] prepared three types of aqueous-based nanofluids and used the traditional Du-Noüy ring method to measure the surface tension. Their results show that both increasing concentration and size can lead to larger surface tension, that is, the opposite of Chen et al.'s results. Khaleduzzaman et al. [76] conducted a systematic study to evaluate the effect of surfactant, concentration, and temperature. They combined both experimental measurements and theoretical correlations. However, many contradictory results were obtained, showing that our understanding of the surface tension of complex fluids is limited.

4.3.2 Effect on nucleation kinetics

How added new composites affect the nucleation kinetics in the drop freezing process can be classified into two groups, either serving as nucleation agents to promote icing nucleation or suppress icing nucleation.

According to Fletcher's theory [77] on heterogeneous nucleation, the generation of a new phase with the assistance of foreign particles requires much less energy, and the efficiency of solid particles depends on their size, shape, and surface chemistry. Among all the additives, silver iodide (AgI) is widely known as an effective icing nucleation agent. Zhang et al. [78] specially prepared nanometer-sized AgI particles and investigated their efficiency in promoting icing nucleation. They found that AgI particles with an average size of 90 nm can exhibit significantly higher icing nucleation efficiency and higher threshold temperature than large AgI particles. Okawa et al. [79] prepared solid particles of different sizes, concentrations, and materials and evaluated their icing nucleation performance in supercooled water. Their results show that the degree of supercooling depends neither on the particle size nor its concentration but instead depends on the total area exposed to the water. Bai et al. [1] identified the critical ice nucleus by studying icing nucleation in water drops containing graphene oxide nanosheets of varying sizes. Their experimental results show that added nanosheets have no impact on icing nucleation if smaller than a certain size that also depends on the degree of supercooling.

Some added composites can hinder icing nucleation and lower the freezing temperature, a phenomenon referred to as freezing point depression as one of the colligative properties of matter. Stefl and George [80] summarized the developments of antifreeze, including different types of antifreezes such as brines, alcohols, and ethylene glycol-based, different applications such as engine coolants and deicing fluids in aviation industries. Many living creatures, polar fishes and many frog species, can survive in supercooling environments due to the presence of biomacromolecules that prevent the intra-cell or tissue growth of ice dendrites. Ampaw et al. [81] introduced the history of biological antifreeze proteins (AFP) and reported four types of AFPs found in 23 documented species. Different from freezing point depression as in solutions of small molecules, AFPs can adhere and occupy the surface of newly emerged ice crystals, preventing them from further growth due to binding more water molecules. Inspired by natural AFPs, many synthetic antifreezes are also reported. Biggs et al. [82] reviewed the rapid growing studies of polymer mimics antifreezes. Polymers designed for antifreezing show no structural similarities to natural AFPs but still have similar macroscopic ice recrystallization inhibition properties.

4.3.3 Morphologies of the freezing front

When the freezing interface approaches a suspended solid particle (or a large molecule), the particle may either get engulfed or rejected into the remaining liquid, accordingly altering the morphologies of the freezing front. The segregation coefficient, defined as the ratio of particle concentrations in the liquid and solid domains near the freezing interface, respectively, is an important factor to evaluate the stability of the freezing interface and the homogeneity of the frozen products. Elliott [83] derived an expression of the nonequilibrium segregation as a function of particle size and freezing rate, providing an explanation for the banding formation when freezing a colloidal suspension. Rempel and Worster [84] characterized an intermolecular force that pushes the suspended particle away from the advancing freezing front. In response to the interfacial premelting, the net force by balancing the intermolecular force and the viscosity resistance determines whether a solid particle is engulfed or rejected. Their results also show the effect of interface curvature on particle behaviors and how morphologies of the freezing interface vary when approaching a solid particle.

For concentrated suspensions, the accumulation of particles in front of the freezing interface may induce structural instabilities and thus damage the integrity of the freezing interface. Peppin et al. [85] studied the unidirectional freezing of a dense hard-sphere colloid and calculated the temperature and concentration profiles ahead of the freezing front. For small particles, Brownian diffusion is stronger, forming a diffusive boundary layer in front of the freezing interface. Constitutional supercooling leads to an instability of the freezing interface. For large particles, Brownian diffusion is negligible, and the accumulated porous layer itself becomes unstable when exposed to large constitutional supercooling. Anderson and Worster [86] examined the unidirectional freezing of a concentrated colloid where a layer of compacted particles together with discrete ice lenses is formed. Three periodic regimes are identified at varying freezing rates. At low freezing rates, a regular periodic sequence of ice lenses is formed with the compact layer thickness and ice lens characteristics depending on the freezing rate. At moderate freezing rates, a diminished compacted layer is formed together with disordered ice lenses. Beyond a critical freezing rate, ice lenses are replaced by a periodic banded structure without any ice lenses or compact layers.

As particles are rejected and accumulated in front of the advancing freezing interface, ice formation of such colloids can be used to prepare composite materials with engineered porosities and structures. Deville et al. [87] were inspired by the freezing of seawater where salt and microorganisms are expelled and entrapped between ice crystals and proposed to apply the same principle to build sophisticated, nacre-like architectures via freezing concentrated suspensions containing ceramic particles.

4.4 Different types of added components

4.4.1 Dissolved gases

Air, as well as other gases, is widely present in water as one dissolved phase. Upon freezing, dissolved gases get rejected from the ice lattice and accumulate to form micrometer-sized bubbles. Chu et al. [43] experimentally characterized the bubble formation process as shown in Fig. 4.6A. They found that bubble formation occurs in a large spectrum of drop size, from micrometer-sized condensed drops to millimeter-sized sessile drops. While the dendritic ice formation in the recalescence stage hardly rejects any gas, most bubbles are formed during the isothermal stage where the planar freezing interface interacts with gas molecules, leaving isolated bubbles either engulfed or rejected. They also proposed a theoretical mode and discussed the correlation between the bubble

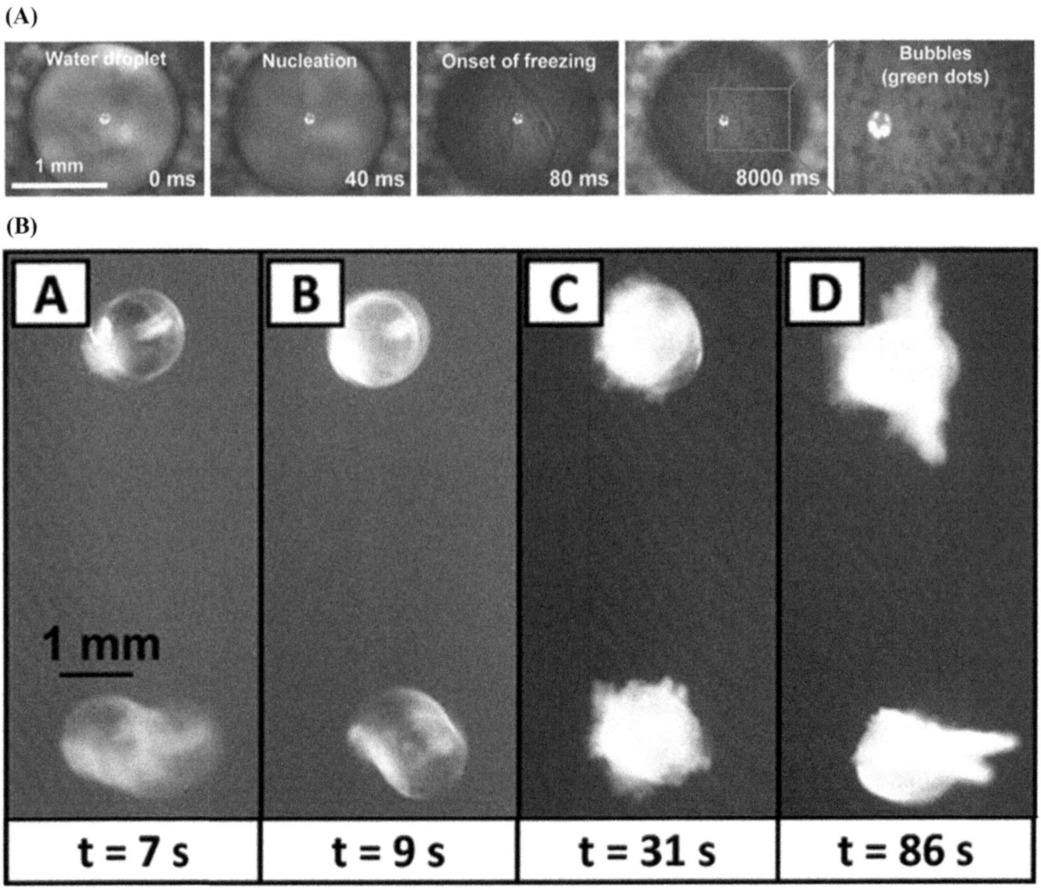

FIGURE 4.6 (A) Bubble formation as the drop freezing proceeds. (B) Morphologies evolution during the solidification of tetrahydrofuran hydrates. Source: *(A) Reprinted with the permission from [43]. ©2019 American Physical Society. (B) Reprinted with the permission from [88]. ©2023 American Chemical Society [43].*

formation rate and its influencing factors, including the degree of supercooling, the drop contact angle, and the gas solubility, among others.

The dissolved gas can act as a heterogeneous medium, promoting icing nucleation and altering the advancement of the freezing interface. Li et al. [89] experimentally investigated the freezing dynamics of water drops at varying gas concentrations. Compared to pure water drops, icing nucleation is triggered at a significantly lower degree of supercooling for drops with dissolved gas. The presence of bubbles also alters the solid–liquid density ratio and accordingly changes the shapes of frozen drops. Additionally, they found that the universal tip angle becomes smaller when increasing the gas concentration.

A special case of drops containing dissolved gas is the hydrate drops, where a crystalline compound is formed instead of ice during the solidification process, with gas molecules trapped in cages of water molecules with hydrogen bonds. McElligott et al. [88] investigated the solidification process of acoustically levitated tetrahydrofuran hydrate drops. A direct digital imaging method was used to capture the solidification process as shown in Fig. 4.6B. Nucleation initiates at the water–air interface where the local temperature is lower due to evaporative cooling. Afterward, the rapid intra-drop growth of hydrates causes the loss of optical clarity. At the last stage of solidification, drops lose more clarity, and protrusions form at the solid–air interface.

4.4.2 Ionic components

A pure water sessile drop freezes into a singularity with the conical angle of approx. 121° in response to the solid–liquid density ratio. Boulogne and Salonen [90] studied the change of this conical angle in the presence of one of the most common types of ionic components, NaCl. They measured the value of the conical angle at extremely small concentrations of 10^{-3}–10^{-8} M and found that the conical angle increases drastically as the salt concentration increases. They attributed such a tendency to the change of electrical conductivity due to the added sodium and chloride ions. Singha et al. [91] investigated the change of frozen drop shapes of considerably larger concentrations. The distinctive two successive freezing stages presenting in the freezing of pure water drops are missing in the freezing of saltwater drops. As the salinity increases, a considerable icing delay occurs as expected due to the freezing point depression. The singularity together with the sharpness of upper drops becomes less noticeable as the salinity increases, and they disappear when the salinity is beyond 0.4 M.

Besides the changes in freezing point and frozen drop shapes, adding salt to pure water drops also alters their other freezing characteristics. Bauerecke et al. [92] monitored the freezing dynamics of acoustically levitated salty water drops of varying concentrations as shown in Fig. 4.7A. For a pure water drop where icing nucleation initiates at −24°C (upper panel), the recalescent freezing occurs with a planer interface of a propagation velocity of 0.12 m/s. For a 1 M NaCl solution drop where icing nucleation initiates at −30°C (intermediate panel), the recalescent freezing occurs as an exploding barre-shaped ice cloud, with the propagation velocity merely a quarter of that of a pure water drop. Further increasing the salt concentration to 3 M (lower panel), the propagation velocity decreases drastically in a similar manner. Lü and Wei [94] compared the supercooling capability of two different solutions, NaCl and KCl, by investigating their freezing behaviors as acoustically levitated

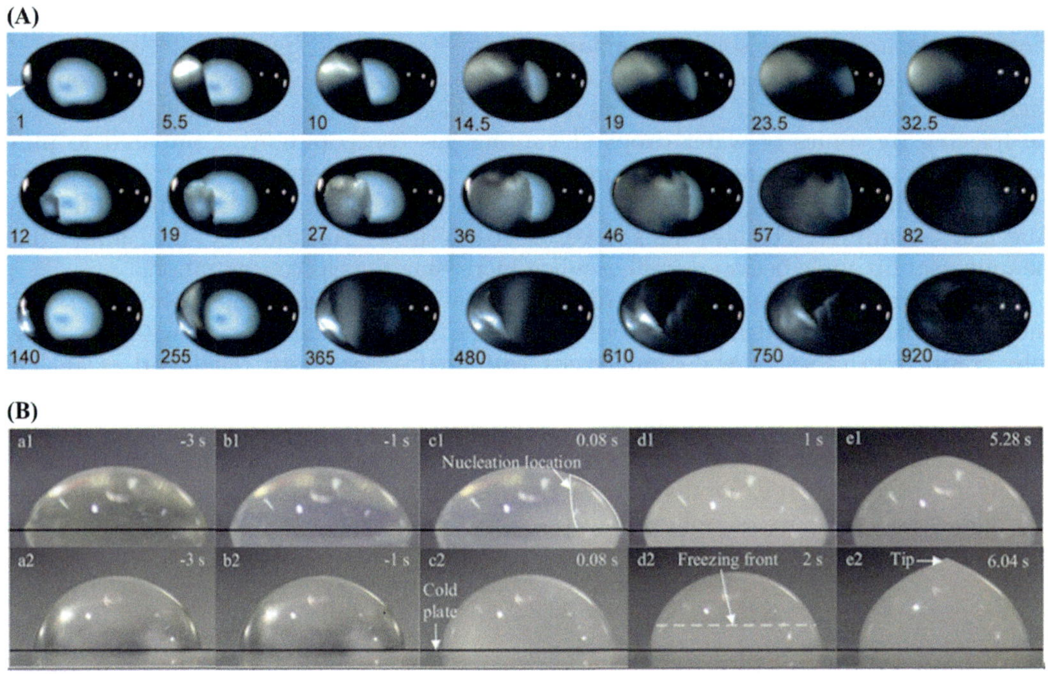

FIGURE 4.7 (A) High-speed snapshots showing the freezing dynamics of a pure water drop (upper panel), 1 M saline solution drop (intermediate panel), and 3 M saline solution drop (lower panel). (B) Comparison of the freezing dynamics of a urine drop (upper panel) and a pure water drop (lower panel) [93]. Source: *(A) Reprinted with the permission from S. Bauerecker, P. Ulbig, V. Buch, L. Vrbka, P. Jungwirth, Monitoring ice nucleation in pure and salty water via high-speed imaging and computer simulations, The Journal of Physical Chemistry C, 112(20) (2008) 7631–7636. ©2008 American Chemical Society. (B) Reprinted with the permission from Q. Dang, M. Song, C. Dang, T. Zhan, L. Zhang, Experimental study on solidification characteristics of sessile urine droplets on a horizontal cold plate surface under natural convection, Langmuir: the ACS Journal of Surfaces and Colloids, 38(25) (2022) 7846–7857. ©2022 American Chemical Society.*

drops. For both salt solutions, their supercooling is apparently enhanced linearly with increasing concentration. However, more detailed studies show that the supercooling also depends on the nature of ions, and NaCl solution is of a 2–3 larger supercooling than KCl solution at the same concentration. Emelyanenko et al. [95] further explored the supercooling capacity of more types of alkali metal chloride solutions.

The freezing of water drops containing other ionic components has been reported in the literature. Dang et al. [93] experimentally studied the freezing characteristics of urine drops which can be treated as a mixture of different types of salts, inorganic or organic. The comparison of freezing dynamics of a urine drop and a water drop is shown in Fig. 4.7B. Urine drops are slightly yellow and have a smaller contact angle due to the presence of polar components. In the supercooling stage, the urine drop gradually becomes white as the temperature decreases in response to the precipitation of inorganic salts. In the recalescence freezing stage, it is very hard to identify the site of nucleation. After the isothermal freezing stage completes, the tip atop the frozen urine drop is more rounded compared to that of the frozen pure water drop.

A selection of solutions such as salty water and glycols can attract water molecules from the surrounding air, termed hygroscopic substances [96]. Sun et al. [97] implanted salt-saturated water and propylene glycol microdrops onto the supercooled substrate to inhibit condensation frosting. These hygroscopic drops can absorb moisture that otherwise condenses and freezes on the substrate, keeping the substrate ice-free for a considerably long duration.

4.4.3 Nanoparticles

Unlike freezing segregation of dissolved gases and ionic components, particle segregation and accumulation can be directly observed using optical microscopy. Ivall et al. [98] experimentally characterized the freezing and melting process of sessile drops containing multiwall carbon nanotubes (MWCNT). They found the MWCNT suspension drop freeze in a relatively slower manner, and a noticeable concentration gradient of particles perpendicular to the track of the freezing interface presents when freezing completes. Even when the whole drop gets melted, a large amount of MWCNT clusters can be observed floating atop the drop. Later, they [99] further increased the volume concentration of MWCNT and experimentally investigated the freezing dynamics of such concentrated MWCNT suspension drops. Their results show two intriguing outcomes: Most MWCNTs are expelled off the solid phase and get dispersed into the remaining liquid domain at low supercoolings; MWCNTs are embedded into the solid phase at high supercooling. Also worth noting is that the advancement of the freezing interface destabilized MWCNTs and yielded large particle clusters due to the loss of dispersion capacity.

Besides the segregation and accumulation of particles as the freezing of nanofluid drops proceeds, more intriguing phenomena were also reported very recently. Zhao et al. [100] studied the freezing characteristics of TiO_2 nanofluid drops as shown in Fig. 4.8A. Even for the diluted case with the particle volume concentration of 0.05%, the pointy singular tip is replaced by a flat plateau. Such plateau size gets larger as the particle concentration increases. They also investigated the freezing outcomes for different material types and sizes of particles, different drop sizes, and contact angles and concluded the formation of this plateau is universal. Zhang et al. [102] developed a lattice Boltzmann (LB) model to simulate the process of how dispersed particles get expelled from the freezing front, accumulate near the tri-junction, and bring the remaining liquid to the edge. Their results also predict that the top plateau size is independent of the initial drop contact angle. Miao et al. [36] compared the freezing processes of a sessile nanofluid drop and a deposited nanofluid drop as shown in Fig. 4.8B. Under the same supercooling and propagation velocity of the freezing front, two distinguished drop shapes are obtained. They found that the top of a frozen nanofluid drop can be altered from a singular tip to a flat plateau by simply changing the thermal condition onset of nucleation. With the assistance of a Hele-Shaw cell experiment, they attribute these two freezing behaviors to the interplay between nanoparticles and ice dendrites. Instead of a set of fundamental studies, these works promise an effective three-dimensional microfabrication strategy.

Physical approaches, such as electrical and magnetic fields, are known to manipulate particles and accordingly the freezing characteristics of nanofluid drops [103,104].

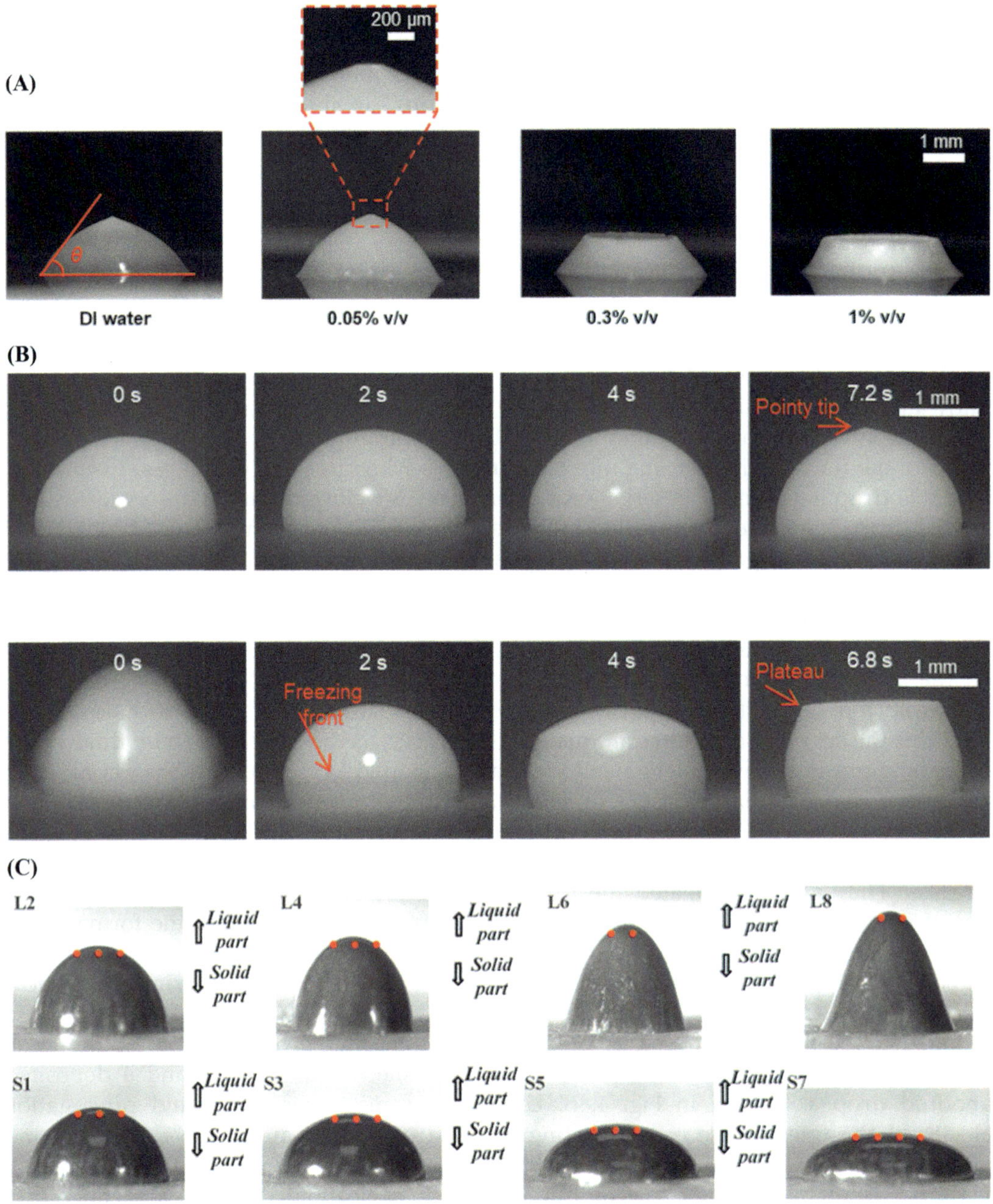

FIGURE 4.8 (A) Freezing dynamics of a nanofluid drop with increasing concentration. (B) Freezing dynamics of a sessile (upper panel) or a deposited (lower panel) nanofluid drop. (C) Freezing dynamics of ferrofluid drops exposed to the magnetic lift condition (upper panel) and the magnetic squeeze condition (lower panel). Source: *(A) Reprinted with the permission from [100]. Copyright 2021 American Institute of Physics. (B) Reprinted with the permission from [36]. Copyright 2022 American Institute of Physics. (C) Reprinted from [101]. Copyright 2020 with permission from Elsevier.*

Zhang et al. [101] proposed to use the magnetic field to change the shape of a ferrofluid drop as shown in Fig. 4.8C. Depending on the nature of the magnetic field, the drop can be either elongated or squeezed during its freezing, and both the freezing time and the frozen drop height are modulated accordingly.

4.4.4 Miscible or immiscible fluids

Liquid mixtures are present ubiquitously in nature and industries, and thus studying the freezing characteristics is meaningful from fundamental and application viewpoints [105]. Miscible fluids can be readily mixed with water of a wide spectrum of volume concentrations. Wang et al. [106] investigated the freezing characteristics of water-ethanol binary drops utilizing high-speed imaging as shown in Fig. 4.9A. When exposed to supercooling, the freezing binary drop exhibits a counterintuitive self-lifting behavior, resulting in a noticeable increase in drop height. They attribute this antigravitational flow transfer to an internal solutal Marangoni flow that elevates the drop as the freezing interface advances. Zeng et al. [107] investigated the freezing of more complex drops which are acoustically levitated ternary mixtures containing hexadecane, ethanol, and diethyl ether as shown in Fig. 4.9B. During the evaporative cooling induced freezing process, three distinct stages occur successively, referred to as "Ouzo state," "Janus state," and "Encapsulating state," respectively.

The freezing of suspension drops in immersible fluids mixed with water with the assistance of surfactants or homogenizers has also attracted lots of research attention.

Kharal and Louf [108] investigated the freezing dynamics of a polymer solution drop. As the polymer concentration increases, the singular tip becomes smaller and even disappears beyond a critical concentration. Meanwhile, the freezing time increases concomitantly when exposed to the same degree of supercooling. Combining the geometrical arguments and heat transfer measurements, they provided a theoretical model to depict the shape evolution of freezing polymer drops. Hauptmann et al. [37] experimentally studied the freezing and thawing characteristics of water-in-oil emulsions as shown in Fig. 4.9C. Initially at the ambient temperature, the water/oil mixture is featureless even at the microscale. In deep supercooling, a dark branch made of solidified water drops is visible. As the reheating proceeds, some sharp structures are missing due to edge-melting. Further heating the system to room temperature, all solid phases are melted and merged into discrete drops.

4.4.5 Special cases

A few special cases of the freezing of multicomponent drops have also been studied recently, where some unique freezing behaviors are revealed, different from both pure water and mixture drops. Ahmadi et al. [109] investigated the freezing dynamics of a soap bubble as shown in Fig. 4.10A. Upon depositing onto the supercooled surface, a plethora of growing ice crystals emerge and swirl around the free interface, reminiscing a snow globe. They attributed this "snow globe effect" to a strong Marangoni flow that detaches and entrains these growing ice crystals from the bottom of the bubble. Hashmi et al. [110]

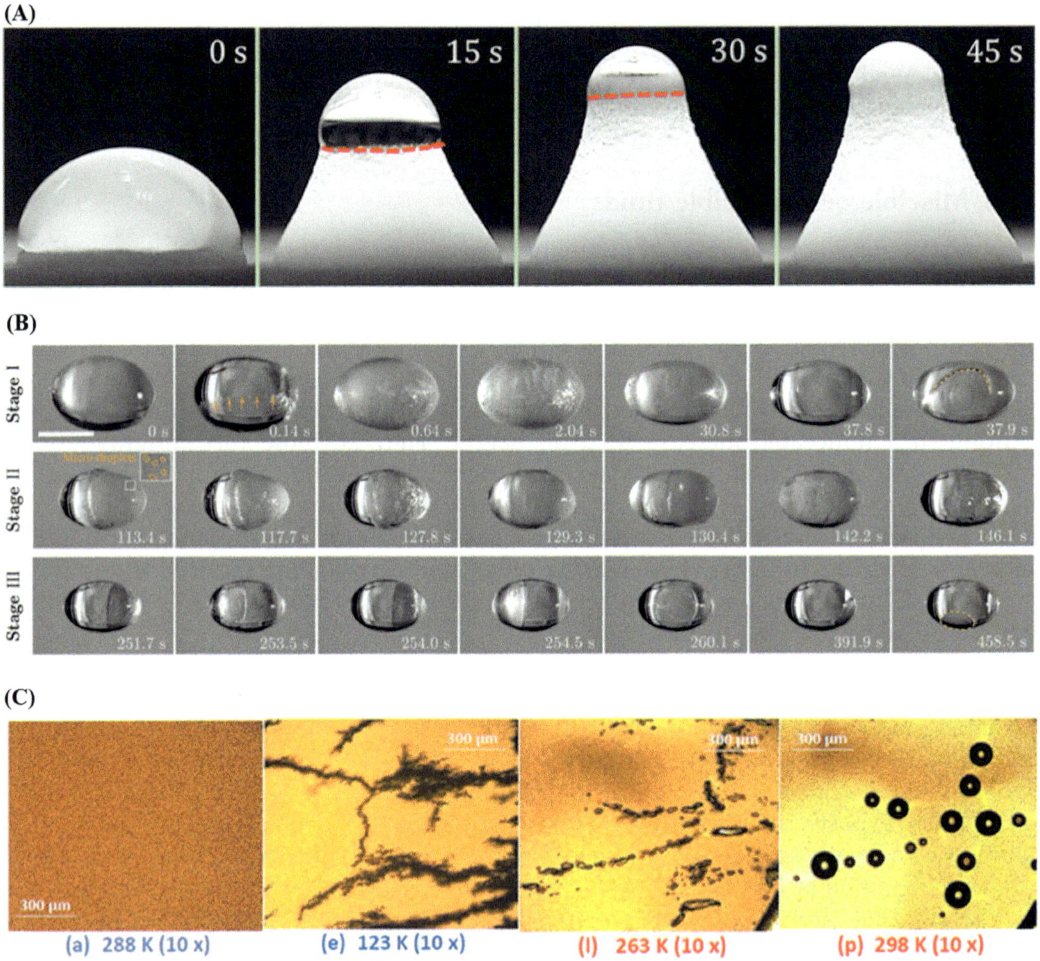

FIGURE 4.9 (A) Snapshots showing the freezing process of a binary drop (the initial volume concentration of ethanol 5%). (B) The lifespan of an acoustically levitated ternary drop. (C) Optical cryomicroscopy investigation of an oil/water mixture as it undergoes supercooling and heating. Source: *(A) Reprinted with the permission from [106]. Copyright 2024 American Physical Society. (B) Reprinted from [107]. Copyright 2023 with permiton from Elsevier. (C) Reprinted with the permission from [37]. Copyright 2016 American Institute of Physics.*

investigated the freezing characteristics of a liquid marble where a liquid-filled core is covered by nanoparticles as shown in Fig. 4.10B. During the freezing process, the liquid marble undergoes a clear shape transition from a truncated sphere to a flying-saucer-shaped morphology. They attributed these shape-changing behaviors to a Marangoni convection that drives the remaining liquid to seep down sideways around the frozen core. Later, Starostin et al. [111] reported three distinct shapes from freezing liquid marbles, namely, "conical tip," "spherical marble," and "flattened marble," respectively.

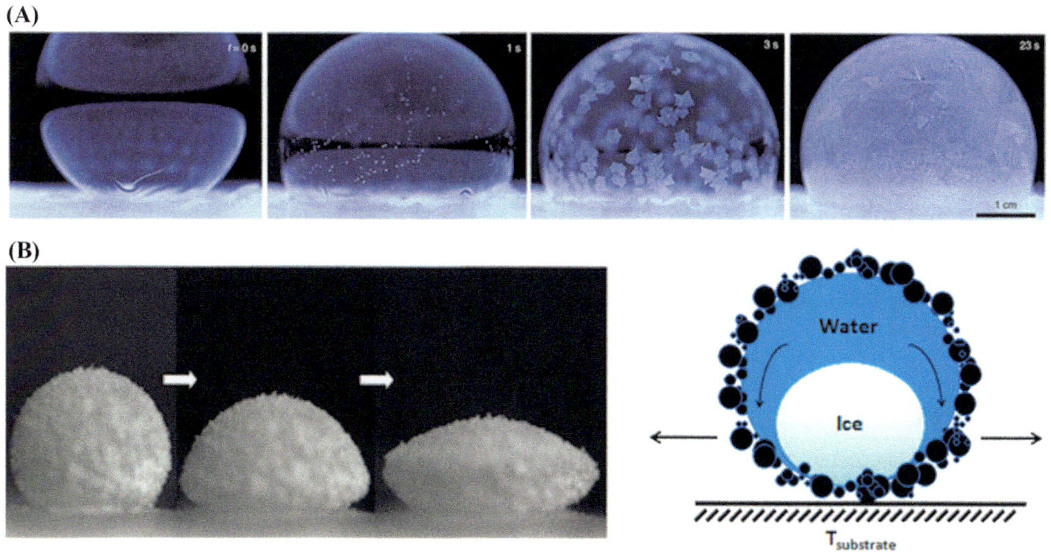

FIGURE 4.10 (A) Freezing of a millimeter-sized bubble deposited on a subcooled surface inside an isothermal freezer. (B) Freezing dynamics of a liquid marble and the schematics showing the proposed mechanism. Source: *(A) Reprinted with the permission from S.F. Ahmadi, S. Nath, C.M. Kingett, P. Yue, J.B. Boreyko, How soap bubbles freeze, Nature Communications, 10 (1) (2019) 2531. ©2019 Springer Nature. (B) Reprinted with the permission from A. Hashmi, A. Strauss, J. Xu, Freezing of a liquid marble, Langmuir: the ACS Journal of Surfaces and Colloids, 28(28) (2012) 10324–10328. ©2012 American Chemical Society.*

4.5 Concluding remarks

Multicomponent drop freezing is closely associated with many industrial processes, and exploring their underlying physics to get a better understanding of their transport mechanisms will significantly improve related applications. In this very recent decade, lots of new phenomena pertaining to this topic have been revealed, making it a true hotspot in the study of heat transfer. This work provides a review of the latest studies on multicomponent drop freezing, with special attention on the characteristics and their mechanisms. As one of the least known pieces in the field of heat transfer and phase change, multicomponent drop freezing needs to be further explored both for in-depth fundamental study and in the breadth of potential applications.

Author contributions

The manuscript was written through contributions of all authors. All authors have given approval to the final version of the manuscript.

Acknowledgment

This study was supported by the National Natural Science Foundation of China (No. 52276079), the Experiments for Space Exploration Program and the Qian Xuesen Laboratory, China Academy of Space Technology (Grant No. TKTSPY-2020–01-01), National Key R&D Program of China (Grant No. 2020YFF0303901), and the Central Guidance on Local Science and Technology Development Fund of Shanghai City Grant No. YDZX20213100003002.

Data availability

The authors declare that the main data supporting the findings of this study are available within the article and its Supplementary Information files. All other relevant data are available from the corresponding author upon reasonable request.

References

[1] G. Bai, D. Gao, Z. Liu, X. Zhou, J. Wang, Probing the critical nucleus size for ice formation with graphene oxide nanosheets, Nature 576 (7787) (2019) 437–441.
[2] D. Rosenfeld, W.L. Woodley, Deep convective clouds with sustained supercooled liquid water down to − 37.5 °C, Nature 405 (2000) 440.
[3] G. John Morris, E. Acton, Controlled ice nucleation in cryopreservation – a review, Cryobiology 66 (2) (2013) 85–92.
[4] G.P. Johari, Water's size-dependent freezing to cubic ice, The Journal of Chemical Physics 122 (19) (2005) 194504.
[5] J.P. Hindmarsh, A.B. Russell, X.D. Chen, Experimental and numerical analysis of the temperature transition of a suspended freezing water droplet, International Journal of Heat and Mass Transfer 46 (7) (2003) 1199–1213.
[6] E.K. Bigg, The formation of atmospheric ice crystals by the freezing of droplets, Quarterly, Journal of the Royal Meteorological Society 79 (342) (1953) 510–519.
[7] Y. Ren, A. Soni, A. Kumar, A.K. Bertram, G.N. Patey, Effects of pH on ice nucleation by the alpha-alumina (0001) surface, Journal of Physical Chemistry C 126 (46) (2022) 19934–19946.
[8] S. Chang, H. Qi, S. Zhou, Y. Yang, Experimental and numerical study on freezing process of water droplets under surfaces with different wettability, Applied Thermal Engineering 219 (2023) 119516.
[9] F. Zheng, Z. Wang, J. Huang, Z. Li, Inkjet printing-based fabrication of microscale 3D ice structures, Microsystems & Nanoengineering 6 (1) (2020) 89.
[10] P. Kamble, R. Hodgir, G. Gote, Y. Mittal, K.P. Karunakaran, Sub-zero additive manufacturing: a review of peculiarities and applications of additive manufacturing at temperatures below 0 °C, Progress in Additive Manufacturing 7 (5) (2022) 993–1008.
[11] H. Asakawa, G. Sazaki, K. Nagashima, S. Nakatsubo, Y. Furukawa, Two types of quasi-liquid layers on ice crystals are formed kinetically, Proceedings of the National Academy of Sciences of the United States of America 113 (7) (2016) 1749–1753.
[12] N. Dhopatkar, A.P. Defante, A. Dhinojwala, Ice-like water supports hydration forces and eases sliding friction, Science Advances 2 (8) (2016) e1600763.
[13] G. Graeber, T.M. Schutzius, H. Eghlidi, D. Poulikakos, Spontaneous self-dislodging of freezing water droplets and the role of wettability, Proceedings of the National Academy of Sciences 114 (42) (2017) 11040–11045.
[14] S. Wildeman, S. Sterl, C. Sun, D. Lohse, Fast dynamics of water droplets freezing from the outside in, Physical Review Letters 118 (8) (2017) 084101.
[15] S. Nath, S.F. Ahmadi, J.B. Boreyko, How ice bridges the gap, Soft Matter 16 (5) (2020) 1156–1161.
[16] Y. Zhao, Z. Yan, H. Zhang, C. Yang, P. Cheng, Promote anti- /de- frosting by suppressing directional ice bridging, International Journal of Heat and Mass Transfer 165 (2021) 120609.
[17] Y. Zhao, F. Zhu, H. Zhang, T.H. New, L. Jin, C. Yang, Triple condensate halo from a single water droplet impacting upon a cold surface, Applied Physics Letters 114 (18) (2019) 183703.

[18] S. Jung, M.K. Tiwari, D. Poulikakos, Frost halos from supercooled water droplets, Proceedings of the National Academy of Sciences 109 (40) (2012) 16073–16078.

[19] Y. Zhao, Z. Zuo, H. Tang, X. Zhang, Ice coverage induced by depositing a water drop onto the supercooled substrate at extreme low vapor pressure, Crystals 11 (6) (2021) 691.

[20] F.-Q. Zhu, W.-Z. Fang, T.H. New, Y. Zhao, C. Yang, Freezing morphologies of impact water droplets on an inclined subcooled surface, International Journal of Heat and Mass Transfer 181 (2021) 121843.

[21] V. Thiévenaz, T. Séon, C. Josserand, Solidification dynamics of an impacted drop, Journal of Fluid Mechanics 874 (2019) 756–773.

[22] P. Kant, R.B.J. Koldeweij, K. Harth, M.A.J. van Limbeek, D. Lohse, Fast-freezing kinetics inside a droplet impacting on a cold surface, Proceedings of the National Academy of Sciences 117 (6) (2020) 2788.

[23] A. Hakimian, M. Mohebinia, M. Nazari, A. Davoodabadi, S. Nazifi, Z. Huang, et al., Freezing of few nanometers water droplets, Nature Communications 12 (2021).

[24] S.L. Gai, Z.B. Peng, B. Moghtaderi, J.L. Yu, E. Doroodchi, Freezing of micro-droplets driven by power ultrasound, Chemical Engineering Science 251 (2022).

[25] C. Huang, Y. Zhao, T. Gu, Ice dendrite growth atop a frozen drop under natural convection conditions, Crystals 12 (3) (2022) 323.

[26] H. Zeng, S. Lyu, D. Legendre, C. Sun, Influence of gravity on the freezing dynamics of drops on a solid surface, Physical Review Fluids 7 (10) (2022).

[27] W.-Z. Fang, F.-Q. Zhu, F. Shen, D. Chai, W.-Q. Tao, Freezing behaviors of impacting water droplets on cold inclined surfaces, Applied Thermal Engineering 219 (2023) 119562.

[28] Z.Y. Ma, W. Xiong, P. Cheng, 3D Lattice Boltzmann simulations for water droplet's impact and transition from central-pointy icing pattern to central-concave icing pattern on supercooled surfaces. Part I: smooth surfaces, International Journal of Heat and Mass Transfer (2021) 171.

[29] Z. Zhu, X. Zhang, Y. Zhao, X. Huang, C. Yang, Freezing characteristics of deposited water droplets on hydrophilic and hydrophobic cold surfaces, International Journal of Thermal Sciences 171 (2022) 107241.

[30] A.J. Meuler, G.H. McKinley, R.E. Cohen, Exploiting topographical texture to impart icephobicity, ACS Nano 4 (12) (2010) 7048–7052.

[31] D. Ehre, E. Lavert, M. Lahav, I. Lubomirsky, Water freezes differently on positively and negatively charged surfaces of pyroelectric materials, Science (New York, N.Y.) 327 (5966) (2010) 672–675.

[32] Q.Y. Deng, H. Wang, Z.T. Xie, Y. Tian, X. Zhu, R. Chen, et al., Fast-freezing behavior of water droplet on subcooled superhydrophobic surface in DC electrostatic field, International Journal of Thermal Sciences 174 (2022).

[33] J.M. Lowe, V. Hinrichsen, M. Schremb, C. Tropea, Ice nucleation in high alternating electric fields: effect of electric field strength and frequency, Physical Review E 103 (1) (2021).

[34] U.G.K. Wegst, M. Schecter, A.E. Donius, P.M. Hunger, Biomaterials by freeze casting, Philosophical Transactions of the Royal Society A: Mathematical, Physical and Engineering Sciences 368 (1917) () (2010) 2099–2121.

[35] G. John Morris, E. Acton, B.J. Murray, F. Fonseca, Freezing injury: the special case of the sperm cell, Cryobiology 64 (2) (2012) 71–80.

[36] Y. Miao, Y. Zhao, M. Gao, L. Yang, C. Yang, Characteristics of a freezing nanosuspension drop in two different schemes, Applied Physics Letters 120 (9) (2022) 091602.

[37] A. Hauptmann, K.F. Handle, P. Baloh, H. Grothe, T. Loerting, Does the emulsification procedure influence freezing and thawing of aqueous droplets? The Journal of Chemical Physics 145 (21) (2016) 211923.

[38] N. Beier, D. Sego, R. Donahue, K. Biggar, Laboratory investigation on freeze separation of saline mine waste water, Cold Regions Science and Technology 48 (3) (2007) 239–247.

[39] L. Vrbka, P. Jungwirth, Brine rejection from freezing salt solutions: a molecular dynamics study, Physical Review Letters 95 (14) (2005) 148501.

[40] J.G. Meijer, P. Kant, D. van Buuren, D. Lohse, Thin-film-mediated deformation of droplet during cryopreservation, Physical Review Letters 130 (21) (2023) 214002.

[41] D. Tsuchida, C. Kang, M. Okada, K. Matsumoto, T. Kawagoe, Ice formation process by cooling water–oil emulsion with stirring in a vessel, International Journal of Refrigeration 25 (2) (2002) 250–258.

[42] B.J. Murray, S.L. Broadley, T.W. Wilson, J.D. Atkinson, R.H. Wills, Heterogeneous freezing of water droplets containing kaolinite particles, Atmospheric Chemistry and Physics 11 (9) (2011) 4191–4207.

[43] F. Chu, X. Zhang, S. Li, H. Jin, J. Zhang, X. Wu, et al., Bubble formation in freezing droplets, Physical Review Fluids 4 (7) (2019) 071601(R).
[44] N.H. Fletcher, Size effect in heterogeneous nucleation, The Journal of Chemical Physics 29 (3) (1958) 572–576.
[45] T. Inada, H. Tomita, T. Koyama, Ice nucleation in water droplets on glass surfaces: From micro- to macro-scale, International Journal of Refrigeration 40 (0) (2014) 294–301.
[46] C. Gurganus, A.B. Kostinski, R.A. Shaw, Fast imaging of freezing drops: No preference for nucleation at the contact line, Journal of Physical Chemistry Letters 2 (12) (2011) 1449–1454.
[47] C. Gurganus, A.B. Kostinski, R.A. Shaw, High-speed imaging of freezing drops: still no preference for the contact line, The Journal of Physical Chemistry C 117 (12) (2013) 6195–6200.
[48] S. Jung, M.K. Tiwari, N.V. Doan, D. Poulikakos, Mechanism of supercooled droplet freezing on surfaces, Nature Communications 3 (2012) 615.
[49] T. Buttersack, S. Bauerecker, Critical radius of supercooled water droplets: on the transition toward dendritic freezing, The Journal of Physical Chemistry. B 120 (3) (2016) 504–512.
[50] Y. Wang, Y. Cheng, New perspectives on the droplet freezing nucleation and early crystal growth mechanisms, International Journal of Heat and Mass Transfer 140 (2019) 1023–1028.
[51] T. Buttersack, V.C. Weiss, S. Bauerecker, Hypercooling temperature of water is about 100 K higher than calculated before, The Journal of Physical Chemistry Letters 9 (3) (2018) 471–475.
[52] D.M. Anderson, M.G. Worster, S.H. Davis, The case for a dynamic contact angle in containerless solidification, Journal of Crystal Growth 163 (3) (1996) 329–338.
[53] G. Chaudhary, R. Li, Freezing of water droplets on solid surfaces: an experimental and numerical study, Experimental Thermal and Fluid Science 57 (2014) 86–93.
[54] T.V. Vu, G. Tryggvason, S. Homma, J.C. Wells, H. Takakura, A front-tracking method for three-phase computations of solidification with volume change, Journal of Chemical Engineering of Japan 46 (11) (2013) 726–731.
[55] A.G. Marín, O.R. Enríquez, P. Brunet, P. Colinet, J.H. Snoeijer, Universality of tip singularity formation in freezing water drops, Physical Review Letters 113 (5) (2014) 054301.
[56] S. Schiaffino, A.A. Sonin, Molten droplet deposition and solidification at low Weber numbers, Physics of Fluids 9 (11) (1997) 3172–3187.
[57] W. Xiong, L. Qi, J. Luo, D. Zhang, J. Liang, H. Yi, Experimental investigation on the height deviation of bumps printed by solder jet technology, Journal of Materials Processing Technology 243 (2017) 291–298.
[58] K.N. Barada, C.G. Mool, W.K. Kurt, Spontaneous formation of nanospiked microstructures in germanium by femtosecond laser irradiation, Nanotechnology 18 (19) (2007) 195302.
[59] M. David, W.K. Kurt, Solidification driven extrusion of spikes during laser melting of silicon pillars, Nanotechnology 17 (11) (2006) 2741.
[60] A. Starostin, V. Strelnikov, L.A. Dombrovsky, S. Shoval, E. Bormashenko, On the universality of shapes of the freezing water droplets, Colloid and Interface Science Communications 47 (2022).
[61] W.Z. Fang, F. Zhu, W.Q. Tao, C. Yang, How different freezing morphologies of impacting droplets form, Journal of Colloid and Interface Science 584 (2021) 403–410.
[62] H. Lambley, G. Graeber, R. Vogt, L.C. Gaugler, E. Baumann, T.M. Schutzius, et al., Freezing-induced wetting transitions on superhydrophobic surfaces, Nature Physics 19 (5) (2023) 649–655.
[63] T.M. Schutzius, S. Jung, T. Maitra, G. Graeber, M. Köhme, D. Poulikakos, Spontaneous droplet trampolining on rigid superhydrophobic surfaces, Nature 527 (7576) (2015) 82–85.
[64] W.-Z. Fang, F. Zhu, L. Zhu, W.-Q. Tao, C. Yang, Self-peeling of frozen water droplets upon impacting a cold surface, Communications Physics 5 (1) (2022) 51.
[65] S.M.S. Murshed, K.C. Leong, C. Yang, Enhanced thermal conductivity of TiO_2—water based nanofluids, International Journal of Thermal Sciences 44 (4) (2005) 367–373.
[66] J. Buongiorno, D.C. Venerus, N. Prabhat, T. McKrell, J. Townsend, R. Christianson, et al., A benchmark study on the thermal conductivity of nanofluids, Journal of Applied Physics 106 (9) (2009) 094312.
[67] L. Fan, J.M. Khodadadi, A theoretical and experimental investigation of unidirectional freezing of nanoparticle-enhanced phase change materials, Journal of Heat Transfer 134 (9) (2012) 092301.
[68] R. Hossain, S. Mahmud, A. Dutta, I. Pop, Energy storage system based on nanoparticle-enhanced phase change material inside porous medium, International Journal of Thermal Sciences 91 (0) (2015) 49–58.

[69] M. Kahani, R.G. Jackson, G. Rosengarten, Experimental investigation of TiO_2/water nanofluid droplet impingement on nanostructured surfaces, Industrial & Engineering Chemistry Research 55 (7) (2016) 2230–2241.
[70] D.T. Wasan, A.D. Nikolov, Spreading of nanofluids on solids, Nature 423 (6936) (2003) 156–159.
[71] S.M.S. Murshed, K.C. Leong, C. Yang, Investigations of thermal conductivity and viscosity of nanofluids, International Journal of Thermal Sciences 47 (5) (2008) 560–568.
[72] L.S. Sundar, M.K. Singh, E.V. Ramana, B. Singh, J. Grácio, A.C.M. Sousa, Enhanced thermal conductivity and viscosity of nanodiamond-nickel nanocomposite nanofluids, Scientific Reports 4 (2014) 4039.
[73] H. Zhang, S. Qing, Y. Zhai, X. Zhang, A. Zhang, The changes induced by pH in TiO_2/water nanofluids: Stability, thermophysical properties and thermal performance, Powder Technology 377 (2021) 748–759.
[74] R.H. Chen, T.X. Phuoc, D. Martello, Surface tension of evaporating nanofluid droplets, International Journal of Heat and Mass Transfer 54 (11-12) (2011) 2459–2466.
[75] M.H.U. Bhuiyan, R. Saidur, M.A. Amalina, R.M. Mostafizur, A. Islam, Effect of nanoparticles concentration and their sizes on surface tension of nanofluids, Procedia Engineering 105 (2015) 431–437.
[76] S.S. Khaleduzzaman, I.M. Mahbubul, I.M. Shahrul, R. Saidur, Effect of particle concentration, temperature and surfactant on surface tension of nanofluids, International Communications in Heat and Mass Transfer 49 (2013) 110–114.
[77] N.H. Fletcher, Nucleation by crystalline particles, The Journal of Chemical Physics 38 (1) (1963) 237–240.
[78] J. Zhang, D. Jin, L. Zhao, X. Liu, J. Lian, G. Li, et al., Preparation of nano-silver iodide powders and their efficiency as ice-nucleating agent in weather modification, Advanced Powder Technology 22 (5) (2011) 613–616.
[79] S. Okawa, A. Saito, R. Minami, The solidification phenomenon of the supercooled water containing solid particles, International Journal of Refrigeration 24 (1) (2001) 108–117.
[80] B.A. Stefl, K.F. George, Antifreezes and deicing fluids, Kirk-Othmer Encyclopedia of Chemical Technology, John Wiley & Sons, Inc, 2000.
[81] A. Ampaw, Antifreeze proteins solve cold problems, Nature Chemistry 14 (11) (2022) 1336. 1336.
[82] C.I. Biggs, T.L. Bailey, G. Ben, C. Stubbs, A. Fayter, M.I. Gibson, Polymer mimics of biomacromolecular antifreezes, Nature Communications 8 (1) (2017) 1546.
[83] J.A.W. Elliott, S.S.L. Peppin, Particle trapping and banding in rapid colloidal solidification, Physical Review Letters 107 (16) (2011) 168301.
[84] A.W. Rempel, M.G. Worster, Particle trapping at an advancing solidification front with interfacial-curvature effects, Journal of Crystal Growth 223 (3) (2001) 420–432.
[85] S.S.L. Peppin, J.A.W. Elliott, M.G. Worster, Solidification of colloidal suspensions, Journal of Fluid Mechanics 554 (2006) 147–166.
[86] M. Anthony, M. Anderson, Grae Worster, Freezing colloidal suspensions: periodic ice lenses and compaction, Journal of Fluid Mechanics 758 (2014) 786–808.
[87] S. Deville, E. Saiz, R.K. Nalla, A.P. Tomsia, Freezing as a path to build complex composites, Science (New York, N.Y.) 311 (5760) (2006) 515–518.
[88] A. McElligott, A. Guerra, A. Denoncourt, A.D. Rey, P. Servio, Interfacial effects during phase change in multiple levitated tetrahydrofuran hydrate droplets, Langmuir: the ACS Journal of Surfaces and Colloids 39 (4) (2023) 1573–1584.
[89] Y. Li, M. Li, C. Dang, X. Liu, Effects of dissolved gas on the nucleation and growth of ice crystals in freezing droplets, International Journal of Heat and Mass Transfer 184 (2022) 122334.
[90] F. Boulogne, A. Salonen, Drop freezing: fine detection of contaminants by measuring the tip angle, Applied Physics Letters 116 (10) (2020).
[91] S.K. Singha, P.K. Das, B. Maiti, Influence of salinity on the mechanism of surface icing: implication to the disappearing freezing singularity, Langmuir: the ACS Journal of Surfaces and Colloids 34 (30) (2018) 9064–9071.
[92] S. Bauerecker, P. Ulbig, V. Buch, L. Vrbka, P. Jungwirth, Monitoring ice nucleation in pure and salty water via high-speed imaging and computer simulations, The Journal of Physical Chemistry C 112 (20) (2008) 7631–7636.
[93] Q. Dang, M. Song, C. Dang, T. Zhan, L. Zhang, Experimental study on solidification characteristics of sessile urine droplets on a horizontal cold plate surface under natural convection, Langmuir: the ACS Journal of Surfaces and Colloids 38 (25) (2022) 7846–7857.
[94] Y.J. Lü, B. Wei, Supercooling of aqueous NaCl and KCl solutions under acoustic levitation, Journal of Chemical Physics 125 (14) (2006).

[95] A.M. Emelyanenko, K.A. Emelyanenko, L.B. Boinovich, Deep undercooling of aqueous droplets on a superhydrophobic surface: the specific role of cation hydration, Journal of Physical Chemistry Letters 11 (8) (2020) 3058–3062.
[96] S. Nath, C.E. Bisbano, P. Yue, J.B. Boreyko, Duelling dry zones around hygroscopic droplets, Journal of Fluid Mechanics 853 (2018) 601–620.
[97] X. Sun, V.G. Damle, A. Uppal, R. Linder, S. Chandrashekar, A.R. Mohan, et al., Inhibition of condensation frosting by arrays of hygroscopic antifreeze drops, Langmuir: the ACS Journal of Surfaces and Colloids 31 (51) (2015) 13743–13752.
[98] J. Ivall, M. Hachem, S. Coulombe, P. Servio, Behavior of surface-functionalized multiwall carbon nanotube nanofluids during phase change from liquid water to solid ice, Crystal Growth & Design 15 (8) (2015) 3969–3982.
[99] J. Ivall, J.-S. Renault-Crispo, S. Coulombe, P. Servio, Ice-dependent liquid-phase convective cells during the melting of frozen sessile droplets containing water and multiwall carbon nanotubes, International Journal of Heat and Mass Transfer 101 (2016) 27–37.
[100] Y. Zhao, C. Yang, P. Cheng, Freezing of a nanofluid droplet: from a pointy tip to flat plateau, Applied Physics Letters 118 (14) (2021) 141602.
[101] H. Zhang, Y. Zhao, W. Fang, C. Zhang, F. Zhu, L. Jin, et al., Active control of the freezing process of a ferrofluid droplet with magnetic fields, Applied Thermal Engineering 176 (2020) 115444.
[102] C. Zhang, S. Yin, H. Zhang, C. Yang, Simulation of a sessile nanofluid droplet freezing with an immersed boundary-lattice Boltzmann model, International Journal of Multiphase Flow 167 (2023).
[103] G.L. Liu, J. Kim, Y. Lu, L.P. Lee, Optofluidic control using photothermal nanoparticles, Nature 5 (1) (2006) 27–32.
[104] W.-Z. Fang, H. Zhang, C.-Y. Zhang, C. Yang, Freezing process of ferrofluid droplets: Numerical and scaling analyses, Physical Review Fluids 5 (5) (2020).
[105] S. Yang, Y. Hou, D. Zhou, X. Zhong, Non-monotonic effect of ethanol concentration on the spreading of an ethanol-water binary droplet impact on a supercooled surface, International Journal of Thermal Sciences 183 (2023).
[106] F. Wang, L. Chen, Y. Li, P. Huo, X. Gu, M. Hu, et al., Self-lifting droplet driven by the solidification-induced solutal marangoni flow, Physical Review Letters 132 (1) (2024) 014002.
[107] H. Zeng, Y. Wakata, X. Chao, M. Li, C. Sun, On evaporation dynamics of an acoustically levitated multicomponent droplet: evaporation-triggered phase transition and freezing, Journal of Colloid and Interface Science 648 (2023) 736–744.
[108] S.P. Kharal, J.-F. Louf, Unidirectional freezing of polymer solution droplets, Langmuir: the ACS Journal of Surfaces and Colloids 40 (1) (2024) 118–124.
[109] S.F. Ahmadi, S. Nath, C.M. Kingett, P. Yue, J.B. Boreyko, How soap bubbles freeze, Nature Communications 10 (1) (2019) 2531.
[110] A. Hashmi, A. Strauss, J. Xu, Freezing of a liquid marble, Langmuir: the ACS Journal of Surfaces and Colloids 28 (28) (2012) 10324–10328.
[111] A. Starostin, V. Strelnikov, L.A. Dombrovsky, S. Shoval, E. Bormashenko, Three scenarios of freezing of liquid marbles, Colloids and Surfaces A: Physicochemical and Engineering Aspects 636 (2022) 128125.

CHAPTER

5

Droplet impact and icing behaviors on flat surfaces

Kailiang Li[1], *Xuan Zhang*[1], *Mengjie Song*[1], *Long Zhang*[1], *Yubo Gao*[1], *Zhongxin Liu*[1], *Yiqing Guo*[2] *and Xiaomin Wu*[2]

[1]Department of Energy and Power Engineering, School of Mechanical Engineering, Beijing Institute of Technology, Beijing, P.R. China [2]Key Laboratory for Thermal Science and Power Engineering of Ministry of Education, Department of Energy and Power Engineering, Tsinghua University, Beijing, P.R. China

Nomenclatures

Symbol	Description	Unit
A	Area of droplet	m^2
a_0	Horizontal half axes of droplet	mm
AR	Aspect ratio	/
Bo	Bond number	/
b_0	Vertical half axes of droplet	mm
D	Diameter of droplet	mm
e	Eccentricity of the ellipse	/
E_k, E_s, E_g	Kinetic, surface, gravitational potential energy	J
g	Gravitational acceleration	m/s^2
H_{max}	Thickness of droplet at maximum spreading	mm
Ra	Surface roughness	nm
Re	Reynolds number	/
S_A	Form factor	/

Frosting and Icing for Efficient Energy Use in Engineering Applications
DOI: https://doi.org/10.1016/B978-0-443-15495-9.00013-7

t	Time	s
T_a, T_s	Air, surface temperature	°C
U	Velocity of impact	m/s
V	Volume of droplet	μL
W_{dis}	Energy of viscous dissipation	J
We	Weber number	/
Greek symbol	**Description**	**Unit**
δ	Thickness of the boundary layer	mm
β_D, $\beta_{D,max}$	Spreading factor, maximum of spreading factor	/
β_H, $\beta_{H,min}$	Height factor, minimum of height factor	/
θ_a, θ_{sta}, θ_r	Advancing, static , receding contact angle	°
μ	Dynamic viscosity	Pa • s
ρ	Density	kg/m^3
σ	Surface tension	N/m^2
ϕ	Viscous dissipation function	J/m^3
Ω	Fluid volume	m^3
Subscript	**Description**	**Unit**
0	Initial	/
a	Advancing	/
D	Length of spreading	/
dis	Dissipation	/
g	Gravitational potential	/
gs	Gas-solid interface	/
H	Height of spreading	/
k	Kinetic	/
ls	Liquid-solid interface	/
max	Maximum	/
min	Minimum	/
r	Receding	/
s	Surface	/
side	Side surface area	/
sta	Static	/
top	Top surface area	/
Y	Young's equation	/

5.1 Introduction

The impact and icing phenomenon of water droplets on cold surfaces is ubiquitous across both natural phenomena and engineering endeavors, closely related to phenomena such as hail formation and frost, and poses a huge threat to the normal operation of aerospace, power production and transportation, and other fields [1–3]. It not only reduces the performance of equipment but also increases safety risks [4–6]. In studying the impact and icing process of water droplets, the first step is to understand the fundamental principles of water droplet impact. The impact process of water droplets is influenced by the complex surrounding environment, which can change the shape of water droplets and affect their impact dynamics [7–9]. So, it is necessary to study water droplet impact on solid surfaces, especially impact dynamics of nonspherical droplets, to gain deeper insights into the interaction between water droplets and surfaces. Second, by investigating the impact and icing characteristics of water droplets and coupling the two processes of impact and icing to better understand the formation and accumulation mechanism of freeze/frost [10–12]. The study of water droplet impact dynamics and the impact and icing characteristics of water droplets is of great significance for improving equipment performance, reducing safety risks, and promoting the development of related technologies [13,14].

The impact behavior of droplets is largely influenced by their inherent physical attributes and the surface's wettability. With the support of various technical conditions, it becomes relatively easy to predict the droplet impact behavior [15]. Lin et al. [16] systematic studies the influence of droplet viscosity and surface wettability on droplet impact behavior. Through experimental observation and theoretical analysis, a universal scaling law for maximum spreading factor during droplet impact has been suggested. Aksoy et al. [17] discussed the negative correlation between maximum spreading time and the factor of droplets and liquid viscosity through comparative analysis of experiments and theoretical modeling. An empirical model for maximum spreading time and spreading factor was established based on energy conservation, taking into account the impact of droplet viscosity on spreading speed. However, the analysis of the results revealed that there is still a certain error in the geometric approximation of droplet shape in the model, and there is room for improvement in modeling viscous dissipation during droplet spreading. Li et al. [18] redefined droplet spreading time ($t_m = 2D_0/U_0$), and the viscous dissipation term has been corrected, establishing a maximum spreading factor model with higher accuracy in predicting the behavior of droplets with high viscosity and low surface tension. Remer et al. [19] propose, through experimental observation and analysis, that the dynamic contact angle induced by fluid flow can more accurately reflect the properties of the surface than values obtained from static and quasistatic measurements. Preliminary results indicate that an increase in the Weber number enhances the discrepancy between advancing contact angles and receding contact angles. Therefore it is necessary to discuss the coupling relationship between the two factors in future research. Xu et al. [20] employed numerical simulation methods, taking into account factors such as Weber number ($We = \rho U_0^2 D_0/\sigma$, where ρ and σ are the droplet density and liquid–gas surface tension), Reynolds number ($Re = \rho U_0 D_0/\mu$, where μ is the droplet viscosity), Ohnesorge number ($Oh = \sqrt{We}/\mathrm{Re} = \mu/\sqrt{\rho\sigma D_0}$), and contact angle to systematically study issues of droplet impact with solid surfaces. They further suggested that during the initial phase of droplet spreading, the role of the contact angle is more pronounced compared to the impact of the Weber number. Liu et al. [21] experimentally explored how surface wettability affects water droplet spreading, specifically when the Weber number is close to zero. According to experimental results and a semiempirical approach to analysis, under the condition of $We < 0.5$, the spreading factor's variation across various surfaces can be modeled using a segmented exponential function and shows a linear relationship. Budakli et al. [22] based on energy conservation and introducing a surface roughness term, a semiempirical model has been provided to determine the maximum spreading factor within the contact angle range from 10° to 150° and the Weber number range of 70–400, with an accuracy of 95%. Yancheshme et al. [23] combine experimental insights with artificial intelligence learning to create predictive models for droplet behavior on diverse wettable surfaces, emphasizing superhydrophobic ones. However, due to the uneven distribution of wettability conditions in the dataset, the model's learning ability for surfaces beyond hydrophobic and superhydrophobic is limited. Most of the above studies simplify the droplet shape as spherical at the moment of impact with the wall surface, without controlling the droplet shape. In reality, the droplet impact process is always affected by factors such as airflow and electromagnetic fields, and the droplet shape at the time of impact may become nonspherical. For shape fluctuations caused by gravity and surface tension, the droplet can be approximately regarded as an ellipsoid with the rotation axis along the gravity direction [24]. Thus further investigation into how droplet shape influences impact behavior and outcomes is crucial [25]. Liu et al. [26] pointed out that at ultra-low Weber numbers, due to the enhanced surface tension, the deformation of the droplet assumes an elliptical shape, which is less significant compared to a spherical shape. Modeling the deformed droplet as elliptical results in a relatively smaller increase in surface energy, and they proposed a theoretical proportional relationship considering the influence of gravity based on energy conservation. Zhang et al. [27] concluded through experiments that elliptical droplets exhibit the same self-similar characteristics as spherical droplets during the initial stages of impact. Under the same impact velocity, the effective stresses after impact of small elliptical water droplets with large spherical water droplets are not significantly different.

As a passive antiicing technology, superhydrophobic surfaces are increasingly attracting attention [28]. It is necessary to discuss the icing experiment of droplets impacting cold surfaces and compare the impact dynamics of room-temperature droplets. In the experiment process of impingement-icing of supercooled droplets on a cold hydrophobic surface, Zhang et al. [29] identified three different forms: complete rebound, partial rebound, and complete adhesion, indicating the competition between fluid dynamics and thermal transport. Yao et al. [30] concluded through numerical modeling and experimental results that, for the case of complete rebound, as the Weber number decreases, the growth rate of contact time with decreasing surface temperature increases. Regarding complete adhesion, alterations in surface temperature have minimal impact on the dynamic response of droplets striking supercold superhydrophobic surfaces (below −40°C). The influence of surface temperature on the maximum spreading factor becomes more pronounced at elevated Weber numbers. Later on, they obtained two different icing patterns, center-concave or center-convex, through numerical simulations of the impacting and icing coupling process on surfaces under various surface temperatures and Weber numbers [31].

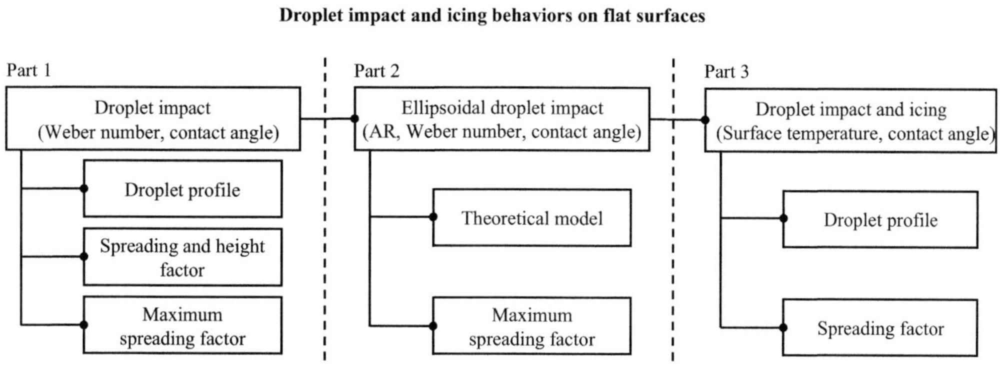

FIGURE 5.1 Outline of droplet impact and icing study.

In terms of droplet impact dynamics, although considerable theoretical, experimental, and simulation works have been conducted, there are still some errors in the predictions of most models. A large number of experimental results are needed to supplement the theoretical deficiencies, providing opportunities for improving existing models. Moreover, in real situations, the initial droplet shape may be affected by environmental factors, which has been given relatively little consideration in its influence on impact characteristics. Regarding the impacting and icing of droplets, due to the challenges in experimentation and simulation, there is limited research on the coupling characteristics. Several qualitative studies have been reported, but the mechanisms of impact and icing are not fully understood and require further discussion.

In this chapter, we commence with an experimental study of droplet impact and icing behaviors on flat surfaces. We investigate the influence of the Weber number and contact angle on the impact process of spherical water droplets. To forecast the maximum spreading factor of water droplets during actual impacts, we formulated a theoretical model considering the initial elliptical form of droplets. Using this model, we further predicted and investigated the influence of the aspect ratio of elliptical water droplets on the impact process, especially regarding the maximum spreading factor, across various Weber numbers and contact angles. Subsequently, we conduct icing tests on the impact of water droplets on cold surfaces. We compare the results of these experiments with the impact dynamics characteristics of the room-temperature droplets and analyze the icing characteristics of the impact droplet. Fig. 5.1 shows the outline of the present chapter.

5.2 Droplet impact behaviors on flat surfaces

Despite previous studies exploring droplet impacts on surfaces of varying wettability, additional experimental evidence is crucial to reinforce and authenticate the behaviors on diverse, particularly superhydrophobic surfaces. Further investigation is essential for a holistic understanding. This section investigates the dynamics of droplet impact on surfaces with varying wettability through droplet impact on flat surfaces experiments.

Specifically, four surfaces ranging from hydrophilic to superhydrophobic were prepared for this study. The influences of Weber number and contact angle on spreading and height change during droplet impact were analyzed, particularly focusing on the effects of these factors on the maximum spreading factor and minimum height factor.

5.2.1 Experimental surfaces and setup

5.2.1.1 *Surfaces preparation*

Four experimental surfaces were prepared utilizing bare aluminum plates. First, the aluminum surface underwent thorough cleaning with ethanol and distilled water for 5 minutes in an ultrasonic cleaner, followed by air drying. Second, a uniformly mixed coating solution was applied to the surface, allowing the solvent to evaporate completely. Subsequently, the coated surface was baked in an oven at a temperature ranging from 140°C to 200°C for 20–30 minutes. Lastly, the wettability and uniformity of surfaces were evaluated by measuring their contact angle and roughness characteristics.

Table 5.1 summarizes the structure, contact angle, and surface roughness of the four surfaces, obtained using scanning electron microscopy (SEM, Merlin VP Compact), contact angle goniometer (JC2000C1, China, ±1° accuracy), and atomic force microscopy (AFM, Bruker Dimension FastScan). These surfaces are labeled Surf. 1 through Surf. 4 based on their wettability and static contact angles, with Surf. 1 being hydrophilic, Surf. 2 and Surf. 3 hydrophobic, and Surf. 4 superhydrophobic. As shown in Table 5.1, Surf. 1 comprises a bare aluminum plate, while Surf. 2, Surf. 3, and Surf. 4 were created by applying distinct coatings onto the aluminum surface. Specifically, Surf. 2 and Surf. 4 employed nano-SiO_2 coatings sourced from Changzhou Nano Coatings Co., Ltd, while Surf. 3 utilized a PTFE coating provided by Daikin Chemical Co., Ltd.

Although using hydrophobic nano-SiO_2 particles (30–50 nm) for both Surf. 2 and Surf. 4, their surface structures, roughness, and contact angles varied due to particle aggregation. Specifically, Surf. 4 exhibited a hierarchical structure, contributing to its superhydrophobic properties. The static contact angle was determined as the average of five measurements using a 2 μL water droplet. Furthermore, advancing and receding contact

TABLE 5.1 Overview of surface properties of test surfaces.

No.	Coating	Static contact angle, θ_{sta} (°)	Advancing contact angle, θ_a (°)	Receding contact angle, θ_r (°)	Contact angle hysteresis, $\Delta\theta$ (°)	Surface roughness, *Ra* (nm)
Surf. 1	Bare aluminum	82	100	39	61	17.4
Surf. 2	Hydrophobic nano-SiO_2	95	114	75	39	2.7
Surf. 3	Hydrophobic PTFE	117	129	88	41	17.6
Surf. 4	Superhydrophobic nano-SiO_2	160	165	155	~10	363.1

angles were obtained through the addition and removal of the liquid, respectively. It was slightly difficult to maintain a stable state of the liquid on the Surf. 4, so their advancing and receding contact angles may not be very accurate, but they can still be used to evaluate the surface wettability.

5.2.1.2 Experimental setup

Fig. 5.2 illustrates the experimental setup employed in this study, comprising a semiconductor-based thermoelectric cooling system, a comprehensive data and image acquisition system, a dedicated droplet generation unit, and the test section itself. This setup can meet the requirements for droplet impact experiments and droplet impact and icing experiments at room temperature. The semiconductor-based thermoelectric cooling system is used to provide a constant cold surface temperature for the test section surface, including a programmable DC power supply from Agilent, model N5766A, made in the USA, with an adjustment accuracy of 0.02 V, a semiconductor cooling plate, TEC1–12706, a cooling water channel, and a cooling circulation pump. By adjusting the output voltage of the DC power supply connected to the semiconductor cooler, the temperature of the test surface pasted on the cold side of the semiconductor cooling plate is set to the desired

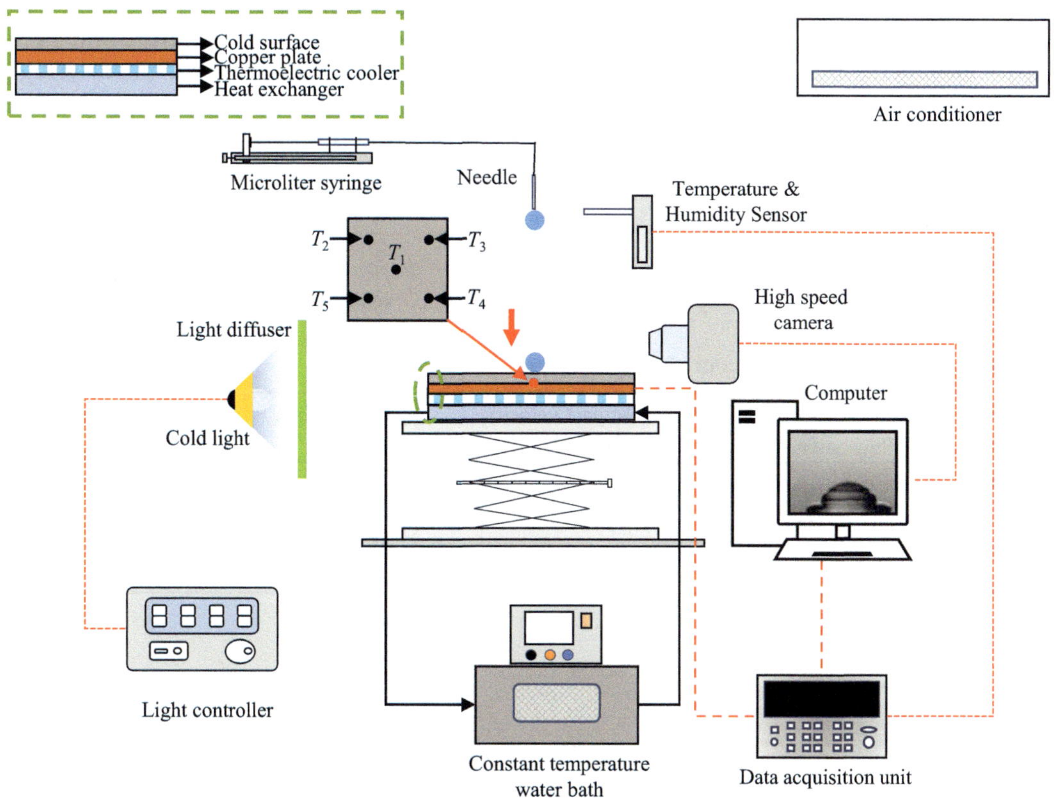

FIGURE 5.2 Schematic diagram of the droplet impact and icing experimental apparatus.

temperature. A copper plate is placed between the test surface and the semiconductor cooling plate to make the temperature more uniform, and thermal grease is used for bonding between the test surface, copper plate, and cold side of the cooling plate to reduce thermal contact resistance. In the experiment, the temperature of the test surface can be adjusted for droplet impact-ice experiments. The data and image acquisition system consists of five T-type thermocouples with a diameter of 0.127 mm, a temperature and humidity sensor, Rotronic HF532 from Switzerland, with temperature accuracy of $\pm 0.1°C$ and humidity accuracy of $\pm 0.8\%$, a data acquisition unit, the Agilent 34970A from the USA, a cold light with a light diffuser, a high-speed camera, the Optronics CP80–3-M-540, with a shooting speed of 2000fps and resolution of 1024pi × 1024pi, equipped with a Nikon AF 105/2.8D lens, and a computer. Temperature and humidity data are recorded by a computer, which is connected to the high-speed camera for control and image storage.

The droplet generation system includes a microsyringe with an accuracy of $\pm 0.2\ \mu L$, an adjustable platform, and a capillary needle. The outer diameter of the needle used in our experiment is 0.70 ± 0.02 mm, and the average droplet volume (V) produced is $12.0 \pm 0.02\ \mu L$, with an average droplet diameter (D_0) of 2.84 ± 0.020 mm. The test section includes an aluminum-based surface measuring 40 mm long by 40 mm wide by 1 mm high. Four surfaces varying in wettability were prepared via coating techniques for the experiments, detailed in Table 5.1.

On the surfaces with different wettability prepared in this chapter, droplet impact experiments were conducted. The experimental conditions were at ambient temperature and standard atmospheric. By finely adjusting the distance between the bottom of the microsyringe and the test surface to different heights, different impact velocities were obtained after the droplets fell freely. The specific experimental conditions are summarized in Table 5.2.

5.2.2 Data processing

Fig. 5.3 shows the morphological changes of water droplets during the impact process on Surf. 2 captured in the experiment. The volume of water droplets is $12\ \mu L$, and the

TABLE 5.2 Experimental conditions of droplet impact.

Volume, V (μL)	Diameter, D_0 (mm)	Initial height, h_0 (mm)	Impact velocity, U_0 (m/s)	Weber number, We	Reynolds number, Re
12.0 ± 0.2	2.84 ± 0.020	10.16	0.44	7.75	1259.33
		25.16	0.70	19.18	1981.77
		50.16	0.99	38.24	2798.20

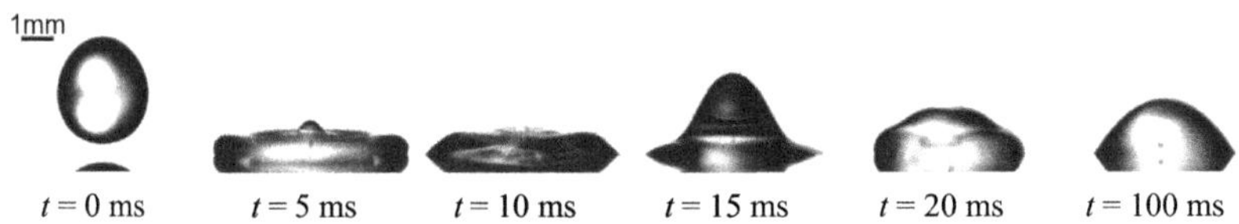

FIGURE 5.3 The process of droplet impact on hydrophilic Surf. 2.

Weber number is 19.18. The impact process of water droplets on the Surf. 2 comprises four distinct phases, that is, spreading, receding, oscillating, and stabilization. It can be clearly seen that the different phases produce different spreading and height factors. In the spreading phase, under the influence of dominant inertial forces, the water droplet's contact line progressively extends, with its kinetic and gravitational potential energies transforming into surface energy. Upon attaining maximum spreading, the surface energy of the water droplet is the largest, and the kinetic energy is zero. In the phase of receding, the contact line of the water droplet recedes gradually under the leading action of surface tension, and the surface energy of the water droplet is converted into kinetic energy and gravitational potential energy. The droplet recedes on the Surf. 2 is obvious (e.g., 10 Ms in Fig. 5.3). During the oscillating phase, due to a certain contact angle hysteresis on the actual surface, the contact line of the water droplet stops moving at a certain position, but the water droplet still oscillates periodically in the height direction. In the stable phase, due to the action of viscous dissipation, the kinetic energy of the water droplet is completely lost, and the water droplet completely stops moving in the height direction. There is a small stable spreading factor on the Surf. 2 (e.g., 100 Ms in Fig. 5.3).

Based on the experimental image in Fig. 5.3, the image recognition method is used to further quantitatively extract temporal variations in spreading and height factors during droplet impact. As depicted in Fig. 5.4, the spreading and height factors vary across the impact phases, and the maximum spreading factor appears at about 5 Ms, which is basically consistent with the morphological shown in Fig. 5.3.

The key features of the droplet impact process are manifested through variations in its spreading and height factors, as evidenced by the above experimental findings. Therefore the following research emphasizes the influence of Weber number and contact angle on the spreading and height factors during the impact process, especially on hydrophobic surfaces.

5.2.3 Experimental results

5.2.3.1 Effect of Weber number

Fig. 5.5A depicts the spreading factor variations of water droplets during the impact process under different Weber numbers on the hydrophobic Surf. 3. As the Weber number rises during

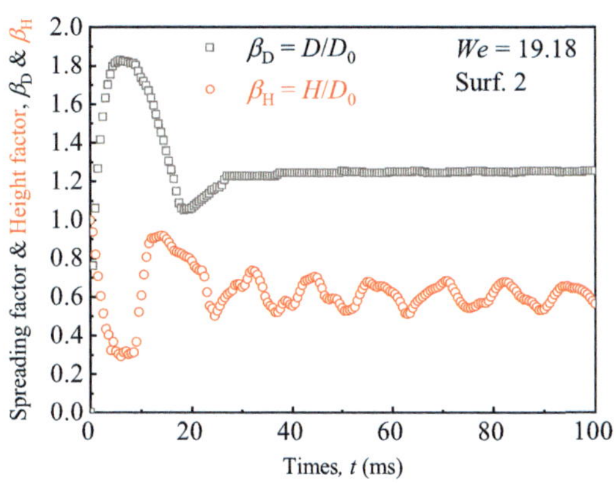

FIGURE 5.4 Temporal evolutions of spreading and height factors of droplet impact.

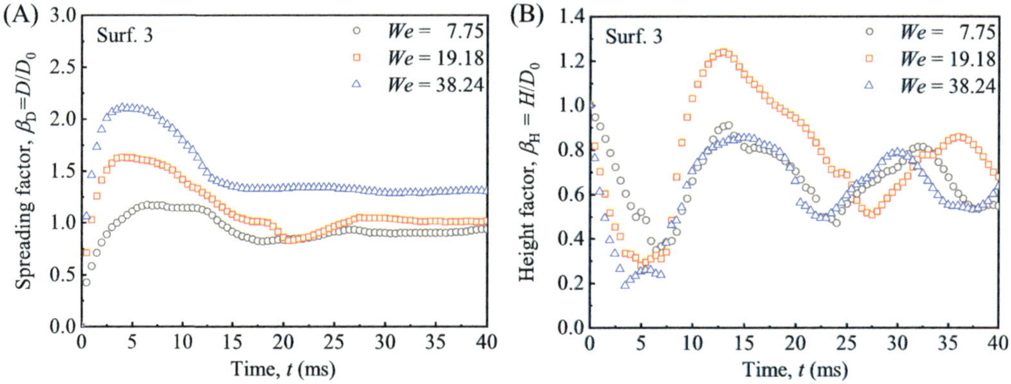

FIGURE 5.5 Evolutions of (A) spreading factors and (B) height factors under different Weber number.

the spreading phase, the water droplets spread more rapidly, leading to a concurrent increase in their spreading factor. This trend is also reflected in the maximum spreading factor, which exhibits an upward tendency. In the subsequent receding phase, an increase in the Weber number also accelerates the dynamic contraction speed of water droplets during the contraction phase, and in the late receding phase, spreading factors change under different Weber numbers tend to be consistent. In the stabilization phase, stable spreading factors under different Weber numbers are anisotropic attributed to the hysteresis of the contact angle.

Fig. 5.5B depicts the height factor variations of water droplets during the impact process on the hydrophobic Surf. 3 under different Weber numbers. As the Weber number rises, the height factor of the water droplet diminishes simultaneously, resulting in a corresponding reduction in the final minimum height factor. The height factor of the droplet is not only affected by the spreading factor but also related to the shape and contour of the droplet. In the actual impact process, due to the inevitable small disturbance, the shape of the water droplets often presents asymmetry, which affects the change of the outline of the water droplets. Therefore in the phase of droplet receding, the influence of Weber number on the droplet height factor is not as obvious as that on the spreading factor, showing a more complex change trend.

5.2.3.2 *Effect of contact angle*

Fig. 5.6A depicts the spreading factor variations of water droplets during the impact process on surfaces with different contact angles under the Weber number of 19.18. The results show that as the contact angle increases, the spreading factor of water droplets decreases, with both maximum and stable spreading factors following this trend. Because the Weber number is fixed, water droplets' impact velocity remains basically constant when impacting different surfaces, resulting in little change in the spreading rate of water droplets at the same time.

Fig. 5.6B depicts the height factor variations of water droplets during the impact process on surfaces with different contact angles under the Weber number of 19.18. At the beginning of spreading, the height factor of water droplets exhibits a similar trend across different surfaces. This similarity indicates the predominant influence of inertial force in the spreading process of water droplets. After reaching the local minimum height factor for the first time,

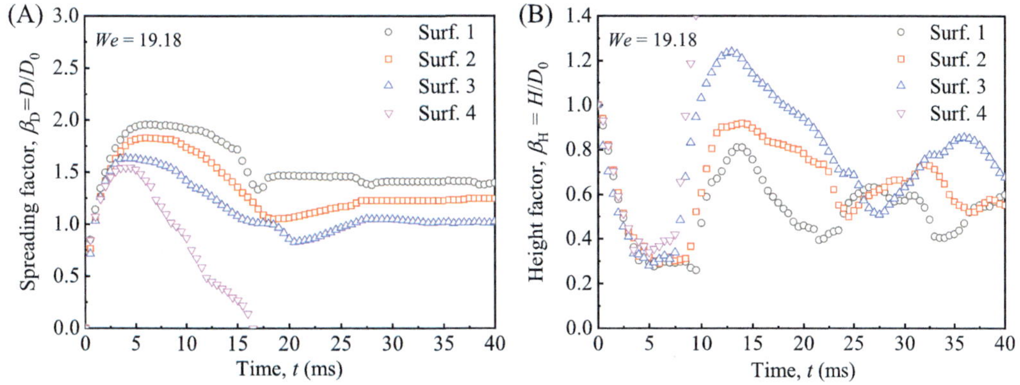

FIGURE 5.6 Evolutions of (A) spreading factors and (B) height factors under different contact angle.

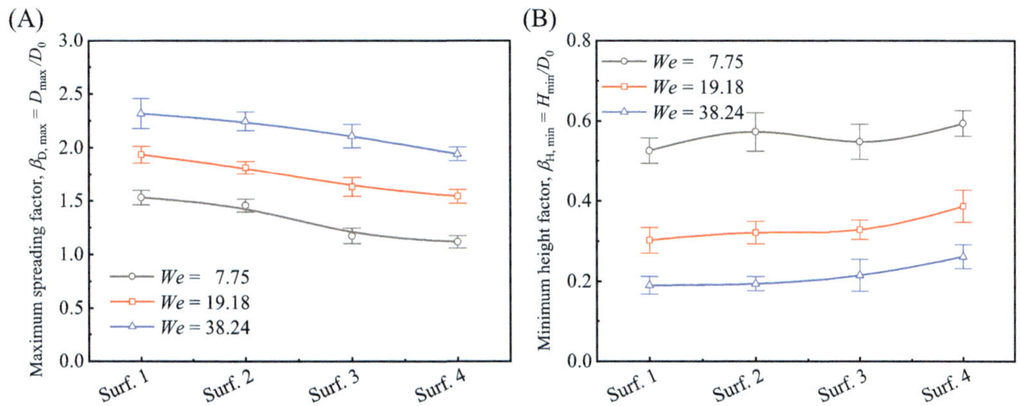

FIGURE 5.7 (A) Maximum spreading factor and (B) minimum height factor for droplet impact.

due to the interaction of viscous force and surface tension, the height factor experiences a brief increase and then continues to decline and reach the global minimum height factor. This process shows that the global minimum height factor of water droplets falls with the rise of contact angle, which is consistent with the variation trend of the spreading factor.

5.2.3.3 Maximum spreading factor and minimum height factor

To further analyze the effect of Weber number and contact angle on the spreading factor and height factor of water droplets, the maximum spreading factor and the first local minimum height factor of water droplets under different conditions were extracted, and the results are shown in Fig. 5.7. As can be seen from Fig. 5.7A, the maximum spreading factor of water droplets gradually decreases with the decrease of Weber number and the increase of contact angle. This is because the driving effect of inertia force on droplet spreading is weakened at a low Weber number, and the hydrophobicity of the surface at a high contact angle is enhanced, thus inhibiting droplet spreading. It can be seen from

Fig. 5.7B that with the increase of Weber number, the first local minimum height factor of water droplets gradually decreases, which is because the inertial force plays a leading role in the deformation of water droplets at the initial phase of impact. The surface hydrophobicity has a weak effect on the height factor under different Weber numbers.

5.3 Maximum spreading of ellipsoidal droplets on flat surfaces

Research on the dynamics of droplet impact on flat surfaces has primarily focused on spherical droplets, examining droplet impact dynamics and analyzing the effects of various factors, particularly Weber number and contact angle, on maximum spreading factor. However, complex environmental factors such as gravity [8], shear flow [9], and electric/magnetic fields [25] can cause droplet deformation, leading to a deviation from sphericity before impact. Changes in the aspect ratio (*AR*) of the droplet during impact can affect subsequent kinetic behavior, especially maximum spreading, thereby altering heat transfer or frictional force between the droplet and surface.

Although we control conditions to minimize these effects, droplet shape fluctuations have been observed in some experiments, as shown in Fig. 5.8. To more accurately study the impact of initial droplet shape on the impact process, especially the effect of aspect ratio on maximum spreading factor, we have analyzed the impact of ellipsoidal droplets by adjusting the term for viscous dissipation in the energy conservation equation. The derived theoretical correlation can predict the maximum spreading factor for droplets with varying aspect ratios. Drawing on the experimental conditions from the previous section, we have verified the predictions for $AR = 1$ and discussed the effect of aspect ratio on the maximum spreading factor under various Weber numbers and contact angles.

5.3.1 Theoretical model

As shown in Fig. 5.9, when in contact with solid surfaces, a nonspherical impact droplet with various kinds of shapes can be reduced to an elliptical droplet with an initial velocity of U_0 and a symmetry axis. Assuming that the initial droplet is at maximum deformation at

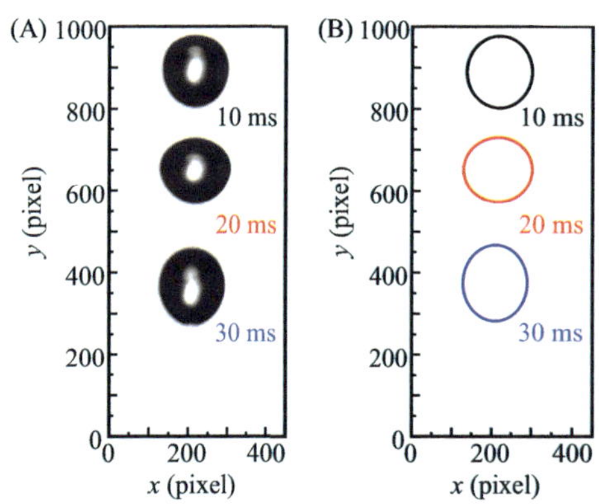

FIGURE 5.8 Shape changes during the droplet falling process from (A) experiment photos and (B) extracted contours.

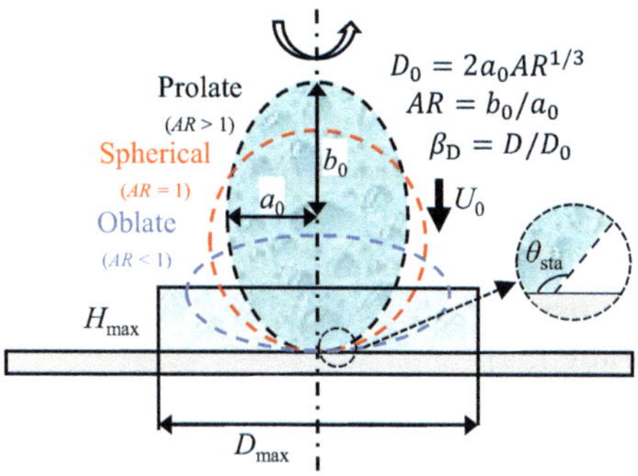

FIGURE 5.9 Shape diagram of ellipsoidal droplets at the initial and maximum spreading.

the contact surface, the internal flow field is ignored. The vertical and horizontal half axes of the droplet are b_0 and a_0, and the droplet's aspect ratio $AR = b_0/a_0$ is obtained. The typical diameter $D_0 = 2(a_0^2 b_0)^{1/3} = 2a_0 AR^{1/3}$ is that of a sphere, which matches the volume of the droplet at its initial state. Owing to the absence of rotation in elliptical droplets upon impact, they display symmetric spreading and retraction behavior on the surface, akin to spherical droplets. The spreading factor is defined as $\beta = D/D_0$, where D is spreading diameter. At the state of maximum spreading ($\beta_{max} = D_{max}/D_0$), the droplet transforms into a thin film with a mean thickness of H_{max}. To theoretically obtain the maximum spreading factor, the transformed droplets that reach the maximum spreading phase are modeled as thin cylindrical films with an enlarged diameter D_{max} and a specified height H_{max}.

According to the law of energy conservation, the correlation formula for the maximum spreading factor of ellipsoidal droplets can be theoretically derived. The droplet impact process includes four energy terms: kinetic energy, gravitational potential energy, surface energy, and viscous dissipation. The energy conservation between the initial and maximum states can be expressed as

$$E_{k,0} + E_{g,0} + E_{s,0} = E_k + E_g + E_s + W_{dis} \tag{5.1}$$

where $E_{k,0}$ and E_k are the kinetic energies of the droplet at two disparate moments, $E_{s,0}$ and E_s represent the surface energy, $E_{g,0}$ and E_g denote the gravitational potential energies, and W_{dis} signifies the energy dissipation during the impact process.

Taking the initial state as the droplet contacting the solid surface, using the solid surface as the benchmark, the energy terms are as follows:

$$E_{k,0} = \frac{\pi}{12}\rho D_0^3 U_0^2, \quad E_{g,0} = \frac{\pi}{12}\rho g D_0^4 AR^{2/3}, \quad E_{s,0} = \sigma A_0 \tag{5.2}$$

where $A_0 = \pi D_0^2 S_A$ represents the initial surface area of the assumed axisymmetric ellipsoidal droplet in Fig. 5.9, and S_A is specified according to [32]

$$S_A = \begin{cases} \dfrac{1}{2AR^{2/3}}\left(1 + \dfrac{AR^2 \text{arctanh} e}{e}\right), e = \sqrt{1 - AR^2}, AR < 1, \\ 1, \qquad e = 0, \qquad AR = 1, \\ \dfrac{1}{2AR^{2/3}}\left(1 + \dfrac{AR \text{arcsin} e}{e}\right), e = \sqrt{1 - \dfrac{1}{AR^2}}, AR > 1, \end{cases} \tag{5.3}$$

where e represents the eccentricity of the ellipse that is defined by the cross-section along the axis of symmetry.

At the state of maximum spreading, based on our previous work [33], in most cases, the remaining kinetic energy of vortices originating from the inside of the thin film constitutes less than 5% of the initial overall energy. Due to the different profiles of droplets under different conditions, accurately determining this energy via theoretical analysis poses a challenge. Furthermore, with the film thickness being substantially thinner than the initial droplet height, gravitational potential energy usually does not play a significant role. Therefore assuming $E_k \approx 0$ and $E_g \approx 0$ is appropriate [34].

The surface energy (E_s) of the entire system includes gas–liquid, gas–solid, and liquid–solid interfaces, expressed as

$$E_s = \sigma \int_A dA + \sigma_{ls} A_{ls} - \sigma_{gs} A_{gs} \tag{5.4}$$

where A, A_{ls}, and A_{gs} are the areas of gas–liquid, liquid–solid, and gas–solid interfaces, respectively, σ_{ls} and σ_{gs} are liquid–solid and gas–solid surface tensions, respectively. During the process of impact, the gas–solid interface is superseded by the liquid–solid interface, which shares the same area, that is,

$$A_{ls} = A_{gs} = \frac{\pi D^2}{4} \tag{5.5}$$

At the state of maximum spreading, $A_{top} = \pi D_{max}^2/4$ and $A_{side} = \pi D_{max} H_{max}$ are the top and side surface areas of the thin film, where $H_{max} = 2D_0^3/3D_{max}^2$. Although Young's equation [35], $\sigma\cos\theta_Y = \sigma_{gs} - \sigma_{ls}$, the spreading phase is predominantly governed by the advancing contact angle [36], θ_Y is considered the θ_{adv} here, and the surface energy is calculated using the following formula, that is,

$$E_s = \pi\sigma D_{max} H_{max} + \frac{\pi\sigma D_{max}^2}{4}(1 - \cos\theta_{adv}) \tag{5.6}$$

The viscous dissipation involved can be obtained from

$$W_{dis} = \int_0^{t_{max}} \int_\Omega \Phi d\Omega dt \approx \Phi\Omega t_{max} \tag{5.7}$$

In which Φ is the viscous dissipation function, Ω is the fluid volume where the viscous dissipation appears, and t_{max} indicates the moment the droplet achieves its maximum spread, with these values calculated according to [37]

$$\phi \sim \mu\left(\frac{\partial U_x}{\partial z}\right)^2 \sim \mu\left(\frac{U_0}{\delta}\right)^2, \Omega = \frac{\pi D_{max}^2 \delta}{4}, t_{max} = \frac{8}{3U_0} 2b_0 = \frac{8D_0}{3U_0} AR^{2/3} \tag{5.8}$$

where δ is the thickness of the boundary layer, formulated as the following equation:

$$\delta = \frac{2D_0}{\sqrt{Re}} AR^{1/3} \tag{5.9}$$

Thus the viscous dissipation is changed to

$$W_{\mathrm{dis}} = \frac{\pi\sigma D_{\max}^2}{3}\frac{We}{\sqrt{\mathrm{Re}}}AR^{1/3} \tag{5.10}$$

The maximum spreading factor can be derived by inserting Eqs. (5.2), (5.6), (5.10) into Eq. (5.1), that is,

$$\left(We + BoAR^{2/3} + 12S_{\mathrm{A}}\right)\beta_{\max} = 8 + \left[3(1-\cos\theta_{\mathrm{adv}}) + \frac{4We}{\sqrt{Re}}AR^{1/3}\right]\beta_{\max}^3 \tag{5.11}$$

where $Bo = \rho g D_0^2/\sigma$ is the Bond number. The results of references [33,38] indicate that the majority of viscous dissipation takes place during the early spreading phase. Therefore using the volume of dissipation at the maximum spreading state to approximate Eq. (5.10) of viscous dissipation may result in significant deviation in the calculated results. The aspect ratio (AR) limited by the shape of the droplet is responsible for the volume of liquid that impacts solid surfaces at any point, so the corresponding dissipative volume can be adjusted proportionally to

$$\Omega(t) \propto \pi a_0^2\cos^2\varphi U_0 dt \propto 1/AR^{\frac{2}{3}} \tag{5.12}$$

where φ represents the parameter extracted from the parametric equation of the ellipse. Therefore Ω can be calculated by

$$\Omega = \frac{\pi D_{\max}^2\delta}{4AR^{2/3}} \tag{5.13}$$

Thus Eq. (5.10) is rewritten as

$$W_{\mathrm{dis}} \approx \frac{\pi\sigma D_{\max}^2}{3AR^{\frac{1}{3}}}\frac{We}{\sqrt{Re}} \tag{5.14}$$

The correlation formula for the maximum spreading factor can be revised to

$$\left(We + BoAR^{2/3} + 12S_{\mathrm{A}}\right)\beta_{\max} = 8 + \left[3(1-\cos\theta_{\mathrm{adv}}) + \frac{4We}{AR^{1/3}\sqrt{Re}}\right]\beta_{\max}^3 \tag{5.15}$$

5.3.2 Model validation

Fig. 5.10 compares maximum spreading factors derived from the experiments in this chapter and reference [38] with those given by the revised theoretical model. Due to the simplified droplet profiles used in the theoretical model for the maximum spreading phase, which differ from those used in the experiments, the model predictions are slightly higher than the experimentally obtained maximum spreading factors for the most part, resulting in a skewed deviation. However, the deviation of the revised model is only -5% to 25%, indicating its strong capability to predict the maximum spreading factor. Therefore the revised correlation equation can better predict the maximum spreading factor of ellipsoidal water droplets with various initial shapes after impact.

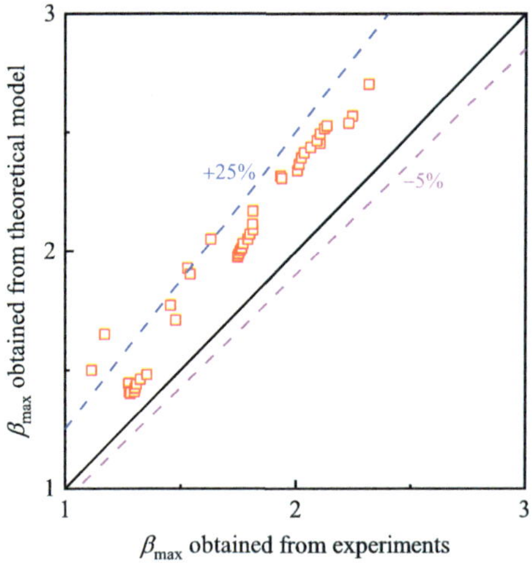

FIGURE 5.10 Comparison of maximum spreading factors from experiments and theoretical model.

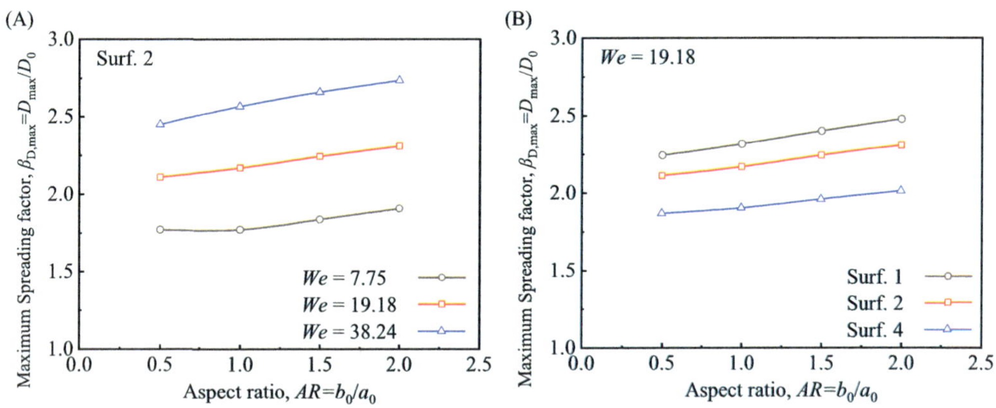

FIGURE 5.11 Variations of maximum spreading factors calculated by theoretical model with aspect ratio (*AR*) under (A) different Weber number and (B) different contact angle.

5.3.3 Results and discussion

Fig. 5.11 further describes the changes in the maximum spreading factor with aspect ratio (*AR*) under various Weber numbers and contact angles. In all cases, the maximum spreading factor tends to increase as the *AR* increases. As the Weber number increases and the contact angle decreases, this increasing trend becomes more pronounced. This is primarily due to the fact that at high Weber numbers and low contact angles, the interaction between the fluid and the wall during the droplet impact is more intense, and the change in aspect ratio affects the energy loss through the impact of the droplet with the wall, thus influencing maximum spreading factor. However, when the contact angle is large and the Weber number is small, the impact of the droplet results in almost no energy loss, and it can even be approximated as an elastic impact [39]. In this case, the change in *AR* has almost no influence on the maximum spreading factor of the droplet.

5.4 Droplet impact and icing behaviors on cold surfaces

The study of impact and icing coupling dynamics of droplets on cold surfaces has always attracted attention, but further research is needed on the interaction mechanism between surface wettability, especially superhydrophobic surfaces, and cold surface temperature. This section conducts experimental work on the impact and icing of droplets on cold surfaces. The influences of contact angle and cold surface temperature during droplet impacting and icing processes have been analyzed comprehensively.

5.4.1 Experimental conditions

In this chapter, we conducted droplet impact and icing experiments on hydrophilic Surf. 1 and superhydrophobic Surf. 4 at room temperature conditions. The experiments were performed with droplets having a diameter of 2.84 ± 0.035 mm and an initial velocity of 0.70 m/s. The surface temperature was controlled at 15°C, −10°C, and −30°C with a precision of ± 0.5°C. The specific experimental conditions are detailed in Table 5.3. By integrating the droplet impact experiments discussed in the previous section, the influence of surface temperature on the droplet impact and icing process can be analyzed.

5.4.2 Experimental results

Fig. 5.12 shows the images of the water droplet impact and icing process under three combined experimental conditions. By comparing the experimental results under conditions (I) and (II), it can be found that the existence of a cold surface mainly affects the receding phase of water droplets at room temperature and has a slight effect on the spreading phase after the impact of water droplets at room temperature, regardless of the hydrophilic Surf. 1 or superhydrophobic Surf. 4. For the hydrophilic Surf. 2, a thin layer of ice is formed at the bottom of the water droplets at room temperature after reaching maximum spreading factor when impacting on the cold surface, so the receding phase similar to that after impacting on the room temperature surface is hardly observed (Fig. 5.12A). The final stable-state contact angle of droplets impacting on cold surfaces is obviously smaller than that of water droplets impacting on a surface at room temperature. At the cold surface temperature of -10°C, water droplets on the Surf. 4 will ice due to the action of the cold surface, and it can be found that an adhesion point begins to form on the wall. However, under the action of surface tension, it can be observed that the water droplets show complete rebound characteristics when they impact on the surface at 15°C (Fig. 5.12B).

TABLE 5.3 Experimental conditions of droplet impact and icing.

Volume, V (μL)	Diameter, D_0 (mm)	Impact velocity, U_0 (m/s)	Weber number, We	Reynolds number, Re	Cold surface temperature, T_s (°C)	Air temperature, T_a (°C)
12 ± 0.2	2.84 ± 0.035	0.70	19.18	1981.77	−10 ± 0.5°C	15 ± 0.5°C
					−20 ± 0.5°C	
					−30 ± 0.5°C	

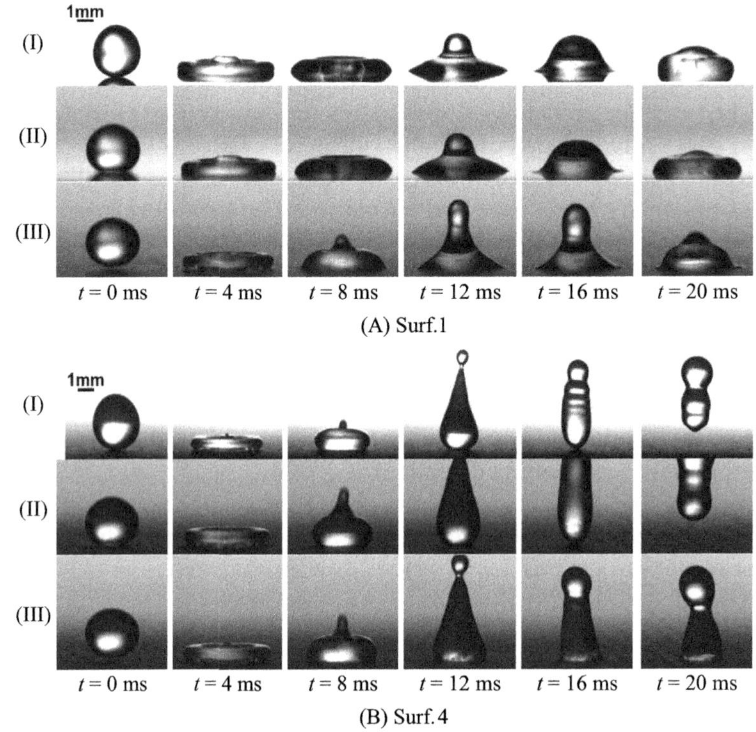

FIGURE 5.12 The process of impact and icing behaviors of droplets on (A) hydrophilic Surf. 1 and (B) superhydrophobic Surf. 4.

By comparing conditions (II) and (III), it can be seen that no matter the hydrophilic Surf. 1 or the superhydrophobic Surf. 4, the difference between various surface temperatures in the impact and icing process of water droplets also mainly exists in the receding phase. For the hydrophilic Surf. 2, from the perspective of water droplet morphology (Fig. 5.12A), the ice thickness at the bottom of the water droplet at the cold surface temperature of −30°C is significantly thicker than that at the cold surface temperature of −10°C during the receding phase, and the ice-water mixture formed inside the water droplet can be obviously seen at the cold surface temperature of −30°C during the receding phase. The inside of the water droplet becomes opaque, and the stable-state profile of the water droplet is also different from that of the water droplet at −10°C cold surface temperature due to the influence of the icing rate. For Surf. 4, at the cold surface temperature of −30°C, the icing rate of water droplets is faster. Compared with the cold surface temperature of −10°C, water droplets cannot completely escape from the cold surface of −30°C after impacting (Fig. 5.12B).

Fig. 5.13 presents the evolutions of the spreading factors when a water droplet impacts under different surface temperatures. For the hydrophilic Surf. 1 shown in Fig. 5.13A, the water droplet reaches a smaller maximum spreading factor due to the icing of the contact line upon impact with a cold surface, and the spreading factor remains almost unchanged in the receding phase.

For the superhydrophobic Surf. 4 displayed in Fig. 5.13B, when the Weber number is 19.18, the surface temperature has little effect on the spreading and maximum spreading

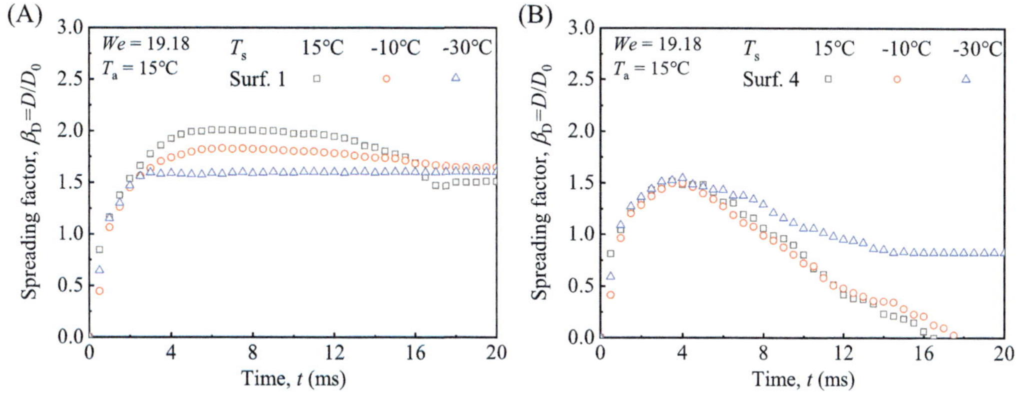

FIGURE 5.13 Evolutions of spreading factors under different surface temperatures on (A) hydrophobic Surf. 1 and (B) superhydrophobic Surf. 4.

factor of the water droplets, with the maximum spreading factors being relatively close. However, the impact of the cold surface on the receding phase is significant, affecting the receding rate. In the receding phase, the water droplets contract more slowly on the cold surface than on the surface at room temperature, and this phenomenon becomes more pronounced at lower cold surface temperatures. Ultimately, the bouncing of droplets on the superhydrophobic Surf. 4 disappears, resulting in a final stable spreading factor. This is primarily due to the increased viscosity of the water droplets caused by the low temperature and the inability of the contact line to move after icing. Therefore if the temperature of the superhydrophobic surface is low enough, its antiicing performance will significantly decline [40]. In general, the final stable spreading factor of the water droplets after impact with Surf. 4 is less than that after impact with Surf. 1.

5.5 Summary

This chapter experimentally investigates the impact process of water droplets and analyzes the influence of Weber number and contact angle on the impact characteristics of water droplets. Then, we study the effect of the initial shape of ellipsoidal water droplets on their impact dynamics characteristics and provide a theoretical model based on theoretical analysis. In addition, the impact and icing process of water droplets with cold surfaces is observed and compared with the impact process of water droplets on room temperature surfaces. The effect of the surface temperature on the impact and icing process was analyzed. The final conclusion is as follows:

1. As the Weber number increases and the contact angle decreases, the spreading speed of water droplets gradually accelerates, and the maximum spreading factor also increases accordingly. The corresponding time for the maximum spreading factor will decrease with the increase of Weber number and contact angle. As the Weber number increases, the decrease rate of droplet height gradually accelerates, and the height factors and the

first local minimum height factor of droplets also decrease accordingly. The variation of droplet height factor is almost unaffected by surface characteristics.

2. Based on the theoretical calculation formula of the maximum spreading factor in the impact process of classical spherical water droplets, by correcting the energy dissipation term in the initial aspect ratio of ellipsoidal water droplets, a more accurate model can be obtained. By applying this model, we ensured that most of the calculated data deviations were within the range of -5% to 25%. Utilizing this model, we further concluded that the maximum spreading factor tends to increase with the increase of the aspect ratio (*AR*) under different Weber numbers and contact angles.
3. Compared to the impact process of water droplets at room temperature, the main difference in the impact and icing process of water droplets lies in the droplet shape and spreading factors during the receding phase. For both hydrophilic and superhydrophobic surfaces, during the impact and icing process of water droplets, a uniform ice-water mixture is formed inside the water droplets. The ice layer at the bottom of the water droplets is thicker, and the stable spreading factor of the water droplets is greater after the contact line stops moving due to icing. The contour of the water droplets is also formed by the effect of icing, forming an irregular ice shell. For superhydrophobic surfaces, due to icing at the contact line, water droplets cannot detach from the cold surface after impacting and cannot show the complete rebound exhibited by water droplets impacting on room temperature surfaces.

Acknowledgments

This work is supported by the National Natural Science Foundation of China (Nos. 52406007, 52076013 and 52306003), Beijing Municipal Natural Science Foundation (Nos. 3244046 and 3232032), Key Laboratory of Icing and Anti/De-icing (Nos. IADL20230113 and IADL20220304), China Postdoctoral Science Foundation (No. 2022M720423), and Young Elite Scientist Sponsorship Program by BAST (No. BYESS2023352).

References

[1] Y. Cao, Z. Wu, Y. Su, Z. Xu, Aircraft flight characteristics in icing conditions, Progress in Aerospace Science 74 (2015) 62–80.
[2] X. Zhang, X.M. Wu, J.C. Min, Aircraft icing model considering both rime ice property variability and runback water effect, International Journal of Heat and Mass Transfer 104 (2017) 510–516.
[3] N. Dalili, A. Edrisy, R. Carriveau, A review of surface engineering issues critical to wind turbine performance, Renewable and Sustainable Energy Reviews 13 (2) (2009) 428–438.
[4] L.Z. Huang, M.J. Song, J. Shen, et al., Experimental study on the effect of inverted aperture plate temperature on frosting characteristics under rising flow considering edge effect, International Journal of Heat and Mass Transfer 224 (2024) 125343.
[5] Y.X. Xu, R.Y. Zhao, K. Wu, et al., Experimental investigation and validation on an air-source heat pump frosting state recognition method based on fan current fluctuation signal and machine learning, Energy 291 (2024) 130372.
[6] K.K. Shao, M.J. Song, Z.Y. Jiang, et al., Experimental study on the formation and distribution of trapped air bubbles in horizontal ice slice, CIESC Journal 74 (S1) (2023) 161–164. In Chinese.
[7] R. Rioboo, M. Marengo, C. Tropea, Time evolution of liquid drop impact onto solid, dry surfaces, Experiments in Fluids 33 (2002) 112–124.
[8] J. Ju, Z. Jin, H. Zhang, Z. Yang, J. Zhang, The impact and icing processes of a water droplet on different cold spherical surfaces, Experimental Thermal and Fluid Science 96 (2018) 430–440.

[9] Z. Xu, T. Wang, Z. Che, Droplet deformation and breakup in shear flow of air, Physics of Fluids 32 (2020) 052109.

[10] X.L. Liu, L.Z. Huang, X. Zhang, et al., A modeling study on icing characteristics of sessile and pendant water droplets1on cold plate surfaces, Applied Thermal Engineering 236 (2024) 121693.

[11] L.Z. Huang, Y.Y. Tian, M.J. Song, et al., An experimental study on edge-affected frosting characteristics on a vertical cold plate at different surface temperatures, Applied Thermal Engineering 236 (2024) 121538.

[12] Y.Y. Tian, X.L. Liu, S.H. Hosseini, et al., An experimental study on edge-affected frosting characteristics on a vertical cold plate at different air velocities, Applied Thermal Engineering 236 (2024) 121882.

[13] M.J. Song, C.B. Dang, Review on the measurement and calculation of frost characteristics, International Journal of Heat and Mass Transfer 124 (2018) 586–614.

[14] S.W. Lei, M.J. Song, W. Dai, et al., Experimental study on the effect of surface temperature on the frost characteristics of a vertical cold plate under natural convection, Experimental Thermal and Fluid Science 137 (2022) 110684.

[15] A.L. Yarin, Drop impact dynamics: splashing, spreading, receding, bouncing, Annual Review of Fluid Mechanics 38 (2006) 159–192.

[16] S.J. Lin, B.Y. Zhao, S. Zou, J.W. Guo, Z. Wei, L.Q. Chen, Impact of viscous droplets on different wettable surfaces: Impact phenomena, the maximum spreading factor, spreading time and post-impact oscillation, Journal of Colloid and Interface Science 516 (2018) 86–97.

[17] Y.T. Aksoy, P. Eneren, E. Koos, M.R. Vetrano, Spreading of a droplet impacting on a smooth flat surface: how liquid viscosity influences the maximum spreading time and spreading ratio, Physics of Fluids 34 (2022) 042106.

[18] J. Li, C. Zhao, C. Wang, Experimental study on the dynamics of droplet impacting on solid surface, Microfluidics and Nanofluidics 27 (2023) 69.

[19] M. Remer, M. Psarski, K. Gumowski, J. Rokicki, G. Sobieraj, M. Kaliush, et al., Dynamic water contact angle during initial phases of droplet impingement, Colloids and Surfaces A: Physicochemical and Engineering Aspects 508 (2016) 57–69.

[20] Y. Xu, S. Vincent, Q.C. He, H. Le-Quang, Spread and recoil of liquid droplets impacting on solid surfaces with various wetting properties, Surface and Coatings Technology 357 (2019) 140–152.

[21] X. Liu, X. Zhang, J. Min, Spreading of droplets impacting different wettable surfaces at a Weber number close to zero, Chemical Engineering Science 207 (2019) 495–503.

[22] M. Budakli, Prediction of maximum spreading factor after drop impact: development of a novel semi-analytical model incorporating effect of surface roughness, Colloid Interface Sci. Commun. 41 (2021) 100384.

[23] A.A. Yancheshme, S. Hassantabar, K. Maghsoudi, S. Keshavarzi, R. Jafari, G. Momen, Integration of experimental analysis and machine learning to predict drop behavior on superhydrophobic surfaces, Chemical Engineering Journal 417 (2021) 127898.

[24] H. Zhang, Z. Jin, M. Jiao, Z. Yang, Experimental investigation of the impact and freezing processes of a water droplet on different cold concave surfaces, International Journal of Thermal Sciences 132 (2018) 498–508.

[25] S. Yun, G. Lim, Ellipsoidal drop impact on a solid surface for rebound suppression, Journal of Fluid Mechanics 752 (2014) 266–281.

[26] Y. Liu, Y. Liu, M. Chen, Maximum spreading diameter of bouncing droplets at ultralow weber numbers, Langmuir: the ACS Journal of Surfaces and Colloids 39 (2023) 7922–7929.

[27] R. Zhang, B. Zhang, Q. Lv, J. Li, P. Guo, Effects of droplet shape on impact force of low-speed droplets colliding with solid surface, Experiments in Fluids 60 (2019).

[28] R. Gao, M. Song, C.C.Y. Hangb, S. Lin, L. Zhang, X. Zhang, Review on condensation frosting and defrosting experiments for superhydrophobic surfaces, Applied Thermal Engineering 236 (2024) 121691.

[29] X. Zhang, X. Liu, X. Wu, J. Min, Impacting-freezing dynamics of a supercooled water droplet on a cold surface: rebound and adhesion, International Journal of Heat and Mass Transfer 158 (2020) 119997.

[30] Y. Yao, K. Wu, R. Yang, H. Zhang, W. Yang, C. Li, Effects of surface temperature and Weber number on the dynamic and freezing behavior of impacting water droplets on a superhydrophobic ultra-cold surface, Applied Thermal Engineering 236 (2024) 121705.

[31] X. Zhang, K. Li, Z. Zhu, W.-Z. Fang, F.-Q. Zhu, C. Yang, Droplet impact and freezing dynamics on ultra-cold surfaces: a scaling analysis of central-concave pattern, Applied Thermal Engineering 239 (2024) 122135.

[32] D. Zwillinger, CRC Standard Mathematical Tables and Formulae, 31st ed., CRC Press, Boca Raton, 2018.

[33] X. Zhang, B.Q. Ji, X. Liu, S.Y. Ding, X.M. Wu, J.C. Min, Maximum spreading and energy analysis of ellipsoidal impact droplets, Physics of Fluids 33 (2021) 1–12.
[34] M. Pasandideh-Fard, Y.M. Qiao, S. Chandra, J. Mostaghimi, Capillary effects during droplet impact on a solid surface, Physics of Fluids 8 (1996) 650–659.
[35] T. Young, An essay on the cohesion of fluids, Philosophical Transactions of the Royal Society of London 95 (1805) 65–87.
[36] H. Huang, X. Chen, Energetic analysis of drop's maximum spreading on solid surface with low impact speed, Physics of Fluids 30 (2018) 1.
[37] X. Liu, X. Zhang, J. Min, Maximum spreading of droplets impacting spherical surfaces, Physics of Fluids 31 (2019) 092102.
[38] S. Yun, I. Kim, Spreading dynamics and the residence time of ellipsoidal drops on a solid surface, Langmuir: the ACS Journal of Surfaces and Colloids 35 (2019) 13062–13069.
[39] T. Mao, D.C.S. Kuhn, H. Tran, Spread and rebound of liquid droplets upon impact on flat surfaces, AIChE Journal. American Institute of Chemical Engineers 43 (1997) 2169–2179.
[40] B. Ding, H. Wang, X. Zhu, R. Chen, Q. Liao, Water droplet impact on superhydrophobic surfaces with various inclinations and supercooling degrees, International Journal of Heat and Mass Transfer 138 (2019) 844–851.

CHAPTER

6

Simulation of the impact and icing processes of water droplets on spheres

Xuan Zhang[1], Xin Liu[2], Jingchun Min[3], Yiqing Guo[4], Xiaomin Wu[4] and Yunda Fei[1]

[1]Department of Energy and Power Engineering, School of Mechanical Engineering, Beijing Institute of Technology, Beijing, P.R. China [2]Chinese Aeronautical Establishment, Beijing, P.R. China [3]Key Laboratory for Thermal Science and Power Engineering of Ministry of Education, Department of Engineering Mechanics, Tsinghua University, Beijing, P.R. China [4]Key Laboratory for Thermal Science and Power Engineering of Ministry of Education, Department of Energy and Power Engineering, Tsinghua University, Beijing, P.R. China

Nomenclatures

Symbol	Description	Unit
A	area	m^2
A_0	cross-sectional area of initial droplet	m^2
A_{mush}	mushy zone constant	/
Ca	Capillary number	/
D	diameter	m
f	function and inverse function	/
F_{st}	volumetric surface tension force	$N \cdot m^{-3}$
g	gravity acceleration	$m \cdot s^{-2}$
h	enthalpy	$j \cdot kg^{-1}$
L	spreading arc length	m
R	radius	m

Frosting and Icing for Efficient Energy Use in Engineering Applications
DOI: https://doi.org/10.1016/B978-0-443-15495-9.00007-1

Re	Reynolds number	/
S_M	momentum source term	$N \cdot m^{-3}$
t	time	s
T	temperature	°C
T_F	freezing temperature (point)	°C
ΔT_{SC}	supercooling degree	°C
U	velocity	$m \cdot s^{-1}$
U_{CL}	contact line velocity	$m \cdot s^{-1}$
We	Weber number	/
Greek symbols	**Description**	**Unit**
α	phase volume fraction	/
β	spreading factor	/
β_A	spreading area factor	/
θ	contact angle	°
θ_{dyn}	dynamic contact angle	°
μ	dynamic viscosity	$Pa \cdot s$
ξ	spreading arc angle	rad
ρ	droplet density	$kg \cdot m^{-3}$
σ	surface tension	$N \cdot m^{-1}$
κ	curvature	m^{-1}
$\kappa*$	curvature ratio	/
Subscript	**Description**	**Unit**
0	initial	/
m	mixture	/
max	maximum	/
adv	advancing	/
rec	receding	/
sta	staic	/
S	sphere/surface	/

6.1 Introduction

The impact and icing phenomenon of the supercooled droplet is widespread in nature and in many fields of production and life such as meteorology [1], food refrigeration [2], wind turbine [3,4], power transmission [5], and aerospace [6,7]. Inappropriate icing in different fields not only seriously affects the operational efficiency of actual production and life but even poses a danger to human life safety.

To avoid icing damage and promote high efficiency and energy saving, the impact and icing process of the supercooled water droplet is studied in this chapter, which is beneficial to provide a theoretical basis for icing prediction and the optimal design of antiicing surfaces. On one hand, with the development and advancement of material science and manufacturing technology, superhydrophobic surfaces have attracted more and more attention because of their unique advantages of retarding or attenuating surface icing in preventing the icing detriment [8–10]. Based on the lotus effect, superhydrophobic surfaces exhibit excellent repulsion to water, which means numerous features that of nucleation barrier, low adhesion force, and self-cleaning features [11–13]. Consequently, it can not only reduce the heat exchange in liquid–solid two-phase but also inhibit the droplet adhesion, so as to achieve the purpose of antiicing. On the other hand, in practical engineering applications, the shapes of various surfaces are complex and diverse, which can be flat or curved, and in most situations are curved, such as the leading edge of the wing, the engine cowling, and so on. Furthermore, the macroscopic freezing process is actually a process in which a single supercooled droplet freezes and gradually accumulates to form ice after interacting with the cold wall. Combining the analyses of the above three points, it is more general and realistic to study the impact and icing process of the individual supercooled droplet on typical and regular curved surfaces such as convex or concave spherical surfaces.

The deformation characteristics of the droplet after impacting on a spherical surface are the result of the combined action of different forces, which mainly include the inertia force, surface tension, and gravity. The evolution process of the droplet morphology usually involves various stages such as the spreading, pinned, retraction, oscillation, splash, or rebound stages [14–17]. The above situation is even more complex when the effects of freezing are considered [18–21], which involves the coupling of phase transition heat transfer and fluid dynamics. To simplify the analysis, the morphology evolution and spreading characteristics during droplet impacting process at room temperature, which only involves fluid dynamics, are studied separately, and then a phase transition heat transfer model is coupled on this basis.

This chapter studies the impact and icing process of the supercooled droplet on a spherical surface from the relatively simple to the complicated and will be divided into three sections: (1) droplet impact process on convex spheres; (2) droplet impact process on concave spheres; and (3) droplet impact and icing processes on cold convex spheres. The outline of the whole chapter is shown in Fig. 6.1.

6.2 Droplet impact process on convex spheres

The schematic of the droplet deformation upon impacting the solid convex sphere is shown in Fig. 6.2. A droplet with an initial diameter of D_0 impacts a convex sphere with a diameter of D_S at an initial velocity of U_0. The initial impacting characteristics of the

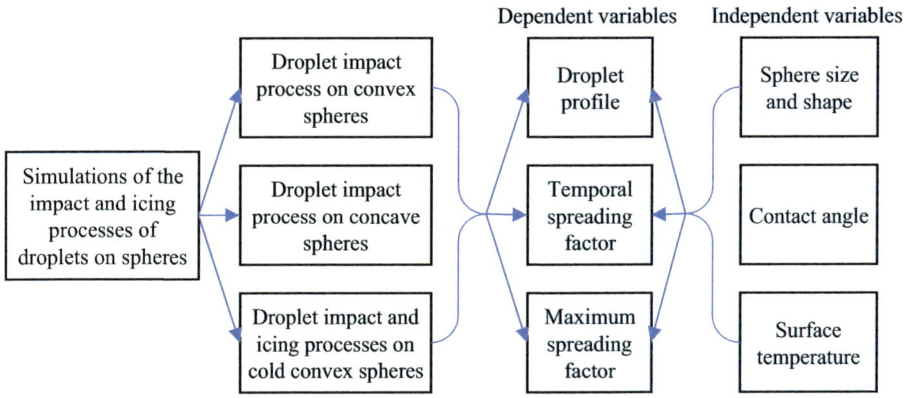

FIGURE 6.1 Outline of the impact and icing processes of water droplets on spheres.

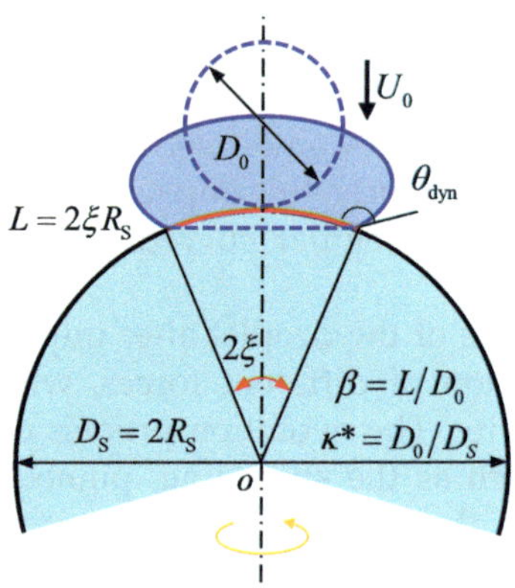

FIGURE 6.2 Schematic of a droplet impacting process on a convex spherical surface.

droplet are described by the dimensionless Weber number and Reynolds number, which are defined as $We = \rho U_0^2 D_0/\sigma$ and $Re = \rho U_0 D_0/\mu$. To quantify the influence of the relative shape of the curved surface, from convex to plane and then to concave, on the droplet impacting process, the sphere-to-droplet curvature ratio is defined as the ratio of spherical surface curvature to initial droplet's curvature, that is, $\kappa^* = (1/D_S)/(1/D_0) = D_0/D_S$. For facilitating subsequent quantitative analysis, the spreading coefficient β of the impacting process is defined as the ratio of the two-dimensional wetting arc length L to the initial diameter of the droplet D_0, that is, $\beta = L/D_0$. ξ is defined as half of the central angle corresponding to the wetting arc, and 2ξ is spreading arc angle accordingly. Apparently, the maximum values of the spreading factor and spreading arc angle are, respectively, expressed as β_{max} and $2\xi_{max}$, which are the two parameters commonly used to analyze the maximum spreading characteristics of impacting droplets. Around the contact line presented in Fig. 6.2, θ_{dyn} is the dynamic contact angle (DCA).

6.2.1 Simulation model

6.2.1.1 Volume of Fluid model

The Volume of Fluid (VOF) method is extensively used to capture the flow at the multiphase interface [22,23]. In this chapter, it mainly refers to the flow tracking at the interface of ice, water, and gas, where ice, solid phase, water, and liquid phase are treated as two states of one phase using the solidification/melt model that will be presented in Section 6.4. Therefore only two-phase fluids of gas and water are included in the VOF model in this section. The volume fraction α_i of each phase in each cell is defined by

$$\alpha_i = \begin{cases} 0 & , \quad \text{if the cell contains no } i\text{th phase} \\ 0 < \alpha_i < 1 & , \quad \text{if the cell contains } i\text{th phase and other phases} \\ 1 & , \quad \text{if the cell is occupied by } i\text{th phase} \end{cases} \tag{6.1}$$

and satisfies

$$\sum \alpha_i = 1 \tag{6.2}$$

As for the two-phase flow in the impacting process, the liquid volume fraction α_{1iq} in each cell can be simplified as

$$\alpha_{1iq} = \begin{cases} 0 & , \quad \text{if the cell is occupied by gas} \\ 0 < \alpha_{1iq} < 1 & , \quad \text{if the cell contains gas and liquid} \\ 1 & , \quad \text{if the cell is occupied by liquid} \end{cases} \tag{6.3}$$

and

$$\alpha_{1iq} + \alpha_{gas} = 1 \tag{6.4}$$

The mass conservation equation of the liquid phase can be expressed by

$$\frac{\partial}{\partial t}\left(\alpha_{1iq}\rho_{1iq}\right) + \nabla \cdot \left(\alpha_{1iq}\rho_{1iq}\boldsymbol{u}\right) = 0 \tag{6.5}$$

while the momentum conservation equation of the fluid flow is as follows:

$$\frac{\partial}{\partial t}(\rho\boldsymbol{u}) + \nabla \cdot (\rho\boldsymbol{u}\boldsymbol{u}) = -\nabla p + \nabla \cdot \mu\left[\nabla\boldsymbol{u} + (\nabla\boldsymbol{u})^{\mathrm{T}}\right] + \rho\boldsymbol{g} + \boldsymbol{F}_{\mathrm{st}} \tag{6.6}$$

in which $\boldsymbol{F}_{\mathrm{st}}$ is the volumetric surface tension force acting on the fluid at the gas–liquid interface, calculated by the continuum surface force (CSF) model [24], that is,

$$\boldsymbol{F}_{\mathrm{st}} = \sigma \frac{\rho\kappa \ \nabla\alpha_{1iq}}{\left(\rho_{1iq} + \rho_{gas}\right)/2} \tag{6.7}$$

where κ is the curvature of the interface and can be obtained from the divergence of the unit normal vector $\boldsymbol{n}$ at the interface, that is,

$$\kappa = \nabla \cdot \boldsymbol{n} \tag{6.8}$$

and $\boldsymbol{n}$ is the gradient of the volume fraction at the interface.

6.2.1.2 Dynamic contact angle model

The contact angle during the dynamic wetting process is not only related to the static contact angle but also concerned with the moving speed of the three-phase contact line due to the hysteresis on a real surface [25–27]. The empirical formula proposed by Kistler [28] is often used to describe the relationship between the DCA and contact line velocity, which is presented by

$$\theta_d = f_{Hoff}\left[Ca + f_{Hoff}^{-1}(\theta_{eq})\right] \tag{6.9}$$

in which $f_{Hoff}(x)$ is the Hoffman function [29], given by

$$f_{Hoff} = \arccos\left\{1 - 2\tanh\left[5.16\left(\frac{x}{1+1.31x^{0.99}}\right)^{0.706}\right]\right\} \tag{6.10}$$

$Ca = \mu U_{CL}/\sigma$ is the Capillary number, in which U_{CL} is the contact line velocity along the tangential vector of the solid surface. θ_{eq} takes the advancing angle θ_{adv} and receding angle θ_{rec} in the spreading and receding stages, respectively.

The above DCA model is applied to the wall contact angle through the User-Defined Function file in Fluent, which can be dynamically and timely adjusted in accordance with the three-phase contact line speed in the simulation.

6.2.1.3 Numerical method

Considering that both the droplet and the convex sphere are geometrically symmetric structures, the no-rotation property of the droplet impacting process indicates that the flowfield structure inside and around the droplet is also symmetrically distributed. Consequently, it is appropriate to adopt a two-dimensional symmetrical computational domain.

A structured mesh is adopted with local refinement to ensure mesh accuracy of the fluid flow area during droplet collision while reducing the calculation cost. Four different minimum mesh sizes of 40 μm × 40 μm, 20 μm × 20 μm, 10 μm × 10 μm, and 5 μm × 5 μm were developed for grid independence verification, and the comparison results of the temporal spreading coefficient and phase interface are shown in Fig. 6.3. As can be seen from

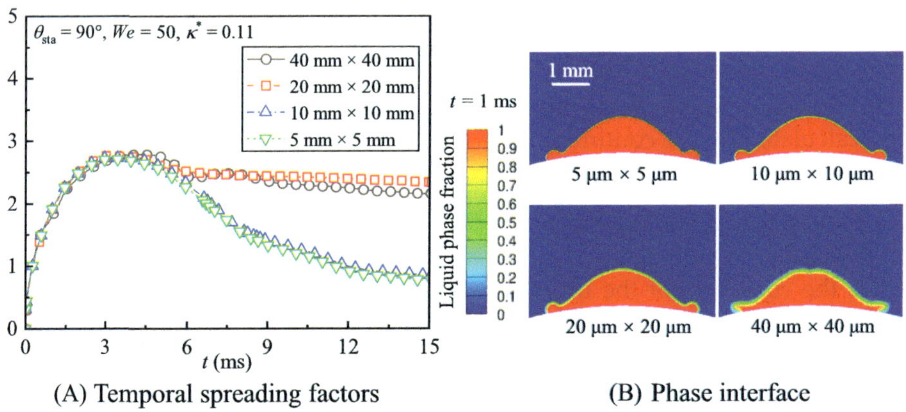

FIGURE 6.3 Grid independence verification of (A) temporal spreading factors and (B) phase interface.

the comparison, simulation results of using minimum mesh sizes of 10 μm × 10 μm and 5 μm × 5 μm grids are basically the same. Therefore considering the simulation accuracy and calculation efficiency, the minimum grid size of 10 μm × 10 μm, which contains 142,420 grids in total, is eventually selected for subsequent numerical calculation.

The boundary condition settings for the sectorial calculation domain with a central angle of ξ_{wall} and a height of $5D_0$ are shown in Fig. 6.4. The left boundary condition is set as the rotational symmetry axis, the upper and right boundary conditions are set as the pressure outlet boundaries, and the no-slip wall boundary is applied as the bottom boundary condition. To eliminate the effect of the domain boundary on the accuracy of the simulation results, the arc length, L_s, corresponding to the bottom wall should meet $L_S > L_{\max}/2 + 5D_0$.

All the involved numerical studies in this chapter are based on the commercial CFD software FLUENT. The initial velocity of the impacting droplet is small, where Mach number is much less than 0.3, so the fluid can be considered incompressible, and the transient solver adopts pressure-based solver. The discrete formats of pressure, momentum, and volume fraction are Body Force Weighted, Second Order Upwind, and Geo-Reconstruct, respectively. The pressure-velocity coupling is achieved with the Pressure Implicit Split Operator (PISO) scheme, and the time discretization format is the First Order Implicit method.

6.2.2 Model validation

Comparison results of droplet morphologies at different moments obtained from the present experiment and simulation, whose numerical and experimental conditions are the same as $We = 19.18$, $\theta_{\text{adv}}/\theta_{\text{rec}} = 108°/78°$, and $\kappa^* = 0.15$, and are embedded in the corresponding spreading-factor evolution diagram, as represented in red line and data points of Fig. 6.5. Comparative analysis shows that temporal droplet profiles of the experiment and simulation are qualitatively in good agreement. However, at 4 and 8 Ms, the numerical and experimental droplet profiles appear to be somewhat less similar, because the numerical result is a two-dimensional outline across the droplet revolving axis, while the experimental result is the outermost outline of the three-dimensional space. In fact, when comparing the

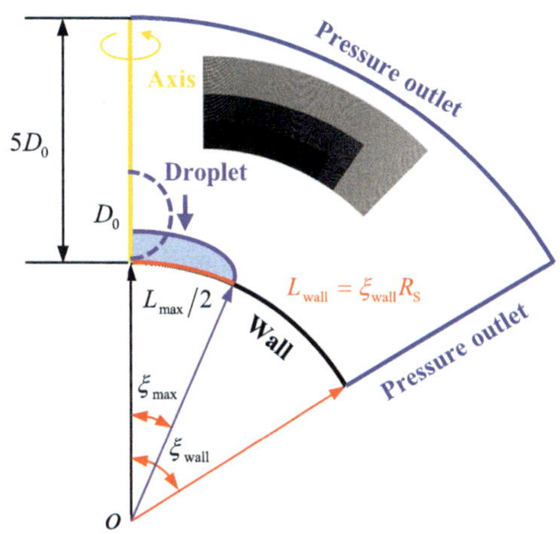

FIGURE 6.4 Computational domain and structured meshes.

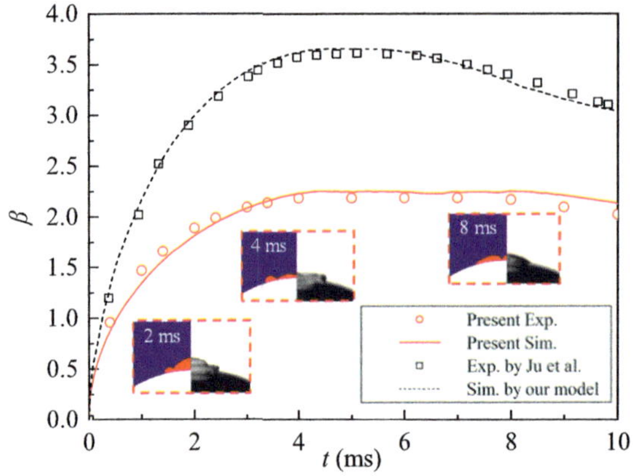

FIGURE 6.5 Evolutions of the spreading factors and droplet profiles.

TABLE 6.1 Calculation conditions.

Droplet volume, V_0 (μL)	Droplet diameter, D_0 (mm)	Weber number, *We*	Contact angle, θ (°)	Sphere-to-droplet curvature ratio, κ^*
5	2.12	1∼90	30, 60, 90, 120, 150	0.35, 0.21, 0.11, 0.04, 0.02

outermost and highest points of the outline, the difference between the two is also very small. In addition, the maximum spreading factors obtained by the experiment and simulation are 1.96 and 2.03, respectively, from quantitatively comparing, generating a deviation of 3.57%. Furthermore, the numerical model in this chapter is also used to numerical simulate the experimental conditions that of $We = 116.53$, $\theta_{sta} = 88°$, and $\kappa^* = 0.07$ in Ju's work [30], and the maximum spreading factors are represented in black line and data points of 3.61 and 3.66, respectively, yielding a deviation of 1.38%. Based on the above analysis, whether compared with the experimental results or the reference data, the deviation of the numerical model adopted in simulation is less than 4%, which is sufficient to demonstrate the reliability and accuracy of the numerical model applied in the chapter.

6.2.3 Results and discussion

A series of simulation cases have been carried out to systematically study the effects of Weber number, contact angle, and the sphere-to-droplet curvature ratio on the droplet impacting process on a convex sphere, especially on the impacting features and the maximum spreading characteristic. The numerical condition parameters are shown in Table 6.1. Moreover, referring to the common contact angle data in the previous experiment [31], the contact angle hysteresis is set as 20 degrees for convenience which is a representative value for a smooth surface. The static contact angle is taken as the average of the advancing and receding contact angles. So, $\theta_{adv} = \theta_{sta} + 10°$ and $\theta_{rec} = \theta_{sta} - 10°$. Different initial impacting velocities with the fixed droplet volume are set to achieve different Weber numbers in the simulation.

6.2.3.1 Impacting behavior

The characteristic of the droplet impacting on a convex spherical surface across varying different Weber numbers under fixed $\theta_{sta} = 120°$ and $\kappa^* = 0.35$ is shown in Fig. 6.6. With the increase of the Weber number, the droplet spreading and shrinking speed accelerates, and the maximum spreading factor increases. Under a small Weber number, $We = 1$, the droplet adheres to the surface completely after impacting the sphere. When the Weber number increases, $We = 10$, the droplet rebounds completely and detaches from the surface at once. With the Weber number further increasing, $We = 50$, due to the large Weber number and inertia force, the droplet tends to overcome the surface tension and break up and finally split into several parts separated from the surface.

The characteristic of the droplet impacting on a convex spherical surface across varying different contact angles under fixed $We = 50$ and $\kappa^* = 0.35$ is shown in Fig. 6.7. With the increase of the contact angle, the droplet spreading speed slows down, while the shrinking

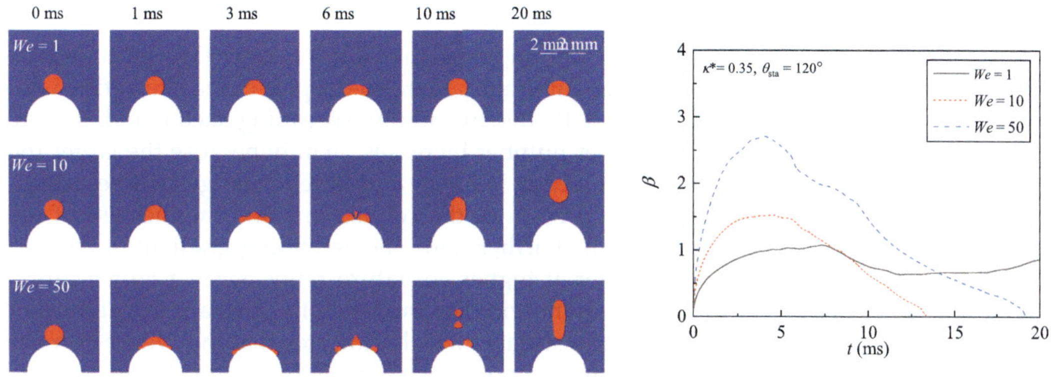

(A) Droplet morphology (B) Spreading factor

FIGURE 6.6 (A) Droplet morphology and (B) temporal spreading factor with different Weber numbers.

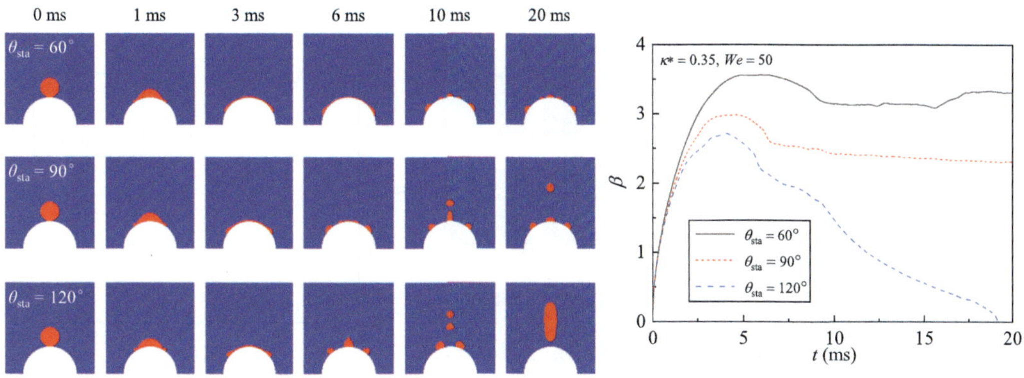

(A) Droplet morphology (B) Spreading factor

FIGURE 6.7 (A) Droplet morphology and (B) temporal spreading factor with different contact angles.

speed accelerates, and the maximum spreading factor decreases, as represented in Fig. 6.7B. When the contact angle is small, $\theta_{sta} = 60°$, the droplet tends to break up after impacting the spherical surface and finally adhere to the surface completely. When the contact angle increases, $\theta_{sta} = 90°$, part of the droplet rebounds from the surface after breakup, and the remaining part adheres to the surface; when the contact angle is further increased, $\theta_{sta} = 120°$, all droplet fragments bounce off the spherical surface after breakup.

The characteristic of the droplet impacting on a convex spherical surface across varying different sphere-to-droplet curvature ratios under fixed $We = 3$ and $\theta_{sta} = 120°$ is shown in Fig. 6.8. As the curvature ratio increases, the droplet spreading speed slows down and the maximum spreading factor decreases, as represented in Fig. 6.8B. Due to the low Weber number of the impacting process, the droplet adheres completely to the surface when the diameter is relatively small ($\kappa^* = 0.11$). When the diameter is relatively large ($\kappa^* = 0.35$), on the contrary, the droplet bounces off the sphere completely.

6.2.3.2 *Maximum spreading characteristic*

The maximum spreading factor and arc angle of the droplet impacting on a convex sphere vary with Weber number under different sphere-to-droplet curvature ratios with fixed $\theta_{sta} = 90°$, as shown in Fig. 6.9. Both the maximum spreading factor and the arc spreading arc angle increase as the Weber number increases, mainly because the larger the Weber number, the larger the initial impacting velocity and kinetic energy will be, resulting in the acceleration of the spreading process of the contact line and the increase of the maximum spreading factor and arc angle. Furthermore, the ascending gradient of the two maximum spreading parameters is larger at higher curvature ratios, since a larger curvature ratio implies greater spherical curvature with the premise of a constant initial droplet curvature, that is, a more convex spherical surface, which indicates that the Weber number has more obvious influence on the maximum spreading characteristics.

The variation trends of the maximum spreading factor and arc angle of the droplet impacting on a convex sphere that vary with the contact angle under different sphere-to-droplet curvature ratios with fixed $We = 50$ are shown in Fig. 6.10. With the increase of the

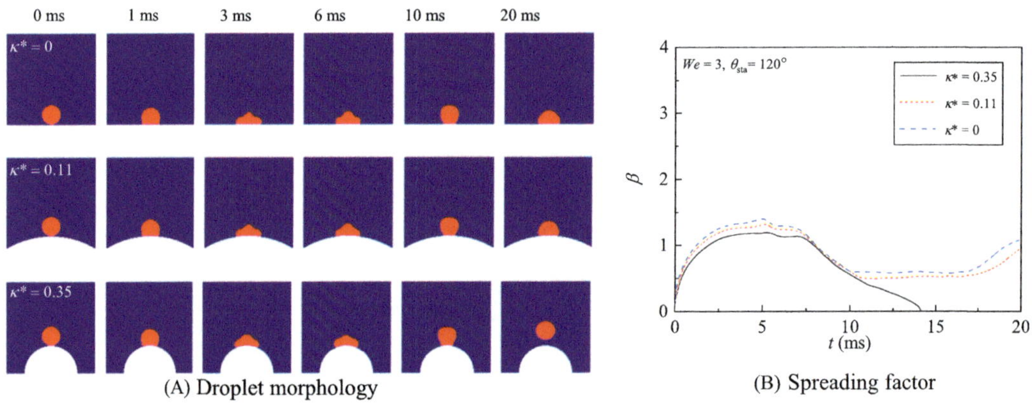

FIGURE 6.8 (A) Droplet morphology and (B) temporal spreading factor with different curvature ratios.

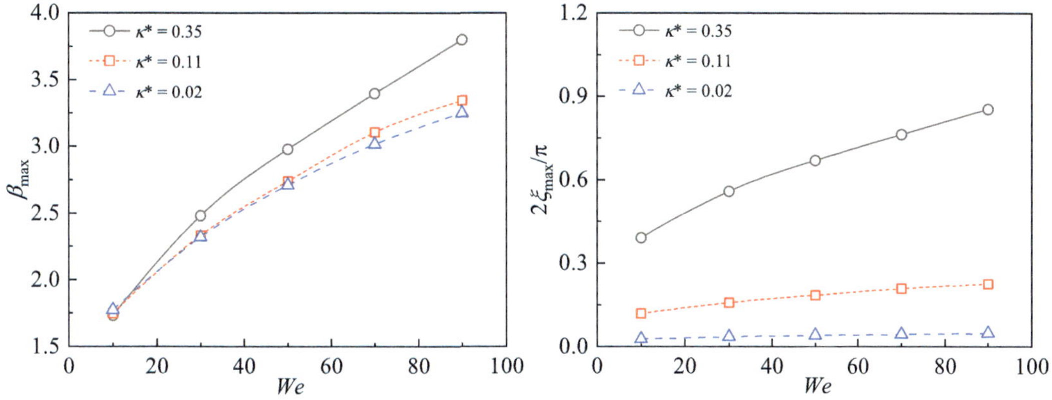

(A) Maximum spreading factor (B) Maximum spreading arc angle

FIGURE 6.9 Effects of Weber number on the maximum spreading (A) factor and (B) arc angle.

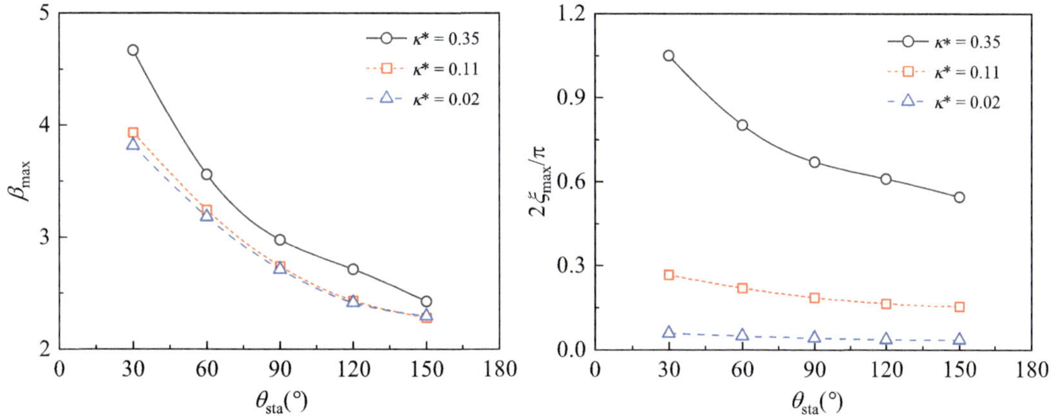

(A) Maximum spreading factor (B) Maximum spreading arc angle

FIGURE 6.10 Effects of contact angle on the maximum spreading (A) factor and (B) arc angle.

contact angle, both the maximum spreading factor and arc angle decrease, which is mainly because the larger contact angle means that the surface is more hydrophobic, leading to inhibit the spreading of the contact line and slow down the spreading speed, resulting in the decrease of the maximum spreading factor and arc angle ultimately. Furthermore, the decreasing slopes of the curves for the two maximum spreading parameters are greater at higher curvature ratios, suggesting that the contact angle has a more pronounced effect on the maximum spreading characteristics.

Fig. 6.11 presents the variations of the maximum spreading factor and arc angle of the droplet impacting on a convex sphere vary with the sphere-to-droplet curvature ratio under different Weber numbers and contact angles. It is observed that the maximum spreading coefficient remains a small amplitude of variation for $\kappa^* \lesssim 0.1$, while it demonstrates an obvious increase as

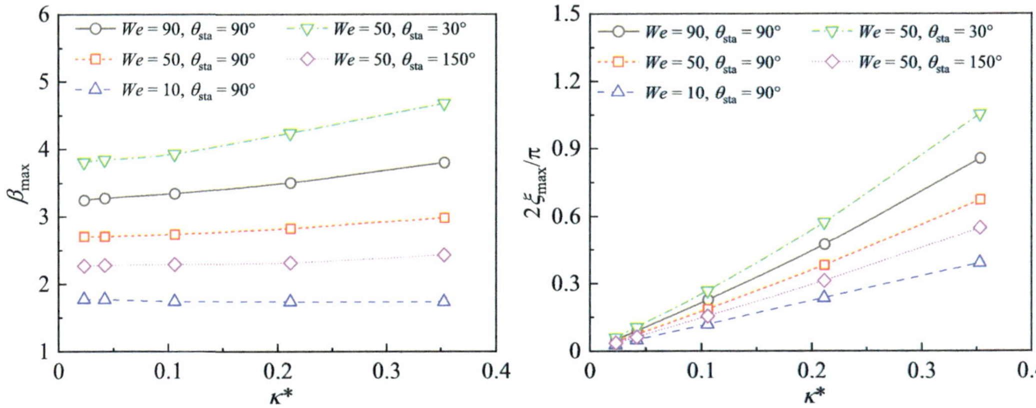

(A) Maximum spreading factor (B) Maximum spreading arc angle

FIGURE 6.11 Variations of the maximum spreading (A) factor and (B) arc angle with different curvature ratios.

the curvature ratio for $\kappa^* \gtrsim 0.1$, as represented in Fig. 6.11A. Besides, the maximum spreading arc angle keeps increasing with increasing curvature ratio. Variation trends in the above two maximum spreading parameters suggest that the bending degree of the spherical surface significantly influences the droplet spreading process and particularly shows enormous levels of effect at larger curvature ratios. Analyzing the causes reveals that the larger the curvature ratio is, the more convex the spherical surface is, consequently, the gravity effect on the droplet spreading stage becomes more and more conspicuous as the curvature ratio increases, which promotes the forward motion of the three-phase contact line and ultimately leads to an increase in the maximum spreading coefficient and arc angle. Furthermore, as the increasing Weber number and the decreasing contact angle, the increasing magnitude of β_{max} and $2\xi_{max}$ becomes obvious, which indicates that the effect of the curvature ratio becomes more remarkable.

6.3 Droplet impact process on concave spheres

When a droplet impacts on a stationary concave sphere, its morphology changes first, and then the droplet gradually spreads out to contraction like on the convex surfaces, which is shown in Fig. 6.12. To facilitate subsequent quantitative analysis, the same parameters are defined as in Section 6.2. Since the curvature of the concave sphere is negative, the sphere-to-droplet curvature coefficient is also negative. Of note, for the purpose of allowing the droplet to enter the sphere and impact the concave surface, a hole larger than the droplet size is reserved at the sphere dome.

6.3.1 Numerical method

Although the types of curved surfaces, convex or concave sphere, impacted by the water droplet are different, the physical essence involved in the impacting process is basically the same. Therefore the simulation model in this section is consistent with that in Section 6.2.

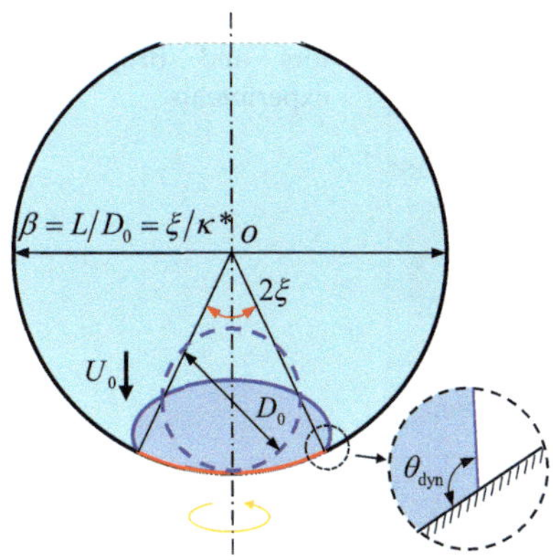

FIGURE 6.12 Schematic of droplet impact process on a concave spherical surface.

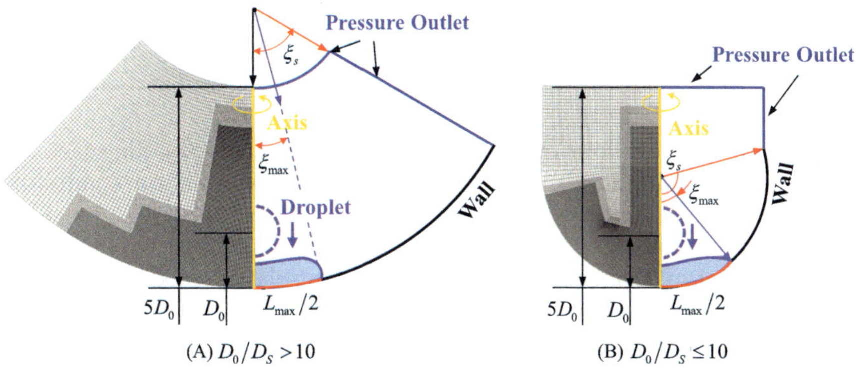

FIGURE 6.13 Computational domains and structured grids for (A) $D_0/D_s > 10$ and (B) $D_0/D_s \leq 10$.

The shape of the computational domain is altered to keep the domain height at $5D_0$ under different D_0/D_S. A sector computational domain is used for $R_S > 5D_0$ with $D_S/D_0 > 10$, as represented in Fig. 6.13A, while a combination of fan-shaped and trapezoidal computational domains is employed for $R_S \leq 5D_0$ with $D_S/D_0 \leq 10$, as shown in Fig. 6.13B. The structured grids are also employed for two kinds of computation domains, and the minimum mesh size is determined as $10 \times 10\ \mu m^2$ after grid independence verification to ensure the solution accuracy. Mesh details are shown in Fig. 6.13. Besides, boundary conditions are also set in accordance with Section 6.2.

6.3.2 Model validation

In the previous section, the simulation results of the droplet impacting characteristic on a convex sphere are in good agreement with the experiment, which proves the accuracy of the simulation model to a certain extent. To further verify its applicability on concave surfaces, the experiment of the droplet impacting on a concave sphere is conducted. Fig. 6.14

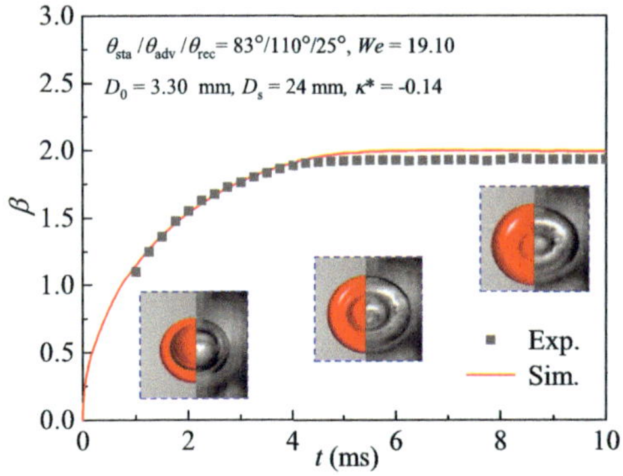

FIGURE 6.14 Comparisons of temporal spreading factors and droplet profiles between simulations and experiments.

TABLE 6.2 Calculation conditions.

Droplet volume, V_0 (μL)	Droplet diameter, D_0 (mm)	Weber number, We	Contact angle, θ (°)	Sphere-to-droplet curvature ratio, κ^*
5	2.12	10∼90	30, 60, 90, 120, 150	−0.35, −0.21, −0.11, −0.04, −0.02

compares the temporal evolutions of droplet morphologies and spreading factors obtained by the experiment and simulation. The experimental and simulation conditions are the same, and the specific details are as shown in Fig. 6.14 legend. It can be seen that the results of the simulation and experiment are basically consistent in two kinds of comparison, which is sufficient to support the accuracy of the simulation model.

6.3.3 Results and discussion

Consistent with the previous section, this section focuses on a comprehensive study of the effects of Weber number, contact angle, and curvature ratio on the impacting process of the droplet on a concave spherical surface. Serials of simulation cases are carried out, and condition details are shown in Table 6.2.

6.3.3.1 Effect of Weber number

Fig. 6.15 illustrates the influence of the Weber number on the maximum spreading factor and arc angle across varying curvature ratios with fixed $\theta_{sta} = 90°$. The analysis of the figure shows that both the maximum spreading coefficient and the maximum spreading arc angle increase with the increasing Weber number, which is compatible with the conclusion of a droplet impacting on a convex sphere. An increase in the Weber number implies an increase in the kinetic energy of the impacting droplet, which can significantly enhance the spreading stage. Furthermore, with the premise of the same Weber number, the two characteristic parameters of the maximum spreading sate increase more obviously under the larger curvature ratio (minus represents the surface bending direction, not the size), which indicates that the influence of Weber number is more conspicuous.

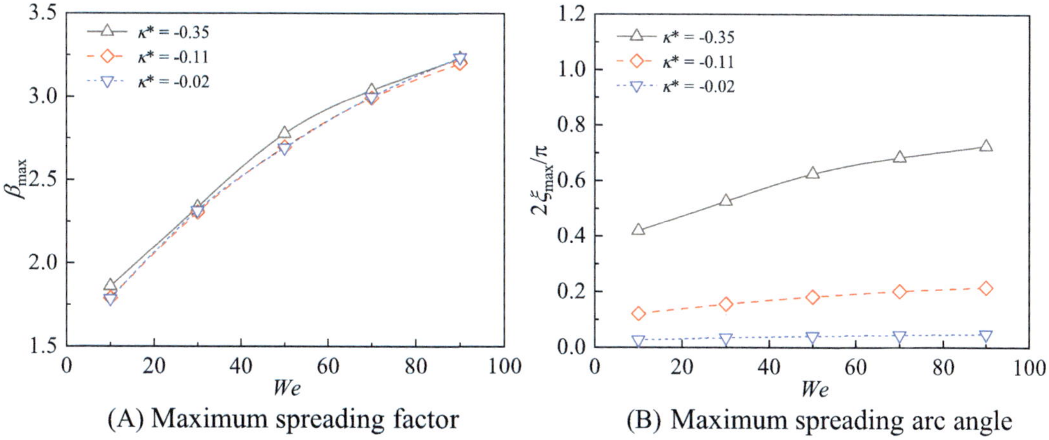

FIGURE 6.15 Effects of Weber number on the maximum spreading (A) factor and (B) arc angle.

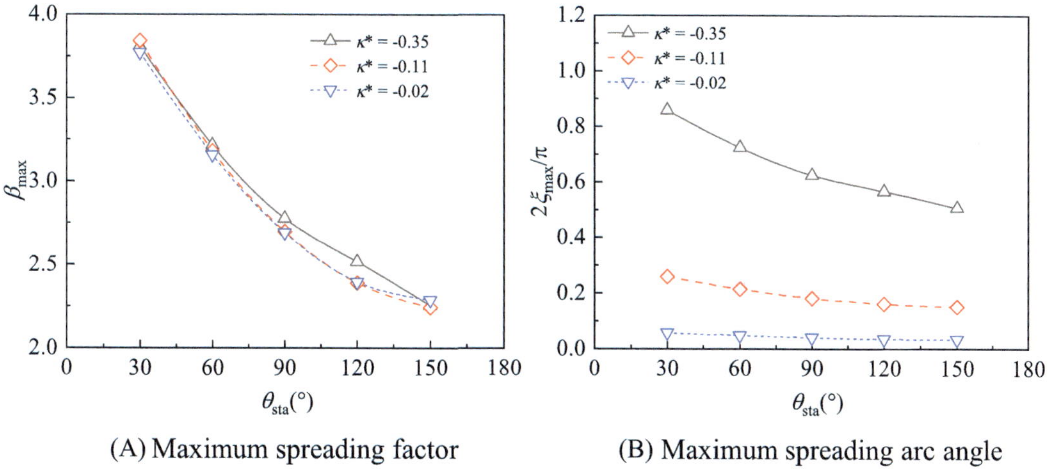

FIGURE 6.16 Effects of contact angle on the maximum spreading (A) factor and (B) arc angle.

6.3.3.2 Effect of contact angle

Fig. 6.16 shows the effects of the contact angle on the maximum spreading factor and arc angle under different curvature ratios with fixed $We = 50$. With the increase of the contact angle, both the maximum spreading coefficient and arc angle decrease, and the decreasing amplitude becomes more severe with the increase of the curvature ratio. Analysis reveals that the larger the contact angle, the stronger the hydrophobicity of the spherical surface, which seriously hinders the advance of the contact line, reduces the spreading speed, and ultimately leads to the weakening of the droplet spreading stage. In addition, the influence of contact angle is more significant at a larger curvature ratio.

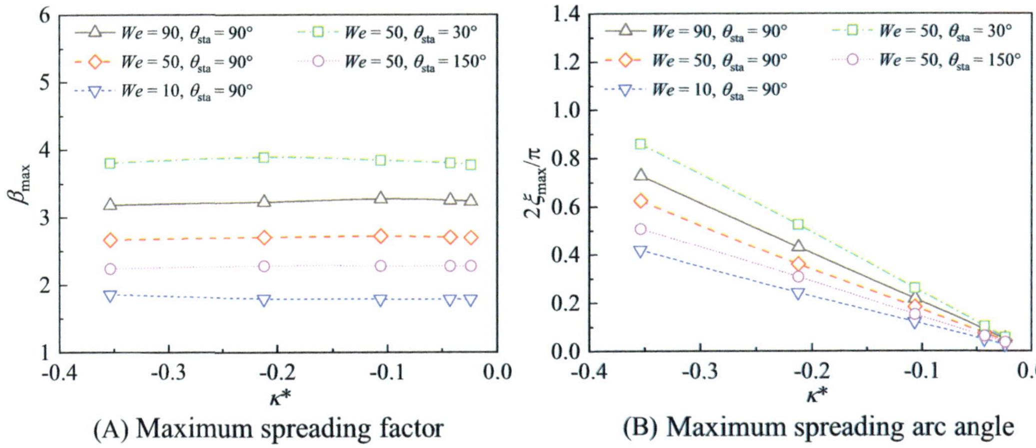

FIGURE 6.17 Effects of curvature ratio on the maximum spreading (A) factor and (B) arc angle.

6.3.3.3 *Effect of sphere diameter*

Fig. 6.17 exhibits the effects of the curvature ratio on the maximum spreading factor and arc angle under different Weber numbers and contact angles. With the increase of the curvature ratio, the maximum spreading factor slightly increases at first and then reduces faintly, but the maximum spreading arc angle keeps decreasing significantly. Analysis reveals that the variation trend of the maximum spreading characteristic parameters is the result of the combined action of gravity and concave shape. Unlike the situation of a convex sphere where gravity promotes the droplet spreading, gravity hinders the advancement of the spreading stage when a droplet impacts on a concave sphere. Moreover, the effect of gravity is more pronounced at a smaller curvature ratio, whereas it can be ignored at a large curvature ratio, since the concave sphere is close to a plane as κ^* approaches 0.

6.4 Droplet impact and icing processes on cold convex spheres

The impact and icing process of the supercooled droplet on the superhydrophobic cooled sphere is shown in Fig. 6.18. Upon the supercooled droplet with the initial temperature of T_0 colliding with the cold sphere, as represented in Fig. 6.18A, the nucleation-recalescence phenomenon occurs; meanwhile, the droplet releases the supercooling degree and changes into a uniform ice-water mixture instantaneously. Soon after, the droplet temperature returns to the freezing point T_F (0 °C), and the volume and diameter turn into V_m and D_m, respectively. In the subsequent spreading (icing) stage, as represented in Fig. 6.18B, the droplet morphology evolves with the freezing front advancing from the bottom to the top inside the droplet and shows a different final state, as represented in Fig. 6.18C.

Considering that the spreading area coefficient during the impacting and icing process can better reflect the heat transfer between the water droplet and the spherical surface, several parameters are defined for analytical convenience, as shown in Fig. 6.18B. A is the spreading area wetted by the liquid calculated by $A = 2\pi R_S^2(1 - \cos\xi)$, and the spreading area factor is defined as $\beta_A = A/A_0$, where A_0 is the cross-sectional area of the initial droplet.

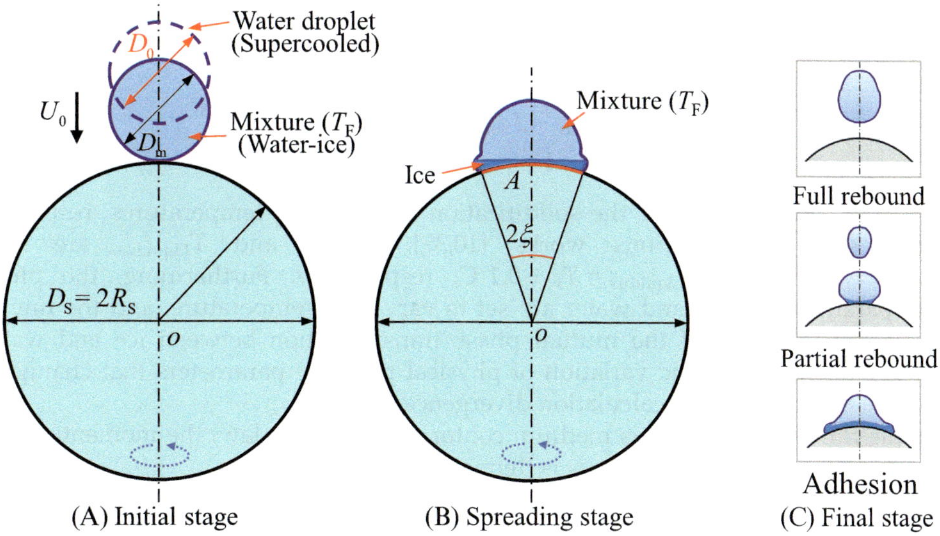

FIGURE 6.18 Schematic of the impacting-icing process of a supercooled droplet on a cold superhydrophobic sphere: (A) Initial, (B) spreading, and (C) final stages.

6.4.1 Simulation model

6.4.1.1 Supercooling effect

Under room temperature, the droplet conventionally experiences stages such as spreading, receding, and oscillating after impacting on the superhydrophobic sphere. However, under the supercooling condition, the impacting process is often accompanied by the phase transition, icing process. Therefore the impact and icing process of the supercooled droplet on the superhydrophobic cooled sphere is a complex process coupling the multiphase fluid flow and heat transfer. To simplify the analysis, the following two assumptions are made: (1) Because collision-induced vibrations contribute significantly to nucleation, the nucleation of the supercooled droplet begins immediately as soon as it contacts with a cold spherical surface [32,33]; (2) the recalescence stage is very fast and completes in a very short time [34–36].

6.4.1.2 Solidification/melting model for icing

The Solidification/Melting model determines whether the second phase is liquid (water) or solid (ice) according to the total enthalpy value of different temperatures. Based on the enthalpy (energy) balance [10,37], the enthalpy-porosity method introduced mush zone is employed to calculate the phase transition process. The total enthalpy is defined as the sum of the sensible enthalpy and latent heat, that is,

$$h = h_{\text{Sensible}} + h_{\text{Latent}} \tag{6.11}$$

and the latent heat is expressed as

$$h_{\text{Latent}} = \alpha_{\text{Liquid}} L_{\text{m}} \tag{6.12}$$

in which the liquid fraction (unfrozen water-ice mixture) α_{Liquid} is calculated by

$$\alpha_{\text{Liquid}} = \begin{cases} 0 & T < T_{\text{Solidus}} \\ \dfrac{T - T_{\text{Solidus}}}{T_{\text{Liquidus}} - T_{\text{Solidus}}} & T_{\text{Solidus}} < T < T_{\text{Liquidus}} \\ 1 & T > T_{\text{Liquidus}} \end{cases} \tag{6.13}$$

where T_{Solidius} and T_{Liquidus} are the solidification and melting temperatures, respectively. Combined with our previous works [10,37], T_{Solidus} and T_{Liquidus} are set to $T_{\text{Solidus}} = T_F - 0.1°C$ and $T_{\text{Liquidus}} = T_F + 0.1°C$, respectively. Furthermore, the physical property parameters of ice and water are set to vary with temperature, and the tiny temperature interval (0.2°C) of the mutual phase transformation between ice and water is applied to smooth the drastic variation of physical property parameters that change with temperature and prevent the calculation divergence.

Since the flow inside a porous medium conforms to Darcy's law, the momentum variation due to phase transition processes is defined as

$$S_M = \frac{\left(1 - \alpha_{\text{Liquid}}\right)^2}{\alpha_{\text{Liquid}}^3 + \varepsilon} A_{\text{mush}} \boldsymbol{u} \tag{6.14}$$

where ε is a minimum value that guarantees the stability of numerical solution. If the second phase is liquid water, then the value of α_{Liquid} is 1, at which point $S_M = 0$ and the momentum equation can be simplified to the N-S equation for a conventional fluid; when the second phase becomes solid ice, the value of α_{Liquid} is 0, at which point S_M tends to infinity and the velocity inside the fluid tends to zero. As for the mushy zone constant A_{mush}, it measures the damping amplitude of the material velocity, with its recommended values lying between 10^4 and 10^7 in the FLUENT help file [38]. According to the comparisons between the experimental and simulation results in our previous works [10,37] and references [39–41], A_{mush} takes different values at different supercooling degrees which are given in Table 6.3. Moreover, the physical properties of air, water, and ice all vary with the temperature in the model calculation, as shown in Table 6.4.

6.4.1.3 *Numerical method*

The same structured grid and local refined methods as in the previous two sections are employed for the present simulations, which will not repeat the presentation here. Considering the effect of icing on the impacting process, the minimum mesh sizes of $40\ \mu m \times 40\ \mu m$, $20\ \mu m \times 20\ \mu m$, $10\ \mu m \times 10\ \mu m$, and $5\ \mu m \times 5\ \mu m$ were applied to reverify the grid independence, and the numerical conditions were selected as follows: contact angle $\theta_{sta} = 150°$, Weber number $We = 50$, the sphere-to-droplet curvature ratio $\kappa^* = 0.11$, supercooling degree $\Delta T = 15°C$. The spreading area factors calculated under different grid

TABLE 6.3 A_{mush} values at different supercooling degrees.

Supercooling degree (°C)	5	10	15	20	25	30	35
A_{mush}	1.0×10^4	5.0×10^4	1.0×10^5	5.0×10^5	1.0×10^6	5.0×10^6	1.0×10^7

TABLE 6.4 Physical properties of air, water, and ice.

Material	Temperature (°C)	Density (kg/m^3)	Specific heat [kJ/(kg · K)]	Thermal conductivity [W/(m · K)]	Viscosity (μPa · s)	Surface tension (mN/m)	Latent heat (kJ/kg)
Air	0.0, −10.0, 20.0, −30.0	1.287, 1.333, 1.380, 1.442	1.006,1.006, 1.006, 1.006	0.0242, 0.0233, 0.0225, 0.0217	17.118, 16.617, 16.117, 15.588	–	–
Water	0.1, −10.0, −20.0, −30.0	999.8, 998.1, 993.6, 983.8	4.220, 4.580, 4.894, 5.132	0.561, 0.525, 0.478, 0.431	1791.1, 2660.0, 4330.0, 5650.0	75.65, 76.98, 78.41, 79.64	333.4
Ice	− 0.1, −10.0, −20.0, −30.0, −50.0	916.7, 918.2, 919.6, 920.9, 923.5	2.10, 2.02, 1.95, 1.88, 1.73	2.16, 2.26, 2.38, 2.50, 2.77	–	–	–

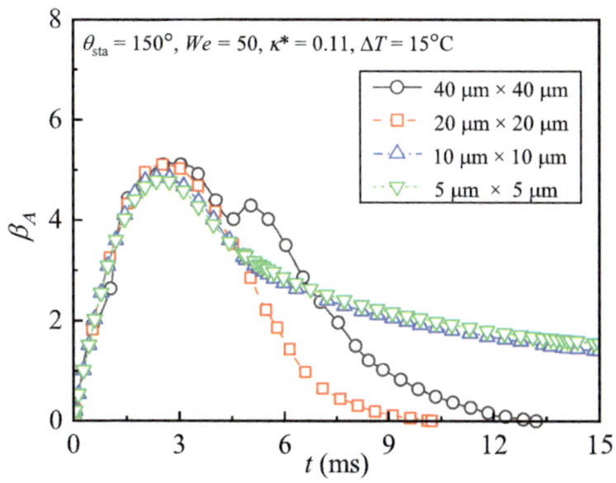

FIGURE 6.19 Grid independence verification.

quantities are shown in Fig. 6.19. The results obtained by employing two kinds of grids, whose minimum sizes are 10 μm × 10 μm and 5 μm × 5 μm, are very close, and the latter is chosen for the subsequent numerical calculation, taking into account the simulation accuracy and calculation efficiency.

The numerical methods of the impact and icing process of the supercooled droplet are consistent with the settings of the water droplet impacting process at room temperature in the previous two sections, and the boundary conditions are almost the same, except that only the boundary condition of the bottom computation domain needs to be set as a cold surface whose temperature is lower than the freezing point. Meanwhile, the Second Order Upwind format is applied to discrete the energy equation.

6.4.2 Model validation

Fig. 6.20 compares the temporal spreading area factor, β_A, obtained from the numerical simulation and the reference experiment [42]. It can be seen that under the same experimental

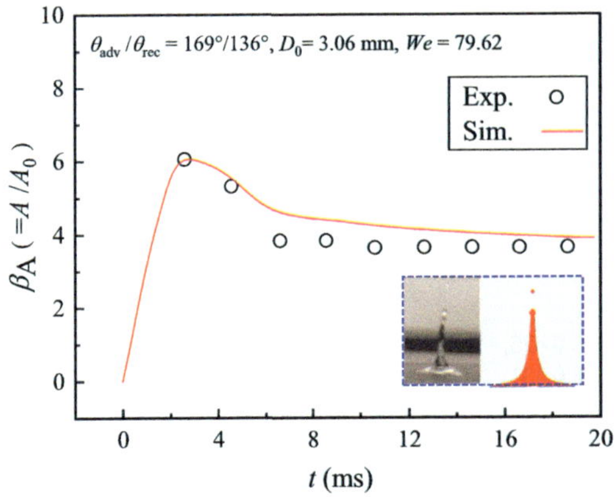

FIGURE 6.20 Comparisons of spreading area factors and droplet profiles between the simulation and experiment.

TABLE 6.5 Calculation conditions.

Droplet volume, V_0 (μL)	Droplet diameter, D_0 (mm)	Static contact angle, θ_{sta} (°)	Weber number, We	Supercooling degree, ΔT_{SC} (°C)	Sphere-to-droplet curvature ratio, κ^*
5	2.12	150	10, 30, 50, 70, 90	5, 7.5, 10, 12.5, 15, 17.5, 20, 22.5, 25, 27.5, 30	0.35, 0.21, 0.11, 0.04, 0.02

conditions in Reference, both the temporal variation trends of spreading area factors and droplet profiles of the simulation and reference experiment are basically the same, and the deviations of the maximum and the steady-state spreading area factor compared with reference experiment are less than 12%. Considering the difference between the contact angle of the droplet surface and that at room temperature under icing conditions, the complexity of the flow and heat transfer coupling during the droplet impact on the spherical surface, and the selection of the empirical value of A_{mush}, the precision of the above simulation is acceptable.

6.4.3 Results and discussion

For comparison with the collision process of the water droplet on a convex sphere at room temperature in Section 6.2, the effect of the supercooling degree, Weber number, and the sphere-to-droplet curvature ratio on the impact and icing process of the supercooled droplet are systematically and comprehensively investigated, and a series of calculation cases listed in Table 6.5 are conducted.

The Weber number is determined by the physical properties of the supercooled droplet and the impacting velocity before nucleation and recalescence occur. Considering that the supercooled water droplet and the cold surface are mostly exposed to the same environment under practical application conditions, the initial temperature T_0 of the supercooled droplet, the temperature T_S of the cold spherical surface, and the temperature T_{air} of the ambient air in the following numerical simulation are set to be equal, that is, $T_0 = T_S = T_{air} = T_F - \Delta T$.

6.4.3.1 Impact and icing behaviors

Fig. 6.21 illustrates the impact and icing behavior of the supercooled water droplet on a superhydrophobic cold sphere with fixed $\kappa^* = 0.11$ and $We = 50$ under different supercooling degrees, and the results of the impacting droplet under the room temperature (*RT*) condition are also presented as a comparison, as represent in black line. It can be seen that although the increase in supercooling degree enhances the heat transfer between the droplet and the cold sphere, the influence on the morphology and spreading area factor of the droplet at the spreading stage is not obvious. When approaching the maximum spreading state, the maximum spreading area factor of the supercooled droplet is marginally smaller than that under the room temperature condition, as represented in Fig. 6.21B, due to the marginal decrease in the spreading velocity of the water droplet hitting the cold spherical surface under supercooled conditions. Compared with the spreading stage, the effect of the supercooling degree on the receding stage is more significant, as shown in Fig. 6.21A. With the increase of supercooling degree, the heat transfer between liquid and solid wall is enhanced [43], and more ice will be formed inside the droplet at the same time, making the contact line tend to freeze in advance and ultimately leading to the droplet adhesion as represented in Fig. 6.21A. Meanwhile, the larger supercooling degree causes the contact line to freeze earlier in the receding stage, resulting in a larger stable spreading area factor, as shown in Fig. 6.21B.

Fig. 6.22 shows the evolution over time of droplet profiles and spreading area factors during the impact and icing process under different Weber numbers, when $\Delta T_{SC} = 10°C$ and $\kappa^* = 0.11$. With the increase of the Weber number, the droplet spreading speed accelerates because of the enlargement in initial kinetic energy and the inertia effect, which is similar to the impacting process at room temperature (Section 6.2). However, it is noteworthy that the corresponding augment of droplet spreading area, as shown in Fig. 6.22B, enhances heat transfer between the liquid–solid phase, resulting in icing and adhering tendency of droplets. After reaching the maximum spreading state, the larger the Weber number, the faster the droplet shrinks, as represented in Fig. 6.22A, while the discrepancy of the final stable spreading area

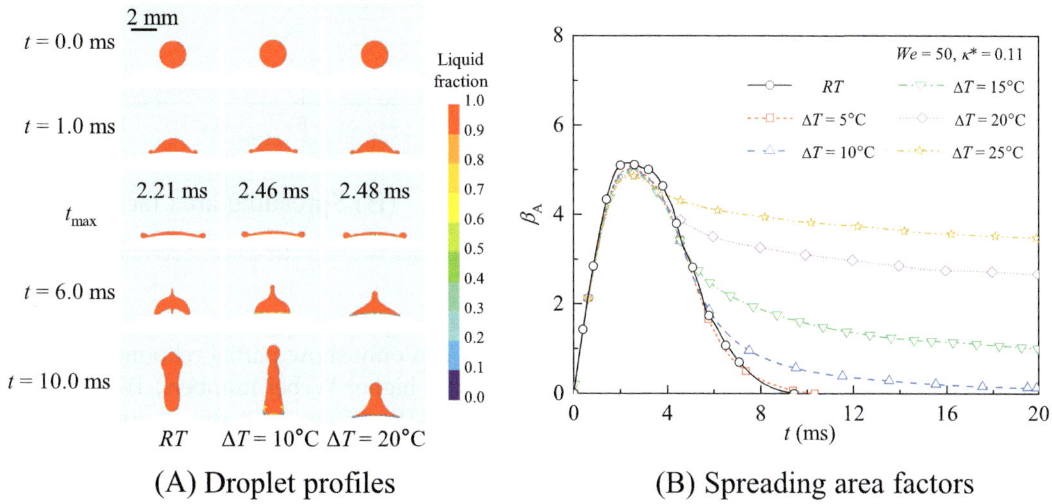

FIGURE 6.21 Evolutions of (A) droplet profiles and (B) spreading area factors under different supercooling degrees.

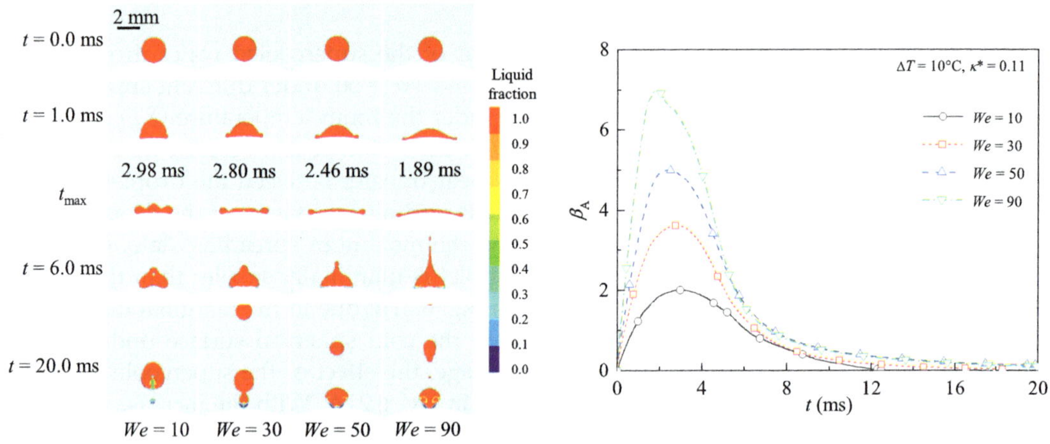

(A) Droplet profiles (B) Spreading area factors

FIGURE 6.22 Evolutions of (A) droplet profiles and (B) spreading area factors under different Weber numbers.

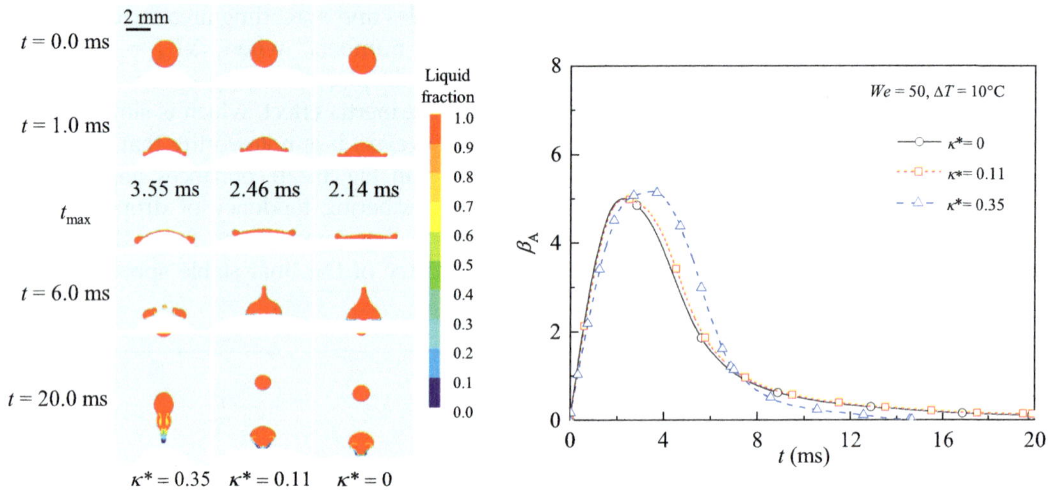

(A) Droplet profiles (B) Spreading area factors

FIGURE 6.23 Evolutions of (A) droplet profiles and (B) spreading area factors under different curvature ratios.

factor is not obvious. Furthermore, the droplet fully bounces off the sphere under lower Weber numbers, $We = 10$ and 30, on the contrary, the droplet can only show partial rebound with the rest sticking to the sphere, as shown in Fig. 6.22A, under higher Weber numbers, $We = 50$ and 90. In addition, even if the droplet is able to fully bounce ($We = 10$ and 30), the rebound time at a larger We is significantly increased due to the occurrence of icing, as represented in Fig. 6.22B.

Fig. 6.23 exhibits the evolution over time of droplet profiles and spreading area factors during the impact and icing process under different curvature ratios, when $We = 50$ and

$\Delta T_{SC} = 10°C$, while the results of a planar surface, $\kappa^* = 0$, are shown for reference. A higher curvature ratio means that the spherical surface is more convex, and under the auspices of gravity effect, the droplet spreading velocity gradually decreases, as represented in Fig. 6.23A, but the time spent approaching the maximum spreading state gradually increases, as shown in Fig. 6.23B, eventually leading to an increase in the maximum spreading area factor. According to the equilibrium relationship between surface energy and kinetic energy, the droplet retraction velocity in the maximum spreading state is approximately as follows [44]:

$$U_{rec} \sim \sqrt{2\sigma(1 - \cos\theta_{sta})/\rho\delta_{max}} \tag{6.15}$$

where δ_{max} is the film thickness at the maximum spreading state. Thus the receding stage is accelerated and a shorter contact time is obtained for a larger κ^*, as represented in Fig. 6.23B. It can also be seen from Fig. 6.23A that the droplet will fully rebound under extreme conditions of very large ($\kappa^* = 0.35$) or very small ($\kappa^* = 0$) curvature ratios but partially rebound under the moderate curvature ratio ($\kappa^* = 0.11$).

6.4.3.2 Maximum spreading characteristic

Fig. 6.24 indicates that the changing law of the maximum spreading area factor during the impact and icing process varies with the supercooling degree under different Weber numbers and curvature ratios. It can be seen that $\beta_{A,max}$ is smaller than that of the droplet at room temperature, which is consistent with the results in Fig. 6.21B. When the supercooling degree ΔT_{SC} is less than 10°C, $\beta_{A,max}$ will increase slowly with the decrease of the supercooling degree, and the larger the Weber number and the curvature ratio, the more conspicuous the influence of the supercooling degree on the $\beta_{A,max}$. When the supercooling degree is greater than 10°C, $\beta_{A,max}$ is almost independent of the supercooling degree.

Fig. 6.25 depicts the variation rule of the maximum spreading area factor during the impact and icing process varying with Weber number under different supercooling degrees and curvature ratios. It can be seen that the maximum spreading area factor increases with the increasing Weber number, which is consistent with the results in Fig. 6.22B. In addition, the effect of the Weber number on the maximum spreading area factor is more significant under the condition of the smaller supercooling degree.

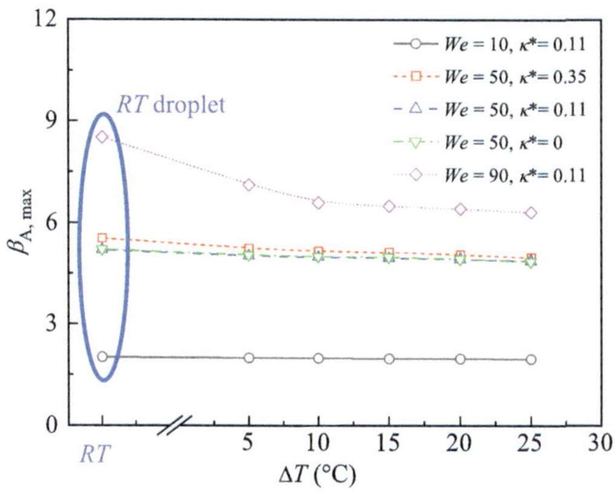

FIGURE 6.24 Effect of supercooling degree on the maximum spreading area factor.

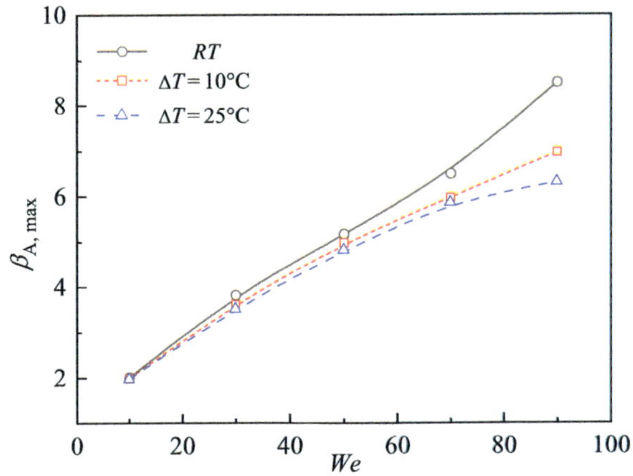

FIGURE 6.25 Effect of Weber number on the maximum spreading area factor.

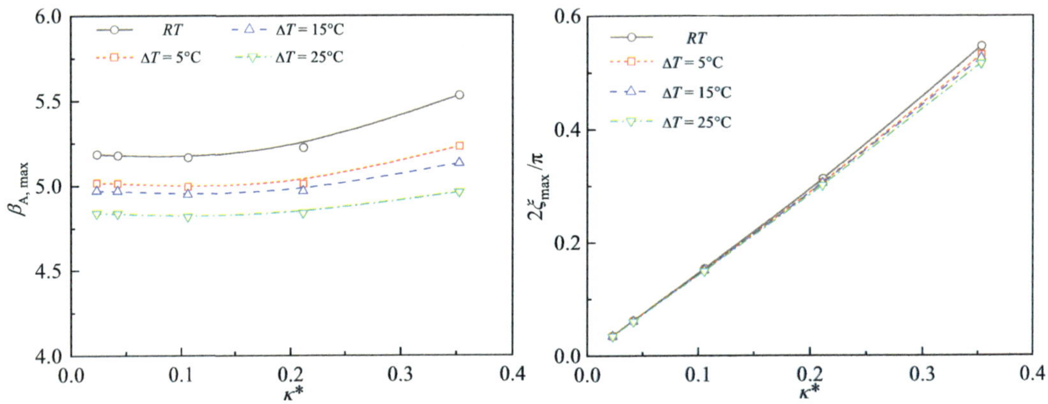

(A) Maximum spreading area factor (B) Maximum spreading arc angle

FIGURE 6.26 Effect of curvature ratio on the maximum spreading (A) area factor and (B) arc angle.

Fig. 6.26 exhibits that the changing law of the maximum spreading area coefficient and arc angle during the impact and icing process varies with the curvature ratio under different supercooling degrees, when $We = 50$. It can be seen that compared with the situation at room temperature, both the change characteristics of the maximum spreading area factor and arc angle with the curvature ratio are similar: When $\kappa^* \lesssim 0.1$, the maximum spreading area coefficient varies slightly with the increase of the curvature ratio, but when $\kappa^* \gtrsim 0.1$, the maximum spreading area coefficient increases significantly. The maximum spreading arc angle keeps increasing with the increase of the curvature ratio. Analysis reveals that the spreading speed slows down and the spreading time extends as increasing κ^*, resulting in the increase of the maximum spreading area factor, the enlargement of the receding speed, and the shortening of the contact time. Furthermore, as the supercooling degree decreases, the effect of the curvature ratio on $\beta_{A,max}$ becomes more pronounced.

6.5 Summary

In this work, numerical simulations are carried out to study the impacting and icing process of the water droplet on a spherical surface. A Volume of Fluid method coupled with DCA model and Solidification/Melting model is established and validated by comparing with experiments or references. The effects of sphere curvature, sphere shape, contact angle, and surface temperature on the droplet impacting morphologies and the maximum spreading characteristics are investigated. The main conclusions are as follows:

On the convex sphere, the maximum spreading factor β_{max} and arc angle $2\xi_{max}$ of the impacting droplet increase with the increasing Weber number, while decreasing with the increasing contact angle. The larger the curvature ratio κ^*, the more significant the effect of Weber number and contact angle on the maximum spreading factor and arc angle. $2\xi_{max}$ increases significantly as the κ^* increases, while the β_{max} increase as κ^* increases especially for $\kappa^* \gtrsim 0.1$, but it remains almost unchanged for $\kappa^* \lesssim 0.1$. The larger the Weber number or the smaller the contact angle, the more pronounced the effect of κ^* on the maximum spreading characteristic.

On the concave sphere, the maximum spreading factor increases slightly and then turns to decrease faintly with increasing κ^* in most cases due to the combined effect of gravity and surface shape. In addition, the variation trend of the maximum spreading factor and arc angle with Weber number and contact angle is similar to that on the convex sphere.

When the droplet is supercooled and the spherical surface is cold, the effect of Weber number and curvature ratio on the maximum spreading characteristics is similar to that of the impacting process at room temperature. The supercooling degree slightly influences the spreading stage of the water droplet but has a substantial effect on the receding stage with a larger stable spreading area factor obtained at a greater supercooling degree. The larger the Weber number and the curvature ratio, the more conspicuous the influence of the supercooling degree on the maximum spreading area factor. As supercooling degree decreases, the effect of the curvature ratio and Weber number on the maximum spreading area factor becomes more pronounced.

6.6 Acknowledgments

This work is supported by the National Natural Science Foundation of China (Nos. 52176079 and 52406007), Beijing Municipal Natural Science Foundation (No. 3244046, Key Laboratory of Icing and Anti/De-icing (No. IADL20230113), China Postdoctoral Science Foundation (No. 2022M720423), and Young Elite Scientist Sponsorship Program by BAST (No. BYESS2023352).

References

[1] H.J. Punge, M. Kunz, Hail observations and hailstorm characteristics in Europe: a review, Atmospheric Research 176–177 (2016) 159–184.

[2] H. Kiani, D.-W. Sun, Water crystallization and its importance to freezing of foods: a review, Trends in Food Science & Technology 22 (2011) 407–426.

[3] N. Dalili, A. Edrisy, R. Carriveau, A review of surface engineering issues critical to wind turbine performance, Renewable and Sustainable Energy Reviews 13 (2009) 428–438.

[4] L. Gao, Y. Liu, H. Hu, An experimental investigation on the dynamic glaze ice accretion process over a wind turbine airfoil surface, International Journal of Heat and Mass Transfer 149 (2020) 119120.
[5] R. Veerakumar, L. Gao, Y. Liu, H. Hu, Dynamic ice accretion process and its effects on the aerodynamic drag characteristics of a power transmission cable model, Cold Regions Science and Technology 169 (2020) 102908.
[6] Y. Liu, K. Zhang, W. Tian, H. Hu, An experimental study to characterize the effects of initial ice roughness on the wind-driven water runback over an airfoil surface, International Journal of Multiphase Flow 126 (2020) 103254.
[7] X. Zhang, X. Wu, J. Min, Aircraft icing model considering both rime ice property variability and runback water effect, International Journal of Heat and Mass Transfer 104 (2017) 510–516.
[8] B. Ding, H. Wang, X. Zhu, R. Chen, Q. Liao, How supercooled superhydrophobic surfaces affect dynamic behaviors of impacting water droplets, International Journal of Heat and Mass Transfer 124 (2018) 1025–1032.
[9] T.M. Schutzius, S. Jung, T. Maitra, G. Graeber, M. Köhme, D. Poulikakos, Spontaneous droplet trampolining on rigid superhydrophobic surfaces, Nature 527 (2015) 82–85.
[10] X. Zhang, X. Liu, X. Wu, J. Min, Impacting-freezing dynamics of a supercooled water droplet on a cold surface: Rebound and adhesion, International Journal of Heat and Mass Transfer 158 (2020)119997.
[11] B. Bhushan, Y.C. Jung, Natural and biomimetic artificial surfaces for superhydrophobicity, self-cleaning, low adhesion, and drag reduction, Progress in Materials Science 56 (2011) 1–108.
[12] F. Chu, X. Wu, B. Zhu, X. Zhang, Self-propelled droplet behavior during condensation on superhydrophobic surfaces, Applied Physics Letters 108 (2016) 194103.
[13] D. Khojasteh, M. Kazerooni, S. Salarian, R. Kamali, Droplet impact on superhydrophobic surfaces: A review of recent developments, J. Ind. Eng. Chem. 42 (2016) 1–14.
[14] C. Josserand, S.T. Thoroddsen, Drop impact on a solid surface, Annual Review of Fluid Mechanics 48 (2016) 365–391.
[15] A.L. Yarin, Drop impact dynamics: splashing, spreading, receding, bouncing…, Annual Review of Fluid Mechanics 38 (2006) 159–192.
[16] Š. Šikalo, M. Marengo, C. Tropea, E.N. Ganić, Analysis of impact of droplets on horizontal surfaces, Experimental Thermal and Fluid Science 25 (2002) 503–510.
[17] D.C.D. Roux, J.J. Cooper-White, Dynamics of water spreading on a glass surface, Journal of Colloid and Interface Science 277 (2004) 424–436.
[18] H. Hu, Z. Jin, An icing physics study by using lifetime-based molecular tagging thermometry technique, International Journal of Multiphase Flow 36 (2010) 672–681.
[19] M. Song, C. Dang, T. Higashi, E. Hihara, Review of experimental data associated with the solidification characteristics of water droplets on a cold plate surface at the early frosting stage, Energy and Buildings 223 (2020) 110103.
[20] M. Strub, O. Jabbour, F. Strub, J.P. Bédécarrats, Experimental study and modelling of the crystallization of a water droplet, International Journal of Refrigeration 26 (2003) 59–68.
[21] X. Zhang, X. Wu, J. Min, Freezing and melting of a sessile water droplet on a horizontal cold plate, Experimental Thermal and Fluid Science 88 (2017) 1–7.
[22] A. Bordbar, A. Taassob, D. Khojasteh, M. Marengo, R. Kamali, Maximum spreading and rebound of a droplet impacting onto a spherical surface at low weber numbers, Langmuir: the ACS Journal of Surfaces and Colloids 34 (2018) 5149–5158.
[23] B. Ji, Q. Song, Q. Yao, Numerical study of hydrophobic micron particle's impaction on liquid surface, Physics of Fluids 29 (2017) 077102.
[24] J.U. Brackbill, D.B. Kothe, C. Zemach, A continuum method for modeling surface tension, Journal of Computational Physics 100 (1992) 335–354.
[25] J.-B. Dupont, D. Legendre, Numerical simulation of static and sliding drop with contact angle hysteresis, Journal of Computational Physics 229 (2010) 2453–2478.
[26] H.B. Eral, D.J.C.M. 'T Mannetje, J.M. Oh, Contact angle hysteresis: a review of fundamentals and applications, Colloid and Polymer Science 291 (2013) 247–260.
[27] T.-S. Jiang, O.H. Soo-Gun, J.C. Slattery, Correlation for dynamic contact angle, Journal of Colloid and Interface Science 69 (1979) 74–77.
[28] O.V. Voinov, Hydrodynamics of wetting, Fluid Dynamics 11 (1977) 714–721.

[29] R.L. Hoffman, A study of the advancing interface, Journal of Colloid and Interface Science 94 (1983) 470–486.

[30] J. Ju, Z. Jin, H. Zhang, Z. Yang, J. Zhang, The impact and freezing processes of a water droplet on different cold spherical surfaces, Experimental Thermal and Fluid Science 96 (2018) 430–440.

[31] X. Liu, X. Zhang, J. Min, Spreading of droplets impacting different wettable surfaces at a Weber number close to zero, Chemical Engineering Science 207 (2019) 495–503.

[32] M. Sun, W. Kong, F. Wang, H. Liu, Impact freezing modes of supercooled droplets determined by both nucleation and icing evolution, International Journal of Heat and Mass Transfer 142 (2019) 118431.

[33] R.R. Czys, Ice initiation by collision-freezing in warm-based cumuli, Journal of Applied Meteorology 28 (1989) 1098–1104.

[34] S. Jung, M.K. Tiwari, N.V. Doan, D. Poulikakos, Mechanism of supercooled droplet freezing on surfaces, Nature Communications 3 (2012) 615.

[35] F. Tavakoli, S.H. Davis, H.P. Kavehpour, Freezing of supercooled water drops on cold solid substrates: initiation and mechanism, Journal of Coatings Technology and Research 12 (2015) 869–875.

[36] X. Zhang, X. Wu, J. Min, X. Liu, Modelling of sessile water droplet shape evolution during freezing with consideration of supercooling effect, Applied Thermal Engineering 125 (2017) 644–651.

[37] X. Zhang, X. Liu, X. Wu, J. Min, Simulation and experiment on supercooled sessile water droplet freezing with special attention to supercooling and volume expansion effects, International Journal of Heat and Mass Transfer 127 (2018) 975–985.

[38] A. Inc, ANSYS Fluent Theory Guide, (2013). http://www.mendeley.com/catalog/ansys-fluent-theory-guide (accessed June 29, 2024).

[39] J. Blake, D. Thompson, D. Raps, T. Strobl, Simulating the freezing of supercooled water droplets impacting a cooled substrate, AIAA J 53 (2015) 1725–1739.

[40] Y. Yao, C. Li, Z. Tao, R. Yang, H. Zhang, Experimental and numerical study on the impact and freezing process of a water droplet on a cold surface, Applied Thermal Engineering 137 (2018) 83–92.

[41] R. Zhang, P. Hao, X. Zhang, F. He, Supercooled water droplet impact on superhydrophobic surfaces with various roughness and temperature, International Journal of Heat and Mass Transfer 122 (2018) 395–402.

[42] L. Mishchenko, B. Hatton, V. Bahadur, J.A. Taylor, T. Krupenkin, J. Aizenberg, Design of ice-free nanostructured surfaces based on repulsion of impacting water droplets, ACS Nano 4 (2010) 7699–7707.

[43] C. Shinan, D. Liang, S. Mengjie, L. Mengyao, Numerical investigation on impingement dynamics and freezing performance of micrometer-sized water droplet on dry flat surface in supercooled environment, International Journal of Multiphase Flow 118 (2019) 150–164.

[44] D. Bartolo, C. Josserand, D. Bonn, Retraction dynamics of aqueous drops upon impact on non-wetting surfaces, Journal of Fluid Mechanics 545 (2005) 329–338.

CHAPTER

7

Growth distribution characteristics of micro-scale trapped air bubbles in ice and their impact on the freezing process

Keke Shao[1,2], *Mengjie Song*[1,2], *Long Zhang*[1], *Xuan Zhang*[1], *Zekang Zhen*[1], *Sirui Yu*[1], *Shiqiang Zhou*[1] *and Runmiao Gao*[1]

[1]Department of Energy and Power Engineering, School of Mechanical Engineering, Beijing Institute of Technology, Beijing, P.R. China [2]School of Mechanical Engineering, Hanyang University, Seoul, South Korea

Nomenclatures

Symbol	Description	Unit
c	Concentration	mol/m^3
d	Mean equivalent diameter	m
g	Gravity acceleration	m/s^2
h	Height	m
k	Distribution coefficient	/
n	Index	/
r	Radius	m
t	Time	s
v	Rate	m/s
x	Distance ahead of an interface	m

Frosting and Icing for Efficient Energy Use in Engineering Applications
DOI: https://doi.org/10.1016/B978-0-443-15495-9.00012-5

y	Position	m
z	Contact radius	m
A	Dimensionless factor	/
C	Change rate of water vapor concentration	/
D	Diffusion coefficient	m^2/s
G	Growth rate	m^2/s
H	Liquid-phase zone length	m
K	Thermal conductivity	W/(m K)
L	Wavelength	m
N	Bubble number density	$/mm^3$
P	Pressure	Pa
R	Migration rate	$\mu m^2/(s\ K)$
T	Temperature	K
V	Volume	m^3
Greek symbol	**Description**	**Unit**
α	Angle between the solid–liquid interface and the gas-solid interface	°
β	Angle between the liquid interface and the horizontal plane	°
γ	Latent heat	kJ/kg
θ	Contact angle between the bubble and ice	°
λ	Amplitude	m
ρ	Density	kg/m^3
μ	Density ratio	/
σ	Surface tension	N/m
φ	Contact angle	
Subscript	**Description**	**Unit**
0	Initial	/
b	Bubble	/
c	Critical	/
d	Concave	/
l	Liquid	/
i	Ice	/
u	Convex	/
w	Water	/

7.1 Introduction

Icing is a complex liquid–solid phase change process of unsteady state, variable density, and moving boundaries. It is widespread in nature and industrial fields such as aviation and navigation, rail transportation, energy and electricity, and refrigeration and often brings adverse effects [1,2]. For example, ice on the surface of an aircraft in high-altitude clouds will affect the attitude and sensor working status, thereby reducing the stability and operability of the aircraft [3,4]. Ice formation on the surface of the outdoor heat exchanger of an air source heat pump will significantly reduce the energy efficiency of the system [5,6]. Ice on the surface of wind turbine blades will reduce power generation efficiency, and ice falling off the blade surface can also cause harm to people and equipment [7,8]. Ice covering the surface of photovoltaic panels will affect light transmittance, thereby reducing power generation efficiency [9,10]. Ice crystals inside frozen beef will pierce the cell membrane and cause the loss of nutrients, affecting its eating quality [11]. The frozen storage of animal organs, tissues, eggs, and sperm also faces the challenge of losing activity due to freezing [12]. The icing problems faced by many of the above fields not only cause large direct economic losses but also waste a lot of time in various anti-icing operations [13,14]. To deal with the problems caused by icing, various mechanical deicing [15], surface processing [16], surface coating [17], and thermal deicing [18,19] methods and technologies have been developed and tested in different industries. However, these methods have shortcomings such as high energy consumption, high cost, short life, and poor stability, and have not been widely used in actual production and daily life. Therefore to fully combine scenarios, develop and optimize various antiicing technologies, and achieve accurate prediction and control of the icing process, it is necessary to conduct research on micro-scale icing processes [20,21].

The solidification process of a single water droplet is a simplification of the icing process. As shown in Fig. 7.1, the solidification process of a single fixed water droplet on the surface of a flat plate can be divided into four stages: precooling or supercooling, recalescence, freezing, and cooling or tempering, according to the temperature fluctuation [22,23]. As the surface temperature of the flat plate decreases, the temperature measured by a thermocouple located inside the water droplet gradually drops from the initial temperature T_0

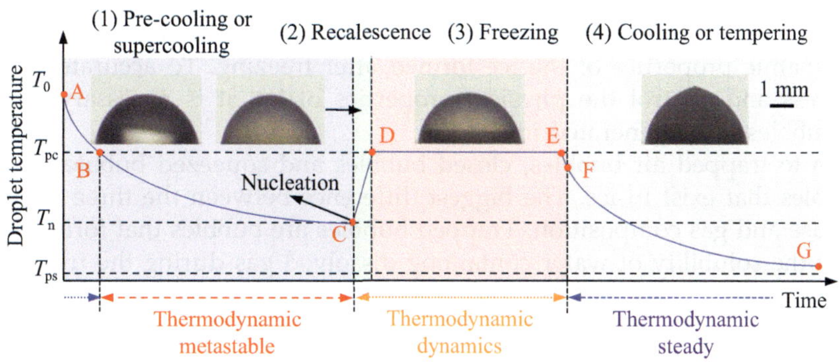

FIGURE 7.1 Stages and morphological changes of solidified water droplets.

of Point A to below the phase change temperature T_{pc} of Point B. At this point, the water droplets remain liquid. When the temperature is as low as the nucleation temperature T_n at Point C, nucleation starts from the triple line on the water droplet surface. The ice crystals that quickly appear on the surface of water droplets after nucleation will reflect and refract light, causing the water droplets to appear white, which is called recalescence. During this stage, the water droplet will change from a liquid state to a coexisting state of solid and liquid, and the duration of this process is generally only 20–50 ms [24,25]. Due to the rapid release of the latent heat of solidification, the internal temperature of the water droplet rises to the phase change temperature T_{pc}. The freezing front will migrate to the top of the droplet, while the temperature remains at the phase changes temperature without changing. This process is the freezing stage. When the freezing front reaches the top of the water droplet, the volume expansion caused by the difference in density of ice and water will eventually form a peach or onion-shaped tip [26,27]. At this time, the temperature at the temperature measuring point has dropped to Point F. In the cooling or tempering stage, the temperature continues to decrease to near the surface temperature T_{ps}. During the entire solidification process, water droplets undergo thermodynamic change processes of thermodynamic steady state, thermodynamic metastable state, thermodynamic dynamic state, and thermodynamic steady state.

In general, the solubility of gases in solids is less than the solubility in liquids because the distance between solid molecules is small. During the freezing stage, air molecules are squeezed out of the ice and converge toward the front of the freezing front [28]. When the air concentration at the front reaches a critical value of supersaturation, bubble nucleation will occur [29]. The tiny bubbles formed after nucleation expand rapidly in water, and the freezing front continues to migrate. During the growth of bubbles on the freezing front, they are mainly affected by the buoyancy force and the adhesion force of the freezing front. When the buoyancy force is greater than the adhesion force, the bubbles will detach from the interface; otherwise, they will continue to grow on the interface [30]. During the growth process, the bubbles are squeezed by the freezing front, causing the shape to gradually change from a spherical shape to an egg-shaped or needle-shaped shape with an aspect ratio of the major axis and the minor axis greater than 1 [31]. The growth of bubbles causes the concentration of air molecules in front of the freezing front to decrease, which in turn causes the growth rate of the bubbles to decrease until they are caught up by the freezing front and form bubbles trapped in the ice [32]. Micro-scale trapped air bubbles not only affect the freezing rate during the freezing but also affect the thermodynamic properties of the ice formed after freezing. To accurately predict the freezing process and control the physical properties of ice, it is necessary to study the trapped air bubbles in ice generated in icing.

In addition to trapped air bubbles, closed bubbles and squeezed bubbles are also two types of bubbles that exist in ice. The biggest difference between the three of them is the formation cause and gas composition. Trapped bubbles are bubbles that form in ice due to differences in the solubility of water containing dissolved gas during the freezing process [29]. Since each gas in the air has different solubility in water, the gas composition in the bubbles is not exactly the same as in air. Gases that are more soluble in water are less abundant in bubbles. Closed bubbles often appear in lakes in high latitude and cold areas. The formation process is that methane and other gases produced by microbial fermentation at the bottom of the lake are blocked by the ice on the lake surface during the process of floating up and are eventually frozen in the ice [33]. Squeezed bubbles are commonly found in polar ice cores. Under the squeeze of the gravity of the ice and snow above, the fluffy snow gradually transforms into solid ice, and the air in it gradually gathers to form bubbles [34]. The main components of the gas in the bubble are approximately the same as those of the Earth's atmosphere when the bubble is formed [35,36]. In addition, boiling bubbles and cavitation bubbles are also two types of bubbles that exist in water [37,38]. Unlike trapped air bubbles that form during relatively mild freezing processes, they tend to form during dramatic energy changes. There are significant differences in their formation time and shape. As shown in Fig. 7.2, not only do trapped air bubbles take a longer time to form, but they also have a wide range of aspect ratios between their major and minor axes. Although the formation process of trapped air bubbles is complex and affected by many factors, its formation conditions are mild, its period is long, its size changes greatly, and its stability is high, which provides convenience for studying its formation process and distribution in the laboratory [39]. In addition, studying trapped air bubbles can also provide references for two other bubbles research.

To clarify the research directions and progress of relevant scholars on freezing, the literature related to freezing is summarized in terms of the research scale. As shown in Fig. 7.3A, statistics on freezing research literature at three scales, namely, microscopic [40,41], mesoscopic, and macroscopic [42,43], are conducted on the CNKI, Science Direct, and Web of Science. The Fig. 7.3B show that there is relatively little research literature on trapped air bubbles in ice at intermediate centimeter scales [44,45]. The frozen ice slice is between micron-scale frozen ice droplets and meter-scale ice and frost. It is a bridge connecting microscopic freezing mechanisms and macroscopic icing/antiicing. However, this part of the research has not been paid attention to, especially the trapped air bubbles in the frozen water film, which will affect the freezing rate on a microscopic level and the adhesion between the ice layer and the surface on a macroscopic level, which is important for preventing deicing. Therefore it is necessary to study the formation and distribution of trapped

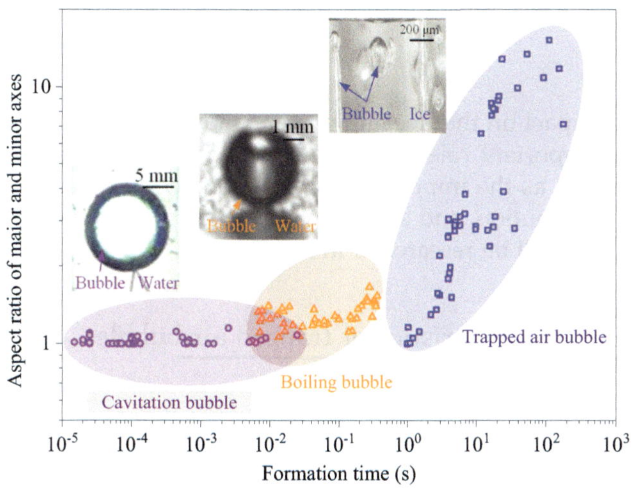

FIGURE 7.2 Aspect ratio comparison of three types of bubbles.

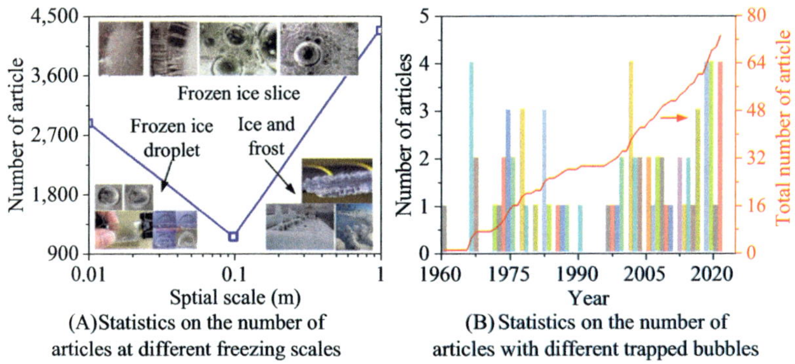

FIGURE 7.3 Number of freezing-related studies at different spatial scales. (A) Statistics on the number of articles at different freezing scales, (B) statistics on the number of articles with different trapped bubbles.

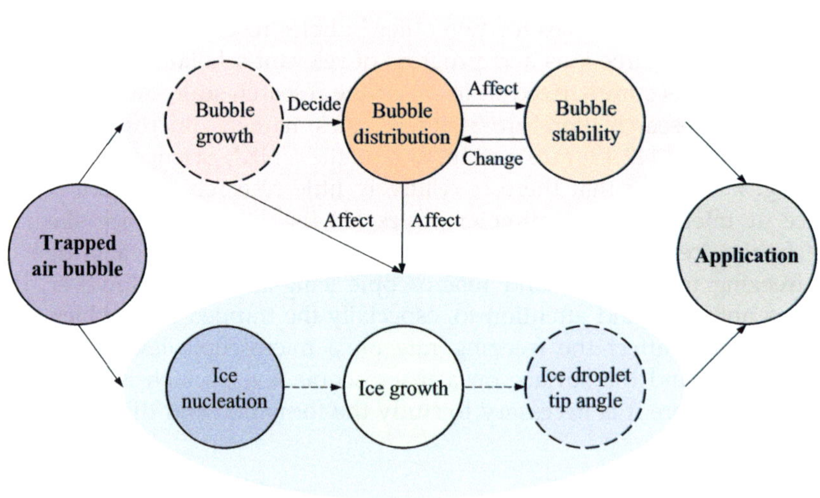

FIGURE 7.4 Outline of the present literature review.

bubbles in the frozen water film and its impact on the freezing process and properties of the entire ice layer after freezing. Given the important role of freezing in the fields of antiicing, medical treatment, food, and sports, as well as the importance of mesoscopic scale freezing research, six aspects of bubble growth, distribution, and stability, and the impact of bubbles on ice nucleation, growth, and morphology will be researched in Fig. 7.4.

7.2 Formation and distribution characteristics of trapped air bubbles

7.2.1 Formation of trapped air bubble in ice

During the freezing, air dissolved in water is squeezed from the ice and converges at the front of the freezing front. The freezing front has a jagged structure, and air concentrations tend to be higher in the jagged depressions. Bubble nucleation occurs when the air concentration in front of the freezing front reaches a critical value [46]. When the bubble size exceeds the critical nucleation size, the bubbles will continue to grow or otherwise dissolve in the water [47,48]. During the bubble growth process, they are affected by the adhesion and buoyancy forces of the freezing front. Among them, a small number of bubbles will detach when their buoyancy is greater than the adhesion force, while a large number of bubbles will adhere to the freezing front. As the freezing front migrates, the high concentration of air around the bubble will diffuse into the bubble and cause the bubble to grow. When the growth rate of the bubble is less than the freezing rate, the freezing front slowly covers the bubble until the bubble is completely frozen in ice [49].

The air concentration in front of the freezing front is an important factor affecting bubble growth. When studying the air concentration distribution in front of the freezing front, early scholars believed that the air distribution in front of the freezing front is uniform [50]. Subsequent research finds that this view is wrong. The diffusion coefficient of air in water is usually 10^{-9} m^2/s, and its diffusion rate is much smaller than the freezing rate of 10^{-6} m/s. Therefore the air concentration in front of the freezing front cannot remain uniform during the freezing process [51,52]. Later, scholars obtained the air distribution expression based on the diffusion equation of solute in liquid. As listed in Table 7.1, most of these equations are related to the diffusion coefficient D and the distribution coefficient k. Lipp et al. obtained an equation related to the physical parameters of air and liquid based on the growth equation of bubbles in a uniformly supersaturated solution [53]. In addition, it can be seen from these expressions that the gas concentration distribution before the freezing front is not only related to constants such as diffusion coefficient and distribution coefficient but also related to the freezing rate. Although these expressions are derived from theoretical analysis, the trends they describe can be well verified in experiments. As the freezing rate decreases, the number of bubbles decreases, and this is also

TABLE 7.1 Expressions for the distribution of gas concentrations in water.

No.	Year	Author	Equation expression
1	1953	Tiller et al. [51]	$c_1(x) = c_{l,0}\left[1 + \frac{1-k}{k}\left(1 - e^{\frac{-kvx}{D}}\right)\right]$
2	1954	Pohl et al. [54]	$c_1(x,t) = c_{1,0}\left\{1 + \frac{1-k}{k}\left[e^{\frac{-vx}{D}} - e^{\frac{-v(1-k)(x+kvt)}{D}}\right]\right\}$
3	1961	Carte et al. [46]	$c_1(x,t) = c_{1,0}\left\{1 + \frac{1-k}{k}\left[e^{\frac{-vx}{D}} - e^{\frac{-(1-k)(x+kvt)}{D}}\right]\right\}$
4	1974	Bari et al. [55]	$c_1(x) = c_{l,0}\left(1 + \frac{1-k}{k}e^{\frac{-vx}{D}}\right)$
5	1981	Geguzin et al. [56]	$c_1(x) = c_{1,0}\left(1 - \frac{1-k}{k}e^{-\frac{x}{H_1}}\right)$
6	1987	Burton et al. [57]	$c_1(x,t) = c_{1,0}\left[1 + \frac{1-k}{k}\left(e^{\frac{-vx}{D}} - e^{\frac{-kv^2t-vx}{D}}\right)\right]$
7	2003	Murakami et al. [58]	$c_1(x) = c_{1,0}\left[\frac{1}{k} - \frac{1-k}{k}e^{-\frac{kvx}{D}}\right]$
8	2008	Yoshimura et al. [32]	$c_1(x,t) = c_{1,0}\left\{1 + \frac{1-k}{k}\left[e^{\frac{-vx}{D}} - e^{\frac{-(1-k)(x+kvt)}{D}}\right]\right\}$

the case with the result that bubbles no longer appear in the ice when the freezing rate is less than 3 μm/s [3]. This experimental phenomenon is consistent with the decreasing trend of gas concentration as the freezing rate decreases described by the expression.

When the air concentration in front of the freezing front reaches the critical value of supersaturation c/c_0, bubbles will nucleate. The degree of supersaturation is related to the freezing rate. When the freezing rate is 2 mm/min, the supersaturation degree of bubble nucleation is 30 [46]. Using classical nucleation theory calculations, it is found that homogeneous nucleation is difficult to occur in frozen CO_2 aqueous solutions [53]. In Eq. 7.1, when the bubble size is 100 nm, the surface tension is 0.01 N/m. However, such a small surface tension is impossible. Therefore some scholars believe that the bubbles trapped in ice are generated by the growth of preexisting micro–nano bubbles in water, rather than the nucleation of high-concentration air. The high stability of MNBs, that is, days or months, can be attributed to the low internal gas dissolution of the surrounding water [59,60]. When freezing a CO_2 aqueous solution, it is found that bubbles will form preferentially in the jagged depressions on the freezing front, and the rougher the surface, the greater the possibility of bubbles forming [53]. Due to the uneven distribution of supersaturation ahead of the freezing front, the supersaturation in depressions may be much higher than that in bulges [56]. In metal solidification research, it is found that heterogeneous substances do not significantly change the nucleation conditions, and when there are gas nuclei such as micro- and nanoscale bubbles in the liquid, bubbles will appear more easily [61]. Studies on the heterogeneous nucleation mechanism using SiO_2 nanoparticles have found a nonmonotonic relationship between the critical gas concentration for nucleation and the radius of the nanoparticles. As the radius of particles increases, the nucleation critical gas concentration first decreases and then increases [62]. These results indicate that bubbles are not prone to homogeneous nucleation on the freezing front and are more likely to undergo heterogeneous nucleation:

$$r_{b,c} = 2\sigma/\Delta P \tag{7.1}$$

where $r_{b,c}$ is the critical radius for bubble nucleation, σ is the surface tension of the liquid, and ΔP is the pressure difference between the inside and outside of the bubble.

In the vertical freezing direction, the bubbles are affected by buoyancy force and interface adhesion force during the growth process. The stress situation is shown in Fig. 7.5A. The adhesion force between the bubble and the freezing front is caused by surface tension and can be expressed as $2\pi x\sigma\sin\theta$. When the buoyancy is greater than adhesion, the bubble will detach from the ice. When the buoyancy force is equal to the adhesion force, the bubbles will adhere to the ice surface and satisfy the following relationship [63]:

$$g(\rho_w - \rho_b)V = 2\pi z\sigma\sin\theta \tag{7.2}$$

where ρ_w and ρ_b are the densities of water and bubbles, V is the bubble volume, y is the radius of the contact point between the bubble and the interface, and θ is the contact angle between the bubble and ice. The relationship between the critical radius when the bubble escapes from the interface and the contact radius can be calculated through the previous formula [64]:

$$\frac{r_{b,c}^3}{z} = \frac{3\sigma\sin\theta}{2g(\rho_w - \rho_b)} \tag{7.3}$$

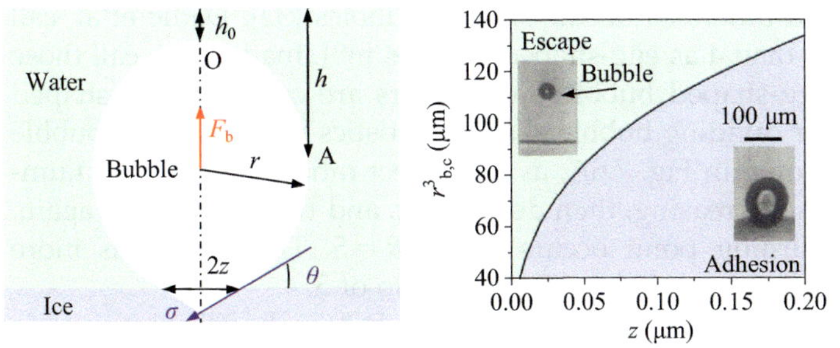

(A) Schematic diagram of the force of trapped bubbles

(B) The relationship between bubble critical escape size and contact radius

FIGURE 7.5 Force analysis and motion of trapped air bubbles. (A) Schematic diagram of the force of trapped bubbles, (B) the relationship between bubble critical escape size and contact radius.

Therefore when x takes a value within a certain range, the critical radius of bubble escape can be obtained, as shown in Fig. 7.5B. When x is 0.01 μm, the critical escape radius of the bubble is approximately 50 μm. When the bubble radius reaches 50 μm, the contact area between the bubble and ice also increases, which shows that it is difficult for the bubble to escape from the interface. However, in vertical freezing experiments, it is found that escape from the freezing front is common, and there is no obvious relationship between the frequency of bubble detachment and the freezing rate [46]. The reason for this phenomenon may be that there is a water film between the bubbles and the ice, causing the adhesion force between the bubbles and the interface to be smaller than the theoretical value [63]. In addition, since the density of water reaches its maximum at 4 °C, this part of the water will move toward the freezing front, causing micro-convection [65], and the movement rate of micro-convection gradually decreases as the freezing rate increases [66]. Therefore micro-convection in front of the freezing front may also be a reason for trapped air bubbles detachment from the interface.

7.2.2 Distribution of trapped air bubble in ice

Trapped air bubbles in ice are constantly squeezed by the freezing front during their formation, so their shape usually does not remain perfectly spherical. The bubble shape in water is determined by the following equation [63]:

$$\frac{2\sigma}{r_{b,0}} = \rho_w(h - h_0)g + \sigma\left(\frac{1}{r_1} + \frac{1}{r_2}\right) \tag{7.4}$$

where r_0 is the radius of curvature of Point O at the top of the bubble, r_1 and r_2 are the curvature radii of any Point A on the bubble surface in two mutually perpendicular surfaces, and h and h_0 are the depths of Points A and O.

Although many scholars have studied the dynamic changes and influencing factors of bubble shapes, there is no unified definition of the classification of shape. There are two main shapes of trapped bubbles: egg-shaped or elliptical and needle-shaped or cylindrical [67]. The length of needle-shaped bubbles is generally much larger than that of egg-shaped bubbles, and their size can even reach several millimeters. To accurately distinguish these two types of bubbles, the shape of the bubbles is generally classified according to the aspect ratio of the major and minor axes. Yoshimura et al. called bubbles with an aspect ratio less than 2.5 as egg-shaped bubbles and others as needle-shaped bubbles [32]. Dadic et al. call bubbles with an aspect ratio of less than 4 as egg-shaped bubbles [68]. Inada et al. call those with an aspect ratio less than 5 egg-shaped bubbles, and others are called needle-shaped bubbles [69]. To clarify the basis for dividing bubble shapes, statistics of trapped air bubble aspect ratios are carried out. As shown in Fig. 7.6A, as the aspect ratio increases, the number of bubbles shows a trend of first increasing, then decreasing, and then increasing again. As Fig. 7.6B demonstrates, this turning point occurs neat $AR = 5$. Therefore it is more appropriate to divide the bubble shape according to the axis ratio of 5.

The growth rate of bubbles is related to the freezing rate. When the freezing rate is larger, the air concentration in front of the freezing front becomes higher, and the bubble growth rate becomes larger. When the freezing rate is small, the air concentration in front of the freezing front becomes smaller, and the bubble growth rate slows down or even stops growing [70]. When the bubble growth rate is equal to the freezing rate, the bubbles will continue to grow in the major axis direction and eventually grow into needle-shaped bubbles. When the freezing rate increases, the major and minor axes of the bubbles decrease due to the shorter growth time [71,72]. The effects of dissolved gas saturation and freezing rate in water on bubble shape are shown in Fig. 7.7A. As the saturation

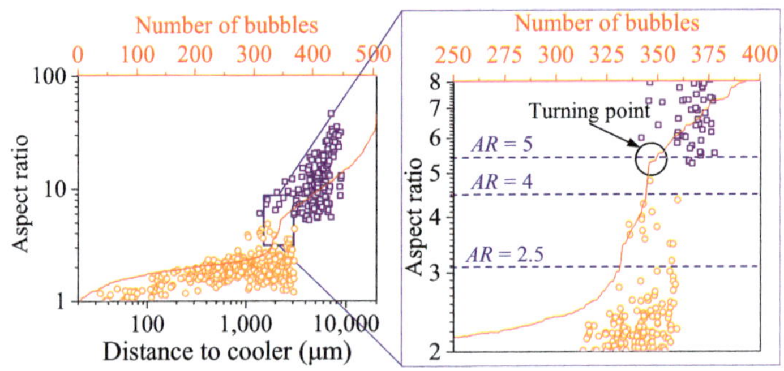

(A) Distribution of bubbles with different aspect ratios in ice **(B)** Classification of bubbles in ice by different aspect ratios

FIGURE 7.6 Distribution of trapped air bubble aspect ratio in ice. (A) Distribution of bubbles with different aspect ratios in ice, (B) classification of bubbles in ice by different aspect ratio.

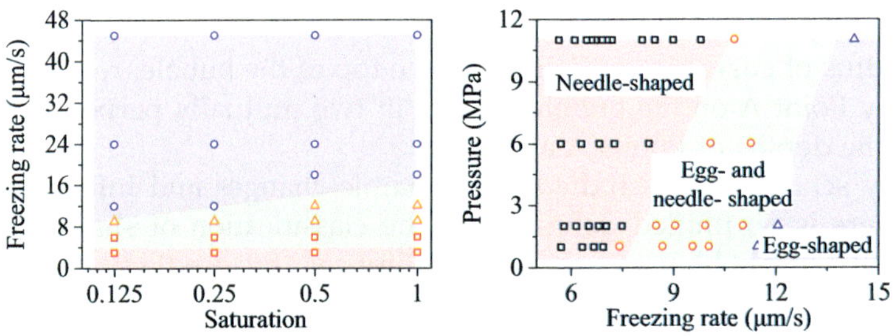

(A) Effect of saturation on bubble shape **(B)** Effect of pressure on bubble shape

FIGURE 7.7 Effect of saturation and pressure on bubble shape. (A) Effect of saturation on bubble shape, (B) effect of pressure of bubble shape.

increases, the bubbles tend to transform into needle-shaped bubbles at the same freezing rate. When the saturation degree is 1, when the freezing rate is greater than 25 μm/s, egg-shaped bubbles form in ice; when the freezing rate is 5–25 μm/s, needle-shaped bubbles form in ice [3]. Furthermore, ambient pressure has a significant effect on the shape of trapped bubbles. As shown in Fig. 7.7B, as the environmental pressure increases, the bubble type tends to transform into a needle-shaped. The increase in environmental pressure will lead to an increase in the concentration of dissolved air in water, which in turn increases the driving force for bubble growth and promotes bubble growth.

In addition to saturation and ambient pressure, the freezing rate also has an impact on bubble shape. Some scholars have studied the growth stages of bubbles from bubble nucleation to complete trapping in ice. In this process, the expansion of the bubble will reduce the surrounding gas concentration, while the increase in freezing rate will increase the air concentration. Therefore the relative relationship between the bubble growth rate and freezing rate will affect the bubble growth process and ultimately the bubble shape. Based on the influence of these two factors on bubble growth, different scholars have divided the bubble growth process into different categories. As listed in Table 7.2, according to the relationship between bubbles and ice, Lipp et al. divided the growth process of bubbles into three stages: spherical growth, partial engulfment, and total encapsulation, and determined each stage based on the relationship between bubble growth rate and freezing rate [53]. Based on the characteristics of bubble growth and the change in contact angle between the bubble and the freezing front, Wei et al. divided the growth process before bubble trapping into two stages: spherical growth and freezing rate-controlled elongation [72,73].

Both division methods regard the first stage of bubble growth as spherical growth, and have also been experimentally verified by other scholars [46]. The reason for the differences in the classification of subsequent bubble growth stages is due to the different types of bubbles they studied. Their bubble growth stage divisions are applicable to egg-shaped bubbles and needle-shaped bubbles, respectively. As shown in Fig. 7.8A, when the lateral growth rate of the bubble is less than 0, the bubble begins to enter the partial engulfment stage. When the freezing front exceeds the center of the bubble, the bubble is completely trapped in ice. This process perfectly describes the growth of egg-shaped bubbles. As shown in Fig. 7.8B, after the spherical growth stage, the bubble diameter reaches the maximum value. At this time, the bubble will not shrink immediately, and while the diameter remains almost unchanged, the length of the bubble will continue to increase until a certain moment when the bubble begins to shrink. This process is described by the growth process of a needle-shaped bubble.

TABLE 7.2 Division of growth stages of trapped air bubbles in ice.

No.	Author	Growth stage	Division criteria
1	Lipp et al. [53]	1. SG	$r_b(t) > 0$; $y_i(t) < y_b(t)$
		2. PE	$r_b(t) > 0$; $y_i(t) < y_b(t)$
		3. TE	$r_b(t) < 0$; $y_i(t) > y_b(t)$
2	Wei et al. [73]	1. SG	$\varphi > \pi/2$
		2. SRCE	$\varphi \leq \pi/2$

PE, Partial engulfment; *SG*, spherical growth; *SRCE*, solidification rate-controlled elongation; *TE*, total encapsulation.

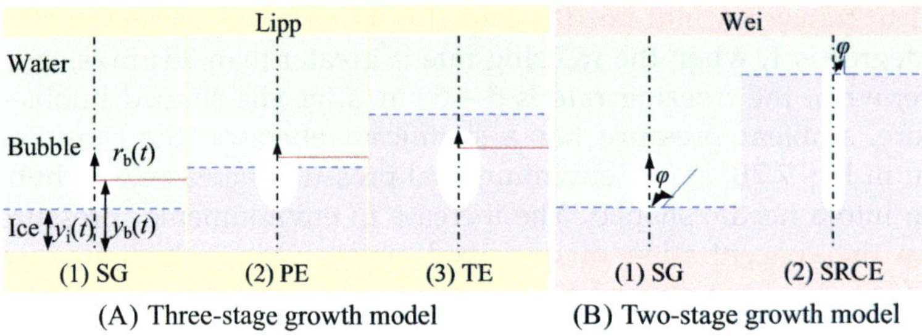

FIGURE 7.8 Schematic diagram of growth stages of trapped air bubbles. (A) Three-stage growth model, (B) two-stage growth model.

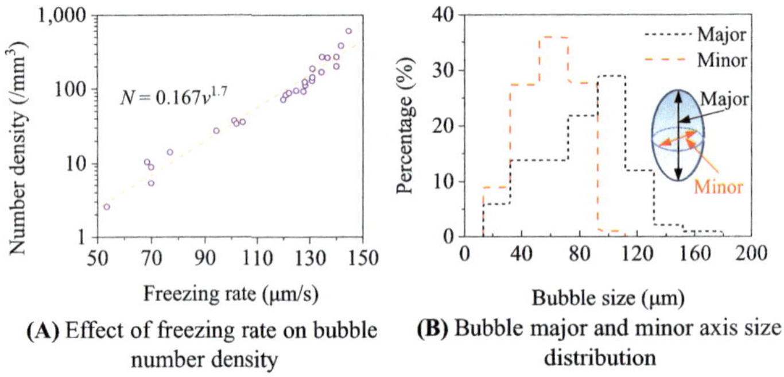

FIGURE 7.9 Number density and distribution of trapped air bubbles in ice. (A) Effect of freezing rate on bubble number density, (B) bubble major and minor axis size distribution.

Trapped air bubbles can be divided into two shapes according to their aspect ratio, and the size of needle-shaped bubbles is usually larger than that of egg-shaped bubbles, but their width does not exceed 300 μm [74]. Within a certain range, bubble size increases with decreasing freezing rate. As shown in Fig. 7.9A, at a freezing rate of 40–160 μm/s, the number of bubbles gradually increases with the increasing freezing rate. The number density of trapped air bubbles in a unit volume of ice is proportional to the 1.7th power of the freezing rate. The distribution statistics of bubble size show that there is also a relationship between bubble size and freezing rate. As shown in Fig. 7.9B, at a freezing rate of 41.67 μm/s, the length and width distribution ranges of bubbles are 10–110 and 10–180 μm, and the medians are 50–70 and 90–110 μm.

The movement of the freezing front increases the concentration of air molecules, while the growth of bubbles decreases the concentration of air molecules. The relative sizes of bubble growth rate and freezing rate affect bubble growth time and shape changes. When the freezing rate is small, the rate of air squeezed out of the ice is less than the bubble consumption rate and air diffusion rate, and bubbles will stop growing and become trapped in ice. As the bubble grows, the air concentration in front of the interface decreases rapidly, and bubble growth rate gradually slows down. At this time, the bubbles will also be caught up by the freezing front, and eventually, the bubbles will also stop growing and become trapped in ice. When there are no bubbles in front of the freezing front, the air concentration increases again, and bubble nucleation and growth will occur again. This periodic alternating increase and decrease in air concentration will cause the periodic formation of bubbles, and a bubble layer structure will be formed.

To analyze the causes and influencing factors of the periodic formation of the bubble layer, the concentration of air in water before and after the formation of the bubble layer is analyzed. Before the bubble layer forms, the air in water gradually decreases from the freezing front to the distance. After bubbles form, the growth of the bubbles will reduce the local air concentration, and the reduction in air concentration will cause the bubble growth rate to slow down until the bubbles are captured by the freezing front and trapped in ice. The movement of the freezing front will gradually increase the air concentration in front of the freezing front until new bubbles form. Since the freezing front is uneven, the distribution of bubble nucleation points is relatively scattered, and all the bubbles in the bubble layer are not in a straight line. In a bubble layer, the maximum distance along the freezing direction in the bubble layer is the amplitude λ, and the distance between two bubble layers is the wavelength L. Physical photos of the bubble layer and the effect of freezing rate on the wavelength of the bubble layer are shown in Fig. 7.10A [51,56].

The formation of bubble layers in ice is related to both freezing rate and air diffusion [66]. To quantitatively study the influencing factors of bubble spacing, a mathematical model was established. As listed in Table 7.3, the distance of the bubble layer is positively related to the reciprocal of the freezing rate. The reason for the difference may be related to the types of dissolved gases in water. Different gases have different solubility in water and differences in their nucleation concentrations. The results of bubble layer distance at different freezing rates obtained from experiments and modeling in the literature are summarized in Fig. 7.10B. It can be seen that as the freezing rate increases, the bubble layer distance gradually decreases, and the modeling results of most bubble layer distances are greater

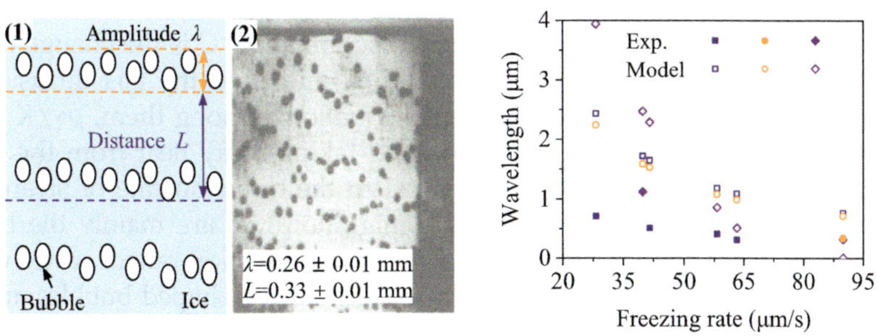

(A) Schematic diagram of bubble layer and experimental photos **(B)** Comparison of bubble layer distance experimental and model values

FIGURE 7.10 Distribution of bubble layers in ice and influencing factors. (A) Schematic diagram of bubble layer and experiment photos, (B) comparison of bubble layer distance experimental and model values.

TABLE 7.3 Expression of wavelength equation for trapped bubble layer in ice.

No.	Author	Expression
1	Bari et al. [55]	$L = 18.14 - 9.79 \cdot \log v$
2	Geguzin et al. [56]	$L = D/kv\ \ln\ \{(1-k)/[1-k(1-c_{l,c}/c_{l,0})]\}$
3	Lipp et al. [53]	$L = D/kv(1-k)\ \ln\ \{[1-k(c_{l,c}/c_{l,0})]/(1-k)\}$

than the experimental values. And there are also large differences in the results between various models. As the freezing rate increases, the model results gradually approach the experimental values. The reason for this phenomenon may be that as the freezing rate increases, the number of trapped bubbles in the ice will also increase, causing the distance of the bubble layer to gradually decrease. Therefore the difference between the experimental results and the model calculation results is small. The two mathematical models independently derive the wavelength of the bubble layer from the gas concentration distribution in ice and the gas concentration distribution in water, but the calculation results have little deviation. In addition, Murakami et al. found that the formation of the bubble layer is related to both the freezing rate and the saturation of dissolved CO_2 in water [58].

7.2.3 Stability of trapped air bubbles in ice

When a temperature gradient is applied to trapped bubbles in ice, the bubbles will migrate slowly toward the warm side [75]. This movement is caused by the sublimation of ice on the warm side and subsequent frost deposition on the cold wall side of the bubble [76]. The migration rate is related to the temperature gradient and bubble shape [68]. Assuming a linear temperature gradient exists far away from the bubble, the theoretical migration rate of the bubble generally takes the form [77]:

$$\frac{1}{R_b} = (1 - A)\frac{\rho_b \gamma}{K} + A\frac{\rho_b}{DC} \tag{7.5}$$

where A is a dimensionless factor related to the shape of the bubble, ρ_b is the bubble density, K is the thermal conductivity, γ is the latent heat of ice sublimation, C is the change rate of water vapor concentration in saturated air with temperature, and D is the gas diffusion constant. There are two terms on the right side of the above equation. Among them, $\rho\gamma/K$ is related to the rate of providing heat to melt the warm side and taking away heat from the cold side to make the water vapor desublimate. ρ/DC is related to the diffusion rate of steam in bubble. This shows that the factors that affect the bubble migration rate are mainly the heat transfer rate at the interface between the bubble and the ice and the diffusion rate of water vapor inside the bubble [78]. The migration rate of the egg-shaped and needle-shaped bubbles are [79],

$$\frac{1}{R_b} = \frac{\rho_b \gamma}{3K} + A\frac{2\rho_b}{3DC} \tag{7.6}$$

$$\frac{1}{R_b} = \frac{\rho_b \gamma}{2K} + \frac{\rho_b}{2DC} \tag{7.7}$$

To verify the accuracy of the equation and explore the factors that affect the migration rate of trapped air bubbles, the experimental values of the migration rates of different types of bubbles under normal pressure are compared with the mathematical model calculation results [79]. As shown in Fig. 7.11A, as the bubble temperature increases, the migration rates of the two types of bubbles show a gradually increasing trend. Whether it is experimental values or mathematical model calculations, the migration rate of needle-shaped bubbles is greater than the migration rate of egg-shaped bubbles. The reason for this difference may be that as the bubbles migrate, the bubbles are gradually elongated,

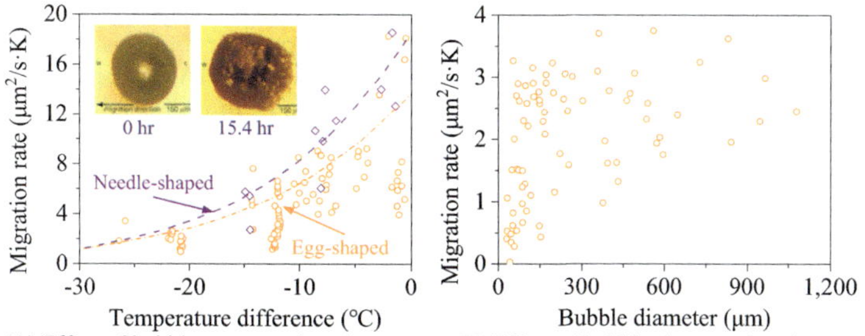

(A) Effect of bubble type on migration rate **(B)** Effect of bubble size on migration rate

FIGURE 7.11 Stability of trapped air bubbles in ice under different conditions. (A) Effect of bubble type on migration rate, (B) effect of bubble size on migration rate.

that is, a/b and m/n gradually increase. The reduction of A weakens the influence of diffusion in bubbles and enhances the influence of conduction. In addition, the longer needle-shaped bubbles lead to a larger temperature difference on both sides, which also accelerates the migration rate.

In addition to the influence of temperature and shape on the migration rate, the bubble size also affects the migration rate of the bubbles [68]. Fig. 7.11B shows the variation trend of migration rate with bubble diameter in trapped bubbles. As the bubble diameter increases, the bubble migration rate shows a gradually increasing trend. However, under the same bubble diameter, there is a big difference in bubble migration rate, and this difference will also decrease as the bubble diameter increases. At larger bubble diameters, the difference in bubble migration rates is smaller. The reason for this phenomenon may be that there are differences in the shape and internal pressure of normal-pressure trapped bubbles obtained from freezing in the laboratory and high-pressure squeezed bubbles in deep polar ice cores. The pressure inside the bubble has a great influence on the migration rate. As the pressure increases, the air molecules in the bubble increase, which will hinder the diffusion of water vapor. On the other hand, the shape of smaller bubbles is variable, which also affects the migration rate.

7.3 Effect of trapped air bubbles on freezing process

7.3.1 Effect of trapped air bubbles on ice nucleation

The growth process of trapped air bubbles that form in ice is also affected by the freezing rate. The size and number of trapped air bubbles in ice are affected by the concentration of air dissolved in water. The influence of trapped air bubbles on the ice nucleation and growth process is shown in Fig. 7.12A. Comparing the temperature changes of the three types of water, the starting times of the main freezing stage are 18, 18.7, and 14.8 seconds, respectively [80]. The difference in cooling rate is mainly caused by the droplet volume and temperature measurement position and has no significant impact on the freezing time. As the air concentration increases, the main freezing stage advances. As the air concentration increases, the duration of droplets from the supercooling stage to the recalescence stage shortens, indicating that the presence of gas may promote the nucleation of supercooled water droplets.

Preexisting air bubbles in water can also impact the nucleation [81]. The results of nucleation time at different water temperatures with varying air saturations after 0.21 W ultrasonic

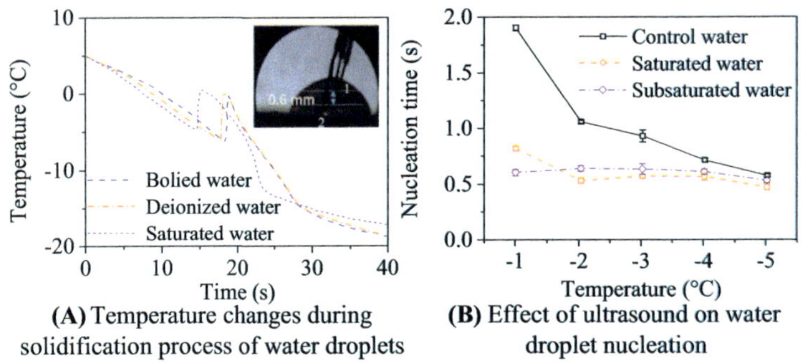

FIGURE 7.12 Effect of trapped air bubbles on ice nucleation. (A) Temperature changes during solidification process of water droplets, (B) effect of ultrasound on water droplet nucleation.

radiation for 3 seconds are shown in Fig. 7.12B. The delay from ultrasonic irradiation to nucleation is shorter for supersaturated water samples at lower ultrasonic application temperatures. Ultrasonic irradiation can effectively trigger the nucleation of water samples, and the nucleation temperature is close to the corresponding irradiation temperature, especially at lower irradiation temperatures. Furthermore, compared to the control water, both the saturated water and the subsaturated water achieved the goal of reducing the lag time, which further confirmed that preexisting bubbles contribute to ultrasound-induced nucleation. It should be noted that for water samples containing bubbles, the reduction in nucleation delay stops after reaching a certain value at ultrasonic irradiation temperature −2°C, and this is also observed for saturated sucrose solution at ultrasonic irradiation temperature −3°C. Similar phenomenon. Saturated water has a shorter delay from irradiation to nucleation at −1°C due to the involvement of more bubbles than saturated water. At −2°C, −3°C, −4°C, and −5°C, the nucleation onset time of saturated water is shorter than that of subsaturated water. As mentioned previously, bubbles initially present in the liquid may become cavitation bubbles, act as nuclei for ice nucleation, and collapse to initiate nucleation [82].

Furthermore, the size of the bubbles has an influence on the crystallization process with the help of ultrasonic irradiation [83]. Generally speaking, oversized bubbles will reduce the number of cavitation bubbles by absorbing surrounding bubbles and turning into larger bubbles. In this study, although the size of preexisting bubbles was not controlled, oversized bubbles should be more numerous in saturated water than in subsaturated water, resulting in a reduced number of cavitation bubbles. To quantitatively study the impact of trapped air bubbles on ice nucleation, a lattice Boltzmann-phase field method is used to study the effect of a single bubble on seawater freezing crystallization [84]. It is found that the presence of bubbles can effectively promote the growth of ice crystals and increase the crystallization rate by 6.78%. This promotion effect increases with the increase of the bubble radius.

7.3.2 Effect of trapped air bubbles on ice crystal growth

During freezing, air molecules dissolved in water are squeezed by the ice, causing gas to accumulate at the freezing front. In most cases, the interaction of bubbles with the freezing front reduces the grain growth rate in polycrystalline materials. Therefore studying the effect of bubbles on grain growth and migration in polycrystalline ice can also provide a better understanding of the ice crystal evolution process. Polycrystalline samples are obtained by recrystallization of a pure ice single crystal [85]. Two types of regions, denoted ZS and ZC, are observed. The ZS zone is initially essentially free of bubbles, while a nonnegligible bubble concentration is always observed in the ZC zone. To study grain growth, the samples are annealed at different temperatures between −2°C and −27°C, their structures are analyzed after annealing times of up to 4000 hours, and the ZS and ZC zones are studied separately [86]. For each temperature, the results are represented in a d^2 versus t plot. The curve in the ZS region shows a sharp decrease in the growth rate after several hundred hours of annealing, so that two growth modes of ZS1 and ZS2 can be distinguished, each satisfactorily adjusted by a straight line. State changes are also observed in the ZC zone, but in this case, the first state of ZC1 is not well defined, while the second state of ZC2 is also well represented by a straight line. Therefore for the three states ZS1, ZS2, and ZC2, the typical grain growth equation can be written as [87]:

$$d^2 = d_0^2 + 4G_i t \tag{7.8}$$

where d is the mean equivalent diameter of the crystals, d_0 is the ice crystal diameter at the initial time, and G_i is the ice crystal growth rate.

The ice crystal growth rates in the three states at different temperatures are listed in Table 7.4. As the temperature decreases, the growth rate of ice crystals gradually slows down, but the growth rate of ice crystals without bubbles is always greater than the growth rate of ice crystals with bubbles [88,89]. This indicates that the bubbles between grain boundaries hinder the migration of grain boundaries, resulting in a reduction in the growth rate of ice crystals.

The ice obtained using the polycrystalline ice preparation method still contains invisible bubbles. To elucidate the influence of the second phase on the normal grain growth of ice, Azuma et al. used phase change method to produce bubble-free fine-grained polycrystalline ice under high hydrostatic pressure and low temperature conditions [90]. This method will be used to prepare bubble-free polycrystalline ice and bubble-containing polycrystalline ice, and the crystal grain distribution before and after standing is statistically analyzed. As shown in Fig. 7.13A and B, the ice crystal diameters in the bubble-free ice and bubble-containing ice before standing are mainly distributed around 0.15 and 0.60 mm, with average values of 0.31 and 0.87 mm. After 840 hours, the ice crystal diameters are

TABLE 7.4 Variation of ice crystal growth rate with temperature.

No.	Temperature/°C	ZS1	ZS2	ZC
1	−2	4.9	1.7	0.15
2	−6	4.0	0.4	0.04
3	−10	1.4	0.16	0.01
4	−13	1.4	0.06	/
5	−20	0.7	0.04	0.01
6	−27	0.5	0.04	/

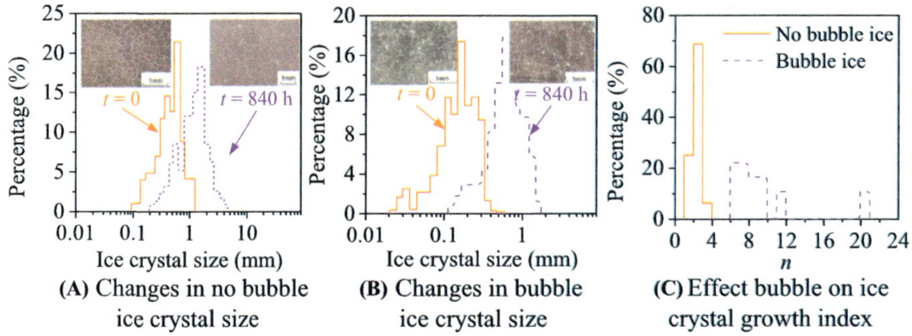

FIGURE 7.13 Effect of trapped air bubbles on ice crystal growth and migration. (A) Changes in no bubble ice crystal size, (B) changes in bubble ice crystal size, (C) effect bubble on ice crystal growth index.

mainly distributed around 1.1 and 0.7 mm, with average values of 2.54 and 1.32 mm. The general form of the grain growth curve is as follows [90]:

$$d^n = d_0^n + 4G_i t \tag{7.9}$$

where n is the ice crystal growth index, which depends on the growth mechanism. In ideal normal grain growth, the theoretical value is 2 [91].

Using Eq. (7.11), e and G_i are determined from the grain size data versus time by varying t and using the least squares method to obtain the best fit to the data. As shown in Fig. 7.13C, the n value of ice samples with bubbles is dispersed between 5.3 and 14.5, with an average value of 8.3, while the n value of ice without bubbles is 2.3, which is very close to the theoretical value of normal grain growth. The n value of the ice sample with bubbles is larger, reflecting the hindering effect of bubbles on grain growth.

7.3.3 Effect of trapped air bubbles on the frozen water droplets

Unfrozen water droplets are approximately spherical under surface tension, while frozen water droplets are ellipsoidal. At the end of freezing, a tip will be formed on the top. The tip is very sensitive to the content in the bubble. The tip angle of the frozen water droplets is related to the impurity concentration [92]. As shown in Fig. 7.14A, the tip angle of frozen water droplets with different air contents is measured and found to be $137.4 \pm 3.0°$, deionized water with an air concentration of 3% is $129.5 \pm 4.1°$, sparkling water with an air concentration of 4.99% is $120.8 \pm 1.3°$, and sparkling water with a gas concentration of 14.66% is $80.3 \pm 9.1°$ [80,81]. The average value of the tip angle of frozen water droplets for 10 times is $63.7 \pm 1.3°$, which is slightly smaller than the theoretically calculated value of 65° [93]. This is because most of the bubbles in the water droplets will not have a significant impact on the formation of tip. Only the bubbles that form in the later stages of freezing will have an impact on the formation of tip. Marin et al. proposed a geometric model of droplet shape based on density changes [94]. Based on theoretical analysis, the unfrozen part is divided into a concave volume part V_d and a convex volume part V_u by a horizontal line ending on both sides of the freeze. As shown in Fig. 7.14B, according to the spherical geometry [95],

$$V_d = r_i^3 f(\alpha - \beta) \tag{7.10}$$

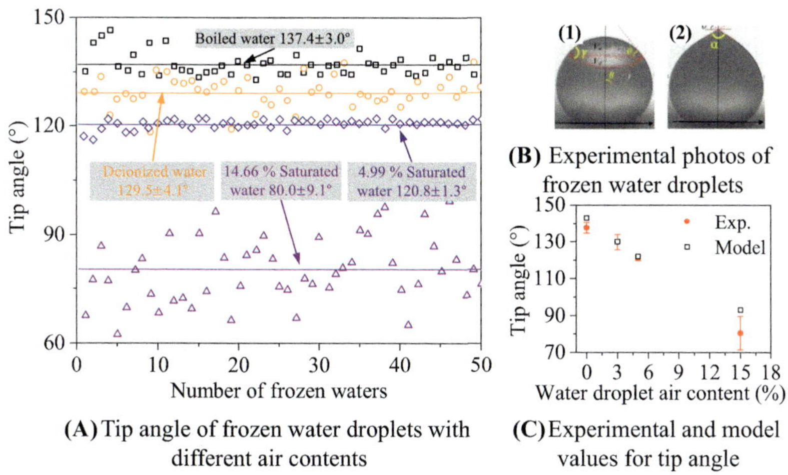

FIGURE 7.14 Trends of tip angles of frozen water droplets. (A) Tip angle of frozen water droplets with different air contents, (B) experimental and model values for tip angle, (C) experimental and model values for tip angle.

$$V_u = r_i^3 f(\beta) \tag{7.11}$$

$$f(\beta) = \frac{\left(2 - 3\cos\beta + \cos^3\beta\right)\pi}{3\sin^3\beta} \tag{7.12}$$

where r_i is the horizontal boundary radius, α is the angle between the solid–liquid interface and the gas–solid interface, and β is the angle between the liquid interface and the horizontal plane. Ignoring the evaporation of water droplets during the freezing process, the mass conservation equation is as follows:

$$\frac{d(V_w + \mu V_i)}{dt} = 0 \tag{7.13}$$

$$V_w = V_d + V_u \tag{7.14}$$

$$V_i = \int_0^z \pi R^2 dz - V_d \tag{7.15}$$

where V_w and V_i are the volumes of water and ice, and μ is the density ratio.

Based on the quasisteady state heat transfer equation, the mass conservation equation can be written as

$$\frac{d(V_w + \mu V_i)}{dz} = 0 \tag{7.16}$$

When R is close to 0, the contact angle β satisfies

$$\mu = \frac{f(\alpha - \beta) + f(\beta)}{f(\alpha - \beta) + \pi\tan(\beta/3)} \tag{7.17}$$

By combining Eqs. (7.10)–(7.17), the tip angles of frozen water droplets under different air concentrations are obtained, as shown in Fig. 7.14C. The trapped air bubbles affect the morphology of the frozen droplets, with their tip angle decreasing as the dissolved air concentration in the water increases.

7.4 Discussion

Trapped air bubble formation is related to air dissolved in water. Air contains a variety of gases, and their solubility in water varies widely. Therefore the gas composition dissolved in water is not the same as the gas composition in air. Gas molecules of different sizes are repelled by ice crystals to varying degrees. Therefore the contents of various gases dissolved in water and the gas contents in bubbles are not the same. When the gas in the ice is collected during melting, some of the gas in the bubbles will be redissolved into the water, causing the composition and content of the collected gas to be different from the gas composition and content in the bubbles. The analysis of gas components during the formation of trapped bubbles and the study of the migration patterns of various gases still need to be carried out in depth.

According to the aspect ratio of the major and minor axes of trapped air bubbles in ice, trapped air bubbles can be artificially divided into egg-shaped bubbles and needle-shaped bubbles. Judging from the statistics of the bubble aspect ratio results, it is reasonable to use an aspect ratio of 5 as the critical point for the two types of bubbles. However, the critical point of aspect ratio 5 has not received enough attention. In addition, scholars have only experimentally studied the effects of freezing rate, saturation of dissolved gases in water, and pressure on bubble shape. There is a lack of in-depth research on the mathematical analysis of bubble growth processes and shape changes. In addition, trapped bubbles will not only undergo metamorphosis under the action of temperature differences, but bubbles in an adiabatic environment will also change shape and position.

Trapped air bubbles in ice will affect the freezing process, causing the freezing rate to slow down and the tip angle of the frozen water droplets to decrease. However, there is a lack of systematic research on the effects of trapped air bubbles in ice on the acoustics, optics, heat, force, electricity, and magnetism of ice formed after freezing. In-depth study of the impact of trapped bubbles on various physical properties of ice is of great significance to optimizing antiicing methods and improving the accuracy of ice detection. To better control and predict the physical properties of ice, it is necessary to further study the effects of bubble content, bubble type, and distribution in ice on the physical properties of ice.

7.5 Summary

Trapped air bubbles are substances formed during the freezing process of microscopic water and have an important impact on the freezing process and the overall physical properties of frozen ice. They are also the link between microscopic water droplet freezing research and macroscopic icing/antiicing research. At present, people have conducted a lot of research on the formation process, growth stages, distribution characteristics, and influencing factors of trapped bubbles in ice. The effects of trapped air bubbles on ice nucleation, growth, and tip angle of frozen water droplets have also been reported.

However, the relevant mechanisms and characteristics of trapped bubble formation still need to be explored in the following aspects:

(1) Bubbles nucleate when the air concentration in front of the freezing front reaches a critical nucleation concentration. At present, many scholars have studied the gas concentration distribution model in front of the freezing front, and there is still controversy about the method of bubble nucleation. Although many scholars believe that the bubble nucleation method is homogeneous nucleation, some scholars point out that the air concentration and pressure in front of the freezing front do not meet the conditions for homogeneous nucleation.
(2) Bubbles on the freezing front are mainly affected by buoyancy and adhesion forces. When these two forces are equal in magnitude, the bubble adheres to the freezing front. When the buoyancy force is greater than the adhesion force, the bubbles will detach from the interface under the action of buoyancy force. However, theoretical calculations have shown that it is difficult for bubbles to detach, but bubble detachment often occurs during the freezing process in the vertical direction. There are currently two main explanations for this phenomenon: micro-convection and the existence of a water film on the bubble boundary, but both lack support from theoretical and experimental phenomena. To better understand the bubble formation process, the bubble nucleation mechanism and stress should be studied.
(3) Trapped air bubble shape and distribution will affect the freezing process and the physical properties of ice after freezing. Depending on whether the aspect ratio of the major and minor axes of bubbles is greater than 5, their shapes can be divided into egg-shaped and needle-shaped. Their distribution in ice is related to the freezing rate, dissolved air concentration in water, and environmental pressure. However, the reasons for the formation of these two shapes are still unclear, and there is a lack of research on mathematical modeling and numerical simulation of the bubble growth process. In addition, adjacent bubbles may inhibit growth and fuse during the growth process, but there are few studies on the interaction between adjacent bubbles.

In summary, trapped air bubbles are formed in ice during the freezing process due to the large difference in the solubility of air in ice and water and have a significant impact on the freezing process and the overall physical properties of the ice after freezing. An in-depth understanding of the basic issues in the formation process of trapped air bubbles is meaningful for deepening the understanding of the freezing process, improving and broadening the application fields of ice, predicting and controlling the freezing process and physical properties of ice, and optimizing and developing antiicing technology. By reviewing research papers on trapped air bubbles, the current research status of basic issues such as the nucleation mechanism, growth process, and distribution characteristics is reviewed, and the impact of trapped air bubbles on ice crystal nucleation and growth is summarized. The future possible research directions are provided to provide references for relevant researchers.

References

[1] K.K. Shao, Z.K. Zhen, R.M. Gao, et al., Comparative experimental study of the effect of loading rate on the typical mechanical properties of bubble and clear ice cubes, Experimental Thermal and Fluid Science 159 (2024) 111264.

[2] Y.X. Xu, R.Y. Zhao, K. Wu, et al., Experimental investigation and validation on an air-source heat pump frosting state recognition method based on fan current fluctuation signal and machine learning, Energy 291 (2024) 130372.
[3] K.K. Shao, M.J. Song, Z.Y. Jiang, et al., Experimental study on the formation and distribution of trapped air bubbles in horizontal ice slice, CIESC Journal 74 (S1) (2023) 161–164 (In Chinese).
[4] D.Y. Fu, H.K. Zheng, W. Sheng, et al., An experimental study on the influence of humidity on ice adhesion strength on superhydrophobic surfaces with microstructures, Applied Thermal Engineering 244 (2024) 122732.
[5] S.W. Lei, M.J. Song, W. Dai, et al., Experimental study on the effect of surface temperature on the frost characteristics of a vertical cold plate under natural convection, Experimental Thermal and Fluid Science 137 (2022) 110684.
[6] M.A. Moradkhani, R.M. Gao, S.H. Hosseini, et al., Estimation of frost layer characteristics during natural convection on vertical and inverted cold surfaces based on smart approaches, Applied Thermal Engineering 239 (2024) 122090.
[7] K.K. Shao, M.J. Song, B.H. Cai, et al., Effects of Environmental Temperature and Trapped Air Bubbles on the Mechanical Properties of Ice Cubes, ACS Applied Materials and Interfaces 16 (46) (2024) 63482–63494.
[8] K.K. Shao, M.J. Song, X. Zhang, et al., Mechanical Analysis of the Critical Conditions for Trapping and Detachment of Microscale Air Bubbles on the Pure Water Freezing Front, Langmuir 40 (47) (2024) 25334–25353.
[9] L.Z. Huang, Y.Y. Tian, M.J. Song, et al., An experimental study on edge-affected frosting characteristics on a vertical cold plate at different surface temperatures, Applied Thermal Engineering 236 (2024) 121538.
[10] L.Z. Huang, Y.Y. Tian, M.J. Song, et al., Experimental study of relative humidity effect on the edge effect of frosting characteristics on a vertical cold plate surface, Applied Thermal Engineering 236 (2024) 121537.
[11] C. Lsygonie, J. Britz, L.C. Hoffman, Impact of freezing and thawing on the quality of meat: review, Meat Science 91 (2) (2012) 93–98.
[12] T. Biron-shental, B. Fishch, R.V. Hurk, et al., Survival of frozen-thawed human ovarian fetal follicles in long-term organ culture, Fertility & Sterility 81 (3) (2004) 716–719.
[13] X.L. Liu, L.Z. Huang, X. Zhang, et al., A modeling study on freezing characteristics of sessile and pendant water droplets1on cold plate surfaces, Applied Thermal Engineering 236 (2024) 121693.
[14] R.M. Gao, M.J. Song, C.C.Y.H. Chao, et al., Review on condensation frosting and defrosting experiments for superhydrophobic surfaces, Applied Thermal Engineering 236 (2024) 121691.
[15] W.Y. Zhang, S.N. Chang, B. Jiang, et al., Numerical simulation of large droplets in bag breakup regime based on VOF method, Acta Aerodynamica Sinica 36 (4) (2018) 605–612.
[16] T.T. Cheng, R. He, Q.H. Zhan, et al., Magnetic particle-based super-hydrophobic coatings with excellent anti-icing and thermoresponsive deicing performance, Journal of Materials Chemistry A 3 (43) (2015) 21637–21646.
[17] H.M. Chan, S. Yasmeen, K. Park, et al., Icephobic coating through a self-formed superhydrophobic surface using a polymer and microsized particles, ACS Applied Materials & Interfaces 14 (2) (2022) 3334–3343.
[18] S.Q. Peng, Y.X. Hu, J. Huang, et al., Surface free energy analysis for stable supercooling of sodium thiosulfate pentahydrate with microcosmic-visualized methods, Solar Energy Materials and Solar Cells 208 (2020) 110390.
[19] Z.K. Zhen, M.J. Song, J. Shen, et al., An experimental study on the effect of CO_2 laser powers on melting characteristics of ice with trapped air bubbles under vertical irradiation, Applied Thermal Engineering 236 (2024) 121533.
[20] M.L. Lu, M.J. Song, X.L. Pang, et al., Modeling study on sessile water droplet during freezing with the consideration of gravity, supercooling, and volume expansion effects, International Journal of Multiphase Flow 147 (2022) 103909.
[21] D.Y. Ma, S.N. Chang, K. Wu, et al., Experimental study on the water film thickness under spray impingement based on planar LIF, International Journal of Multiphase Flow 130 (2020) 103329.
[22] M.J. Song, C.B. Dang, Review on the measurement and calculation of frost characteristics, International Journal of Heat and Mass Transfer 124 (2018) 586–614.
[23] L. Dong, W.D. Zhang, X.B. Wang, et al., Visual observation of water droplet freezing on superhydrophobic cold surface, Journal of Nanjing Tech University (Natural Science Edition) 38 (06) (2016) 116–123 (In Chinese).
[24] Q. Dang, M.J. Song, X. Zhang, et al., Modelling study on freezing process of water droplet on inclined cold plate surface with droplet dynamic behavior considered, International Journal of Heat and Mass Transfe 197 (2022) 123327.

[25] X. Zhang, Y.B. Gao, X. Liu, et al., Modelling of the icing processes of sessile supercooled water droplets on cold spheres, Applied Thermal Engineering 243 (2024) 122596.
[26] X. Zhang, X.M. Wu, J.C. Min, Freezing and melting of a sessile water droplet on a horizontal cold plate, Experimental Thermal and Fluid Science 88 (2017) 1–7.
[27] L. Chen, Y.N. Huo, J. Shen, et al., Research on simulating the freezing phenomenon of refrigerator freezer drawers, Journal of Appliance Science & Technolog (2021) 188–190 (S1).
[28] S.N. Chang, L. Ding, M.J. Song, et al., Numerical investigation on impingement dynamics and freezing performance of micrometer-sized water droplet on flat dry surface in supercooled environment, International Journal of Multiphase Flow 118 (2019) 150–164.
[29] F.Q. Chu, X. Zhang, S.K. Li, et al., Bubble formation in freezing droplets, Physical Review Fluids (2019). 071601(R).
[30] P.S. Wei, S.Y. Hsiao, Pore formation from bubble entrapment by a solidification front, American Journal of Heat and Mass Transfer 2 (2) (2015) 76–88.
[31] S.Y. Hsiao, P.S. Wei, Case study of ambient pressure effects on pore shape in solid, Journal of Thermophysics and Heat Transfer 31 (4) (2017) 1–9.
[32] K. Yoshimura, T. Inada, S. Koyama, Growth of spherical and cylindrical oxygen bubbles at an ice-water interface, Crystal Growth & Design 8 (7) (2008) 2108–2115.
[33] K.M. Walter anthony, D.A. Vas, L. Brosius, et al., Estimating methane emissions from northern lakes using ice-bubble surveys, Limnology and Oceanography: methods 8 (2010) 592–609.
[34] S.H. Fraia, I. Weikusat, N. Azuma, The microstructure of Polar ice. Part I: Highlights from ice core research, Journal of Structural Geology 61 (2014) 2–20.
[35] J.M. Barnola, D. Ratnaud, Y.S. Korotkevich, et al., Vostok ice core provides 160,000-year record of atmospheric CO_2, Nature 329 (1987) 408–414.
[36] S.H. Faria, J. Freitag, S. Kipfstuhl, Polar ice structure and the integrity of ice-core paleoclimate records, Quaternary Science Reviews 29 (1-2) (2020) 338–351.
[37] Z.Y. Jiang, M.J. Song, C.B. Dang, et al., Two mathematical models of flow boiling and flow instability in rectangular expanding microchannel heat exchangers and structure optimization, Applied Thermal Engineering 228 (2023) 120483.
[38] S. Azizifar, M.J. Song, C.Y.H. Chao, et al., A numerical study of multiphase flow boiling heat transfer of nanofluids in the horizontal metal foam tubes, International Journal of Thermofluids 22 (2024) 100605.
[39] S.H. Hong, C.Z. Li, C.B. Dang, et al., Experimental research on cooling performances of radial expanding-channeled heat sinks applied for multiple heat sources, Applied Thermal Engineering 237 (2024) 121789.
[40] Q. Dang, M.J. Song, X. Zhang, et al., Experimental study about solidification processes of sessile deformed water droplets on the biaxial inclined cold plate surface under natural convection, Applied. Thermal Engineering 236 (2024) 121883.
[41] M.J. Song, C.N. Dang, T. Higshi, et al., Review of experimental data associated with the solidification characteristics of water droplets on a cold plate surface at the early frosting stage, Energy and Buildings 223 (2020) 110103.
[42] M. Mousavi, F. Sotoudeh, B. Chun, et al., The potential for anti-icing wing and aircraft applications of mixed-wettability surfaces—a comprehensive review, Cold Region Science and Technology 217 (2024) 104042.
[43] Y.H. Cao, W.Y. Tan, Z.L. Wu, Aircraft icing: an ongoing threat to aviation safety, Aerospace Science and Technology 75 (2018) 353–385.
[44] K. Matsumoto, K. Samehima, Y. Teraoka, et al., Formation of ozone ice by freezing water containing ozone micro-bubbles, International Journal Refrigeration 36 (3) (2013) 842–851.
[45] N.A. Mishchuk, The effect of water cooling conditions on the mechanisms of porous ice formation, International Journal of Heat and Mass Transfer 165 (2021) 120612.
[46] A.E. Carte, Air bubble in ice, Proceedings of the Physical Society London 77 (495) (1961) 757–768.
[47] Z.X. Wang, Z.X. Li, L. Bai, et al., Formation and stability of nanobubble at solid/liquid interface, CIESC Journal 72 (7) (2021) 3466–3477.
[48] H.G. Zhang, Z.J. Guo, X.R. Zhang, Surface enrichment of ions leads to the stability of bulk nanobubbles, Soft Matter 16 (23) (2020) 5470–5477.
[49] P.S. Wei, S.Y. Hsiao, Effects of supersaturation on pore shape in solid, Journal of Crystal Growth 460 (2017) 126–133.

[50] C.D. Ho, H.M. Yeh, T.L. Yeh, Optimal zone lengths in multi-pass zone-refning processes, Separations Technology 6 (4) (1996) 227–233.
[51] W.A. Tilier, K.A. Jackson, J.W. Rutter, et al., The redistribution of solute atoms during the solidification of metals, Acta Metallurgica 1 (4) (1953) 428–437.
[52] A. Widdicombe, P. Ravindrarajah, A. Sapelkin, et al., Measurement of bitumen viscosity in a room-temperature drop experiment: student education, public outreach and modern science in one, Physics Education 49 (4) (2014) 406–411.
[53] G. Lipp, C.H. Korber, E.U. Hartmann, et al., Investigation of the behavior of dissolved gases during freezing, Cryobiology 24 (1987) 489–503.
[54] R.G. Pohl, Solute redistribution by recrystallization, Journal of Applied Physics 25 (9) (1954) 1170–1178.
[55] S.A. Bari, J. Hallet, Nucleation and growth of bubbles at an ice-water interface, Journal of Glaciology 13 (69) (1974) 489–520.
[56] Y.E. Geguzin, A.S. Dzuba, Crystallization of a gas-staurated melt, Journal of Crystal Growth 52 (1981) 337–344.
[57] J.A. Burton, R.C. Paim, W.P. Slichter, The distribution of solute in crystals grown from the melt. Part I. Theoretical, The Journal of Chemical Physics 21 (11) (1953) 1987–1991.
[58] K. Murakami, H. Nakajima, Formation of pores during unidirectional solidification of water containing carbon dioxide, Materials Transactions 43 (10) (2002) 2582–2588.
[59] M. Sakr, M.M. Mohamed, M.A. Maragq, et al., A critical review of the recent developments in micro-nano bubbles applications for domestic and industrial wastewater treatment, Alexandria Engineering Journal 61 (8) (2022) 6591–6612.
[60] N. Nirmalkar, A.W. Pacek, M. Barigou, On the existence and stability of bulk nanobubbles, Langmuir: the ACS Journal of Surfaces and Colloids 34 (37) (2018) 10964–10973.
[61] Z. Radovic, M. Lalovic, Numerical simulation of steel ingot solidification process, Journal of Materials Processing Technology 160 (2) (2005) 156–159.
[62] X.L. Deng, Y. Shan, X.H. Meng, et al., Direct measuring of single-heterogeneous bubble nucleation mediated by surface topology, Proceedings of the National Academy of Sciences of the United States of America 119 (29) (2022). e2205827119.
[63] N. Maenon, Air bubble formation in ice crystals, Physics of Snow and Ice: Proceesings 1 (1) (1967) 207–218.
[64] T. Temsgen, T.T. Bui, M. Han, et al., Micro and nanobubble technologies as a new horizon for water-treatment techniques: A review, Advances in Colloid and Interface Science 246 (2017) 40–51.
[65] R. Viskanta, Heat transfer during melting and solidification of metals, Journal of Heat Transfer 110 (4b) (1988) 1205–1219.
[66] Y.Z. Wang, L.L. Regel, W.R. Wilcox, Can propagation of gas bubbles lead to detached solidification? Experiments on freezing of water, Crystal Growth & Design 2 (5) (2002) 453–461.
[67] P.S. Wei, C.C. Hsiao, Microbubble or pendant drop control described by a general phase diagram, International Journal of Heat Mass Transfer 52 (5-6) (2009) 1304–1312.
[68] R. Dadic, B. Light, S.G. Warren, Migration of air bubbles in ice under a temperature gradient, with application to "Snowball Earth, Journal of Geophysical Research 115 (2010) D18125.
[69] T. Inada, T. Hatakey, F. Takemura, Gas-storage ice grown form water containing microbubbles, International Journal of Refrigeration 32 (2009) 462–471.
[70] K.K. Shao, M.J. Song, X. Zhang, et al., Growth and distribution characteristics of trapped air bubbles in ice slices, Physics of Fluids 35 (2023) 113319.
[71] K.F. Vsaconcellos, J. Beech, The development of blowholes in the ice/water/carbon dioxide system, Journal of Crystal Growth 28 (1) (1975) 85–92.
[72] P.S. Wei, Y.K. Kuo, S.H. Chiu, et al., Shape of a pore trapped in solid during solidification, International Journal of Heat and Mass Transfer 43 (2) (2000) 263–280.
[73] P.S. Wei, C.C. Huang, Z.P. Wang, et al., Growths of bubble/pore sizes in solid during solidification - An in situ measurement and analysis, Journal of Crystal Growth 270 (3) (2004) 662–673.
[74] K.K. Shao, M.J. Song, J. Shen, et al., Experimental study on the distribution and growth characteristics of trapped air bubbles in ice slices at different freezing temperatures, Applied Thermal Engineering 244 (2024) 122600.

[75] R. Dadic, M. Schneebeli, M. Wiese, et al., Temperature-driven bubble migration as proxy for internal bubble pressures and bubble trapping function in ice cores, JGR Atmospheres 124 (17-18) (2019) 10264–10282.
[76] J. Mariti, K. Mauersberger, A survey and new measurements of ice vapor pressure at temperatures between 170 and 250 K, Geophysical Research Letters 20 (5) (1993) 363–414.
[77] R.L. Shreve, Migration of air bubbles, vapor figures, and brine pockets in ice under a temperature gradient, Journal of Geophysical Research 72 (16) (1967) 4093–4100.
[78] B. Light, R.E. Brandt, S.G. Warren, Hydrohalite in cold sea ice: Laboratory observations of single crystals, surface accumulations, and migration rates under a temperature gradient, with application to "Snowball Earth.", Journal of Geophysical Research 114 (2009) C07018.
[79] N.S. Stehle, Migration of bubbles in ice under a temperature gradient, Physics of Snow and Ice: Proceedings 1 (1) (1967) 219–232.
[80] Y.L. Li, M.X. Li, C.B. Dang, et al., Effects of dissolved gas on the nucleation and growth of ice crystals in freezing droplets, International Journal of Heat and Mass Transfer 184 (2022) 122334.
[81] F. Hu, D.W. Sun, W.H. Gao, et al., Effects of pre-existing bubbles on ice nucleation and crystallization during ultrasound-assisted freezing of water and sucrose solution, Innovative Food Science & Emerging Technologies 20 (2013) 161–166.
[82] D.W. Sun, B. Li, Microstructural change of potato tissus frozen by ultrasound-assisted immersion freezing, Journal of Food Engineering 57 (4) (2003) 337–345.
[83] J. Cárcel, J. Benedito, C. Rossello, et al., Influence of ultrasound intensity on mass transfer in apple immersed in a source solution, Journal of Food Engineering 78 (2) (2007) 472–479.
[84] D. Zhang, Y.N. Luo, Y. Zhao, et al., LBM–PFM simulation of directional frozen crystallization of seawater in the presence of a single bubble, Desalination 542 (2022) 116065.
[85] O.B. Nasello, C.L. Di Prinzio, P.G. Guzmán, Temperature dependence of "pure" ice grain boundary mobility, Acta Materialia 53 (18) (2005) 4863–4869.
[86] H.J. Frost, C.V. Thompson, Computer simulation of microstructural evolution in thin films, Journal of Electronic Materials 17 (1988) 447–458.
[87] L. Arena, O.B. Nasello, L. Levi, Effect of bubbles on grain growth in ice, The Journal of Physical Chemistry. B 101 (32) (1997) 6109–6112.
[88] T. Ohuchi, M. Nakamura, Grain growth in the forsterite–diopside system, Physics of the Earth and Planetary Interiors 160 (1) (2007) 1–21.
[89] T. Hirage, C. Tachibana, N. Ohshi, et al., Grain growth systematics for forsterite ± enstatite aggregates: effect of lithology on grain size in the upper mantle, Earth and Planetary Science Letters 291 (1–4) (2010) 10–20.
[90] N. Azuma, T. Miyakoshi, S. Yokayama, et al., Impeding effect of air bubbles on normal grain growth of ice [J], Journal of Structural Geology 42 (2012) 184–193.
[91] B. Evans, J. Renner, G. Hirth, A few remarks on the kinetics of static grain growth in rocks, International Journal of Earth Sciences 90 (2001) 88–103.
[92] F. Boulogne, A. Salonen, Drop freezing fine detection of contaminants by measuring the tip angle, Applied Physical Letter 116 (2020) 103701.
[93] A. Schetnikov, V. Matiunin, V. Cheronov, Conical shape of frozen water droplets, American Journal of Physics 83 (2015) 36–38.
[94] A.G. Marin, O.R. Enriquez, P. Brunet, et al., Universality of tip singularity formation in freezing water drops, Physical Review Letter 113 (2014) 054301.
[95] K.K. Varanasi, T. Deng, J.D. Smith, et al., Frost formation and ice adhesion on superhydrophobic surfaces, Applied Physical Letter 97 (2010) 234102.

CHAPTER

8

Research on mechanism and measurement technology of frost formation of micro water frosting

Xiaoqin Zhi, Yinghe Qi, Wenhan Shu and Limin Qiu

Institute of Refrigeration and Low Temperature, Zhejiang University, Hangzhou, P.R. China

Nomenclatures

Symbol	Description	Unit
a	Diffraction dark fringe series with narrow and wide fringe spacing	/
d	Thickness	m
h	Thickness	m
k	Number of fringe	/
m	Mass	kg
n	Refractive index	/
r	Electromagnetic vector of the reflection part of the light wave	/
t	Electromagnetic vector of the transmission part of the light wave	/
u	Velocity	m/s
x	X-axis coordinate	/
y	Y-axis coordinate	/
B	Width of the single slit	m
C	Calibration factor	/
I	Measured light intensity	/
L	Distance from the observation screen to the single slit plane	m

Frosting and Icing for Efficient Energy Use in Engineering Applications
DOI: https://doi.org/10.1016/B978-0-443-15495-9.00009-5

R	Reflectivity	/
S	Scale factor	/
T	Temperature	K
X	Distance from the dark fringe center of the corresponding series to the center of the zero-order fringe	m
Greek symbol	**Description**	**Unit**
ω	Water content	ppmv
λ	Wave length	m
θ	Angle	°
δ	Frost layer thickness	m
α	Wave vector	/
σ	Standard deviations	/
τ	Time	s
μ	Elastic stiffness coefficient	$kg/(m \cdot s^2)$
ρ	Density	kg/m^3
Subscript	**Description**	**Unit**
a	Air phase	/
c	Change	/
ca	Calibration value	/
f	Freezing	/
i	Ice phase	/
in	Inlet	/
n	Nucleation	/
q	Quartz crystal	/
avg	Average value	/
cr	Critical probing depth	/
exp	Experimental value	/
theo	Theoretical value	/

8.1 Introduction

Frost formation conditions can be classified according to temperature and water vapor content. Byun et al. [1] defined the temperature range of frosting and systematically divided frosting temperature into general cold zone ($T > -30°C$), low-temperature zone ($T \leq -30°C$), and transition zone from $-30°C$ to $-160°C$. Depending on the water content conditions, there are wet air frosting (water content above 1000 ppmv) and micro water frosting (water content below 1000 ppmv). According to the above conditions, frost formation can be divided into four categories, as shown in Fig. 8.1. Because general cold and micro water content usually do not meet the conditions of frost phase transition, the actual frost formation scene can be divided into three categories: general cold and wet air, cryogenic and wet air, and cryogenic and micro water content.

Compared to the common wet air frost, the frost mechanism under cryogenic and micro water content conditions is seldom studied, but it is very important in some fields. Micro water frosting will have a bad effect on cryogenic wind tunnel, space detectors and telescopes, micro refrigerator, and so on. In a large cryogenic wind tunnel, in a temperature change chamber, frost will form on the surface of the testing model due to insufficient drying. In the test section, due to insufficient replacement and water absorption of insulation materials, the residual water vapor content can reach tens of ppmv [2], and there is also a risk of frost formation. In experiments at the National Transonic Facility (NTF), it was found that residual water in the test section and the greenhouse would condense and settle on the model during the test, affecting its aerodynamic data. Even the small scale distributed roughness caused by frost formation will have a significant impact on the aerodynamic characteristics of the model, resulting in test misalignment and even failure [3]. During the on-orbit operation of the optical system in the space probe, due to the venting of materials, micro amounts of water and other impurities are deposited on the optical surface, causing pollution, resulting in changes in the transmittance, reflectivity, and scattered light of the optical surface. It will cause the optical system to fail to meet the requirements of resolution and sensitivity [4]. During the operation of the micro Joule–Thomson (JT) refrigerator, water molecules will be diffused with the airflow, and the narrow pipes encountered are prone to forming frost deposits. It is found that water molecules easily produce ice blockage in the counter flow heat exchanger (CFHX) and the throttle valve in the refrigerator. At the same time, water molecules in the vacuum chamber may also be deposited on the surface of the cooler. This will have a negative impact on the micro-chillers [5]. The above problems of frost pollution show that with the progress of technology, the research of low-temperature and trace water frosting has become urgent.

There are two mechanisms of frost formation: condensation and freezing, desublimation and deposition [6,7]. Frost formation under general cold and wet air conditions is usually a condensation and freezing mechanism, while frost formation under cryogenic and micro

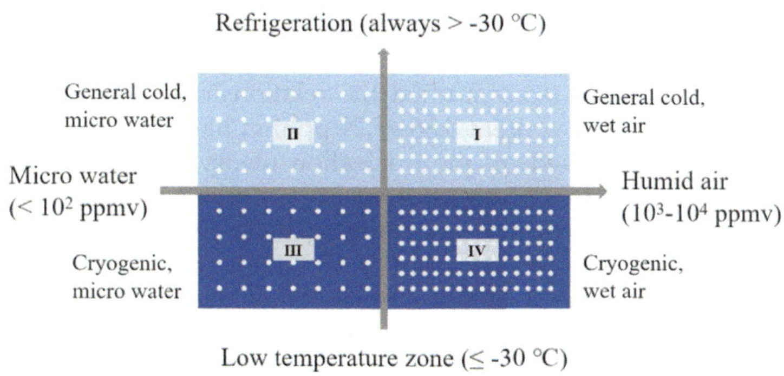

FIGURE 8.1 Frosting conditions diagram.

water content is usually a desublimation and deposition mechanism. Different frost formation mechanisms lead to significant differences between frost thickness, density, thermal conductivity, surface roughness, and other aspects [8]. Compared with the general cold zone (243K–273K), the low-temperature zone (<243K) has a large temperature span, which makes frost formation at low temperatures more challenging in both theoretical and experimental research studies. First of all, in terms of theoretical research, the large supersaturation caused by large temperature differences makes the calculation of mass transfer rate more difficult. In addition, the difficulties such as the sharp change in thermal and physical properties of mixed air and solid water under large temperature difference also increase the challenge of theoretical calculation and analysis of frost formation at low temperatures. In the experimental research, the temperature control and monitoring of low-temperature cold surface, the system sealing under low-temperature conditions, and the realization of visual observation of low-temperature closed system all have great difficulties.

Compared with the wet air region, the water vapor content of the micro water region is one to three orders of magnitude smaller. Under such a large water content gap, the numerical and experimental methods used in the frost formation of the wet air and the obtained physical properties of the frost layer are no longer applicable, which also brings great theoretical and experimental research challenges. First of all, in terms of theoretical research, the volume fraction of ice phase in the frost layer formed by micro water frosting is very small, and capturing the gas–solid interface during the growth of the frost layer is a key theoretical difficulty. In addition, the diffusion property of micro water in low-temperature airflow and the relative lack of dew point temperature correlation of micro water vapor also increase the difficulty of theoretical research. In terms of experimental research, unlike the humidity and humidity regulation of wet air through mature enthalpy wet chamber technology, the preparation and measurement of mixed gas-carrying micro water are extremely challenging. In addition, the extremely thin frost layer increases the difficulty of visual observation. How to achieve high precision frost layer capture is the key difficulty in the research of micro water frosting.

In recent years, the frost pollution problem of large low-temperature wind tunnels and space equipment has been widely concerned, and the research of frost formation of low-temperature micro water has been promoted. On the mechanism of frost formation under cryogenic and micro water content condition, Cao et al. [9] from the University of Twente studied the clogging phenomenon caused by water molecule deposition in micromechanical JT coolers. A numerical model is developed to calculate the temperature distribution and the water deposition rate of a countercurrent heat exchanger. The deposition process takes into account the diffusion of water molecules in nitrogen and the dynamics of water molecules on the wall. Using this model, the team studied the relationship between the frosting of very small amounts of water (water content 0.86–7.29 ppb) in high purity nitrogen and gas impurities and gas pressure. Based on the nucleation theory, Ye et al. [10] from Zhejiang University obtained the specific conditions of nucleation type (condensation or desublimation) and nucleation mode (heterogeneous or homogeneous) under variable parameter conditions through theoretical calculation in MATLAB® and further analyzed the frost formation mechanism and path under different surface temperature, water content, and surface contact angle conditions. Qi et al. [11] improved the CFD numerical model of previous frost formation studies by taking into account changes in frost layer thermal conductivity and introducing classical nucleation theory and applying it to frost formation of low-temperature micro water to obtain the distribution of frost layer density and thermal conductivity along the airflow. In terms of the measurement of low-temperature micro water frosting, Qi et al. [12] combined the low-temperature experimental chamber with the microphotography system to collect images of the frost formation process of micro water. It was found that the frost growth pattern changed from needle vertical growth to grass circular growth in the early stage of frost formation, revealing the difference in frost formation mechanism and characteristics under different temperature conditions. From 2012 to 2013, the AEDC at Arnold Air Force Base, Tennessee, United States, used the experimental platform [13] for measuring frost layer thickness by multi-beam thin film interferometry built in the ultrahigh vacuum (UHV) chamber to carry out high vacuum and low-temperature frost formation experiments. The frost formation thickness and rate of low-temperature micro water were measured by using calcium sulfate dihydrate ($CaSO_4$–$2H_2O$) to provide micro water vapor under pressure of 10^{-7} Torr and temperature of 30K–65K. At the same time, quartz crystal microbalance (QCM) was applied to frost formation experiment to monitor the quality of frost layer, which was used to compare and verify the optical measurement results. The experimental results show that the frost thickness of low-temperature micro water is about 45 microns (error is less than 120 nm), and the calculated mass area density measurement value is less than 10% of the operating range of QCM. Guo et al. [14] of Beijing Satellite Environmental Engineering Institute conducted frost formation experiment under the vacuum of 250 Pa through the frost formation visualization experimental device, obtained the frost formation image of low-temperature micro water, and found that the frost layer existed as a granular solid, the growth rate was slow, and would be evenly distributed on the cold surface, and cracking would occur during the growth process. Yuan et al. [4] used QCM to detect pollutant deposition on the optical surface of spacecraft, and the pollutants also included frost formed by micro water in space, which further promoted the application of QCM in the measurement of low-temperature frost formation quality. Ye et al. [15] improved the QCM structure, used twin crystal vibration to form a contrast, excluded the effect of temperature on crystal frequency, and successfully measured the cooling and frosting quality of water content of 80, 44, and 24 ppmv.

This chapter will introduce the frost formation mechanism of micro water, and the progress of measurement methods of micro water frost formation. The frost formation mechanism includes frost nucleation condition of low-temperature water, growth characteristics of frost layer, relationship between frost formation condition, and frost layer morphology. The measurement methods mainly include low-temperature visualization measurement, optical measurement, and low-temperature micromass measurement based on QCM.

8.2 Research of micro water frosting mechanism

8.2.1 Research on nucleation mechanism

The frost formation process begins with water vapor nucleation [16–18], and the nucleation process needs to overcome the minimum barrier of nucleation, which means that not only saturation conditions need to be satisfied but also a certain degree of saturation. In the study of water vapor nucleation, the following questions are usually concerned: (1) the preferred type of nucleation and (2) critical nucleation conditions.

The preferred nucleation method is condensation or desublimation, depending on which nucleation pressure is lower under certain conditions. At higher cold surface temperatures, the nucleation pressure of desublimation is usually lower than that of condensation, which favors condensation as the preferred type of nucleation. When the cold surface temperature is lower than the critical transition temperature (T_c), the nucleation pressure of condensation is lower than that of desublimation, which is favorable for desublimation to become the preferred nucleation type. In addition, to achieve nucleation, the critical nucleation temperature (T_n) must be satisfied. Therefore the type of initial nucleation depends on the relationship between the critical transition temperature (T_c) and the critical nucleation temperature (T_n). If the critical nucleation temperature is higher than the critical transition temperature, condensation nucleation is preferred; otherwise, condensation nucleation is preferred. It is worth noting that both T_c and T_n vary greatly with the surface contact angle.

As shown in Fig. 8.2, the critical transition temperature (T_c) decreases rapidly as the contact angle increases. In the region above the critical transition temperature curve, the nucleation type is condensation, while in the region below the critical transition temperature curve, the nucleation type is desublimation. Nath et al. [19] provide a similar preferred nucleation phase diagram with a temperature range of −30°C to 0°C, as shown in the blue solid rectangular box. Considering the chemical limit of hydrophobicity, the critical transition temperature with contact angle range of 1−120 degrees is given.

For a given water vapor content, it is necessary to further determine whether its nucleation temperature is above or below the critical temperature to determine the type of nucleation. Fig. 8.3 shows the critical nucleation temperature (T_n) at atmospheric pressure (0.1 MPa) with different water vapor contents, in which six orders of magnitude of water

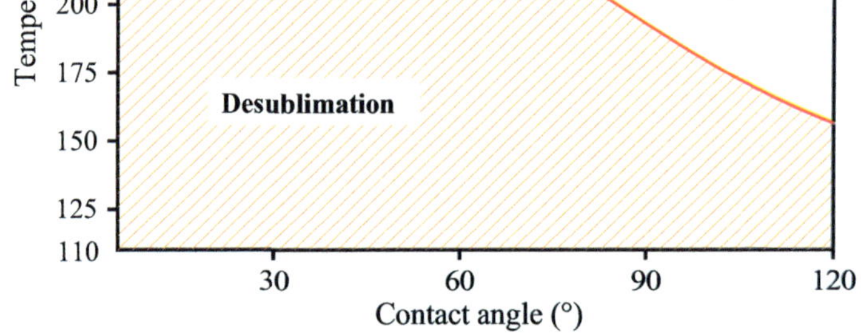

FIGURE 8.2 Phase diagram for the preferred nucleation type divided by the critical transition temperature.

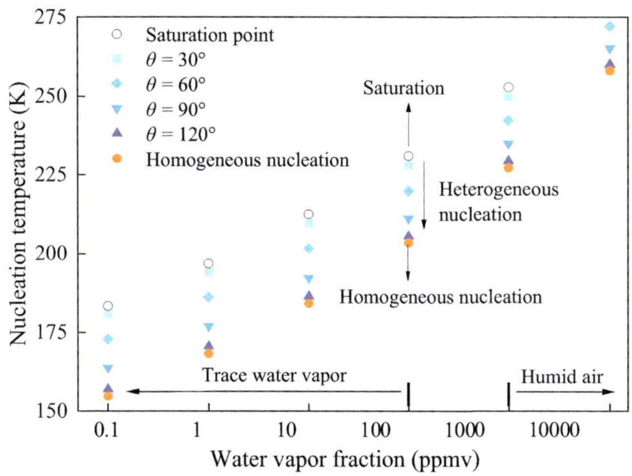

FIGURE 8.3 Nucleation temperatures at different water vapor contents and contact angles.

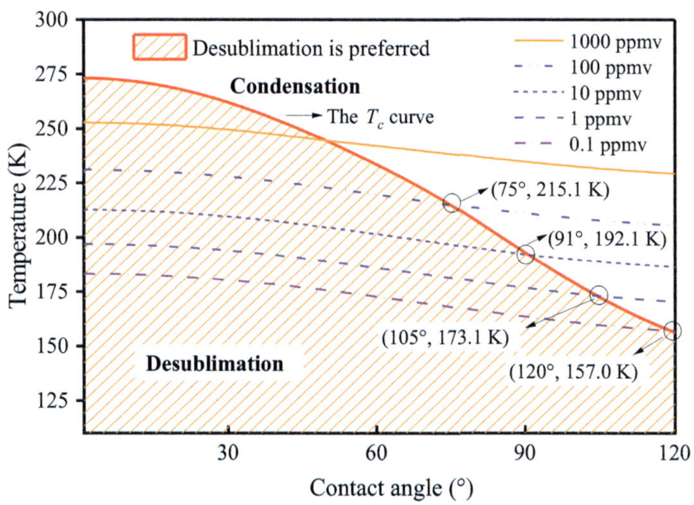

FIGURE 8.4 Phase diagram for the initial nucleation at different water vapor contents (ppmv).

vapor content are selected, ranging from 0.1 to 10,000 ppmv. First, it is observed that at a certain contact angle, the lower the water vapor content, the lower the nucleation temperature. At the same time, with the increase of the contact angle, a lower nucleation temperature is required due to the need to overcome the larger nucleation barrier. The uniform nucleation temperature can be used to determine whether nucleation occurs in gas flow. For example, at water vapor concentrations of 1000, 100, and 0.1 ppmv, the corresponding uniform nucleation temperatures are 227.0K, 203.2K, and 154.5K, respectively.

From the point of view of nucleation, some interesting conclusions are drawn that are different from those under saturated conditions. Fig. 8.4 shows the preferred nucleation phase diagram under different water vapor contents. The solid red line is the critical transition temperature curve, and the remaining lines are the nucleation temperature curves under different water vapor content. Under the condition of saturation phase transition, that is, when the equivalent contact angle is 0 degrees, the nucleation temperature is equal to the saturation temperature. Therefore when the water vapor content is less than 100 ppmv, desublimation nucleation is assumed to occur because the nucleation temperature is lower than the critical nucleation temperature. On the other hand, since the corresponding water vapor pressure of 100 ppmv at atmospheric pressure (10 Pa) is less than the triple point pressure (611 Pa), it is often misunderstood that only a desalination phase transition occurs.

Similar to the critical transition temperature, the critical transition water vapor content also changes at different contact angles, as shown in Fig. 8.5. Even under the same surface contact angle conditions, changes in water vapor content can alter the preferred nucleation type. If the water vapor content exceeds the critical transition content, condensation nucleation is preferred; otherwise, desublimation nucleation is preferred. That is, the critical transition content of 75 and 120 degrees is 100 and 0.1 ppmv, respectively. The critical transition water vapor content for uniform nucleation (surface contact angle 180 degrees) is less than 0.01 ppmv, and achieving this at such low water vapor content is often challenging. As a result, homogeneous nuclei almost always condense into nuclei and then freeze into ice grains.

8.2.2 Frost growth characteristics

Qi et al. [20,21] established a CFD numerical model to simulate the frost formation process of micro water vapor in nitrogen flow on a low-temperature surface by introducing the source term of volume mass transfer rate with saturation criterion into the Euler multiphase flow model.

The distribution of ice phase volume fraction, mass transfer rate, and water vapor content during frost growth was simulated. The maximum ice phase volume fraction increases with

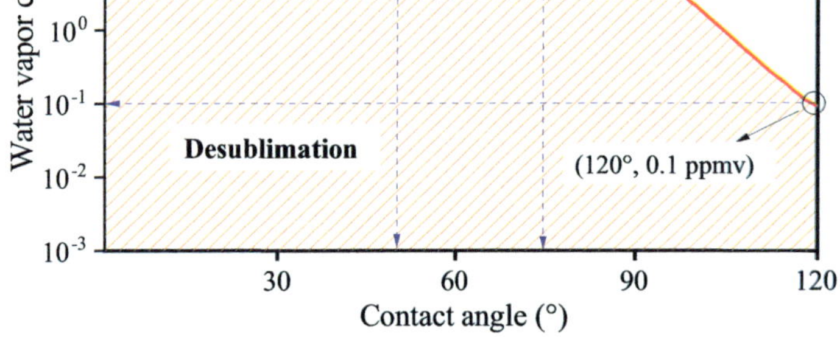

FIGURE 8.5 Phase diagram for the initial nucleation type divided by the critical transition water vapor content.

the frost-forming time, and the thickness of the region with the ice phase volume fraction greater than zero also increases with the time. On the other hand, the maximum mass transfer rate over the entire computing domain decreases over time. Since the mass transfer rate and the gas phase volume fraction are proportional to the dimensionless phase transition driving force, the ice phase volume fraction in the control volume increases, while the gas phase volume fraction decreases with the increase of frost formation time. Meanwhile, the thermal resistance of the frost layer leads to the increase of the control volume temperature, which leads to the decrease of the phase transition driving force. Therefore the maximum mass transfer rate in the calculation domain decreases with time.

In addition, the density and mass transfer rate of the frost layer are maximized near the gas inlet and become thinner and thinner toward the gas outlet. In the vertical direction, the density and mass transfer rate of the frost layer in the calculated domain are maximum near the cold side and gradually decrease along the positive direction of the y axis. This is because the control volume closer to the gas inlet has a larger water vapor content, that is, a higher saturation temperature Ts, while the control volume closer to the cold side has a lower control volume temperature T, both of which result in a larger phase transition driving force and therefore a larger mass transfer rate and a higher frost layer density.

At the same time, the frost growth process under the condition of water vapor content in different inlet gas mixtures was simulated by CFD, and the influence of water vapor content on the physical properties of the frost layer was quantitatively studied. With other operating conditions unchanged ($T = 77K$, $u_{in} = 1\ m/s$, $T_{in} = 260K$), five working conditions in which the water vapor content of the inlet gas varied between 5 and 1000 ppm were selected for calculation. The result of ice phase volume is shown in Fig. 8.6. According to the simulation

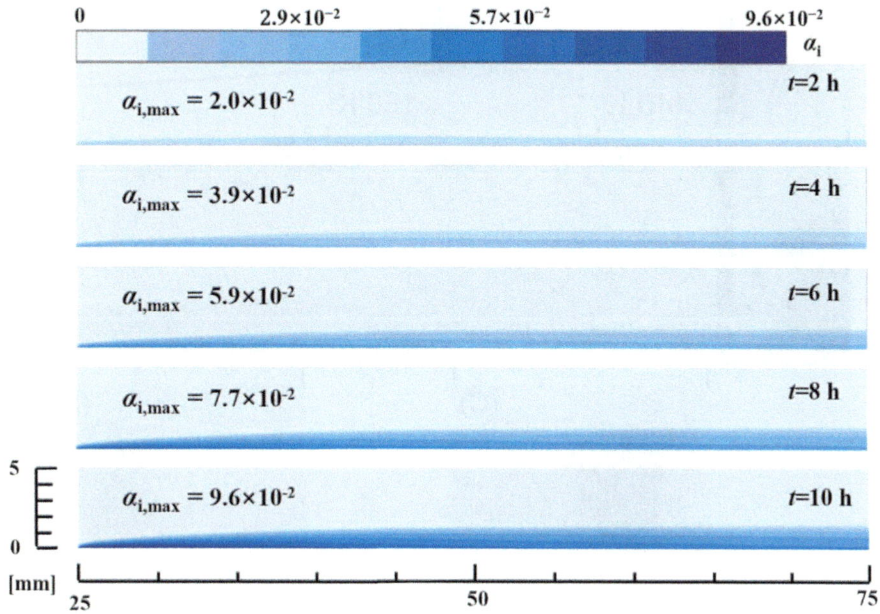

FIGURE 8.6 Volume of ice phase [20].

results, the thickness and density of frost layer increase rapidly with the increase of ω. When ω is less than 10 ppm, the ice phase volume fraction in the frost layer becomes very small [20]. Under the condition of $\omega = 5$ ppm, the average ice phase volume fraction of the frost layer accumulated for 10 hours is only 0.003, and the frost layer with such an order of ice phase volume fraction is extremely difficult to capture in experimental research.

8.2.3 Micro water deposition and sublimation process

Micromechanical JT coolers can be used to cool small electronic devices to low temperatures, provide superconducting electrons, and more. This is done by precooling the incoming high-pressure gas in a countercurrent heat exchanger (CFHX), which amplifies its cooling effect. However, impurities in the working fluid can be deposited in the pipeline, causing blockage, which limits the long-term performance of the microcooler. The reason for the pipe blockage is that, due to the gas flow, the deposited ice separates from the CFHX surface and migrates, depositing in the narrow area; thus blocking the H. Cao's team, who conducted microscopic observations of the deposition and subsequent sublimation of water molecules within the restricted slit and further understood the mechanism of microchannel clogging [5].

Fig. 8.7A shows the restricted slit cross section where ice formation is observed. Fig. 8.7B–E shows the deposition positions of water molecules in the observed limiting

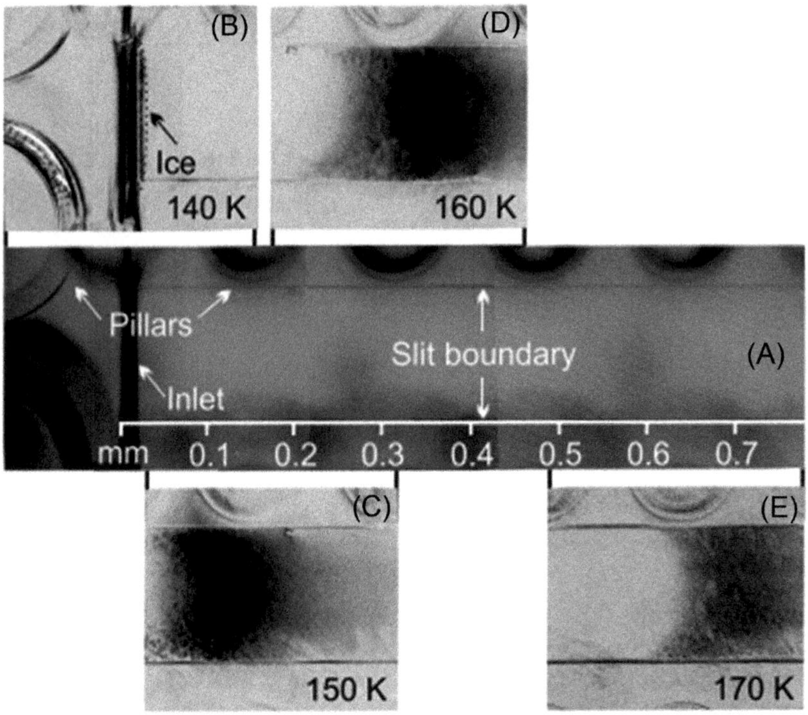

FIGURE 8.7 Image of ice formation [5]. (A) Image of a part of the observed restriction slit made by combining three images taken along the restriction slit. The positions of the figures (B), (C), (D), and (E) show their relative locations in the restriction.

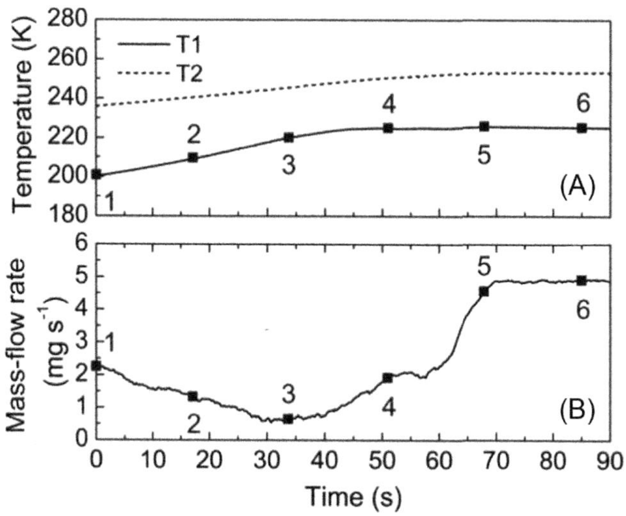

FIGURE 8.8 Deposition and subsequent sublimation of water molecules [5]. Temperature (A) and mass-flow rate (B) recorded during clogging (points 1-3) and subsequent declogging (points 3-6).

slit when the cold end temperature is controlled for gradually increasing temperatures (140K, 150K, 160K, and 170K, respectively).

The ice growth rate along the restricted slit at different cold end temperatures was simulated. Due to the blockage, the actual mass flow rate will be reduced. This decrease in mass flow affects the pressure drop in the low-pressure CFHX channel, which affects the location of the blockage in the restricted zone. To keep the cold end temperature of the microcooler at a certain value, the minimum mass flow rate is about 75% of the initial maximum. In this case, the mass flow reduction has less effect on the blockage position within the 1.15 mm length limit than 0.01 mm. At 140K, the ice mainly formed at the entrance of the restricted area, and the ice peak moved to the exit of the restricted area with the increase of the cold end temperature. This is consistent with the experimental results.

Fig. 8.8 shows a series of images showing the deposition and subsequent sublimation of water molecules in the observed limiting slit. From point 1 to point 2, the mass flow rate drops from 2.3 to 1.4 mg/s due to the blockage of the slit. At point 3, when the cold end temperature is 218K, the minimum mass flow rate is 0.6 mg/s. At this point, sublimation occurs, and the mass flow rate begins to increase. From points 4 to 5 and 6, the deposited ice gradually sublimates into steam, and the mass flow rate increases to 4.9 mg/s. As the restricted area heats up, the ice does not migrate but sublimates, leaving no micro in the restricted area.

8.3 Measurement methods of micro water frosting

In the 21st century, experimental testing technology has been further developed. In the field of general cold and wet air, the thickness of frost layer can be measured by optical methods such as interference [22], diffraction [23], or image processing [24,25]. For the density of the frost layer, the total mass after melting of the frost layer can be measured using a digital balance to further calculate the density of the frost layer ***[66–68]. On the basis of the experimental data, some new empirical correlation equations and simple one-dimensional frost formation models are proposed to predict the physical property parameters of frost layer thickness [26–28], density [26,27], thermal conductivity [29,30], and heat transfer efficiency [29,31,32].

However, the accuracy of the above methods is insufficient for cryogenic and micro water frosting, so it is very important to develop new measurement methods. At present, the microphotography technique has been successfully applied to the measurement of the shape and thickness of water frosting. Surface plasmon resonance (SPR) imaging technology breaks through the limitations of traditional optical technology and has a broad application prospect in frost crystal scale measurement. QCM is the only feasible method for cryogenic micromass measurement.

8.3.1 Visualization measurement technology

Qi et al. [11] from Zhejiang University designed and built a cryogenic micro water frosting experimental system, which consists of a cryogenic experimental chamber, a gas distribution system, and a testing system. The microscopic imaging method often uses a high-precision microscopic system to capture the process of frosting at a specific angle. After the experiment, image processing techniques are used to analyze the information in the images and calculate parameters such as the frost layer thickness and surface roughness at different frosting stages.

Currently, a combination of Adobe Photoshop and MATLAB® is commonly used for image processing. Adobe Photoshop can easily make overall changes to images, such as highlighting the measured part by increasing contrast. MATLAB® focuses on mathematical analysis of images, such as extracting coordinate information of different pixels, fitting contour curves, and finding the motion trajectories of the same point in multiple images [21].

The microscopic imaging system [11], as shown in Fig. 8.9, consists of a Navitar PL-D795CU high-performance CMOS camera, a $12\times$ zoom lens, coaxial illumination fiber optic cable, an LVT-250T-5600K ultra-bright LED microscope dedicated light source (lightbox), and a support platform with a precision micro-adjustment of ± 25 mm travel in the XYZ direction. The working distance of the microscopic imaging system is 165 mm, and the minimum resolution is 6.66 μm.

Assuming that the physical properties of the frost layer images taken from the side of the test section obtained by the microscopic imaging method can represent the average properties of the frost layer on the entire cold surface. In the image processing process, Adobe Photoshop is used to enhance the sharpness and texture of the image, improving the contrast of the image to highlight the surface area of the frost layer. This processing makes the subsequent conversion of the image into grayscale and binary images more accurate. Using MATLAB® to extract the coordinates of the frost layer surface pixels in the image, the frost surface curve is fitted in the coordinate system and integrated to obtain the frost layer side view area. The area is further divided by the x-axis range to calculate the average frost layer thickness δ_{avg}, and the arithmetic average of the absolute

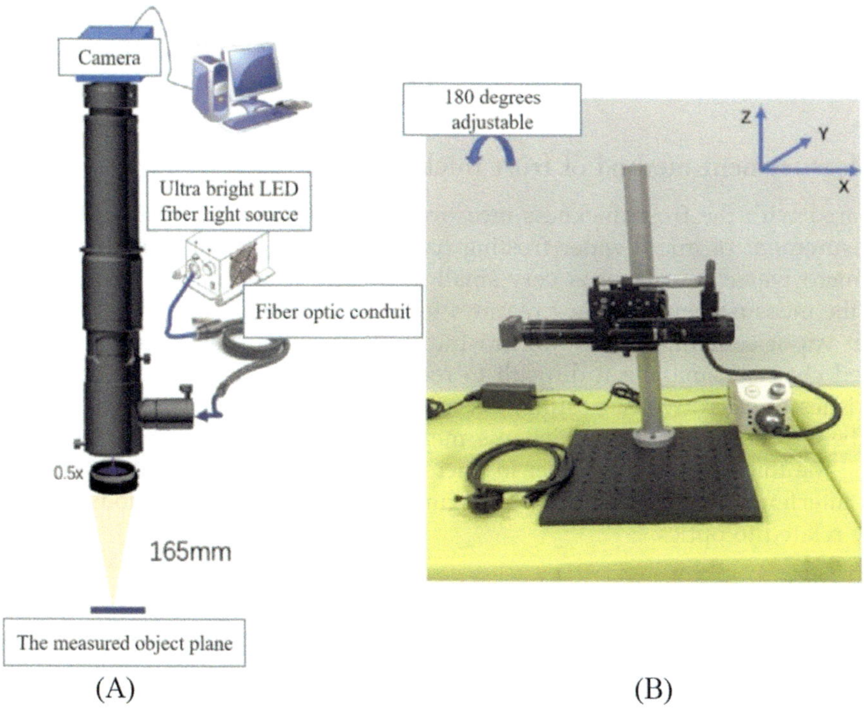

FIGURE 8.9 Microscopic measurement system [11].

deviations of the height of each point on the frost surface curve from δ_{avg} is calculated to obtain the average frost surface roughness R_a [11]:

$$(x_n - x_1)\delta_{avg} = \int_{x_1}^{x_n} y dx \tag{8.1}$$

$$nR_a = \sum_{i=1}^{n} |y_i - \delta_{avg}| \tag{8.2}$$

By using Eqs. (8.1) and (8.2), the calculation of the frost layer thickness and frost surface roughness is further realized. The total uncertainty of the frost layer thickness measurement is $\pm 2.80\%$. The microscopic imaging and computer image processing methods can obtain several frost layer properties such as thickness and surface roughness. However, since only images from specific angles are captured, the three-dimensional information of the frost layer cannot be obtained. As a result, when the environmental conditions of the experiment are unevenly distributed, resulting in significant differences in the overall morphology of the frost layer, accurate frost layer thickness cannot be obtained through this method.

When processing the images, it is necessary to judge the pixels on the surface of the frost layer based on the grayscale, and the critical grayscale value setting will directly affect the subsequent calculations. The determination of this value has not yet been systematically studied and standardized, so there is still room for improvement in this method in practical applications.

8.3.2 Measurement method of frost thickness

Compared with the frost thickness measurement of humid air frosting, the frost thickness measurement of micro water frosting has more requirements. First of all, the thickness of micro water frost layer is very small, generally in the micron or even nanometer level, so the measurement accuracy requires to be very high. Second, due to the extremely low water vapor content, micro water frosting experiments usually need to be carried out in a sealed chamber, making it difficult to realize in situ measurement. According to the characteristics of micro water frosting, the measurement methods based on optical principles are better choices for frost thickness measurement. In this section, two methods for measuring the thickness of frost layer in micro water are introduced, including multibeam thin film interference method and SPR imaging technology. These methods are directly or indirectly related to optics.

8.3.2.1 Multibeam thin film interference method

Moeller et al. used this method to measure the thickness of frost layer deposited in the high vacuum cryogenic experimental chamber in 2012 and 2013, respectively. During the experiment, the thickness of the frost layer deposited on the surface of horizontal optical gold in high vacuum (10^{-7} Torr) and cryogenic (60K) was measured. The source of water is the bound water carried by calcium sulfate dihydrate. By heating the effusion pool, the water moleculars are desorbed and led to the cold surface. The frost thickness is measured by the multibeam interference of the reflected He–Ne laser beam.

The measurement principle of multibeam thin film interference method is shown in Fig. 8.10. When the light wave passes through the film with a thickness of *d*, it will undergo multiple refraction and reflection, where *t* represents the electromagnetic vector

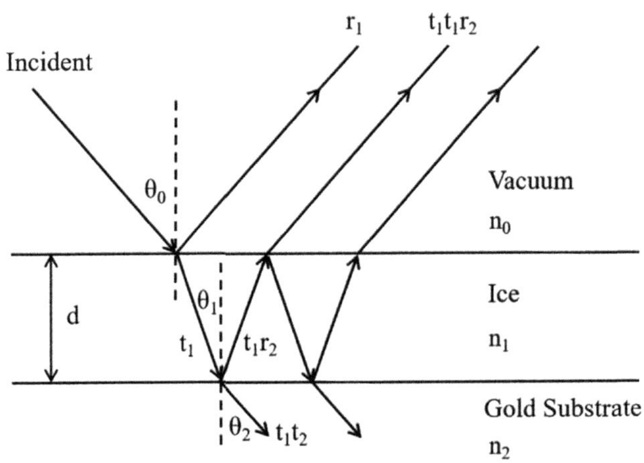

FIGURE 8.10 Optical path diagram of gold surface film.

of the transmission part of the light wave, and r represents the electromagnetic vector of the reflection part of the light wave [33,34].

For a single plane, the relative magnitude of the reflected and transmitted electric and magnetic vectors of s-polarized light perpendicular to the incident plane are given by the Fresnel equation:

$$r_{1s} = \frac{n_0 \cdot \cos\theta_0 - n_1 \cdot \cos\theta_1}{n_0 \cdot \cos\theta_0 + n_1 \cdot \cos\theta_1}$$

$$t_{1s} = \frac{2 \cdot n_0 \cdot \cos\theta_0}{n_0 \cdot \cos\theta_0 + n_1 \cdot \cos\theta_1} \tag{8.3}$$

In the formula, n_0 and n_1 are the complex refractive index of vacuum and thin film, respectively, r_1 and r_2 are the normalized amplitude of reflected beam and incident beam, respectively, and t_1 and t_2 are the normalized amplitude of transmitted beam, respectively. Combined with the complex Snell's law ($n_0\sin\theta_0 = n_1\sin\theta_1$), the phase change of light passing through the film is considered:

$$\delta_1 = 2\pi n_1 d_1 \cos\theta_1 \tag{8.4}$$

The reflectivity of the film light intensity caused by multibeam interference is presented in an infinite series:

$$R = r_1 + t_1 t_1^- r_2 e^{-2i\delta_1} - t_1 t_1^- r_1 r_2^2 e^{-4i\delta_1} + \cdot\cdot\cdot = r_1 + \frac{t_1 t_1^- r_2 e^{-2i\delta_1}}{1 + r_1 r_2 e^{-2i\delta_1}} \tag{8.5}$$

By using the law of conservation of energy, the above expression can be changed into:

$$R = \frac{r_1 + r_2 e^{-2i\delta_1}}{1 + r_1 r_2 e^{-2i\delta_1}} \tag{8.6}$$

From this, the reflectivity $|R|^2$ [34] can be obtained, which can also be calculated by the ratio of the intensity of the reflected part to the intensity of the transmitted part. Therefore after measuring the intensity of the reflected and transmitted light, the thickness of the film to be measured can be calculated by the following formula [34]:

$$|R|^2 = \left| \frac{r_1 + r_2 e^{-2i\delta_1}}{1 + r_1 r_2 e^{-2i\delta_1}} \right|^2 \tag{8.7}$$

The experimental platform for measuring frost thickness using multibeam thin film interferometry is located in the UHV chamber of AEDC at Arnold Air Force Base, Tennessee. The experimental platform is mainly composed of optical measurement system, QCM, UHV experimental chamber, and effusion pool [35]. The optical measurement system is composed of a He–Ne laser, a light-emitting diode group (PD1 and PD2), a lens group, a mirror group, a beam splitter, and a gold-plated mirror. The crystal oscillator in the QCM is Inficon's 6 MHz quartz crystal oscillator with a Maxtek shell. The UHV chamber is roughly cylindrical, with an outer diameter of about 0.6 m and a length of about 1 m. There are transparent windows at each end for optical measurement. The liquid pool is 76 mm in diameter and 57 mm in length. It is connected to the pressure sensor

(Baratron) through a copper tube to monitor the water vapor pressure. An electric shutter is placed at the front end of the effusion pool to start and stop liquid deposition [35].

The effusion pool is bolted to the internal guide rail of the UHV room. The He–Ne laser is mounted on an optical platform outside the chamber. The beam emitted by the He–Ne laser is directly reflected by the reflector group and received by PD1. The other one is incident on the gold-plated mirror through one of the end windows at a nearly vertical angle, and the reflected beam is received by PD2.

The water source in the UHV chamber is 3 Angstrom KA zeolite molecular sieve outside the chamber, which can release water into the chamber in the form of molecules. Water molecules enter the liquid pool through a 16 mm diameter pipe. The liquid pool is aligned upward at an angle of 30 degrees to a customized mounting seat.

The condition of high vacuum cryogenic frosting experiment is kept at the pressure of 10^{-7} Torr and the temperature of 30K–65K in the UHV chamber. In the experiment, calcium sulfate dihydrate ($CaSO_4$–$2H_2O$) was used as a desiccant for water storage. Since one molecule of calcium sulfate can carry two water molecules, it can store a large amount of water in low-mass samples (water accounts for about 20% of the total mass).

Optical data such as light intensity are obtained by two photodiodes (Thorlabs DET10A). The beam with a wavelength of 632.8 nm is emitted by the He–Ne laser He–Ne laser and is divided into two channels by the beam splitter at a ratio of 50:50. One of them is guided to PD1 as the value of laser output power; the other beam enters the chamber through the transparent window, is reflected by the gilded mirror after passing through the frost layer, and returns to PD2 through the chamber window. The digital data recorded include experimental time, the output of photodiode PD1, and PD2 during the experiment.

After three experiments in the above way, the collected data are analyzed, and the following results and conclusions can be obtained. Through the experiment, it is found that under the cryogenic and high vacuum condition, the frost layer can grow to about 20 μm after 13 hours. By comparing the experiment and the best single-period reflectivity curve, the visual observation of this data simplification method shows that the function form is very consistent. This corresponds to the error bar of ± 17 nm in the thickness measurement, which is almost an order of magnitude higher than the ± 120 nm in the initial experiment [36].

8.3.2.2 *Surface plasmon resonance imaging technology*

Since the first application of the SPR phenomenon for sensing almost two decades ago, this method has made great strides both in terms of instrumentation development and applications. SPR sensor technology has been commercialized, and SPR biosensors have become a central tool for characterizing and quantifying biomolecular interactions [37].

SPR is a physical optical phenomenon. By using the vanishing wave of total reflection at the glass interface, free electrons on the metal surface can be induced to produce surface plasmons. Under the condition that the incident angle or wavelength is a certain appropriate value, the frequency and wave number of the surface plasmon and the vanishing wave are equal, and the two will resonate. The incident light is absorbed, so that the reflected light energy decreases sharply, and the resonance peak appears on the reflection spectrum (i.e., the minimum reflection intensity). When the refractive index of the medium close to the surface of the metal film is different, the resonance peak position will be different [37].

Based on the Maxwell equation, the reflectivity of single-wavelength light incident on a thin layer can be predicted. The following Eq. 8.8 can theoretically predict the reflectivity of incident light on a four-layer system:

$$R = \frac{r_1[1+\exp(-2i\alpha_2 d_2)] + [r_1 r_2 + \exp(-2i\alpha_2 d_2)]r_3\exp(-2i\alpha_3 d_3)}{1 + r_1 r_2\exp(-2i\alpha_2 d_2) + [r_2 + r_1\exp(-2i\alpha_2 d_2)]r_3\exp(-2i\alpha_3 d_3)} \quad (8.8)$$

where r_i represents the reflection coefficient between layer i and layer $i+1$, α_i represents the wave vector of layer i, and d_i represents the thickness of layer i. Subscripts 1–4 represent glass (BK7), gold layer (50 nm), test medium, and bulk medium in the experimental conditions, respectively. The reference optical path is shown in Fig. 8.11. If the thickness of the test medium is greater than the measurable thickness range of the SPR signal, the medium itself appears to be infinite, and the configuration is simplified to a three-layer system, thereby simplifying Eq. (8.8) [38].

The incident angle of the light is adjusted so that the incident angle at the interface between the prism and the dielectric film is greater than the critical angle of total reflection of the interface, which satisfies the condition of excitation SPR. The reflectivity of the incident light is measured, and the thickness of the film to be measured is obtained by the relationship between the reflectivity and the film thickness.

The SPR imaging device consists of 50 nm gold-plated BK7 cover glass (Edmund Optics, Dove prism), CCD camera (JAI CM-141MCL), lens array (including polarizer), band-pass filter (600 ± 5 nm, Edmund Optics), LED light source (Edmund Optics, SugarCUBETM LED Illuminators), data recorder, prism, thermocouple, injection pump, and workstation. The specific structure is shown in Fig. 8.12.

After passing through the band-pass filter, the light from the LED light source is converted into a single-wavelength beam, and the subsequent collimating lens array converts the divergent beam into a parallel beam. Finally, the p-polarized collimated light enters the prism. The prism is composed of BK7 glass, and the corresponding refractive index at $\lambda = 600$ nm is 1.516. A cover glass (Platypus Tech., Au.0500.CSS) with a 50 nm gold thin layer is attached to the prism using a matching liquid (Norland, IML150). The matching liquid needs to have the same refractive index as the BK7 prism.

The CCD camera is connected to a magnification lens with a magnification of 7.9 times, and the pixel resolution is 810 nm. The CCD camera captures single-wavelength light reflected

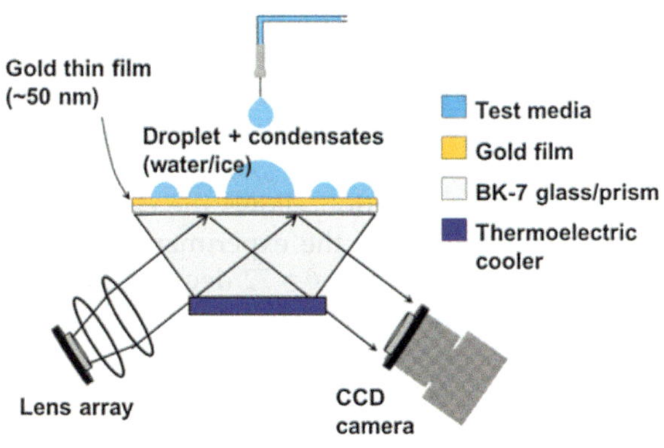

FIGURE 8.11 SPR measurement optical path diagram [38]. *SPR*, Surface plasmon resonance.

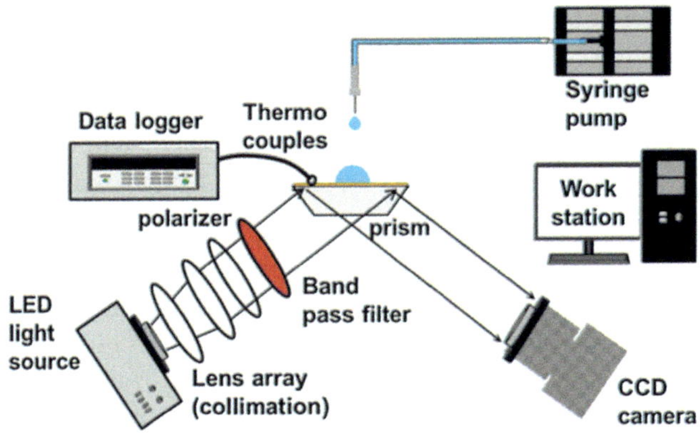

FIGURE 8.12 SPR measuring equipment [38]. *SPR*, Surface plasmon resonance.

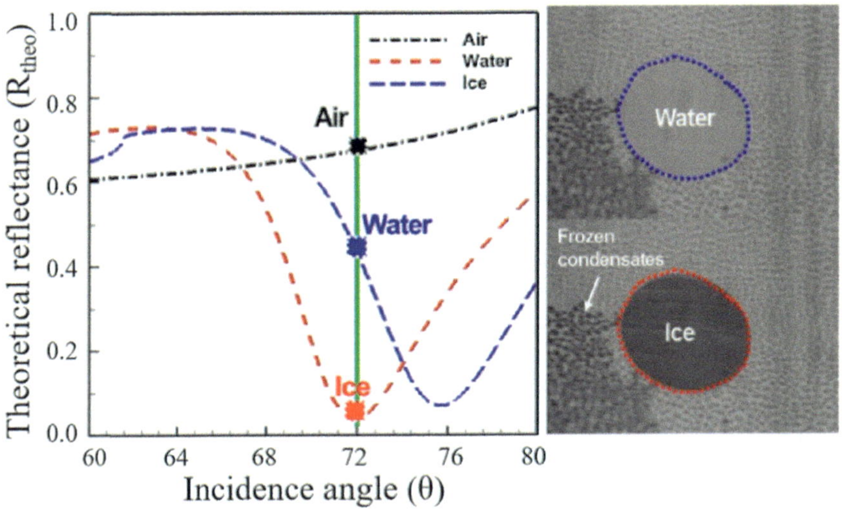

FIGURE 8.13 Experimental results obtained by SPR equipment [38]. *SPR*, Surface plasmon resonance.

from multiple thin layers at a speed of 0.2 frames per second. When the test medium is composed of air, water, or ice, the equation can be used to predict the reflectivity value based on the SPR angle. As shown in Fig. 8.13, the most sensitive SPR angle (ni ~ 1.3070, −4.0°C) to the refractive index of ice corresponds to an incident angle of 72 degrees. In Fig. 8.13, the bright, gray and dark areas correspond to air, water, and ice, respectively [38].

For quantitative analysis, calibration is required. The calibration allows the calculation of the corresponding reflectivity based on the pixel intensity of the experimental image. First, under the current experimental conditions of $\lambda = 600$ nm and $\theta = 72$ degrees, the factor S correlates the image intensity with the reflectivity as follows:

$$S = \frac{R_{a,\text{theo}} - R_{i,\text{theo}}}{I_{a,\text{exp,avg}} - I_{i,\text{exp,avg}}} \tag{8.9}$$

where S, R, and I represent the scale factor, reflectivity, and measured light intensity, respectively. The subscripts theo, exp, and μ represent the theoretical, experimental, and average values, respectively. The subscripts of a and i represent the air and ice phases. For the case of ice and air as the medium, the corresponding reflectivity curve is determined from Fig. 8.13. In addition, the measured intensities corresponding to air/ice are determined to be the mean values of 1879 (I_a, exp, avg) and 528 (I_i, exp, avg), and their standard deviations (σ) are calculated to be 112 (σ_a) and 41 (σ_i), respectively.

The reflectivity is obtained by multiplying the intensity by the scale factor (S). The reflectivity value always contains systematic errors, which may come from light source, angle control, or surface uniformity. The system error can be corrected by defining a calibration factor (C), which is the average of the experimental reflectivity (the product of the average of the measured intensity I and the scale factor S)—the theoretical reflectivity:

$$C = I_{i,\text{exp,avg}} \times S - R_{i,\text{theo}} \tag{8.10}$$

Finally, the specific pixel position and one calibration reflectivity can be obtained from the following relationship:

$$R(x, y, \tau)_{ca} = S \times I(x, y, \tau) - C \tag{8.11}$$

where x, y, and τ represent pixel coordinates and time. The subscript c represents the calibration value [38].

After the calibration is completed, the cooling and frosting experiment can be promoted, and the LED light source can be adjusted to make the incident light angle at the interface between the prism and the dielectric film greater than the critical angle of total reflection at the interface, so as to meet the conditions for exciting SPR, thus exciting SPR on the surface of the film to be tested. The reflectivity at different times was measured, and the thickness of the film to be measured was obtained by using the formula inversion calculation, and the characteristics of the frosting process were analyzed.

Jeong et al. used the SPR method to study the characteristics of the frost layer and detected the existence of nanoscale thin frost layer on the cold surface before the visible frost layer appeared. Good spatial resolution can be obtained by testing below the critical detection thickness δ_{cr} (270 nm). This shows that the current SPR system can successfully visualize the frosting process and can measure the thickness of nanoscale thin frost layer in situ. Fig. 8.14 is the relationship between the reflectivity and the frost thickness.

The frosting process was also analyzed. The frost formation process is usually divided into four different stages: initial stage, frost seed formation and growth stage, phase transition stage, and frost growth stage. These stages can be clearly observed in Fig. 8.15, and Fig. 8.16 shows the average calibration reflectivity of the reflected SPR signal over time. The black line a represents the area of the air and condensate region, and the red line b represents the fixed droplet region. Based on the reflectivity signal, the frost formation process can be divided into four stages. The initial stage corresponds to (1) in the reflectivity diagram (from 0 to 4 minutes), and the reflectivity remains unchanged because there is no phase transition and condensate formation. After 4 minutes, the thermoelectric cooler is started to cool the experimental samples, prisms, and gold-plated surfaces. Frost seed formation and growth stages correspond to the portion from 4 to 17.4 minutes (2). Since the liquid condensate is formed on the surface, and the condensate continues to grow and aggregate, a slow continuous change in reflectivity is observed.

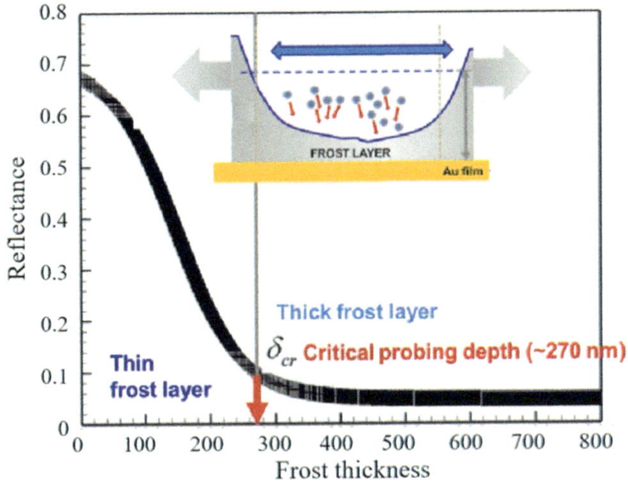

FIGURE 8.14 Relationship between reflectivity and frost thickness [38].

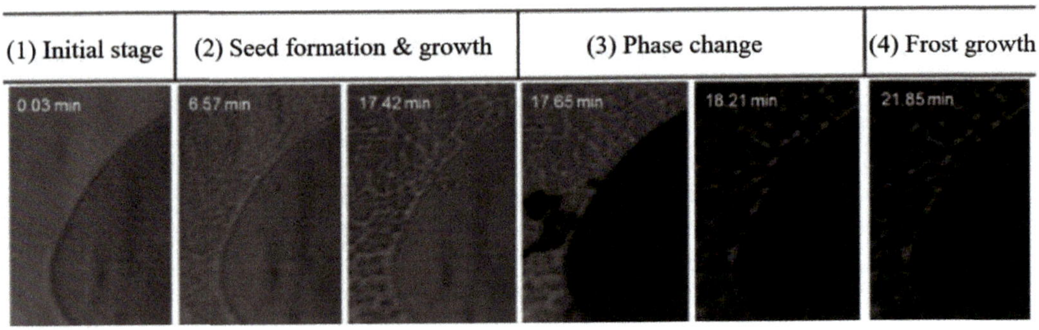

FIGURE 8.15 Frost growth stage [38].

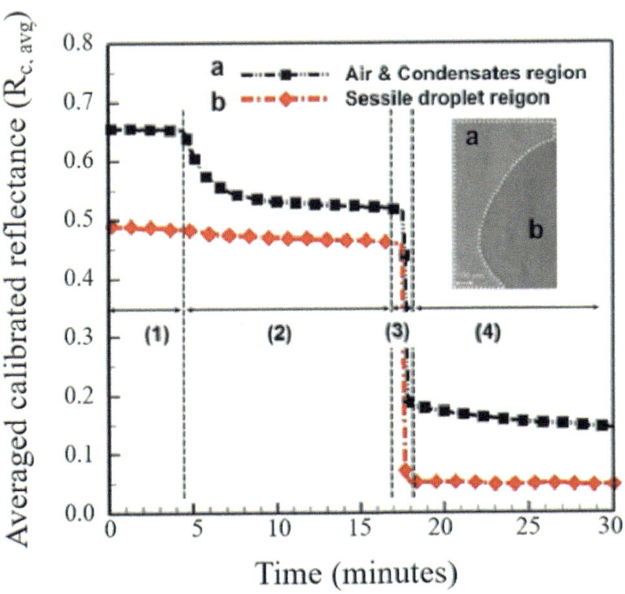

FIGURE 8.16 Variation trend and analysis of reflectivity [38].

In Fig. 8.15, (3) represents the phase transition stage. At this stage, the condensate liquid gradually freezes into a solid. This solid–liquid phase change is reflected in the rapid decrease of reflectivity detected. During the freezing process, the condensed liquid between the frozen regions plays a bridge-like role in frost propagation [39]. The frost propagation phenomenon ended in one minute, and the freezing time (τ_f) was defined as the end time of the significant change in reflectivity. In this experiment, $\tau_f = 18.21$ minutes. The last stage corresponds to the frost growth stage, as shown in (4). The water vapor in the air cools between the solid frost and the condensate, which contributes to the gradual increase of the frost thickness until it reaches a dynamic equilibrium state.

The SPR imaging method overcomes the traditional optical limitations, dynamically visualizes the nanoscale thin frost layer, and achieves the measurement of the microscale frost layer. In addition, through the spatial distribution of reflectivity related to frost thickness, it is shown that local information about frost thickness can be obtained by SPR imaging method rather than only the average value of the overall frost thickness. Collecting real-time reflectivity signals can grasp the frosting process and help to more intuitively reflect the characteristics of different frosting stages.

The potential of SPR in film feature extraction and monitoring of metal interface processes was recognized in the late 1970s. Up to now, SPR biosensor has become a key tool for characterizing and quantifying the interaction of biomolecules [40]. Therefore related devices are relatively mature. Detection and analysis of chemical and biochemical substances are required in many important fields, including medicine, environmental monitoring, biotechnology, drug and food monitoring. SPR sensor technology has application potential in these fields, and the research on micro water frosting is expected to become the next application field of this technology.

8.3.3 Measurement method for frost layer weight

The QCM is a precision detection instrument that uses the piezoelectric effect of quartz crystals to convert the mass of adsorbed substances and changes in liquid viscosity on the surface of the quartz crystal electrode into measurable frequency signals [41]. Due to its nanogram-level precision and advantages of small size and light weight of its core component, the quartz crystal resonator, the QCM has been widely used in various fields, such as electrochemistry, air hygiene quality testing, and biomolecular research. Professor Qiu Limin's research group at Zhejiang University proposed to introduce the QCM into the frosting experimental system to realize the mass measurement of micro water frosting, which is a further expansion of the application field of the QCM.

8.3.3.1 Measurement principle

In the 1880s physicists Pierre Curie and Jacques Curie first discovered the piezoelectric effect through experiments: applying external forces in a certain direction to a particular type of medium can cause it to undergo mechanical deformation. In this process, the internal structure of the medium changes due to the crystal structure, which causes the upper surface of the medium to produce two opposite electrical charges, called polarization. When the external force is removed, the phenomenon disappears, and the crystal returns to its original state without deformation. When the direction of the external force changes, the polarity of the equal charges on the upper and lower surfaces also changes, and when the magnitude of the external force changes, the density of the equal charges changes positively [42].

The Curie brothers further discovered in their experiments that applying an electric field in the polarization direction of the medium can cause it to undergo mechanical deformation. When the applied electric field disappears, this phenomenon also disappears, and the crystal returns to its original state without deformation. This phenomenon corresponds completely to the piezoelectric effect and is named the inverse piezoelectric effect, while the electric effect produced by external forces is renamed the direct piezoelectric effect. Fig. 8.17 shows a schematic diagram of the piezoelectric effect. Crystals that exhibit the piezoelectric effect are called piezoelectric crystals, and quartz crystals are a common type of piezoelectric crystal.

In the 1960s Sauerbery discovered the relationship between the deposited mass and the resonant frequency based on the piezoelectric effect, which formally opened the era of quartz crystal as a microbalance research method. The Sauerbery equation holds under the following three assumptions [42]:

1. The deposited material is a rigid medium and does not exhibit viscoelasticity.
2. The deposited material is uniformly distributed on the surface of the quartz crystal resonator with a random non-Gaussian distribution.
3. The mass of the deposited material is much smaller than the mass of the quartz crystal resonator itself, typically less than 2%.

The equation expressing the relationship between the deposited mass and resonant frequency is as follows:

$$\Delta f = -\Delta m \frac{2f_0^2}{A\sqrt{\mu_q \rho_q}} \tag{8.12}$$

where Δf represents the change in frequency, Δm represents the amount of material deposited on the surface of the quartz crystal resonator, f_0 represents the original resonant

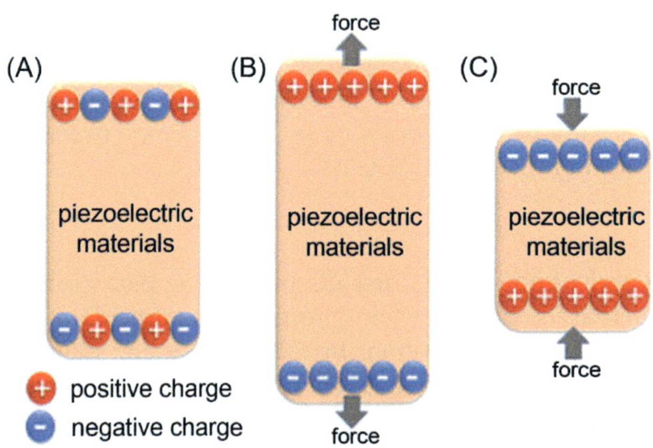

FIGURE 8.17 Principle of piezoelectric effect [43]. (A) Piezoelectric materials with no strain; and (B, C) Piezoelectric materials under different types of strain.

frequency of the quartz crystal resonator, A represents the surface area of the quartz crystal resonator, μ_q represents the elastic stiffness coefficient of the crystal, which is 2.947×10^{11} g/cm^3 • s^2, and ρ_q represents the density of the quartz crystal, which is 2.643 g/cm^3. The negative sign indicates that the resonant frequency decreases as the mass is deposited [42]. With the above data, it can be calculated that for every 1 nanogram of material deposited on the surface of the quartz crystal resonator, the fundamental frequency will decrease by approximately 9 Hz.

8.3.3.2 Single crystal vibration structure microbalance

Moeller et al. used single crystal vibration structure microbalance to measure the frosting rate in 2012 and 2013, respectively, and compared the results with those of frost layer thickness measurement. Under the assumption that the ice density is 0.934 g/cm^3, the mass area density measurements calculated from QCM and interferometer results are within 10% of the operating range of QCM.

Table 8.1 shows the results of cryogenic experiments conducted by Moeller et al. in 2012, each lasting approximately 5, 10, 1.5, and 13.3 hours, respectively. At the beginning of Experiment 1, it was found that the laser was severely misaligned, and therefore no useful optical data was obtained. Due to the lack of optical data, QCM data collection was terminated after 1 hour. During Experiment 2, 4.1 hours of optical data were collected, after which the experiment was terminated due to frost layer rupture. However, QCM data collection was stopped due to QCM failure after 1.3 hours. In Experiment 3, which lasted 1.5 hours, QCM and optical data were almost continuously collected throughout the entire duration, but the QCM produced anomalous readings near the end of the run. Due to QCM failure at the end of the third run, no data was collected using the QCM during Experiment 4 [13]. Based on the above, it can be concluded that the QCM in Experiments 2 and 3 failed at around 1.5 hours when the water content and environmental temperature were basically the same, and there is reason to suspect that this failure was due to the temporary failure of the oscillator in the cooling experiment of this study caused by reaching its range limit.

8.3.3.3 Twin crystal structure microbalance

Due to the sensitivity of quartz crystal resonators to environmental factors such as temperature, the Sauerbery equation needs to be corrected during actual measurements, with the addition of temperature and pressure correction factors. Generally, these correction factors are complex polynomials, and the coefficients and degrees may also change with temperature, which brings considerable difficulty to the measurement. To eliminate the

TABLE 8.1 Experimental results of the Tennessee Laboratory quartz crystal microbalance (QCM).

No.	Nominal duration (h)	Temperature (K)	QCM data acquisition time (h)
1	5	30	1
2	10	33	1.3
3	1.5	31	1.2
4	13.3	66	None

influence of environmental factors, Q. Hu et al. from Zhejiang University designed a new QCM based on the Twin-QCM structure, using two quartz crystal resonators as control and measurement groups, respectively, to perform cryogenic micromass measurements.

To achieve resonance in a quartz crystal resonator, two circuit elements are required: an oscillator circuit element and a receiver circuit element. The quartz crystal resonator is a good insulator and cannot output a current signal on its own. Therefore a layer of metal with a uniform thickness is deposited on the upper and lower surfaces of the resonator. The metal used should have excellent conductivity, be easy to obtain and process, and have stable properties. Generally, gold, silver, or chromium gold is used. However, silver is prone to oxidation when exposed to air and is difficult to remove after adsorbing substances, so quartz crystal resonators with gold electrodes are more preferred. When the crystal resonates, the resonant signal is fed back to the data receiver through the frequency signal transmitted by the metal layer on both ends, and the resonant frequency of the crystal under the electrical excitation can be obtained. Fig. 8.18 shows a complete schematic diagram of the metal-coated quartz crystal resonator and a deposition diagram. The gray part in the middle is the quartz crystal, the gold part is the metal coating, and the blue part in the right figure represents the deposited material [42].

To reduce the frequency deviation caused by environmental factors such as temperature, a set of cryogenic structural components for QCM was designed, as shown in Fig. 8.19. The main body is a cylindrical ceramic piece (high thermal conductivity ceramic made of aluminum nitride, with a thermal conductivity of 170 W/m K at room temperature). Two

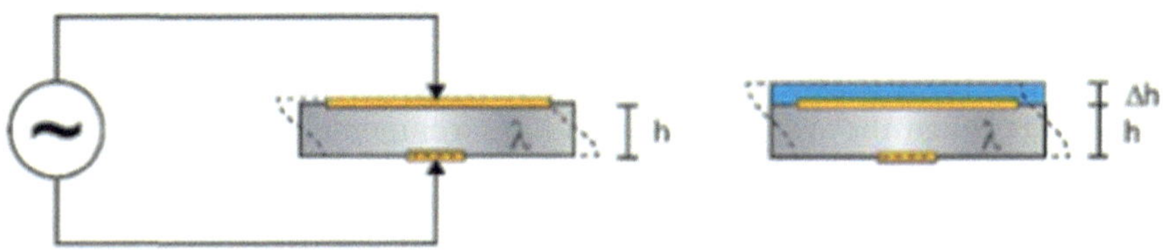

FIGURE 8.18 Schematic diagram of metal coating electrodes and deposition for quartz crystal resonator [42].

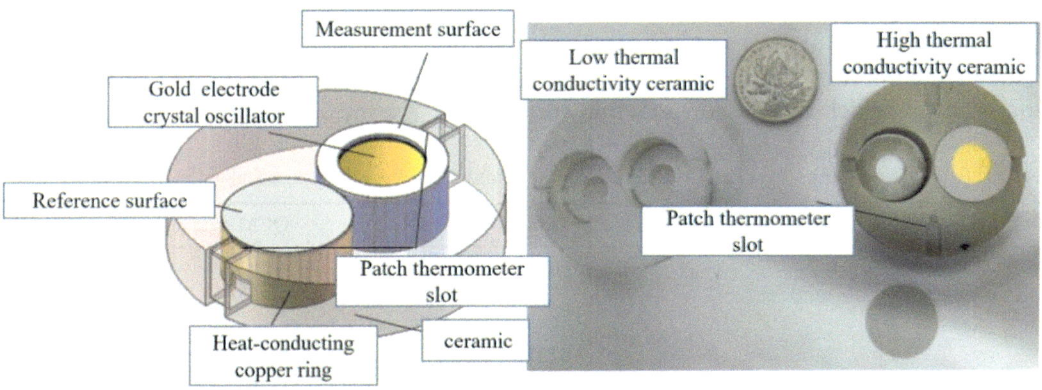

FIGURE 8.19 Design and physical diagram of quartz crystal microbalance [42].

symmetrical cylindrical grooves are used to support the support parts of the quartz crystal resonators and the copper heat-conducting rings used to fix the quartz crystal resonators (oxygen-free copper, with a thermal conductivity of 401 W/m K at room temperature). The support parts and copper heat-conducting rings also serve as conductors. A patch temperature sensor for measuring the surface temperature of the quartz crystal resonator can be glued to the copper heat-conducting ring using the small grooves on both sides of the ceramic piece. Due to the high thermal conductivity of the copper ring, its surface temperature can be approximately equal to the surface temperature of the resonator. The ceramic piece can serve as a bonding material to prevent the resistance connected directly to the resonator from being too large, which may cause the resonant frequency of the designed oscillator circuit to deviate.

In Fig. 8.19, the quartz crystal resonator on the left side of the ceramic piece is completely covered, and micro water vapor does not come into contact with the metal electrode on the resonator surface, so it is not affected by flow rate and water vapor condensation and is only affected by temperature due to heat dissipation. Its frequency is only related to temperature and serves as the "control surface." On the right side of the ceramic piece, the surface is exposed, and the quartz crystal resonator is partially in contact with the experimental chamber. The surface diameter of the 6 MHz crystal resonator electrode is 13.57 mm, which can be completely covered by the 10 mm inner diameter ceramic ring. At this time, the resonator is affected by flow rate, pressure, temperature, and changes in deposited mass caused by different concentrations, resulting in frequency changes, and serves as the "measurement surface."

Based on the STM32 microcontroller, the quartz crystal oscillator circuit was designed using the open-source design scheme openQCM. For data acquisition, the printed circuit board (PCB) has a built-in micro USB interface, which is connected to the USB port on the computer's chassis using a 3-m USB-MicroUSB cable, and data is collected using the open-source software openQCM (which can be searched and downloaded on GitHub).

The assembled QCM is fixed to the cold surface inside the main body of the cryogenic experimental chamber using cryogenic thermal adhesive, and high-purity nitrogen gas with a certain water content is introduced to begin cooling and data acquisition. Liquid nitrogen is used as the cooling source. When the cooling starts, the liquid nitrogen valve is opened to inject liquid nitrogen into the liquid nitrogen heat exchange pool on the cold surface to lower its temperature.

The data to be collected in the experiment includes the temperature and frequency of the oscillator for the reference surface and the measurement surface, as well as the water content of the gas. The frequency of the oscillator is collected using the openQCM software, while its temperature is collected by using the Keithley 2700 data acquisition card to collect the raw signal, which is then processed using a LabVIEW program on the computer.

Cooling and frosting experiments were carried out on a 6 MHz quartz crystal oscillator under conditions of high-purity nitrogen gas with water contents of 80, 40, 20, 2, and 0.1 ppm. During the experiment, it was found that the frequency change trend of the reference group and the measurement group was basically the same, and in the first four groups of experiments, the frequency of the reference group was always higher than that of the measurement group, which is consistent with the theory and demonstrates the feasibility of using two oscillators for cryogenic micromass measurement.

When the experiment is performed for a long time, the frequency of the quartz crystal oscillator will almost vertically drop to 0 Hz. The time of sudden frequency mutation is closely related to the water content, and the higher the water content, the shorter the time from the beginning of the experiment to the frequency mutation. This mutation is caused by the range of the quartz crystal oscillator itself, that is, when the frequency of the oscillator drops to a certain extent, it will lose the ability to measure mass, but it will return to normal after defrosting. This is consistent with the experimental results of Moeller et al.

8.3.3.4 Advantages and disadvantages of the method

The application of the QCM in cryogenic micromass measurement expands the range of mass measurement to the nano-kilogram level. This means that even in the nucleation stage of frost that cannot be captured by microscopic imaging techniques, the QCM can still provide mass data on the frost layer, which helps to advance experimental research on the nucleation stage of frost. However, temperature has a great influence on the crystal frequency. If QCM wants to be used in cryogenic environment, cryogenic micromass calibration is needed. In addition, the quartz crystal oscillator itself has a limited range, and the theoretical mass limit of a 6 MHz crystal oscillator is about 60 μg. Therefore QCM cannot be used in high water content or long time frosting experiments.

8.4 Conclusion

This chapter focuses on low-temperature micro water frosting and introduces the mechanism of micro water frosting and the measurement method of micro water frosting. The mechanism of frost formation mainly involves the optimum nucleation conditions of frost formation of micro water, the influence of water content on the growth characteristics of frost layer, and the deposition and sublimation of micro water. The summary of the measurement methods of micro water frosting is shown in Table 8.2.

At present, the research on micro water frosting is not mature, especially in the measurement method, which is relatively lacking. Although the thickness of the frost layer can be determined by photographing, optical measurement, and other methods, its volume is difficult to calculate due to the loose structure of the frost. Micromass measurement techniques, such as QCM, have not yet been calibrated at low temperatures. With the deepening of research on low-temperature wind tunnel and carbon dioxide capture, the problem of frost formation in low-temperature micro water will receive more and more attention, and there will be broad research prospects in the future.

TABLE 8.2 Summary of micro water frost measurement methods.

Detection quantity	Detection method	Accuracy level	Section
Frost layer morphology	Microscopic imaging method	10 μm	Section 8.3.1
Frost layer thickness	Multibeam thin film interference method	Error: 10^1 nm	Section 8.3.2
	Surface plasmon resonance technology	Nanometer level	Section 8.3.2
Frost layer weight	QCM mass measurement method	Nanogram level	Section 8.3.3

QCM, quartz crystal microbalance.

References

[1] S. Byun, H. Jeong, H. Son, et al., Frost formation from general-low to ultra-low temperatures: a review, International Journal of Heat and Mass Transfer 195 (2022) 123164.

[2] B. Gloss, R. Bruce, A solution to water vapor in the National Transonic Facility, 27th Aerospace Sciences Meeting, 1989.

[3] F. Payne, C. Bosetti, G. Gatlin, et al., Progress in flaps down flight Reynolds number testing techniques at the NTF; in: Proceedings of the 45th AIAA Aerospace Sciences Meeting and Exhibit, F, 2007.

[4] X. Yuan, C. Zhou, D. Yang, Pollution control technology for aerospace optical system, Spacecraft Environmental Engineering 168 (2012).

[5] H. Cao, S. Vanapalli, H.J. Holland, Clogging in micromachined Joule-Thomson coolers: mechanism and preventive measures, Applied Physics Letters 103 (2013) 34107.

[6] H. Jeong, S. Byun, D.R. Kim, et al., Frost growth mechanism and its behavior under ultra-low temperature conditions, International Journal of Heat and Mass Transfer 169 (2021) 120941.

[7] Z. Long, S. Mengjie, D. Shiming, et al., Frosting mechanism and behaviors on surfaces with simple geometries: a state-of-the-art literature review, Applied Thermal Engineering 215 (2022) 118984.

[8] J. Lee, K. Lee, The behavior of frost layer growth under conditions favorable for desublimation, International Journal of Heat and Mass Transfer 120 (2018) 259–266.

[9] H. Cao, S. Vanapalli, H.J. Holland, et al., Numerical analysis of clogging dynamics in micromachined Joul-Thomson coolers, International Journal of Refrigeration 81 (2017) 60–68.

[10] H. Ye, H. Jiang, W. Shu, et al., Analysis of frost formation mechanism and path under different parameters, in: 16th Cryogenic Engineering Conference, 2023.

[11] Y. Qi, Study on frosting mechanisms of extra-low content water vapor on a cryogenic surface. Hangzhou: Zhejiang University, 2023.

[12] Y. Qi, H. Ye, W. Shu, et al., Frost layer growth behavior study on a cryogenic surface with water vapor content <1000 ppmv, International Communications in Heat and Mass Transfer 148 (2023) 107027.

[13] T.M. Moeller, L.M. Smith, F.G. Collins, et al., Measurement of the accumulation of water ice on optical components in cryogenic vacuum environments, Optical Engineering (2012) 1–9.

[14] Q. Guo, D. Qie, J. Wang, et al., Experimental study of frost formation on vertical cryogenic surface under different pressures, Spacecraft Environmental Engineering 35 (2018) 283.

[15] H. Ye, Q. Hu, Y. Qi, et al., Real-time mass detection of trace water vapor frosting on a cryogenic surface based on Quartz, Applied Thermal Engineering (2023) 120785.

[16] M. Song, C. Dang, Review on the measurement and calculation of frost characteristics, International Journal of Heat and Mass Transfer 124 (2018) 586–614.

[17] Y. Hayashi, A. Aoki, S. Adachi, et al., Study of frost properties correlating with frost formation types, Journal of Heat Transfer-Transactions of the Asme 99 (1977) 239–245.

[18] Y. Tao, Characteristics of frost growth on a flat plate during the early growth period, ASHRAE Transactions (1993).

[19] S. Nath, J.B. Boreyko, On localized vapor pressure gradients governing condensation and frost phenomena, Langmuir: the ACS Journal of Surfaces and Colloids 32 (2016) 8350–8365.

[20] Y. Qi, X. Zhi, H. Zhang, et al., Mechanisms of trace water vapor frosting on a cryogenic surface in nitrogen gas flow, International Journal of Heat and Mass Transfer 169 (2021) 120898.

[21] Y. Qi, H. Ye, Q. Hu, et al., Study of frosting on a cryogenic surface from a perspective of thermal resistance analysis, Applied Thermal Engineering 208 (2022) 118158.

[22] K.Y. Qu, S. Komori, Y. Jiang, Local variation of frost layer thickness and morphology, International Journal of Thermal Sciences 45 (2006) 116–123.

[23] J. Dong, W. Wang, Z. Dong, Real-time measurement of frosting rate in evaporator, Laser Technology 24 (2000).

[24] J.H. Wu, G.A. Ouyang, P.X. Hou, et al., Experimental investigation of frost formation on a parallel flow evaporator, Applied Energy 88 (2011) 1549–1556.

[25] B. Xu, C. Zhang, Y. Wang, et al., Experimental investigation of the performance of microchannel heat exchangers with a new type of fin under wet and frosting conditions, Applied Thermal Engineering 89 (2015) 444–458.

[26] H. Kim, D. Kim, H. Jang, et al., Microscopic observation of frost behaviors at the early stage of frost formation on hydrophobic surfaces, International Journal of Heat and Mass Transfer 97 (2016) 861–867.

[27] T. Benitez, S.A. Sherif, Modeling spatial and temporal frost formation with distributed properties on a flat plate using the orthogonal collocation method, International Journal of Refrigeration-Revue Internationale Du Froid 76 (2017) 193–205.
[28] M. Kandula, Frost growth and densification on a flat surface in laminar flow with variable humidity, International Communications in Heat and Mass Transfer 39 (2012) 1030–1034.
[29] C.J.L. Hermes, R.O. Piucco, J.R. Barbosa, et al., A study of frost growth and densification on flat surfaces, Experimental Thermal and Fluid Science 33 (2009) 371–379.
[30] M. Kandula, Frost growth and densification in laminar flow over flat surfaces, International Journal of Heat and Mass Transfer 54 (2011) 3719–3731.
[31] W. Wang, Q.C. Guo, W.P. Lu, et al., A generalized simple model for predicting frost growth on cold flat plate, International Journal of Refrigeration-Revue Internationale Du Froid 35 (2012) 475–486.
[32] A. El Cheikh, A. Jacobi, A mathematical model for frost growth and densification on flat surfaces, International Journal of Heat and Mass Transfer 77 (2014) 604–611.
[33] Z. Wu, Z. Chen, Z. Huang, University Physics, Zhejiang University Press, 2001.
[34] M.S. Westley, G.A. Baratta, R.A. Baragiola, Density and index of refraction of water ice films vapor deposited at low temperatures, Journal of Chemical Physics 108 (1998).
[35] T.M. Moeller, L.M. Smith, F.G. Collins, et al., Further results in measuring water ice buildup on optical components in cryogenic vacuum chambers, Proc. of SPIE (2013) 8841.
[36] J.M. Labello, Water Ice Films in Cryogenic Vacuum Chambers, University of Tennessee, Knoxville, 2011.
[37] X. Zhao, H. Chen, X. Song, et al., Surface plasmon resonance sensor I: basic principles, Analytical Instrument 4 (2000) 1–8.
[38] C.H. Jeong, D.H. Shin, V. Konduru, et al., Quantitative measurements of nanoscale thin frost layers using surface plasmon resonance imaging, International Journal of Heat and Mass Transfer 124 (2018) 83–89.
[39] S. Kishi, K. Asaoka, S. Inoue, et al., Freezing propagation between condensed droplets: visualization and formulation of governing equation from ice bridge growth velocity, International Journal of Refrigeration 160 (2024) 1–8.
[40] J. Homola, S.S. Yee, G.N. Gauglitz, Surface plasmon resonance sensors: review, Sensors and Actuators B-Chemical 54 (1999) 3–15.
[41] G. Su, Design of Quartz crystal Microbalance Drive Circuit, Suzhou University of Technology, Suzhou, 2013.
[42] Q. Hu, Mass measurement method research of micro water vapor desublimation based on quartz crystal sensor under cryogenic conditions, Zhejiang University, Hangzhou, 2022.
[43] Z. Jiang, X. Tan, Y. Huang, Piezoelectric effect enhanced photocatalysis in environmental remediation: state-of-the-art techniques and future scenarios, Science of the Total Environment 806 (2022) 150924.

CHAPTER

9

Frost formation and growth on cold surfaces: a review on experimental studies

Yuan Wang, Xiaofei Yue and Duo Zhang

College of Aerospace Science and Engineering, National University of Defense Technology, Changsha, Hunan, P.R. China

9.1 Introduction

Frosting begins when moist air blows over a surface whose temperature is lower than the freezing temperature of the water and the dew point of moist air. Surface frosting is likely to take place in several engineering applications, from aviation to HVAC&R (heating, ventilation, air conditioning, and refrigerating), impacting not only equipment performance but also aspects of operation such as safety and cost. For instance, with the rapid development of aerospace propulsion technology, air-breathing combined cycle engines have drawn growing attention. High heat dissipation heat exchangers are fabricated in the precooled turbojet engine to deeply cool the incoming air for higher thrust and specific impulse. However, the atmospheric humidity is relatively high at a near-earth amplitude (usually <5 km), which will be encountered during the air-breathing mode of the engines. Water vapor in the deeply cooled air will unavoidably be condensed and frozen on the precooler external surfaces immediately. Consequent deleterious effects will occur such as increased pressure loss, obstructed flow passages, reduced heat transfer rate and heat transfer uniformity due to high thermal resistance, undesirable changes in material properties, etc. [1]. Frost formation during air precooling degrades the heat transfer efficiency and can result in catastrophic engine failure in the air-breathing mode [2]. Therefore it is of significant importance to understand the frost formation processes and develop anti-frosting techniques to ensure the high efficiency of the precooling systems.

Due to the complexity of frost formation, the existing literature review is dedicated to a particular aspect of frost. For example, some papers reviewed the frost in mechanism [3]

Frosting and Icing for Efficient Energy Use in Engineering Applications
DOI: https://doi.org/10.1016/B978-0-443-15495-9.00006-X

and calculation [4,5]. Some introduced defrosting methods in air sour heat pumps [6,7] and microchannel heat exchangers [8]. In this chapter, we hope to completely present the frost phenomenon research in experiments. The experimental investigations on frost formation and growth on cold surfaces under natural convection and forced convection will be surveyed. The influential factors on frost growth are concentrated on. Moreover, surface topographical textures and the corresponding impact on frosting are discussed particularly, showing its potential usage for antifrosting.

It is worth mentioning that technically, icing and frosting represent different types of solidification of water [9]. In the present review, we describe the solidification process of a single sessile droplet of water on cold surfaces as icing and the solidification process of water vapor in moist air as frosting. The surfaces showing antifrosting and antiicing characteristics are all called surfaces with icephobicity.

9.2 Criteria and classification of frost formation on cold surfaces

According to [9], the dew point of the air T_d and the surface temperature T_{suf} determine the onset of frost formation. The phase change criteria of water vapor can be written as follows:

$T_{suf} > T_d$	No vapor condense
$T_d > 0°C$, $0°C < T_{suf} < T_d$	Vapor condenses into liquid
$T_d > 0°C$, $T_{suf} < 0°C < T_d$	Condensed liquid freeze
$T_{suf} < T_d < 0°C$	Vapor changes into solid directly (desublimation)

When the temperature of the cold surface, which is lower than the freezing point, is below the dew point of the environment, frost formation will be initiated. When the dew point is higher than the freezing point of water, moisture will be condensed into liquid before freezing. On the other hand, water vapor will directly switch into the solid phase when the dew point is below the freezing point. The present study shows interest in the situation when water vapor condenses into liquid droplets and the droplets freeze into a solid ice/frost layer.

According to nucleation theory, the phase-changing process is driven by the Gibbs free energy difference between the saturated vapor state and the state of the phase-changing substance at the cold surface temperature, which can be considered a potential barrier for nucleation initiation. Therefore the frost formation process is closely related to the topographical texture of the substrate, which largely influences the surface energy barrier.

The frost formation and growth process can be classified into three main steps, namely, crystal growth period, frost layer growth period, and frost layer full growth period. This classification was first published by [10] and has been widely accepted and cited in the investigations on frost formation. The frost growth process has also been divided into two stages, including nucleation and crystal growth [11,12], and the early stage, or one-dimensional growth of ice column, and multidimensional growth of porous structure [13]. Based on theoretical analysis and experimental measurements, frost formation and growth have been found to be related to many environmental parameters such as the surface temperature, air velocity, air temperature, air humidity, surface structure, and additional particles [14], the impacts of which on the frost characteristics will be discussed in the following sessions.

9.3 Summary of experimental studies on frost formation and growth on cold surfaces

Frost nucleation commencement, frost thickness, and mass changing rates are usually acquired to evaluate the heat transfer performances in applications where frost accumulates. The study on frost accumulation on cold surfaces can be traced back to more than 80 years ago [15]. In the present review, literature published in 2000–17 are surveyed, the experimental parameters of which are summarized in Table 9.2. To quantitatively examine the frosting characteristics, various experimental methodologies have been employed. The frost nucleation commencement is mostly captured via visualization methods, for example, CCD and ESEM (environmental scanning electron microscope). There are several methods for frost layer thickness measurements, including CCD and microscopics [16–19], laser source measurement [20], micrometer measurement [21–25], and imaging processing technique [26,27]. Various ways of frost mass measurement methods have also been utilized, including simultaneous measurement using high precision balance or scales as the frost formation is proceeding [28–30], measurement after the frost has been scraped from the cold surface [24], and posttesting measurement by removing the frost-covered coil and measuring the increased weight [25,31].

Growing attention has been drawn to the frost formation and propagation mechanisms on cold surfaces with various topographical textures. However, the tested air velocities are mainly below 20 m/s. Besides, the incoming air temperatures are usually lower than 50°C. Very limited studies on frost growth characteristics on cold surfaces under high temperatures (hundreds of degrees Celsius) and high-speed (hundreds of meters per second) airflow conditions have been published so far. Therefore the experimental database is waiting to be enlarged before developing the antifrosting methods for precooled air-breathing combined cycle engines.

9.3.1 Frost formation on cold surfaces under natural convection

Even though airflow is commonly encountered on most occasions, it is still necessary to find out the frosting characteristics under natural convection conditions or when airflow velocity can be neglected, for example, in cryogenic transportation systems. Investigations have been conducted to study the influential factors of frost formation and growth when there is no externally forced airflow.

With the aid of a CCD camera and magnifier lens, the frost formation process was captured by [57], revealing a typical four-stage crystal pregrowth process, including vapor condensation, droplet coalescence, supercooled droplets freezing, and crystal nucleation on iced droplets. This crystal pregrowth process is then followed by the three-stage crystal growth process as described by [10].

Precise measurement of frost growth showed a linearly growing frost mass as frost accumulation is proceeding [21]. In the study of [21], the frost thickness was found to be a function of the product of time and the temperature difference between ambient air and the cold surface, which indicated the determining roles of ambient and surface temperatures during frost accumulation. For instance, the frost thickness growth rate increased sharply as the surface temperature decreased from −8.2°C to −20.5°C [58]. Higher thickness growth rate at lower surface temperatures was also captured by [26,27]. Nevertheless, the tested ambient temperature (20°C and 35°C) was evidenced to have minor effects on the frost density and frost surface temperature in the study of [26] on the natural frost formation over a cold horizontal cylinder. One can believe that the cold surface temperature is relatively more dominant compared with the ambient temperature during frost formation and growth processes.

The significant effect of air relative humidity on frost formation and growth has been highlighted in various studies in Table 9.1. Air relative humidity can serve as a boundary for frost formation, under a certain value of which the frost on a cold surface may not even happen. In the study of Liu et al. [58], frost deposition was not observed when humidity was below or at 30%. Generally, the thickness growth rate increased drastically with increasing air relative humidity. However, it was found that the impact of air humidity is more evident at the early stage of frost growth [26].

Moreover, among the influential factors of frosting, cold surface structure where the heat and mass transfer mainly occur during frost formation and growth, is of conspicuous importance. Based on the surface static contact angle, Sommers et al. [82] proposed a correlation formula to predict the average frost thickness. It was indicated that the surface structure was an important factor in affecting the frost formation process. The study of Li et al. [34] revealed that as the frost grows, the superhydrophobic structures of the surfaces are covered by the deposited ice particles, which weakens the ability of antifrosting. The effect of surface texture will be discussed in session 4 in detail.

In recent years, with the application of high-resolution microscopic observation and digital processing technology, scholars have focused on frost propagation and frost crystal growth characteristics at the early frosting stage. These studies can help deeply understand the frost formation mechanism. Yang et al. [55] stabilized the droplets in a Cassie state through topological design, exhibiting good hydrophobicity. The droplets formed during condensation on different topological surfaces exhibit a Cassie or Wenzel wetting state, while the frozen droplets formed after nucleation and freezing remain in a Cassie or Wenzel ice state. Lo et al. [42] observed the formation of Cassie ice on the hydrophobic surface through ESEM electron microscopy. It was found that an air pocket existed in the interface between the ice and surface, which leads to low adhesion strength, and the strength of ice adhesion is the key element in determining the antiicing ability of the surfaces.

The essence of frost propagation on cold surfaces is to realize interdroplet nucleation between condensed droplets. Since the saturated vapor pressure on the supercooled droplet is always greater than that on the ice under the constant wall temperature condition [83], the droplet will evaporate and transfer to the ice through mass diffusion to form an ice bridge. When the ice bridge grows and contacts the droplet, the droplet will nucleate and start to freeze [78], thus forming frost propagation through a chain reaction [84]. The frost propagation velocity is determined by the freezing of a single droplet and the formation of the ice

TABLE 9.1 Experimental investigations on frosting characteristics.

Authors	Year	Geometry	Flow condition	Examined influential parameters					
				Surface temperature (°C)	Air humidity	Air velocity	Air temperature (°C)	Surface treatment	Geometry
Lüer and Beer [59]	2000	Horizontal parallel plate channel	Laminar flow	−16 to 2	40%–90%	Re 700–2300	12–30	–	–
Fossa and Tanda [21]	2002	Plate in vertical channels	Natural convection	−13 to −4	31%–58%	–	26–28	–	–
Cheng and Shiu [60]	2002	Horizontal plate	Forced convection	−18 to 0	40%–70%	2.0–8.0 m/s	18–30	–	–
Lee and Ro [61]	2002	Vertical plate	Hydrodynamically and thermally developing flow	−20 to −10	Air humidity ratio 0.0030–0.0055	Re 1000–3000	5–20	–	–
Cheng and Wu [62]	2003	Flat surface	Atmospheric airflow	−13 to −2	40%–80%	2–13 m/s	20–35	–	–
Lee et al. [63]	2004	Flat surface	Forced convection	−28.4 to −11.6	40%–80%	Re 7000–17,000	10–20	√	–
Liu et al. [58]	2006	Plate	Natural convection	−20.5 to −8.2	30%–73%	–	15.4–21	√	–
Fossa and Tanda [64]	2006	Plate in vertical channels	Natural convection	−40 to −4	31%–85%	–	27	–	√
Kwon et al. [58]	2006	Horizontal plate	Laminar flow	–	–	1.5 m/s	2/1	–	–
Wu et al. [57]	2007	Horizontal plate	Natural convection	−20, 0	15%–85%	–	22	√	–
Liu et al. [65]	2007	Vertical plate	Natural convection	−12 to −7.1	51%–68%	–	20.4–24	√	–

(Continued)

TABLE 9.1 (Continued)

				Examined influential parameters					
Authors	**Year**	**Geometry**	**Flow condition**	**Surface temperature (°C)**	**Air humidity**	**Air velocity**	**Air temperature (°C)**	**Surface treatment**	**Geometry**
Liu et al. [66]	2008	Horizontal plate	Natural convection	−10.1	40%	–	18.4	√	–
Hermes et al. [28]	2009	Flat surface	Forced convection	−18 to −5	50%–80%	0.7–1.0 m/s	16–22	–	–
Fossa and Tanda [24]	2010	Plate in vertical channels	Natural convection	−40 to −4	31%–85%	–	27	–	√
Tahavvor and yaghoubi [26]	2010	Horizontal cylinder	Natural convection	−10 to 4	30%–60%	–	20–35	–	–
Barzanoni et al. [25]	2011	Horizontal cylinder	Forced convection	−22 to −16	Absolute humidity 0.005552–0.009319 kg/kg_{air}	Re 3300–11,800	21.5–24	–	–
Kandula [67]	2012	Flat surface	Laminar flow	−17.75	Continuously changing from 81% to 54%	1.77 m/s	21.95°C	–	–
Moallem et al. [29]	2012	Louvered fin microchannel heat exchanger	Forced convection	−4, −7, −9	Absolute humidity 0.0029–0.0041 kg/kg_{air}	0.7–1.5 m/s	1.7	–	–
Rahman and Jacobi [68]	2013	Microgrooved surface	Natural convection	−8, −18	30%–70%	–	22 ± 2	√	–
Jing et al. [69]	2013	Horizontal plate	Natural convection	−20	55%	–	10	√	–
Kim et al. [70]	2014	Folded-louvered-fin microchannel heat exchanger	Forced convection	Coolant temperature: −8.0, −5.5	40%–95%	2.5–5.0 m/s	−10 to 50	–	√

Qin et al. [30]	2014	Tube-fin air cooler	Forced convection	Evaporation temperature: −25, −20, −15, −10	70%, 80% and 90%	Fan motor frequency: 40, 50 Hz	6	√	√
Amini et al. [27]	2014	Fin-and-tube heat exchanger	Natural convection	Coolant temperature: −10, −15, −20	50%–70%	–	20–30	–	–
Y. Chen et al. [71]	2015	Flat surface	Natural convection	−9.6, −14	Absolute humidity 16.7 g/kg_{air}, and 17.6 g/kg_{air}	–	24.5–25.5	–	–
M.A. Rahman and A.M. Jacobi [72]	2015	Microgrooved metal surface	Natural convection	−8, −10, −13, −18	50%–90%	–	20	√	√
Liu et al. [73]	2016	Plate	Natural convection	−191.3°C to room temperature	35%–95%	–	16–30	–	–
Wu et al. [74]	2016	Finned surfaces	Forced convection	−19 to 10	80%	0.3–1.1 m/s	−2 to 7	√	√
Sommers et al. [75]	2016	Vertical plate	Natural convection	−4 to −11	60%, 80%	–	–	√	–
Kim et al. [76]	2016	Horizontal plate	Forced convection	−14 to −8	60%, 75%, 90%	2–4 m/s	4, 6, 8	√	–
Wu et al. [77]	2017	Grooved surfaces	Forced convection	−16 to −10	75%–85%	0.92–1.16 m/s	2	√	–
Zhao and Yang [78]	2017	Patterned surfaces	Natural convection	−5 to −30	40 ± 5%	–	20 ± 0.2	√	–
Ma et al. [79]	2018	Wavy plates	Forced convection	−5 to −20	85%	0.31–0.92 m/s	−5 to 2	–	√
Wang et al. [80]	2019	Plate	Natural convection	−7.4 to −21.3	39.5% ± 5.4%	–	26.7 ± 2.3	√	–

(Continued)

TABLE 9.1 (Continued)

Authors	Year	Geometry	Flow condition	Examined influential parameters					
				Surface temperature (°C)	Air humidity	Air velocity	Air temperature (°C)	Surface treatment	Geometry
Lee et al. [81]	2018	Plate	Forced convection	−30 to −50	30%−70%	0.75−2.25 m/s	8−12	–	–
Sommers et al. [82]	2018	Plate	Natural convection	−13 to −5	40%−80%	–	12	√	–
Hermes et al. [32]	2019	Plate	Forced convection	−20 to −10	50%−80%	1−2 m/s	5−16	√	–
Yonezawa et al. [33]	2019	Grooved surfaces	Natural convection	−20 to −30	50%−60%	–	25	√	√
Li et al. [34]	2019	Plate	Natural convection	−30 to −70	30.2%	–	20.1	√	–
Kim et al. [35]	2020	Plate	Forced convection	−10 to −20	50%−90%	1 m/s	15	–	–
Qin et al. [36]	2020	Plate	Natural convection	−5 to −26	5%−95%	–	25	–	–
Zhang et al. [37]	2020	Plate	Forced convection	−15 to −25	50%−75%	0.5−1.5 m/s	6−16	–	–
Koninck et al. [38]	2020	Cable	Natural convection	−10	8%	–	20−22	√	√
Chu et al. [39]	2020	Plate	Natural convection	−16 to −18	20%−30%	–	21−25	√	–
Yao et al. [40]	2020	Plate	Natural convection	−12	25%	–	23.5	√	√
Zhao et al. [41]	2021	Plate	Natural convection	−5	60%	–	22	√	√
Lo et al. [42]	2021	Plate	Natural convection	−25	–	–	25	√	–

Jeong et al. [43]	2021	Plate	Forced convection	−100 to −160	5.85–7.66 g/kg	1–3 m/s	15–25	–	–
Byun et al. [44]	2021	Plate	Forced convection	−100 to −160	30%–70%	1–3 m/s	15–25	√	–
Jeong et al. [45]	2022	Plate	Forced convection	−180	6.08–9.14 g/kg	3–5 m/s	15–25	–	–
Lei et al. [46]	2022	Plate	Natural convection	−10 to −30	59.1%	–	23.0	–	–
Yang et al. [47]	2022	Plate	Forced convection	−10	85%	3 m/s	15	√	√
Wei et al. [48]	2022	Microtubule	Forced convection	0 to −260	1.8, 6.4 g/kg	10 m/s	50	–	–
Song et al. [49]	2022	Plate	Natural convection	−10 to −30	53.8%–64.2%		23.4–24.6	–	–
David et al. [50]	2022	Plate	Natural convection	−20	80%	–	24		–
Wang et al. [51]	2023	Plate	Natural convection	−10	65%	–	12	–	√
Zhang et al. [52]	2023	Plate	Forced convection	−10	50%	0.5–2.5 m/s	20		–
Li et al. [53]	2023	Plate	Natural convection	−20	85%	–	25.3	√	–
Zhang et al. [54]	2023	Plate	Forced convection	−7.4 to −13.0	50%	1–6 m/s	20	√	–
Yang et al. [55]	2023	Plate	Natural convection	−8 to −12	50%	–	20	√	–
Zuo et al. [56]	2023	Plate	Natural convection	−8 to −25	55%	–	20	–	–

bridge. Experiments have verified that the ice bridge formation rate was constant, independent of the droplet radius [50]. Further research has found that interdroplet nucleation can be divided into two modes: "contact-nucleation" and "absorption-nucleation" [51]. Increasing the surface contact angle or reducing the contact angle hysteresis can increase the occurrence frequency of the "absorption-nucleation" mode, thus hindering frost propagation. Droplet jumping on superhydrophobic surfaces is also a key factor in affecting frost propagation [39]. The droplet jumping ability and frost propagation velocity can be evaluated quantitatively by analyzing the average distance and proximity between the adjacent droplets.

By imitating the micro-morphology on the leaf surface, the antifrosting surface with serrated structures on the millimeter scale was carried out [40]. Experimental research has found that even on the superhydrophilic surface, due to the lower water vapor pressure surrounding the ice, the condensed droplets at the serrated valley were evaporated, resulting in a frost-free zone in the valley. This frost-free zone can resist frost propagation and reduce frost coverage. Yonezawa et al. [33] designed the microscale lattice-patterned grooves on glass and copper. It was observed that the droplets at the groove valley evaporated during the frost crystal growth process, and the groove structure could effectively increase the porosity of the frost layer, thus inhibiting the growth of the frost layer. The study of Zhao et al. [41] had the same conclusion that superhydrophobic surfaces with groove structures could effectively intercept the formation of ice bridges and reduce frost coverage. Furthermore, due to the low ice adhesion strength, the melting ice drops would detach from the surface during the defrosting process. Koninck et al. [38] designed a cable with a ring groove array. By utilizing the moisture absorption characteristic of ice, a dry overlap zone was created between adjacent ice rings to prevent continuous frost accumulation on the cable. Zuo et al. [56] constructed ice walls to form a discontinuous frost layer on the surface, which reduced the frost coverage effectively.

It is generally believed by the present research that columnar and stacked stellar plate shapes are the primary frost crystal morphologies. By changing cold surface temperature and relative humidity, different frost crystal morphologies can be obtained, such as needle-like and dendritic [46,49,80]. Wang et al. [80] obtained the frost crystal growth rate by analyzing a single frost crystal growth. It was reported that the frost crystal growth rate increases at the early stage and is reduced after reaching a maximum value, which is related to the balance of the sensible heat removal rate and latent heat release rate. Lei et al. [46] and Song et al. [49] revealed that the time of the dendrite touch stage is shorter compared with that of the reversed melting stage, and the periodic melting of frost crystals causes fluctuations in surface roughness and frost density of the frost layer.

In addition to frost suppression and prevention, the study of frost formation can also be applied to the self-cleaning of the surface. It was found that the superhydrophobic surface [53] had a self-cleaning function during the defrosting process. The dust particles could be wrapped in the curly frost layer and fall together with the defrosting ice–water mixture. The dust removal rate can reach 32.7%. The frost formation process can also be affected by the physical field. Qin et al. [36] studied the effect of a magnetic field on frost coverage. It was reported that as the magnetic flux density increased, frost coverage slowly decreased, indicating that the presence of a magnetic field would reduce frost formation.

Additionally, the effects of experimental parameters are closely related to the cold surface geometry. Fossa and Tanda [24,64] are continuously dedicated to investigating the frost characteristics, including the local frost thickness, frost surface temperature, air-to-frost heat flux, and the frost mass in vertical channels under natural convection. Rectangular plexiglass channels with variable cross-sectional depth were utilized. It was first pointed out that the heat/mass transfer features were similar to those on the vertical cold plates. However, the impact of channel depth on the frost formation has been emphasized, and the channel confinement was found to remarkably alter the frosting process. Thus the correlations developed for frost growth on isolated plates should not be applied to the cooling surfaces in narrow channels. Moreover, the frost formation on a fin-tube heat exchanger was different from that of the flat surface or in narrow channels [27]. Different frost formation stages were believed to be attributed to the heat transfer characteristics of the heat exchanger. Besides, it is worth mentioning that there exist some coupling impacts among different influential factors of frost formation. Tahavvor and Yaghoubi [26] described that the frost retardation effect of surface structure is related to surface temperature and air humidity.

Among the reviewed literature, there are limited studies on frost formation under natural convection conditions, a part of which specifically focused on the influence of surface structure. Few studies have reported the visualization results of the frost growth processes, which calls for more high-quality visualization observations in future studies. Nevertheless, as we are more interested in the air precooling system, more attention is paid to the frost formation processes under forced convection conditions.

9.3.2 Frost accumulation on cold surfaces under forced convection

Frost formation and deposition under forced convection are the major interests in the open literature as airflow is frequently encountered during the cooling processes in various applications such as air precoolers of combined cycle engines, airplane airfoils, air source heat pumps, and so on. Apparently, the influence of airflow velocity is emphasized under forced convection conditions. Thus the effect of airflow should be considered when exploring the frosting mechanisms and building comprehensive models of frost accumulation.

The frost layer exhibits different characteristics under forced convection compared with those obtained under natural convection conditions. Different types of frost patterns were detected. A three-stage frost growth (Frost pattern I) was observed according to the frost distribution streamwise [59], including a frost thickness-reducing stage, a constant thickness stage, and a thickness rerising stage (or the frost layer fully growth period). Similar frost thickness distribution was reported by [60]. Specifically, the early stage frost accumulation process on a cold flat surface has been observed [62]. Three stages of frost growth were identified (Flow pattern II), including a relatively short period of crystal growth, a frost layer growth period, and a frost layer full growth period when air velocity was elevated. Besides, the leading and trailing edges induced by the airflow were captured, which were one of the differences of the frost layer between natural and forced convections. Due to the dual effects of airflow and gravity, the inclination of the cold surface can change the trajectory of scattered frost particles. It was indicated that the increase in the inclination angle could lead to an increase in frost thickness and a decrease in frost density [35]. Zhang et al. [37] used digital induction tracking laser measurement technology to measure the surface roughness of the frost layer. It was found that the airflow could promote the evolution of the frost crystal morphology from feather shape to needle shape, thereby affecting the surface roughness of the frost layer.

Concerning the influential parameters of frost growth, the impact of air temperature on the frost layer thickness was pointed out to rely on the air velocity [60]. This agrees with the conclusion drawn in natural convection where the coupling effects of various parameters exist. The effects of air humidity and surface temperature were more considerable than those of air velocity and temperature [61]. It was explained that the increase in air humidity and the decrease in cold surface temperature would largely increase the concentration of driving force at the frost and air boundary. Consequently, the frost growth process was intensified. The increasing trend of frost thickness with increasing air humidity and decreasing plate temperature was also captured in Ref. [63]. However, this differs from [59,60], which emphasized the relatively more important influence of air velocity. The decreasing surface temperature was found to cause an increased frost thickness [62]. However, the frost thickness was slightly changed by the increased air temperature. Similarly, cold plate surface temperature showed the most conspicuous impact on frost growth rate [28]. Different flow conditions are possibly attributed to the discrepancy. Further experiments are needed to comprehensively analyze the impact of each influential factor on frost formation and growth. The frost thickness was also found to be a decreasing function of increasing surface temperature [25]. Hermes et al. [32] conducted a series of experiments to analyze the influence proportion of different factors on frost growth and obtain the correlation equation of frost average thickness. It was revealed that the proportion of cold surface temperature and supersaturation were significantly larger than other factors.

In addition, as the air velocity rose from 0.6 to 1.1 m/s, the frost growth promotion and thickness augmentation were captured. The increase in airflow velocity leads to an increase in phase change driving force and water vapor flux, thus promoting frost growth [79]. However, the impact of flow velocity was insignificant in the experiments of [28], which unveiled the different frost growth mechanisms between flat surface and cylinder. Interestingly, the effects of air relative humidity and absolute humidity on frost growth were just the opposite due to the opposite influences on the humidity driving potential near the cold surface. Actually, it was realized that the frost thickness along the cylinder requires more detailed measurement due to the nonuniformity of the frost layer. Wei et al. [48] conducted an experimental study on liquid nitrogen low-temperature heat transfer in a heat exchanger microchannel under airflow velocity of 10 m/s. The results showed that high humidity and low surface temperature would bring greater pressure loss. Zhang et al. [54] reveal the influence of airflow velocity on the frost formation process through the perspective of droplet freezing. It was found that increasing airflow velocity causes the droplet freezing to transit from peak mode to peakless mode.

When the supercooling temperature reaches a lower value, the water vapor in the air will directly condense into frost, which is different from the process of condensation and freezing to form frost. Lee et al. [81] pointed out that no obvious condensation phenomenon was observed in the frosting early stage when the cold surface temperature was below −30°C. The feather-shaped frost crystals formed on the surface could maintain one-dimensional growth. The decrease in surface temperature (−30°C to −50°C) had no significant effect on the frost accumulation, and the increased airflow velocity increased frost density without increasing frost thickness. The study of Jeong et al. [43] revealed that compared to ordinary cold conditions, the frost formation mechanism under ultralow temperature (−100°C to −160°C) conditions was completely different. There was a turning point in the average frost layer thickness over time. The appearance of this point indicated a transition in the frost formation mechanism. Byun et al. [44] reveal that under ultralow temperature (−100°C to −160°C) conditions, superhydrophobic surfaces have no obvious advantages on antifrosting. They also proposed functions to predict the frost thickness and density concerning Reynolds number, Fourier number, absolute humidity, and dimensionless temperature, and the maximum error of the prediction is within 6% [85]. Jeong et al. [45] observed the frost formation behavior on ultralow temperature (−180°C) surfaces under intense forced convective conditions and divided the frosting process into frost film formation period, seeding period, seed growth period, and coalescence period.

Most of the studies in open literature were conducted at constant humidity during each test run. However, Kandula [67] focused on the effect of variable humidity, while frost is accumulating. The surface roughness of the flat aluminum surface ranged from 0.8 to 3.2 μm. It is worth mentioning that it is of great necessity to provide the surface roughness condition because the surface wetting characteristics are closely related to the frost formation processes (as discussed in session 4). It was argued that there is a joint relationship between relative humidity, frost thickness, and density. So far it was confirmed that the frost growth model for constant relative humidity [86] can be applied to variable humidity conditions. Aiming at the air-precooling applications, more efforts are needed to enlarge the database of frost growth in varying humidity.

The impacts of frost growth influential factors show some differences among flat surface, cold cylinder, and fin-tube heat exchangers. Moallem et al. [29] probed into the dominant impact factors of the frosting time and growth rate on a louvered microchannel heat exchanger coil. Actually, it is debatable to consider the heat exchanger channels as microscale because the cross-sectional dimension of the coolant tubes is 5 mm × 1 mm. It was argued that the frost growth rate was primarily affected by the fin surface temperature and air humidity, while the influence of air velocity was less dominant. Zhang et al. [52] pointed out that the freezing duration of the condensate droplet under the edge effect is significantly shorter than that without the edge effect.

The geometry of the heat exchangers also plays an important role in the frosting processes. Frost accumulation in folded-louvered-fin microchannel heat exchangers was studied by [70]. The frost uniformity on the fins, which was influential to the heat exchanger thermal performances, was enhanced by larger fin pitch and smaller coil depth. Additionally, the effects of fin space and evaporating temperature on frost characteristics were demonstrated by [30]. Increasing fin space and airflow rate hindered the commencement of frosting, which could be attributed to the increased surface temperature of the tube and fins, resulting in a smaller potential barrier and relayed frosting occurrence.

Generally, the influence of air relative humidity is usually exhibited on the change of temperature difference between dew point and surface temperature and the corresponding change of mass transfer rate and frost density. The discrepancies among different investigations on the influences of air temperature and velocity on frost thickness should be probed extensively. The change in the frost density should be considered when trying to explain the trends. The reviewed investigations on frost formation characteristics are mainly encountered in regular heat exchangers such as air conditioners, refrigerators, and heat pumps. The conclusions of the influential factors on frost formation and the methods for frost control will be more instructive in the design of frosting countermeasures in turbojet air precooling systems by widening the experimental condition ranges. The combination of active and passive frost prevention technology is the future development direction [47].

9.4 Hydrophobicity versus icephobicity: effects of surface topographical textures on frost growth

Since icing/frosting has caused serious problems and become a significant issue in many industrial fields. Many of the antiicing or antifrosting methods and materials have been developed [87–90]. Current de-icing techniques, such as spraying of snow-melting agents, may lead to an excess of the threshold of chronic toxicity for freshwater organisms, negatively affecting the environment [87]. The method of electric heating or mechanical force tends to be low-efficiency and of high energy consumption [91]. Alternatively, as one of the passive antiicing techniques, surface coating is more economical and environment-friendly without consuming external energy to prevent or remove frost accumulation. Surface hydrophobicity is crucial for frost nucleation and frost deposition mainly because of the change of surface energy compared with regular uncoated surfaces. Fig. 9.1 is the

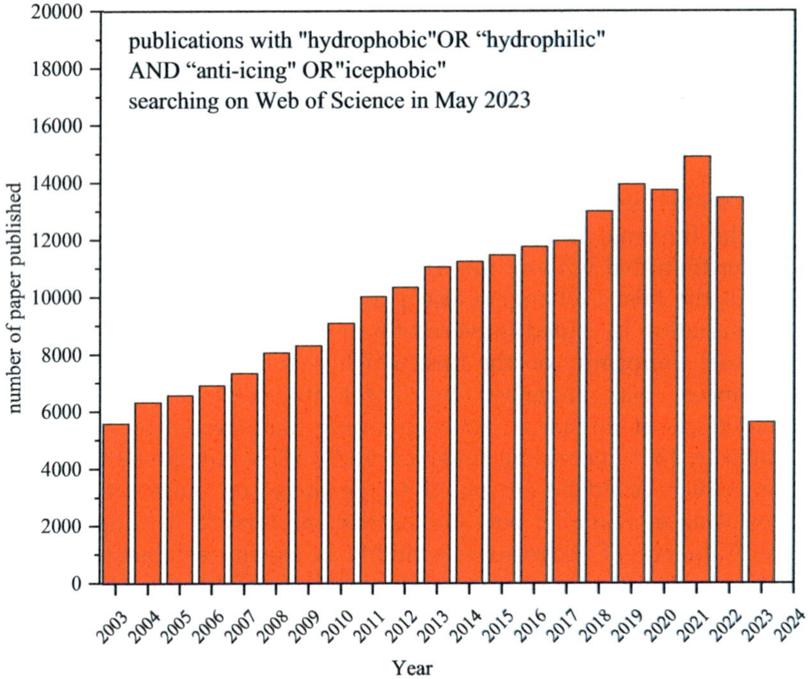

FIGURE 9.1 Recent search results of the publications related to hydrophobic/hydrophilic and antiicing or icephobic in Web of Science.

recent search results of the publications related to hydrophobic/hydrophilic and antiicing or icephobic in Web of Science. Increasing attention has been paid to the icephobicity/anti-icing effects of hydrophobic/hydrophilic surfaces.

Controversial conclusions were published on the frost-retarding effects of different surface conditions. Some claimed delayed frosting on hydrophilic surfaces [29,30,42,77,92], while some described frost retardation on hydrophobic surfaces [1,93–96] the possible reasons of which include frost crystal propagation restriction, reduced water vapor diffusion flux [77], reduced contact area between condensed droplets and surface [69,97], droplet coalescence [98,99], and mostly, larger energy barrier [65,69,100–102].

However, the surface hydrophobicity/hydrophilicity sometimes showed no remarkable impact on frost formation. The icephobicity was not necessarily maintained on the hydro-/superhydrophobic surfaces [24,103–105] or merely affected the initial stage of frost accumulation [65,106]. Meanwhile, the frost growth was either enhanced or inhibited on different surface structures. Finding out the relation between icephobicity and hydrophobicity/hydrophilicity is still not a closed topic.

Table 9.2 shows some of the surveyed research studies particularly on the effect of surface structure and wettability on antifrosting characteristics. Typical topographical textures are presented with corresponding research remarks.

There are several types of typical surface textures developed to explore the icephobicity in the open literature, including surfaces with postarray [44,108], aperiodic surface structure fabricated via chemical etching [123], coated with nanopowers [104,105,124], and the combination of different surface treatments with hybrid structural characteristics.

Unfortunately, lots of studies on the effect of surface treatment on frosting failed to provide the surface texture images, which therefore are not listed in Table 9.2. For a fin-tube heat exchanger fabricated by [103], hydrophilic surface and hydrophobic surface with contact angles of 12° and 124° showed weak impact on the thermal and the frosting performances. What is worse, the surface treatments even exhibited frost acceleration rather than retardation. Lee et al. [63] proposed frost maps according to the frost crystal structure. Lower frost thickness and higher frost density were obtained on the hydrophilic surface, while frost structure showed substantial similarity on both of the surfaces.

The frost retardation effect of surface hydrophobicity was sometimes only detected at the onset of frost incipient [65], after which the frost thickness growth rate was hardly restrained by surface structure treatment. Given the typical frost growth stages [10,11,13,125], it is understandable that the surface structure is dominant only during the early stage of frost growth where the crystal growth is rigidly controlled by surface energy. Varanasi et al. [108] monitored the dynamic frost formation process on a superhydrophobic surface with the aid of an ESEM. High-quality visualization results of the frost accumulating on the microscale posts were captured, which clearly demonstrated that the frost appears on the surface textures without any special preference, revealing that the superhydrophobicity showed no outstanding icephobic performances from practical point of view. Similar trends were found by [29], in the study of which five surface coatings were applied on the microchannel heat exchanger, including uncoated, hydrophilic, hydrophobic, and highly hydrophobic surfaces, the contact angle range of which was less than 5° to 105°. Frosting time and growth rate were hardly changed by different surface hydrophobicity. Only the hydrophilic surface showed slightly longer frosting time and lower pressure drop.

TABLE 9.2 Experimental studies on the effects of topographical textures on antifrosting characteristics.

Authors	Year	Surface treatment	Contact angle	Typical topographical feature	Surface icephobicity
Wang et al. [101]	2007	Copper surfaces coated with nano-sized calcium carbonate particles	155°	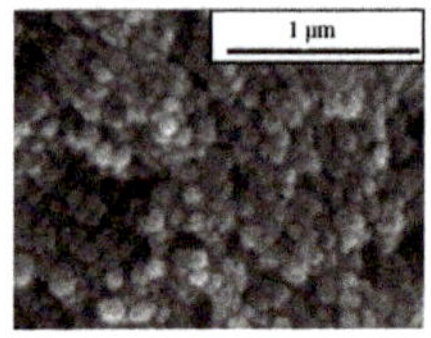	The frost formation on the tested superhydrophobic surface was greatly retarded compared with that on bare copper surface
Liu et al. [66]	2008	Silicon substrate fabricated with hydrophobic a-C films, and a further fluorinated process with CF_4 plasma; plain copper plate	162°, 72° (for reference)	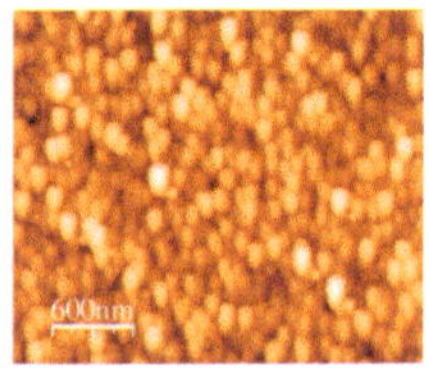	For the superhydrophobic surface, a 55 min delay of frost retardation was captured, and an obvious depression of frost growth was found
Mishchenko et al. [107]	2010	Flat aluminum (hydrophilic), smooth fluorinated Si (hydrophobic) and microstructures fluorinated Si (superhydrophobic)	0–5°, 115°, 145–170°	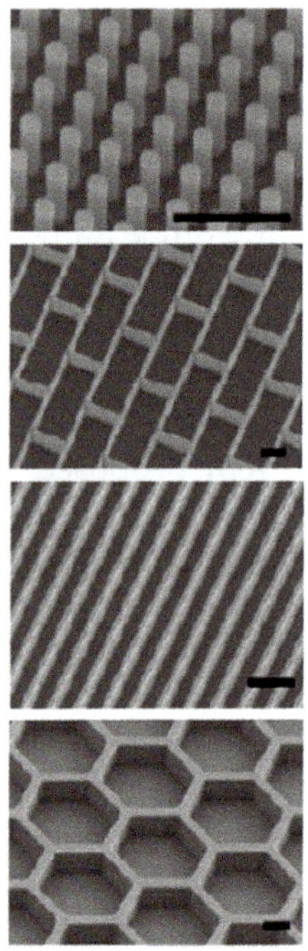 scale bars: 10 μm	Highly ordered superhydrophobic surfaces could be completely ice free when surface temperature was above −25°C to −30°C, beyond which the ice could be easily removed. The fabricated surface textures were promising candidates for antiicing.

(Continued)

TABLE 9.2 (Continued)

Authors	Year	Surface treatment	Contact angle	Typical topographical feature	Surface icephobicity
Varanasi et al. [108]	2010	Surface with hydrophobic posts array which was coated with a hydrophobic layer to achieve superhydrophobicity	145°–163°		Frost appeared on all available surfaces of the test samples and deteriorated the surface hydrophobicity. The icephobic properties were weaken as frost forming on the superhydrophobic surfaces due to increased ice adhesion strength. Superhydrophobicity showed no outstanding icephobic performances from practical point of view.
Farhadi et al. [109]	2011	Aluminum substrates modified via various techniques including spin-coating and etching.	152.4 ± 2.6°, 153.2 ± 2.4°, 155.1 ± 1.8° 154.8 ± 2.1°	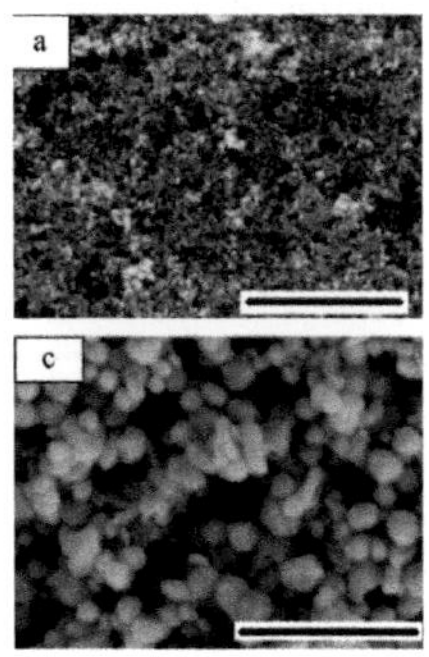	Lower ice adhesion strengths were detected on the superhydrophobic surfaces than that on the polished aluminum. However, antiicing efficiency is negatively affected by the air humidity. The tested surface structures were not suggested for antiicing in humid environments.
Huang et al. [110]	2012	Copper surfaces treated via chemical etching and fluorination modification	97.2°–156.2°	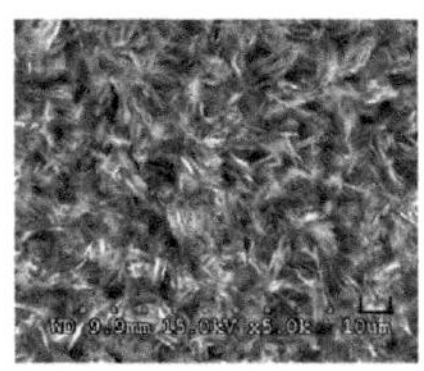	Though the droplet freezing time increases with increasing surface contact angle, the frost growth is accelerated. Thus the tested superhydrophobic surfaces showed poor icephobicity.

(Continued)

TABLE 9.2 (Continued)

Authors	Year	Surface treatment	Contact angle	Typical topographical feature	Surface icephobicity
Wang et al. [111]	2012	Superhydrophobic surface with a nano-fluorocarbon film	163.01°, 84.24° (plain copper surface for reference)	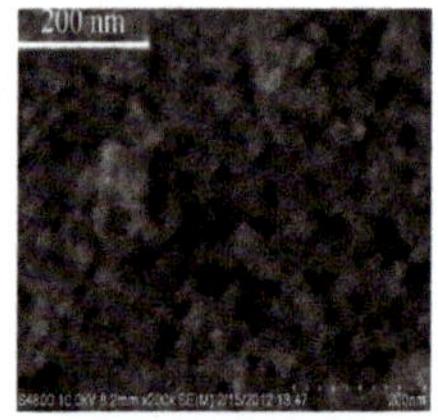	For the superhydrophobic surface, the onset of droplet icing was evidently delayed. Besides, the entire icing period of a droplet was 120 s longer than that on the plain copper substrate.
Rahman and Jacobi [68]	2013	Microgrooved brass surfaces with groove depth range 27–122 μm, pillar width 26∼187μm and groove width 130μm	125°–149°	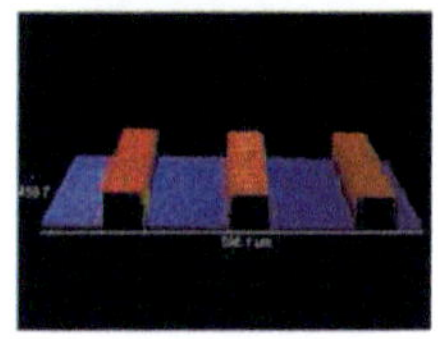	Microgrooved surface with the largest groove depth and pillar width showed the best performance in frost growth suppression and frost densification
Jing et al. [69]	2013	Aluminum plates with various treatments using anodization, HDFS coating and nanoporous replication	0°–169°	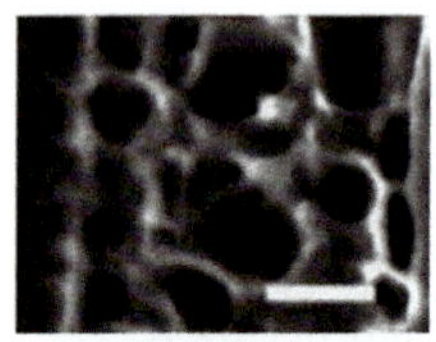 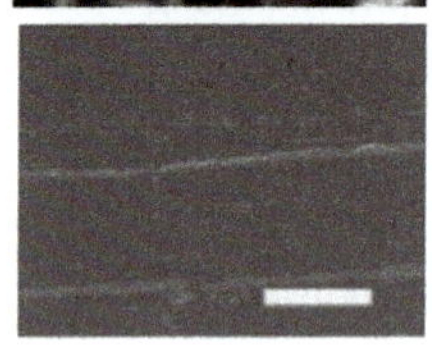 HDFS coating 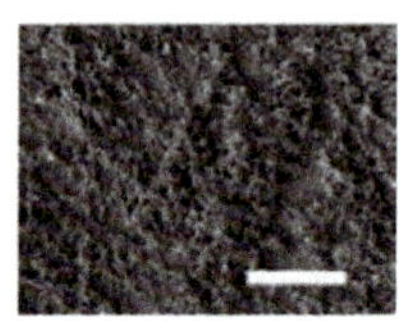 Anodization and replication of nanoporous AAO 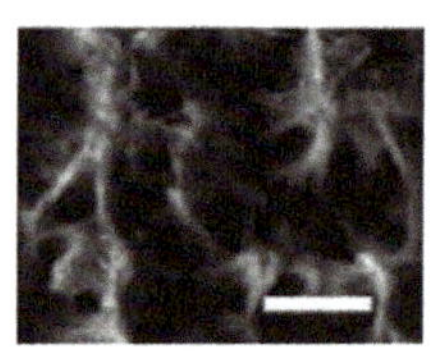 Anodization and HDFS coating	Superhydrophobic surface showed remarkable advantage of antifrosting, the frost weight per unit area on which was largely reduced compared with hydrophilic surfaces

(Continued)

TABLE 9.2 (Continued)

Authors	Year	Surface treatment	Contact angle	Typical topographical feature	Surface icephobicity
Boinovich et al. [112]	2013	Coated stainless steel surface	168.7 ± 4.8°, 169.7 ± 4.2°, 166.1 ± 2.5°		Compared with the untreated stainless steel, the superhydrophobic surface effectively prevented the snow accumulation, and prolonged the freezing time of rain droplets. The designed coating was suggested for antiicing on stainless steel substrate.
Wang et al. [97]	2013	PTES coated aluminum plat, etched aluminum plat, and etched aluminum plat with PTES coating	15°–164°	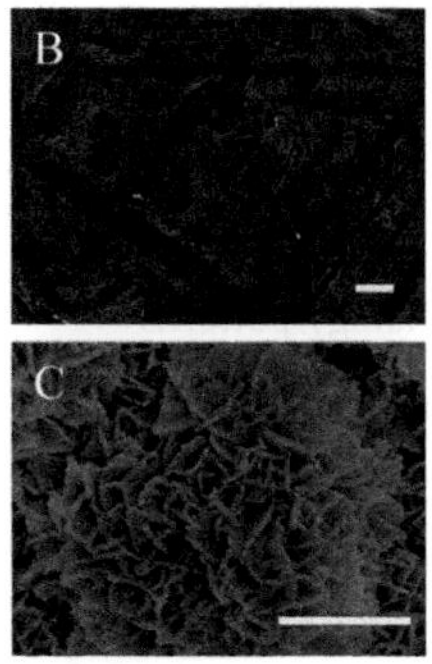 Scale bar: 500 nm	Very low water sliding angle was achieved on the superhydrophobic surface, resulting in low water droplet adhesion. The ice adhesion on the superhydrophobic surface was only 13% of that on the superhydrophilic surface. Besides, excellent durability was reported. The superhydrophobic surface structure was suggested as an active icephobic texture.
Yue et al. [113]	2016	Black silicon, two-stage black silicon surfaces and uncoated silicon	76°–146.645°	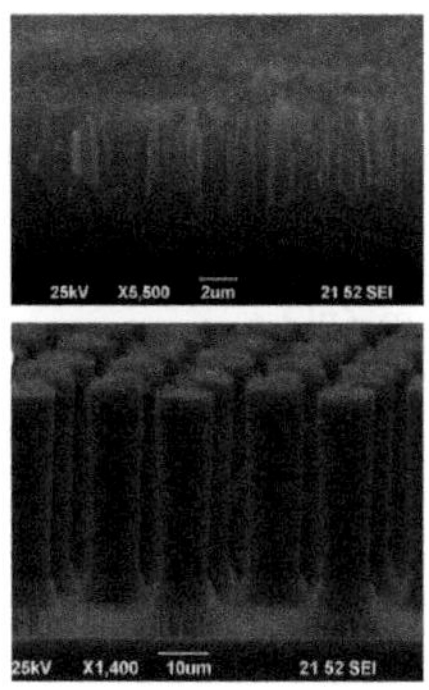	Black silicon (CA = 126.495) and two-stage black silicon surfaces largely increased the surface hydrophobicity, reduced the freezing thickness deformation and showed antifrosting performance

(Continued)

TABLE 9.2 (Continued)

Authors	Year	Surface treatment	Contact angle	Typical topographical feature	Surface icephobicity
Wu et al. [77]	2017	Al plates with parallel microgrooves and crossed grooves creating uniform square pillars	Aluminum surface: 80° ± 1°; the contact angles of the grooved surfaces were not provided	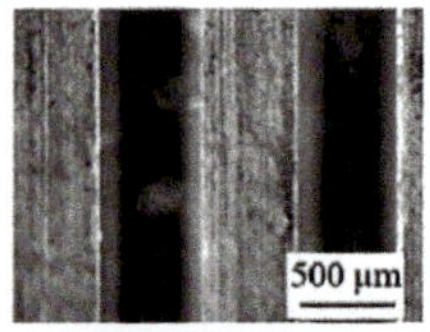 	Droplet condense nucleation was suppressed and the frost accumulation was retarded on grooved surfaces compared with bare aluminum surface
Zhao and Yang [78]	2017	Five different surfaces, including smooth hydrophilic/ hydrophobic surfaces and surfaces with wettability and morphology patterns	67.5° ± 2.2°, 109.2° ± 1.4°, 107.6° ± 1.7°, 152.3° ± 5.1°, and 148.6° ± 4.9°	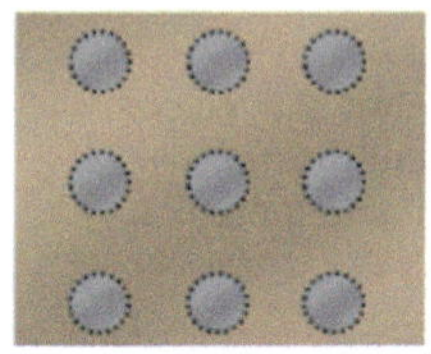 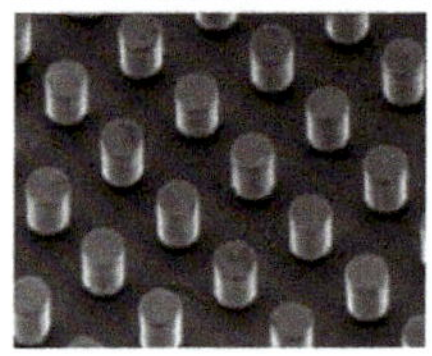	Surfaces with morphology patterns were suggested for antiicing over a wide range of surface temperatures
Zuo et al. [114]	2017	A superhydrophobic surface with ZnO nanorods	167.1°	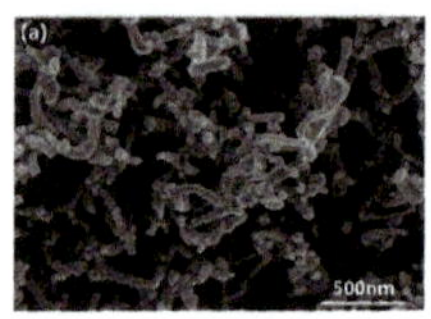	A 140 min delay of frost formation was achieved on the superhydrophobic surface, and the frost propagation rate was remarkably reduced
Wang et al. [115]	2018	Micro-ratchet structured PDMS substrate coated by uniform ZnO nanorods	160° (from Fig. S1 in supplementary data)	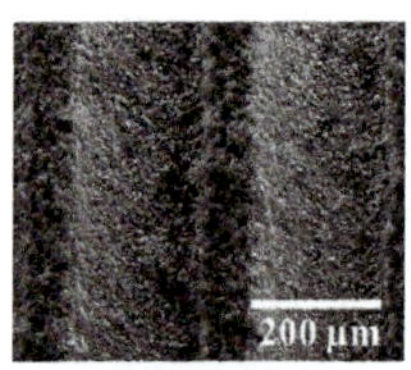 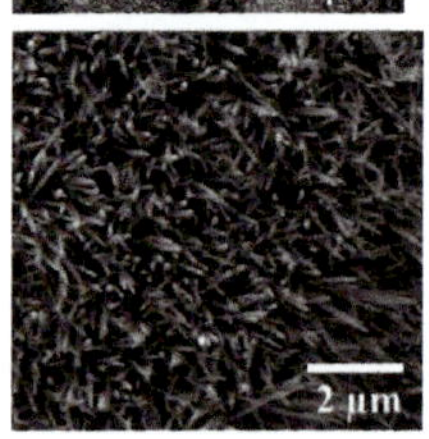	The surface performs excellent icephobicity at higher than -30 °C and fast recovery of the robust performance from freezing state due to the elasticity of the basement material.

(*Continued*)

TABLE 9.2 (Continued)

Authors	Year	Surface treatment	Contact angle	Typical topographical feature	Surface icephobicity
Kim et al. [116]	2019	Glass fiber reinforced composites prepared with different sand paper grits (from neat to 2000CC)	62°–118°		The rate of frost formation, starting at the rate on the neat surface, decreased to a minimum (optimal) value for the surface produced using 320 CC sand paper and then increased to a value very near that of the neat surface for the surface produced using 2000 grit sand paper
Haque et al. [117]	2019	An atomic smooth graphene coating and a 3D nanostructured inkjet-printed coating	87.96 ± 2.56° 138.26 ± 0.44°	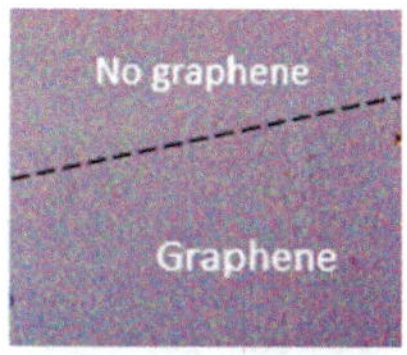 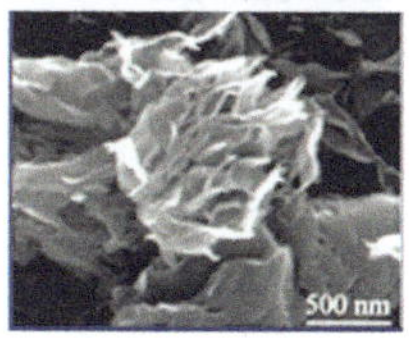	Both hydrophobic and hydrophilic graphene surfaces delay freezing compared to a plain silicon surface under same operating conditions. A maximum of ~3.65 × freezing delay was observed for InkJet-printed graphene (IPG) surface compared to hydrophilic graphene surface at the same freezing temperature
Wang et al. [118]	2020	Coated aluminum fins	140.0 ± 1.0° 160.6 ± 1.0°	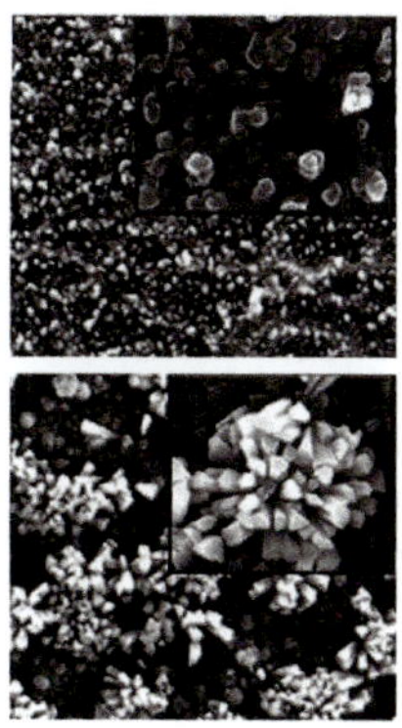	Defrosting temperature could be reduced by optimizing the nanostructure size of superhydrophobic surface to save defrosting energy consumption

(Continued)

TABLE 9.2 (Continued)

Authors	Year	Surface treatment	Contact angle	Typical topographical feature	Surface icephobicity
Mahvi et al. [119]	2021	Unaltered, superhydrophilic, and superhydrophobic fin heat exchangers which substrates are aluminum. The hydrophilic and hydrophobic have the same structures, while the only difference is the presence of low surface energy coating on the hydrophobic surface.	$77 \pm 1.5^\circ$ $\sim 5^\circ$ $160 \pm 1^\circ$	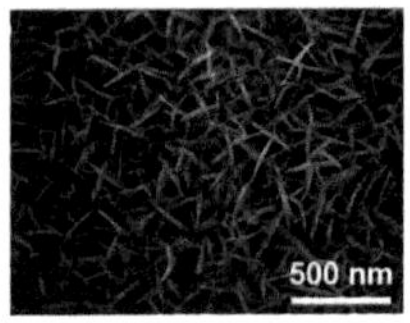	Superhydrophobic surfaces can delay frost formation and can substantially improve the performance of heat pumps when applied to the external surface of an evaporator. Although the benefit diminishes as the system cycles between frosting and passive defrosting, the higher heat transfer rates and COPs are maintained long enough to for most automotive cabin heating applications.
Fan et al. [120]	2021	Hydrophobic glass surface which is prepared by sol–gel method	$128 \pm 3^\circ$	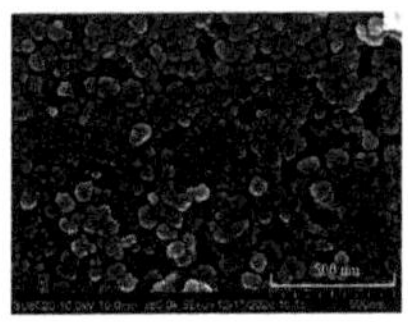	The frost crystal appearance time on the hydrophobic surface is more than 660 s, which is 540 s longer than that on the unmodified glass surface. The frost coverage of the hydrophobic surface is significantly smaller than that of the unmodified plain glass surface, the speed of frost propagation is only 47.5 μm/s, which is 29.6% lower than that of unmodified surface.
Chen et al. [121]	2022	A macroscopic aluminum honeycomb structure with a superhydrophobic coating	161.1°	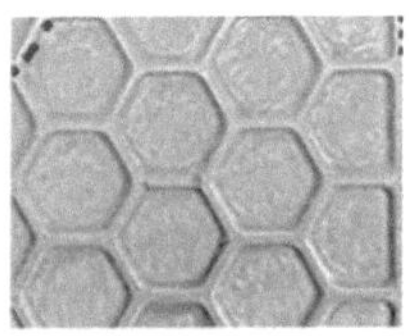	The total ice-up time of single droplet was prolonged by 190 s on a platform at 15°C. After 300 wear tests, the frosting amount increased by only 0.03 g when the platform was tilted to 60% at 20°C and 75 ± 5% humidity. In the outdoor ice test, no large area of ice was observed on the surface.

(Continued)

TABLE 9.2 (Continued)

Authors	Year	Surface treatment	Contact angle	Typical topographical feature	Surface icephobicity
He et al. [122]	2023	Aluminum-based superhydrophobic by the spray method	161.5 ± 1° 159.7 ± 1°	 Superhydrophobic nano surface  Superhydrophobic micro-nano surface	The frost inhibition ability of superhydrophobic surface can be effectively improved with the increase of the density of nanostructures.

For those studies that reported a noticeable impact of surface structures on frosting, both hydrophobic and hydrophilic surfaces exhibited antifrosting effects but through different mechanisms.

Kim and Lee and their group conducted continuous investigations on the frosting characteristics of surface-treated fins [126–129]. In Ref. [126], moderate frost retardation (less than 10 min) was observed on the hydrophobic surfaces (CA = 142°), which agreed with [100,130]. However, the frost thickness was diminished on the hydrophilic surfaces (CA = 2.5°) with a higher frost density. Therefore the hydrophilic surfaces were suggested to detract from the block-ratio and thermal resistance of the frost layer. In the comparisons among three types of surface treatments of a two-row louvered-fin heat exchanger unit, including hydrophilic (CA = 3°), hydrophobic (CA = 130°), and dual-coated surfaces [69,127], remarkable frost retardation was captured on the hydrophobic surfaces. Subsequently, the hydrophobic-treated heat exchanger was suggested under frosting conditions because of the lower air-side pressure drop and better heat transfer performance. Moreover, two extra hybrid coatings were employed [129]. Applying hybrid coating on the heat exchanger can help enhance the initial heat transfer rate thanks to the hydrophilic side while diminishing the heat transfer rate reduction as a result of the frost retardation effects of the hydrophobic side.

Specifically, the effect of microgroove geometry on the frost formation performances was examined [68]. Microgrooved samples with the largest groove depth and pillar width showed the best performance in frost growth suppression and frost densification. Moreover, the dependency of frost properties on the grooved structure was clarified and will be instructive in antifrosting surface design.

Wang et al. [131] qualitatively examined the frost thickness and mass on the superhydrophobic heat exchanger fins, which were 17.1% and 28.8% lower than those of the bare surfaces, respectively. The frost nucleation retardation and frost growth suppression effects of the hydrophilic surface were also evidenced but were weaker than those of the superhydrophobic surface. However, the surface contact angles of the tested surfaces were not provided. Wang et al. [1] innovatively utilized an aluminate coupling agent on an aluminum surface (with a contact angle of 62°) and obtained a hydrophobic surface with a contact angle of 147°. At a surface temperature of −10°C and the ambient air relative humidity of 60%, both frost nucleation and frost growth were remarkably hindered on the hydrophobic surface.

Apparently, no inclusive conclusion can be drawn concerning the icephobicity of hydrophobic/hydrophilic surfaces so far. More surface structures are waiting to be developed to achieve better frost retardation performance. The aim is not to simply find out one single surface structure that is most effective in antifrosting but to seek out better antifrosting strategies.

Based on the survey, it is noticed that limited studies have been conducted on the droplet freezing process. However, we believe that the droplet freezing process is of great significance since it corresponds to the frost crystal nucleation period, which is the very first stage of frost deposition.

The freezing process of a water droplet and the frost formation on a plain copper plate (CA = 72°) and a superhydrophobic treated silicon plate (CA = 162°) were experimentally observed [66]. A 55 minutes delay of frost retardation was captured, and an obvious depression of frost growth was found on the superhydrophobic surface. It was believed that the frost deposition started from the condensed water droplets on the cold surface, the characteristics of which were largely related to the surface contact angle. The increase in contact angle would substantially raise the potential barrier for water condensation and, subsequently, retard the frost nucleation.

The frost nucleation period on surfaces with CA ranges from 72.6° to 129.5° was particularly examined by 0. Surface roughness ranged from 0.05 to 6.92 μm. The frost nucleation limit subcooling degree was found to be the lowest on the roughest plate with the largest contact angle. However, a relatively large contact angle but a low surface roughness induced higher nucleation subcooling. Accordingly, high contact angle surface was suggested for frost retardation. The joint effect of contact angle and surface roughness needs to be quantified in detail. However, surface roughness is not provided in most of the studies on the effects of surface structure. Therefore we suggest conducting the surface roughness measurement in future experimental investigations on the frosting characteristics of various surface structures.

Recently, we experimentally observed the droplet freezing and frost accumulation processes under natural convection conditions on surfaces with different topographical textures. As shown in Fig. 9.2, test sections, including a copper plate with resin spray, ICP-etched black silicon surfaces (aperiodic structure), and periodic black silicon surfaces (including one-stage and two-stage fabrications), were used. A water droplet with a volume of 10 μL was positioned on the test section, the temperature of which was regulated by a low-temperature thermostat bath. The profile of the droplet was monitored via a drop shape analyzer. Temporal droplet height and diameter are measured (as shown in Fig. 9.3). In Fig. 9.4, comparisons of the droplet deformation coefficients, defined as the ratio of temporal and initial diameters or heights on four types of surface textures, reveal

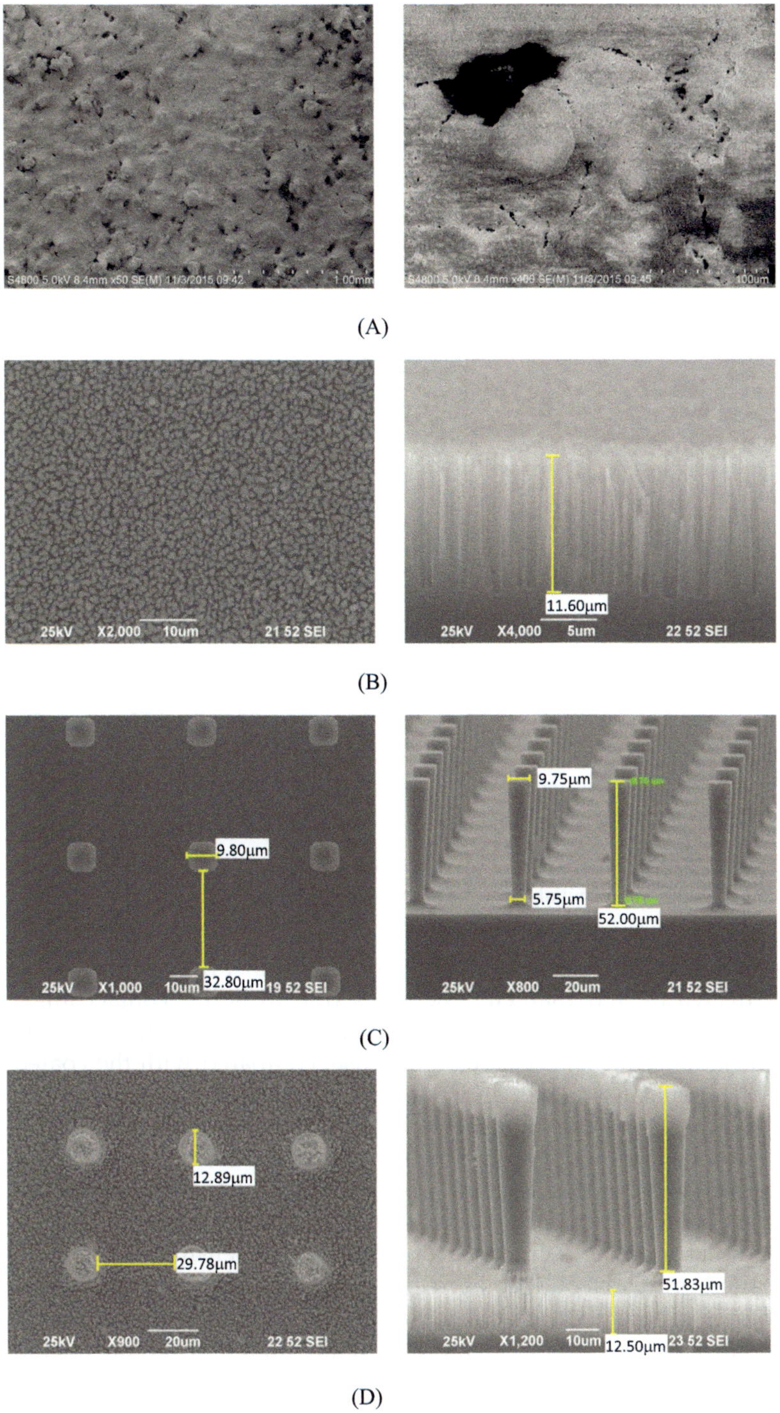

FIGURE 9.2 Scanning electron microscope (SEM) images of the tested surface structures in our study: (A) copper plate sprayed with resin, (B) Inductively coupled plasma (ICP)-etched black silicon, (C) one-stage silicon microstuds, and (D) two-stage silicon microstuds with aperiodic structures tops.

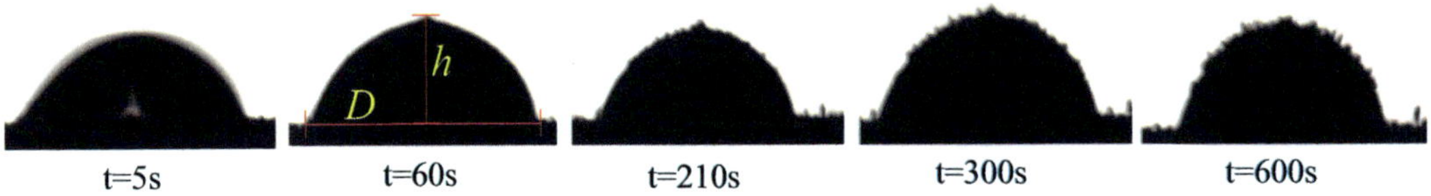

FIGURE 9.3 Droplet profile variation during freezing at surface temperature of -6°C on two-stage silicon microstuds with aperiodic structures top.

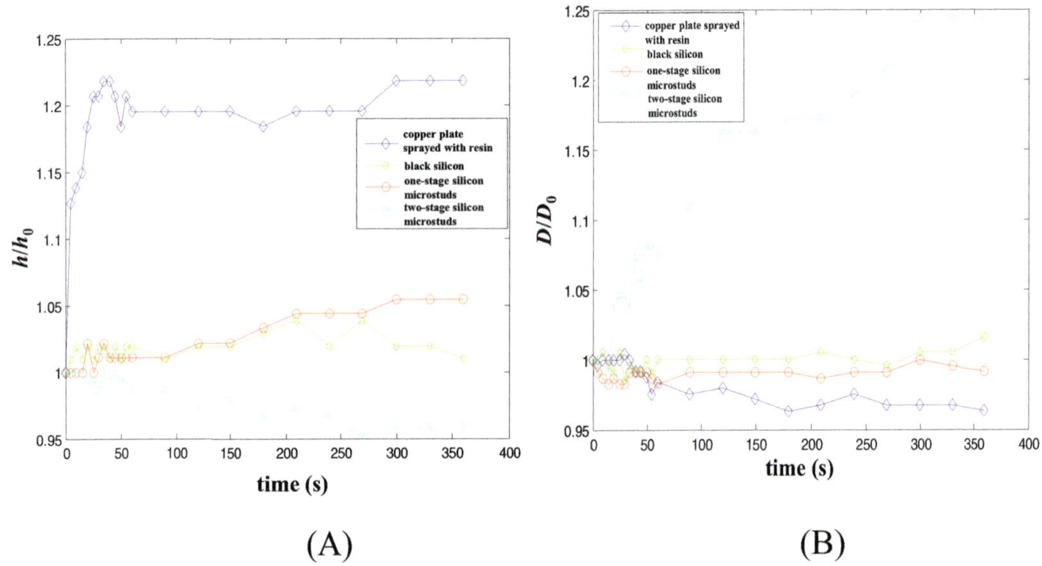

FIGURE 9.4 Droplet deformation coefficients of different surface structures at 0°C surface temperature: (A) temporal-to-initial height ratio and (B) temporal-to-initial diameter ratio.

that the two-stage silicon microstructure can evidently delay the droplet freezing onset, on which the droplet height decreases and diameter increases due to liquid contact angle decrease. Meanwhile, droplets on other surfaces have already begun freezing and show slight height or diameter variation. This is as expected because, compared with the coated copper plate, ICP-etched black silicon surface, and one-stage silicon microstuds surface, the aperiodic protrusions of the two-stage morphology have more air pockets trapped under the water droplet, reducing the heat transfer to the cold surface. The unique topographical texture is suggested to be applied in air precooler antifrosting. Some of the results with two-stage black silicon surfaces have already been published in Ref. [113].

Since frost usually accumulates on fully frozen droplets after condensing on the cold surfaces, exploring the detailed processes of droplet freezing is important for us to analyze the icephobic characteristics of different structured surfaces. Based on a series of experiments, we have captured clear images of the shape and temperature evolution during droplet cooling and freezing as shown in Figs. 9.5 and 9.6. Some of the results with droplet

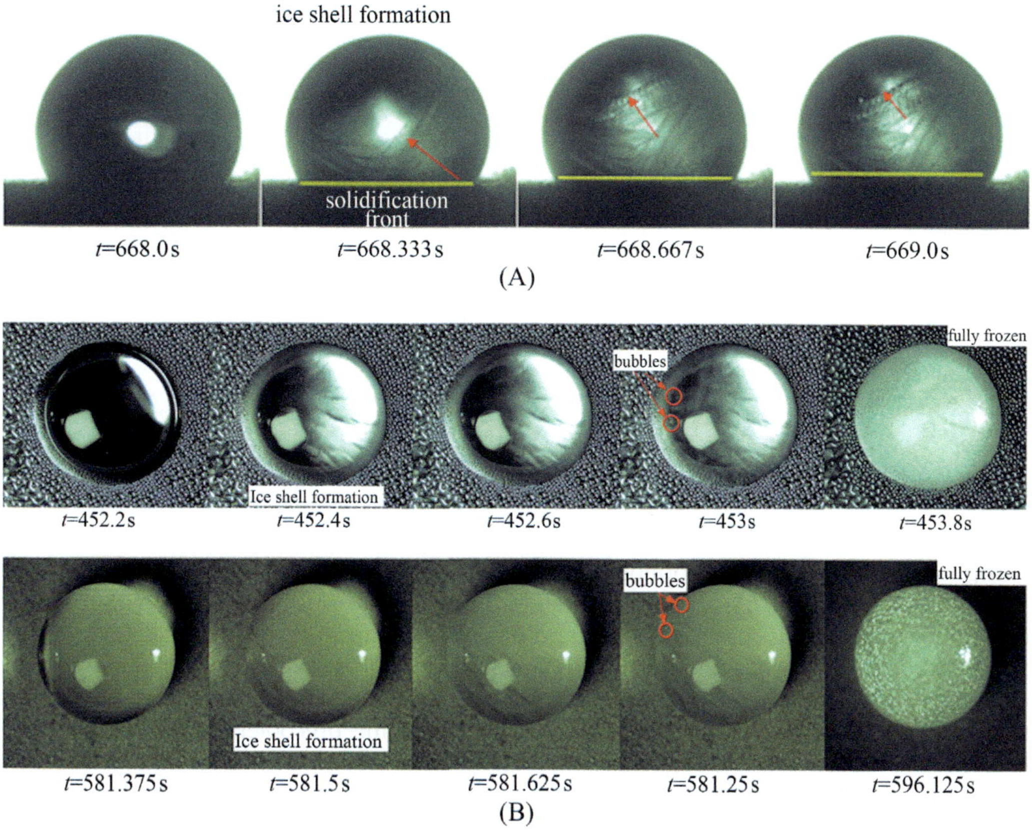

FIGURE 9.5 Droplet shape evolution during cooling and freezing: (A) lateral view, (B) top view. (A) Lateral view of the droplet ice nucleation, (B) top view of the droplet ice[133] nucleation.

freezing processes have already been published in Ref. [132–134]. More outcomes are expected in our further experimental investigations.

There are also several investigations particularly introducing the design and preparation of antifrosting surface structures and coating materials [107,112,123,131,135–137] the details of which are beyond our interests in the present review. Efforts have been made to develop a new painting material that can both delay the frost nucleation and minimize the frost growth on the cold surface [58,138]. The fabricated hydrophilic painting can postpone the frost nucleation for at least 15 minutes, and the frost thickness and mass were reduced by more than 40%. Astonishing results were reported that when air humidity was lower than 60% and the plate surface temperature was higher than −10°C, the surface could be frost-free for 3 hours. This is very appealing for antifrosting applications. However, the optimum design of the coating thickness should be accomplished before bringing the material into practical applications.

Generally, we still need to sig further to clarify the relation between the surface wettability and the surface ice-phobic capacity. Most of the studies show the possibility of using

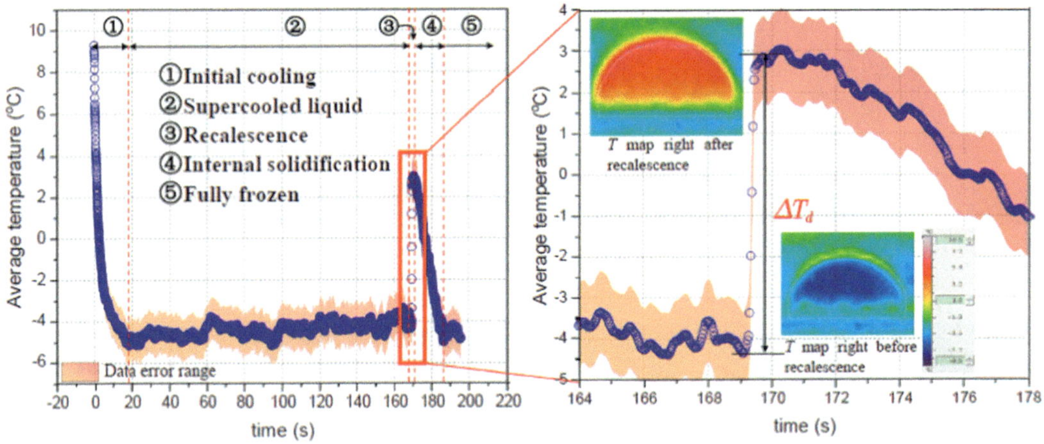

FIGURE 9.6 Average temperature of the droplet surface and the temperature jump (zoomed-in on the right) of the averaged surface temperature (ΔT_d) which is used for crystal production calculation.

hydrophobic and/or superhydrophobic surfaces for frost retardation. More importantly, the use of surface treatment for antifrosting can simplify the propulsion system without introducing redundancies. It is a potential candidate for future antifrosting strategies in precooled turbojets. However, the frosting characteristics under extreme conditions such as high temperature (hundreds of degrees Celsius) and high-speed (hundreds of meters per second) airflow conditions in deeply cooled turbojet engine precoolers still require further exploration. Most of the existing experimental data are obtained at air velocity less than 20 m/s and mostly in laminar flow condition with the ambient air temperature lower than 100°C. Wide ranges of experiments are urgently waiting to be conducted before the hydrophobic/hydrophilic surface structures can be used in precooler antifrosting techniques.

9.5 Summary and prospect

Frost nucleation and deposition are commonly encountered in deeply cooled turbojet air precoolers. The high antifrosting capacity of the precooling system is the prerequisite for maintaining the engine performance. An understanding of the frost formation process and the influencing factors of the frost characteristics is of significant importance to develop better antifrosting techniques.

Experimental investigations on frost formation and growth on cold surfaces (e.g., flat surfaces, cylinders, and heat exchangers) under natural convection and forced convection have been surveyed, which show that among the main influencing parameters of frosting, cold surface temperature, air relative humidity, and surface structures are the most influential ones. However, the coupling effects of different influential factors on the frosting characteristics should be carefully examined.

However, it is realized that among the existing experimental studies, the tested air velocities are mainly below 20 m/s, and the incoming air temperatures are usually lower than 50°C. Very limited studies on frost growth characteristics on cold surfaces under high temperature and high-speed airflow conditions have been published so far. Therefore the experimental database is waiting to be enlarged before developing the antifrosting methods for precooled air-breathing combined cycle engines.

Particularly, the effects of surface topographical texture on frost formation and growth show in-conclusive trends. Controversial conclusions are found on the icephobicity of hydrophobic and/or hydrophilic surfaces even though most of the studies show the possibility of using hydrophobic and/or superhydrophobic surfaces for frost retardation. Moving forward, it is in the opinion of the authors that impending research shall focus on, but is not limited to, the following areas:

1. With many available fabrication methods, successfully producing hydrophobic and/or icephobic coatings is no surprise; however, mechanical and weather durability is still a major concern and hence should be addressed and standardized.
2. Most of the fabrication methods reported in the literature are very expensive and are not suitable for large-scale production. These reasons have prevented hydrophobic and/or icephobic surfaces from being commercially successful so far. Hence, laboratory preparation of these surfaces should aim to foresee commercial aspects, for example, costs, durability, and ease of large-scale production.
3. It is recommended that images of test sample topographical textures be provided instead of simply giving the contact angle. Images obtained via SEM, STF, AFM, and others can provide detailed surface textures. ESEM can be used to monitor the real time frost growth process. Besides, surface roughness measurement is also suggested.

In light of the present overview, we believe that surface treatment should be utilized to improve the existing antifrosting techniques in air-breathing combined cycle engines. It should also be realized that the air deep precooling technique is in its infancy in China. Experimental investigations on ice-phobic surface structures and the corresponding frosting characteristics are essential to develop more energy-efficient and environment-friendly antifrosting systems and to bridge the gap.

References

[1] Z.-J. Wang, et al., Frost formation and anti-icing performance of a hydrophobic coating on aluminum, Experimental Thermal and Fluid Science 60 (2015) 132–137.
[2] T. Sato, et al., Development study of precooled-cycle hypersonic turbojet engine for flight demonstration, Acta Astronautica 61 (1) (2007) 367–375.
[3] Saurabh Ahmadi, et al., A review of condensation frosting, Nanoscale & Microscale Thermophysical Engineering 21 (2017) 81–101.
[4] S.-M. Li, C.-C. Wang, Predictive models on the frost formation for plain surface - a review and comparative study, International Communications in Heat and Mass Transfer 129 (2021) 105670.
[5] M. Song, C. Dang, Review on the measurement and calculation of frost characteristics, International Journal of Heat and Mass Transfer 124 (2018) 586–614.
[6] W. Su, X. Zhang, Review of anti-frosting and defrosting methods on air source heat pump, Journal of Engineering-Physics 42 (2021) 2195–2215.
[7] Y. Zhang, et al., Frosting phenomenon and frost-free technology of outdoor air heat exchanger for an air-source heat pump system in China: an analysis and review, Energies 11 (10) (2018).

[8] M. Xiong, et al., Review on research status of frosting characteristics of microchannel heat exchanger, Journal of Refrigeration 41 (2020) 22–30.
[9] M. Massoud, Engineering thermofluids, Thermodynamics, Fluid Mechanics, and Heat Transfer, Springer, 2005.
[10] Y. Hayashi, et al., Study of frost properties correlating with frost formation types, Journal of Heat Transfer 99 (2) (1977) 239–245.
[11] N.H. Fletcher, Vol. Chapter 4 and 5 The Chemical Physics of Ice, Cambridge University Press, 1970.
[12] C.T. Sanders, The influence of frost formation and defrosting on the performance of air coolers, Delft Technical University, 1974.
[13] A.R. Tahavvor, M. Yaghoubi, Analysis of early-stage frost formation in natural convection over a horizontal cylinder, International Journal of Refrigeration 32 (6) (2009) 1343–1349.
[14] J Shin, A V Tikhonov, C Kim, Experimental study on frost structure on surfaces with different hydrophilicity: density and thermal conductivity [J], J. Heat Transfer 125 (1) (2003) 84–94.
[15] W. Piening, Der Wärmeübergang an Rohren bei freier Strömung unter Berücksichtigung der Bildung von Schwitzwasser und Reif, Gesundheits-Ingenieur 42 (1933) 493–497.
[16] L.A. Kennedy, J. Goodman, Free convection heat and mass transfer under conditions of frost deposition, International Journal of Heat and Mass Transfer 17 (1974) 477–484.
[17] C.C. Wang, Some observations of the frost formation in free convection: with and without the presence of electric field, International Journal of Heat and Mass Transfer 47 (2004) 3491–3505.
[18] J.-T. Kwon, An experimental study on frosting of laminar air flow on a cold surface with local cooling, International Journal of Refrigeration 29 (2006) 754–760.
[19] E. Moallem, et al., Experimental measurements of the surface coating and water retention effects on frosting performance of microchannel heat exchangers for heat pump systems, Experimental Thermal and Fluid Science 39 (2012) 176–188.
[20] Q. Kaiyang, S. Komori, J. Yi, Local variation of frost layer thickness and morphology flow, International Journal of Thermal Sciences 45 (2006) 116–123.
[21] M. Fossa, G. Tanda, Study of free convection frost formation on a vertical plate, Experimental Thermal and Fluid Science 26 (2002) 661–668.
[22] B. Na, R.L. Webb, New model for frost growth rate, International Journal of Heat and Mass Transfer 47 (5) (2004) 925–936.
[23] D.K. Yang, K.S. Lee, Dimensionless correlations of frost properties on a cold plate, International Journal of Refrigeration 27 (2004) 89–96.
[24] M. Fossa, G. Tanda, Frost formation in vertical channels under natural convection, International Journal of Multiphase Flow 36 (3) (2010) 210–220.
[25] Y. Barzanoni, et al., Experimental investigation of frost formation on a horizontal cold cylinder under cross flow, International Journal of Refrigeration 34 (4) (2011) 1174–1180.
[26] A.R. Tahavvor, M. Yaghoubi, Experimental and numerical study of frost formation by natural convection over a cold horizontal circular cylinder, International Journal of Refrigeration 33 (7) (2010) 1444–1458.
[27] M. Amini, A.R. Pishevar, M. Yaghoubi, Experimental study of frost formation on a fin-and-tube heat exchanger by natural convection, International Journal of Refrigeration 46 (2014) 37–49.
[28] C.J.L. Hermes, et al., A study of frost growth and densification on flat surfaces, Experimental Thermal and Fluid Science 33 (2) (2009) 371–379.
[29] E. Moallem, et al., Experimental investigation of the surface temperature and water retention effects on the frosting performance of a compact microchannel heat exchanger for heat pump systems, International Journal of Refrigeration 35 (1) (2012) 171–186.
[30] Q. Haijie, et al., Experimental study of the characteristic of frosting on low-temperature air cooler, Experimental Thermal and Fluid Science 55 (2014) 106–114.
[31] K. Kim, K.-S. Lee, Frosting and defrosting characteristics of a fin according to surface contact angle, International Journal of Heat and Mass Transfer 54 (13) (2011) 2758–2764.
[32] C.J. Hermes, et al., A study of frost build-up on hydrophilic and hydrophobic surfaces under forced convection conditions, Experimental Thermal and Fluid Science 100 (2019) 76–88.
[33] S. Yonezawa, et al., Suppression of frost layer growth on glass or copper surfaces with micro-scale lattice-patterned grooves, International Journal of Refrigeration 106 (2019) 33–40.
[34] L. Li, et al., The theoretical analysis and experimental study on anti-frosting performance of surface characteristics, International Journal of Thermal Sciences 137 (2019) 343–351.

[35] J. Kim, et al., Frost growth behavior according to the cold surface inclination angle, International Journal of Heat and Mass Transfer 146 (2020) 118841.
[36] Y. Qin, B. Dong, W. Li, Experimental study of the frosting characteristic of water on a cold surface in the magnetic field, Experimental Thermal and Fluid Science 114 (2020) 110044.
[37] T. Zhang, D.L. O'Neal, S.T. McClain, Analysis of frost thickness and roughness growth from the perspective of frost crystal structure, International Journal of Refrigeration 112 (2020) 314–323.
[38] L.H. De Koninck, S.F. Ahmadi, J.B. Boreyko, Passive anti-frosting cables, International Journal of Heat and Mass Transfer 146 (2020) 118808.
[39] F. Chu, et al., Quantitative relations between droplet jumping and anti-frosting effect on superhydrophobic surfaces, Energy and Buildings 225 (2020) 110315.
[40] Y. Yao, et al., Frost-free zone on macrotextured surfaces, Proceedings of the National Academy of Sciences 117 (12) (2020) 6323–6329.
[41] Y. Zhao, et al., Promote anti-/de-frosting by suppressing directional ice bridging, International Journal of Heat and Mass Transfer 165 (2021) 120609.
[42] C.-W. Lo, J.-X. Li, M.-C. Lu, Frosting and defrosting on the hydrophilic nylon-6 nanofiber membrane–coated surfaces, Applied Thermal Engineering 184 (2021) 116300.
[43] H. Jeong, et al., Frost growth mechanism and its behavior under ultra-low temperature conditions, International Journal of Heat and Mass Transfer 169 (2021) 120941.
[44] S. Byun, et al., Frost layer growth behavior on ultra-low temperature surface with a superhydrophobic coating, International Communications in Heat and Mass Transfer 128 (2021) 105641.
[45] H. Jeong, et al., Optical investigation of cryogenic frost formation under forced convection, Applied Thermal Engineering 202 (2022) 117887.
[46] L.E.I. Shangwen, et al., Experimental study on the effect of surface temperature on the frost characteristics of a vertical cold plate under natural convection, Experimental Thermal and Fluid Science 137 (2022) 110684.
[47] J.B. Yang, et al., Frosting and defrosting characteristics of multi-layer coated aluminum surfaces, International Communications in Heat and Mass Transfer 139 (2022) 106460.
[48] H. Wei, et al., Experimental investigation on effects of freestream humidity on the frosting and defrosting characteristics of microtubule precooler, Applied Thermal Engineering 206 (2022) 118104.
[49] S. Mengjie, et al., An experimental study on the effect of horizontal cold plate surface temperature on frosting characteristics under natural convection, Applied Thermal Engineering 211 (2022) 118416.
[50] D. Paulovics, et al., Dynamics of frost propagation on breath figures, Langmuir 38 (9) (2022) 2972–2978.
[51] F. Wang, et al., Freezing propagation of condensate droplets at early stage of frosting on vertical hydrophobic surface, Case Studies in Thermal Engineering 41 (2023) 102617.
[52] L. Zhang, et al., An experimental study on the dynamic frosting characteristics on the edge zone of a horizontal copper plate under forced convection, International Journal of Heat and Mass Transfer 200 (2023) 123541.
[53] Y. Li, M. Li, C. Dang, Effect of surface wettability on dust removal characteristics during frosting cleaning process, Applied Thermal Engineering 229 (2023) 120592.
[54] D. Zhang, et al., Freezing modes of water droplet on cold plate surface under forced convection, Applied Thermal Engineering 223 (2023) 120025.
[55] S. Yang, et al., Efficient anti-frosting on discrete nanoclusters via spatiotemporal control of condensation frosting dynamics, Chemical Engineering Journal 465 (2023) 142991.
[56] Z. Zuo, et al., Suppressing condensation frosting using micropatterned ice walls, Applied Thermal Engineering 224 (2023) 120099.
[57] X. Wu, et al., Mesoscale investigation of frost formation on a cold surface, Experimental thermal and fluid science 31 (8) (2007) 1043–1048.
[58] Z. Liu, et al., An experimental study on minimizing frost deposition on a cold surface under natural convection conditions by use of a novel anti-frosting paint. Part I. Anti-frosting performance and comparison with the uncoated metallic surface, International Journal of Refrigeration 29 (2) (2006) 229–236.
[59] A. Lüer, H. Beer, Frost deposition in a parallel plate channel under laminar flow conditions, International journal of thermal sciences 39 (1) (2000) 85–95.
[60] C.-H. Cheng, C.-C. Shiu, Frost formation and frost crystal growth on a cold plate in atmospheric air flow, International Journal of Heat and Mass Transfer 45 (21) (2002) 4289–4303.
[61] Y.B. Lee, S.T. Ro, Frost formation on a vertical plate in simultaneously developing flow, Experimental Thermal and Fluid Science 26 (8) (2002) 939–945.

[62] C.-H. Cheng, K.-H. Wu, Observations of early-stage frost formation on a cold plate in atmospheric air flow, Journal of Heat Transfer 125 (1) (2003) 95–102.
[63] H. Lee, et al., Frost formation on a plate with different surface hydrophilicity, International Journal of Heat and Mass Transfer 47 (22) (2004) 4881–4893.
[64] G. Tanda, M. Fossa, Free convection frost growth in a narrow vertical channel, International Journal of Heat and Mass Transfer 49 (2006) 1946–1957.
[65] Z. Liu, et al., Influences of surface hydrophilicity on frost formation on a vertical cold plate under natural convection conditions, Experimental Thermal and Fluid Science 31 (7) (2007) 789–794.
[66] Z. Liu, et al., Frost formation on a super-hydrophobic surface under natural convection conditions, International Journal of Heat and Mass Transfer 51 (25) (2008) 5975–5982.
[67] M. Kandula, Frost growth and densification on a flat surface in laminar flow with variable humidity, International Communications in Heat and Mass Transfer 39 (8) (2012) 1030–1034.
[68] M.A. Rahman, A.M. Jacobi, Effects of microgroove geometry on the early stages of frost formation and frost properties, Applied Thermal Engineering 56 (1–2) (2013) 91–100.
[69] T. Jing, et al., Frosting and defrosting on rigid superhydrohobic surface, Applied Surface Science 276 (2013) 37–42.
[70] K. Kim, et al., Thermal performance of microchannel heat exchangers according to the design parameters under the frosting conditions, International Journal of Heat and Mass Transfer 71 (2014) 626–632.
[71] Y. Chen, et al., Experimental study on frost formation on a cold surface in low atmospheric pressure, Applied Thermal Engineering 90 (2015) 86–93.
[72] M.A. Rahman, A.M. Jacobi, Experimental study on frosting/defrosting characteristics of microgrooved metal surfaces, International Journal of Refrigeration 50 (2015) 44–56.
[73] Z. Liu, Y. Dong, Y. Li, An experimental study of frost formation on cryogenic surfaces under natural convection conditions, International Journal of Heat and Mass Transfer 97 (2016) 569–577.
[74] X. Wu, S. Hu, F. Chu, Experimental study of frost formation on cold surfaces with various fin layouts, Applied Thermal Engineering 95 (2016) 95–105.
[75] A.D. Sommers, et al., Densification of frost on hydrophilic and hydrophobic substrates – examining the effect of surface wettability, Experimental Thermal and Fluid Science 75 (2016) 25–34.
[76] H. Kim, et al., Microscopic observation of frost behaviors at the early stage of frost formation on hydrophobic surfaces, International Journal of Heat and Mass Transfer 97 (2016) 861–867.
[77] X. Wu, et al., Frost formation and frost meltwater drainage characteristics on aluminum surfaces with grooved structures, Applied Thermal Engineering 118 (2017) 448–454.
[78] Y. Zhao, C. Yang, Frost spreading on microscale wettability/morphology patterned surfaces, Applied Thermal Engineering 121 (2017) 136–145.
[79] Q. Ma, et al., Experimental and numerical investigations of frost formation on wavy plates, Applied Thermal Engineering 138 (2018) 627–632.
[80] Y. Wang, Y. Cheng, Early stage condensation frosting characteristics on plain and nano Al2O3-epoxy mixture-coated brass, Applied Thermal Engineering 160 (2019) 113971.
[81] J. Lee, K.-S. Lee, The behavior of frost layer growth under conditions favorable for desublimation, International Journal of Heat and Mass Transfer 120 (2018) 259–266.
[82] A.D. Sommers, C.W. Gebhart, C.J. Hermes, The role of surface wettability on natural convection frosting: Frost growth data and a new correlation for hydrophilic and hydrophobic surfaces, International Journal of Heat and Mass Transfer 122 (2018) 78–88.
[83] S. Nath, J.B. Boreyko, On localized vapor pressure gradients governing condensation and frost phenomena, Langmuir 32 (33) (2016) 8350–8365.
[84] Y. Zhao, et al., A review of recent literature on icing phenomena: transport mechanisms, their modulations and controls, International Journal of Heat and Mass Transfer 159 (2020) 120074.
[85] S. Byun, et al., Quantitative analysis of frosting characteristics at ultra-low temperatures under forced convection conditions, Energy and Buildings 248 (2021) 111186.
[86] M. Kandula, Frost growth and densification in laminar flow over flat surfaces, International Journal of Heat and Mass Transfer 54 (2011) 3719–3731.
[87] A. Kenzhebayeva, et al., A mini-review on recent developments in anti-icing methods, POLYMERS 13 (23) (2021).
[88] Z. Wang, et al., Bioinspired anti-icing hydrogel enabled by ice-nucleating protein, CCS Chemistry 4 (1) (2022) 104–111.
[89] G. Bai, et al., Probing the critical nucleus size for ice formation with graphene oxide nanosheets, Nature 576 (7787) (2019) 437–441.

[90] K. Liu, et al., Janus effect of antifreeze proteins on ice nucleation, Proceedings of the National Academy of Sciences 113 (51) (2016) 14739–14744.
[91] W. Zheng, et al., Magnetic responsive and flexible composite superhydrophobic photothermal film for passive anti-icing/active deicing, Chemical Engineering Journal 427 (2022) 130922.
[92] J. Zhou, H. Xu, L. Tang, Facile fabrication of high performance hydrophilic anti-icing polyurethane methacrylate coatings cured via UV irradiation, Progress in Organic Coatings 182 (2023) 107657.
[93] S. Noormohammed, D.K. Sarkar, Rf-sputtered Teflon®-modified superhydrophobic nanostructured titanium dioxide coating on aluminum alloy for icephobic applications, Coatings 11 (4) (2021) 432.
[94] A. Allahdini, R. Jafari, G. Momen, Transparent non-fluorinated superhydrophobic coating with enhanced anti-icing performance, Progress in Organic Coatings 165 (2022) 106758.
[95] C.H. Moon, et al., Icephobic coating through a self-formed superhydrophobic surface using a polymer and microsized particles, ACS Applied Materials & Interfaces 14 (2) (2022) 3334–3343.
[96] S.S. Latthe, et al., Recent developments in air-trapped superhydrophobic and liquid-infused slippery surfaces for anti-icing application, Progress in Organic Coatings 137 (2019) 105373.
[97] Y. Wang, et al., Verification of icephobic/anti-icing properties of a superhydrophobic surface, ACS Applied Materials & Interfaces 5 (8) (2013) 3370–3381.
[98] J.B. Boreyko, C.P. Collier, Delayed frost growth on jumping-drop superhydrophobic surfaces, ACS Nano 7 (2) (2013) 1618–1627.
[99] F. Chu, X. Wu, Q. Ma, Condensed droplet growth on surfaces with various wettability, Applied Thermal Engineering 115 (2017) 1101–1108.
[100] B. Na, R.L. Webb, A fundamental understanding of factors affecting frost nucleation, International Journal of Heat and Mass Transfer 46 (20) (2003) 3797–3808.
[101] H. Wang, et al., Fabrication and anti-frosting performance of super hydrophobic coating based on modified nanosized calcium carbonate and ordinary polyacrylate, Applied Surface Science 253 (22) (2007) 8818–8824.
[102] X. Wu, et al., Visual and theoretical analyses of the early stage of frost formation on cold surfaces, Journal of Enhanced Heat Transfer 14 (3) (2007).
[103] S. Jhee, K.-S. Lee, W.-S. Kim, Effect of surface treatments on the frosting/defrosting behavior of a fin-tube heat exchanger, International Journal of Refrigeration 25 (8) (2002) 1047–1053.
[104] S.A. Kulinich, et al., Superhydrophobic surfaces: are they really ice-repellent? Langmuir 27 (1) (2010) 25–29.
[105] D. Yu, J. Tian, Superhydrophobicity: is it really better than hydrophobicity on anti-corrosion? Colloids and Surfaces A: Physicochemical and Engineering Aspects 445 (0) (2014) 75–78.
[106] X. Yue, W. Liu, Y. Wang, Freezing delay, frost accumulation and droplets condensation properties of micro-or hierarchically-structured silicon surfaces, International Journal of Heat and Mass Transfer 126 (2018) 442–451.
[107] L. Mishchenko, et al., Design of ice-free nanostructured surfaces based on repulsion of impacting water droplets, ACS Nano 4 (12) (2010) 7699–7707.
[108] K.K. Varanasi, et al., Frost formation and ice adhesion on superhydrophobic surfaces, Applied Physics Letters 97 (23) (2010).
[109] S. Farhadi, M. Farzaneh, S.A. Kulinich, Anti-icing performance of superhydrophobic surfaces, Applied Surface Science 257 (14) (2011) 6264–6269.
[110] L.H.A. B, et al., Effect of contact angle on water droplet freezing process on a cold flat surface - ScienceDirect, Experimental Thermal and Fluid Science 40 (1) (2012) 74–80.
[111] H. Wang, G. He, Q. Tian, Effects of nano-fluorocarbon coating on icing, Applied Surface Science 258 (18) (2012) 7219–7224.
[112] L.B. Boinovich, et al., Durable icephobic coating for stainless steel, ACS applied Materials & Interfaces 5 (7) (2013) 2549–2554.
[113] X. Yue, W. Liu, Y. Wang, Effects of black silicon surface structures on wetting behaviors, single water droplet icing and frosting under natural convection conditions, Surface and Coatings Technology 307 (2016) 278–286.
[114] Zhiping Zuo, et al., Anti-frosting performance of superhydrophobic surface with ZnO nanorods, Applied Thermal Engineering 110 (2017) 39–48.
[115] M. Wang, et al., A novel flexible micro-ratchet/ZnO nano-rods surface with rapid recovery icephobic performance, Journal of industrial and engineering chemistry 62 (2018) 52–57.
[116] J.-H. Kim, et al., Evaluation of surface roughness and frost retardancy of a glass fiber/unsaturated polyester composite, International Journal of Heat and Mass Transfer 130 (2019) 282–289.

[117] M.R. Haque, S.R. Das, A.R. Betz, Experimental investigation of condensation and freezing phenomena on hydrophilic and hydrophobic graphene coating, Applied Thermal Engineering 160 (2019) 113987.
[118] F. Wang, et al., Effects of frost thickness on dynamic defrosting on vertical hydrophobic and superhydrophobic fin surfaces, Energy and Buildings 223 (8)) (2020)110134.
[119] A.J. Mahvi, et al., Superhydrophobic heat exchangers delay frost formation and enhance efficency of electric vehicle heat pumps, International Journal of Heat and Mass Transfer 172 (2021) 121162.
[120] P. Fan, et al., Preparation and performance of nanoparticles-based anti-frosting transparent hydrophobic surfaces, International Journal of Refrigeration 2 (2021).
[121] X. Chen, L. Hu, Y. Du, Anti-icing and anti-frost properties of structured superhydrophobic coatings based on aluminum honeycombs, Materials Chemistry and Physics 291 (2022) 126683.
[122] H. He, et al., Effect of micro-nano structure on anti-frost and defrost performance of the superhydrophobic fin surfaces, Experimental Thermal and Fluid Science 144 (2023) 110878.
[123] J. Yang, W. Li, Preparation of superhydrophobic surfaces on Al substrates and the anti-icing behavior, Journal of Alloys & Compounds 576 (2013) 215–219.
[124] S.A. Kulinich, M. Farzaneh, Ice adhesion on super-hydrophobic surfaces, Applied Surface Science 255 (18) (2009) 8153–8157.
[125] C.T. Sanders, The influence of frost formation and defrosting on the performance of air coolers, PhD Thesis, Delft University of Technology, 1974.
[126] K. Kim, K.S. Lee, Frosting and defrosting characteristics of a fin according to surface contact angle, International Journal of Heat & Mass Transfer 54 (13-14) (2011) 2758–2764.
[127] K. Kim, K.S. Lee, Characteristics and performance evaluation of surface-treated louvered-fin heat exchangers under frosting and wet conditions, International Journal of Heat & Mass Transfer 55 (23-24) (2012) 6676–6681.
[128] K. Kim, K.S. Lee, Frosting and defrosting characteristics of surface-treated louvered-fin heat exchangers: effects of fin pitch and experimental conditions, International Journal of Heat & Mass Transfer 60 (2013) 505–511.
[129] K. Kim, D.R. Kim, K.S. Lee, Local frosting behavior of a plated-fin and tube heat exchanger according to the refrigerant flow direction and surface treatment, International Journal of Heat & Mass Transfer 64 (2013) 751–758. sep.
[130] R.O. Piucco, et al., A study of frost nucleation on flat surfaces, Experimental Thermal & Fluid Science 32 (8) (2008) 1710–1715.
[131] F. Wang, et al., Effects of surface characteristic on frosting and defrosting behaviors of fin-tube heat exchangers, Applied Thermal Engineering 75 (2015) 1126–1132.
[132] Y. Wang, Y. Cheng, New perspectives on the droplet freezing nucleation and early crystal growth mechanisms, International Journal of Heat and Mass Transfer 140 (2019) 1023–1028.
[133] Y. Wang, Z.-G. Wang, Sessile droplet freezing on polished and micro-micro-hierarchical silicon surfaces, Applied Thermal Engineering 137 (2018) 66–73.
[134] Y. Wang, P. Zhao, Temperature-based analysis of droplet cooling and freezing on femtosecond laser textured surfaces, Applied Thermal Engineering 206 (2022) 118046.
[135] M. He, et al., Superhydrophobic surface at low surface temperature, Applied Physics Letters 98 (9) (2011).
[136] X.S. Chaudhuri, Design of anti-icing coatings using supercooled droplets as nano-to-microscale probes, Langmuir 28 (9) (2012) 4434–4446.
[137] P. Tourkine, M. Le Merrer, D. Quéré, Delayed freezing on water repellent materials, Langmuir 25 (13) (2009) 7214–7216.
[138] Z. Liu, et al., An experimental study on minimizing frost deposition on a cold surface under natural convection conditions by use of a novel anti-frosting paint. Part II. Long-term performance, frost layer observation and mechanism analysis, International Journal of Refrigeration 29 (2) (2006) 237–242.

CHAPTER

10

A sustainable defrosting technology for air source heat pump systems

Long Zhang[1], Longping Wu[1], Jiatai Jiang[1], Mengjie Song[1,2], Xuan Zhang[1], Keke Shao[1,2], Han Shi[1] and Yiqiang Jiang[3]

[1]Department of Energy and Power Engineering, School of Mechanical Engineering, Beijing Institute of Technology, Beijing, P.R. China [2]School of Mechanical Engineering, Hanyang University, Seoul, South Korea [3]Department of Building and Thermal Engineering, School of Architecture, Harbin Institute of Technology, Harbin, P.R. China

Nomenclatures

Symbol	**Description**	**Unit**
L_w	The latent heat of melting water	kJ/kg
L_{PCM}	The latent heat of PCM	kJ/kg
m_w	The mass of the melting water	kg
Q	The amount of energy supplied by the phase change ESC for defrosting	kJ
V_{PCM}	The volume of the PCM	L
Greek symbol	**Description**	**Unit**
ρ_{PCM}	The liquid density of PCM	kg/m^3
Abbreviation	**Description**	
ASHP	Air source heat pump	
DSC	Differential Scanning Calorimetry	
ESC	Energy storage component	
PCM	Phase change material	
RCD	Reverse cycle defrosting	
TESD	Thermal energy storage defrosting	

Frosting and Icing for Efficient Energy Use in Engineering Applications
DOI: https://doi.org/10.1016/B978-0-443-15495-9.00010-1

10.1 Introduction

Air source heat pump (ASHP), as a sustainable energy-saving technology, has been widely utilized all over the world [1]. It can be used for space heating and hot water supply in many fields, such as construction, industry, agriculture, and vehicle [2]. With the intensification of the energy crisis and environmental issues, ASHP technology has once again entered the stage of rapid development. In general, low ambient temperature and frosting problems are key bottlenecks for the further development of ASHP. Currently, the low ambient temperature problem has greatly been improved with the help of quasi-two-stage compression technology, also named vapor injection technology [3]. By contrast, the frosting problem of ASHP is still not well improved due to the complexity of the frosting process [4]. In general, frosting may significantly decrease the energy efficiency, heating capacity, and stability of ASHP and lead to frequent start and stop [5]. To resolve the frosting problem, a large number of studies on frosting mechanism and characteristic, antifrosting technology, and improving defrosting technology have been conducted for decades.

Studies on the frosting mechanism and characteristic can be further divided into three parts in terms of the research scale. The first part mainly aims at the frost growth on simple cold surfaces, such as flat plate, cylinder, and fin [6]. These small-scale studies are more likely to explain the mechanism of frost formation. For example, the three-dimensional structure of frost layer was first reconstructed using X-ray CT and light-curing resin methods, and the frost density and tortuosity were precisely measured [4,7]. Zhang et al. [8–11] quantitatively investigated the temporal and spatial characteristics of frost growth on cold flat plates with frosting stage division and plate edge effect considered. They also claimed that frost crystal lodging and reverse melting occur frequently during the frost layer growth stage and remarkably affect the frost layer morphology. The second part mainly aims at the intermediate-scale research on frost growth on finned tube heat exchangers, where the thermodynamic and frost distribution characteristics are the priority. It has been demonstrated that the reduction of air flow, rather than the thermal resistance, is the main reason for the decrease in heat transfer rate of the heat exchanger [12–14]. Moreover, the frost layer accumulated on the edges of windward fins accounted for up to 13.7% of the total frost amount, and the frost distribution on the surfaces of the finned tube heat exchanger along the air flow direction was significantly uneven [15–17]. Accordingly, the frosting model was further improved based on this uneven distribution characteristic [18,19]. The last part mainly aims at the large-scale frosting performances of ASHP, which shows that the reduction in energy efficiency for ASHP caused by frosting is up to 20% [20–22], and the uneven frost distribution among multicircuits also brings notable negative impacts on system heating performances [23].

For the antifrosting technology of ASHP, it can also be further divided into five parts in terms of the antifrosting mechanism. The first mainly aims at utilizing surface treatment technology, such as hydrophobic [24], hydrophilic-hydrophobic [25], superhydrophobic [26], and lubricant-impregnated surfaces [27]. Although surface treatment technology can significantly improve the antifrosting performances of ASHP, it is not widely used due to the durability of the treated surface and the complex operating environment in real projects [28]. The second part mainly aims at structure optimization of evaporator, such as bowl-shaped evaporator [29], stripe concavo-convex fins [30], uneven fin pitch [31], uneven fin structure [32], and smaller evaporator depth in air flow direction [33]. Some of these antifrosting technologies are in the laboratory research stage, while others have been applied in practice. The third part mainly aims at optimizing system control strategy based on condensation-frosting maps [34]. For a variable-speed ASHP, the frequency of compressor and the outdoor coil fan have a significant effect on the frosting performances of an ASHP [35]. In some situations, reducing the compressor frequency and increasing the outdoor coil fan frequency can prolong the heating period by up to 157.68% [36,37]. The fourth part mainly aims at utilizing dehumidifying technology to realize frost-free operation [38,39]. Although the desiccant device can effectively reduce the moisture content of inlet air of ASHP and hence realize frost-free operation, the desiccant needs periodic regeneration. The regeneration energy consumption and cyclic properties of desiccant impend the application of this technology [40]. Different from the above that can inhibit frost growth, the last one can improve the heating performances of ASHP under frosting conditions by promoting frost growth. This is because the air flow passing through the outdoor coil of the novel ASHP can be reversed, and the rapid reduction in heating performance caused by the frost on the windward side is retarded due to the even frost distribution [5,41].

Given that the defrosting of ASHP is avoidable, several studies on improving the defrosting technologies have been conducted [42]. Currently, the most widely used defrosting method is reverse cycle defrosting (RCD), which has the advantages of simple structure and low investment [43]. However, the RCD method faced the problems of long defrosting time and high energy consumption [44]. Accordingly, some researchers focused on the optimization of the RCD method from the aspect of improvements in control strategy [45] and components, such as outdoor coil [46], indoor coil [47–49], gas–liquid separator [50], electric expansion valve [51], and air fan [52]. To resolve the problem of insufficient defrosting energy, on the one hand, multiple heat exchangers [53] or multiple circuits [54,55] defrosting method is proposed. The core idea of this method is to defrost multiple heat exchangers or multiple circuits in turn. In addition to the defrosting heat exchangers or circuits, other heat exchangers or circuits serve as evaporators to provide energy for defrosting [56,57]. Although the multiple heat exchanger or multiple circuit defrosting method can improve the defrosting energy efficiency by up to 17.5% [53], it has the disadvantages of a complex system and control strategy, which impends its application. On the other hand, the thermal energy storage defrosting (TESD) method is proposed, and the core idea of this method is to add an energy storage component (ESC) used for defrosting [58–60]. The TESD method can significantly improve the defrosting energy efficiency and even realize uninterrupted indoor heating during the defrosting process. However, the energy stored by the ESC usually occupies a part of the system heating capacity, which may have some negative impacts on the heating performance [58].

Therefore in this chapter, a sustainable defrosting technology based on the TESD method is proposed, which has few negative impacts on the heating performances of an ASHP under the normal heating mode. The outline of this chapter is shown in Fig. 10.1. First, the development of the sustainable defrosting ASHP, such as the design of the phase change ESC and defrosting modes. Then the thermal stability of the selected phase change material (PCM) is experimentally investigated. This is followed by an optimization of the sustainable defrosting method in terms of a comparison between the RCD performances and the sustainable defrosting performances. Finally, concluding remarks are given, including conclusions of this work and expectations.

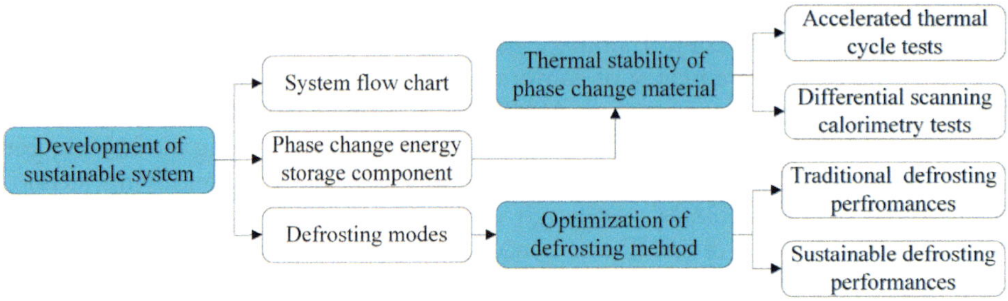

FIGURE 10.1 The outline of this chapter.

10.2 Development of a sustainable defrosting air source heat pump system

As mentioned in Section 10.1, the TESD method is a promising defrosting technology for ASHPs. However, the energy source for the ESC usually occupies a part of the system heating capacity, which may have some negative impacts on the heating performance. Accordingly, a sustainable defrosting technology based on the TESD method is proposed, which uses the heat dissipated by the compressor of an ASHP as the energy source of ESC. The designs of the sustainable defrosting ASHP system, phase change ESC, and defrosting modes are presented in this section.

10.2.1 Design of a sustainable defrosting air source heat pump system

Fig. 10.2A shows the schematic of the prototype of the ASHP system, and Fig. 10.2B shows that of the sustainable defrosting ASHP system. As seen, the novel system adds a phase change ESC to the conventional ASHP system, which stores the heat dissipated by the compressor to the surroundings during the normal heating process and serves as a low-level heat source of the system during the defrosting process. Besides, four valves are also added to realize the intended refrigerant cycles, such as refrigeration, heating, and defrosting cycles.

For the refrigeration cycle, Valve 1 is opened, Valves 2–4 are closed, and the four-way valve is reversed. The refrigerant with high temperature and pressure comes from the compressor, passes through the four-way valve, and then enters into the outdoor coil to condensation. Thereafter, the refrigerant is throttled by the electronic expansion valve and capillary tube and becomes the status of low temperature and pressure. Finally, the refrigerant passes through the indoor coil to evaporation and then comes back to the compressor. On the other hand, the heating cycle is opposite to the refrigeration cycle. The four-way valve is not reversed, and the on-off status of the added four valves in the heating cycle is the same as that of the refrigeration cycle. Moreover, two kinds of defrosting cycles in terms of whether to heat the indoor room during the defrosting process will be presented in detail in Section 10.2.3.

10.2.2 Design of phase change energy storage component

To design the phase change ESC, PCM should be selected first. In this experiment, the phase change temperature of PCM should be between the compressor wall temperature and the

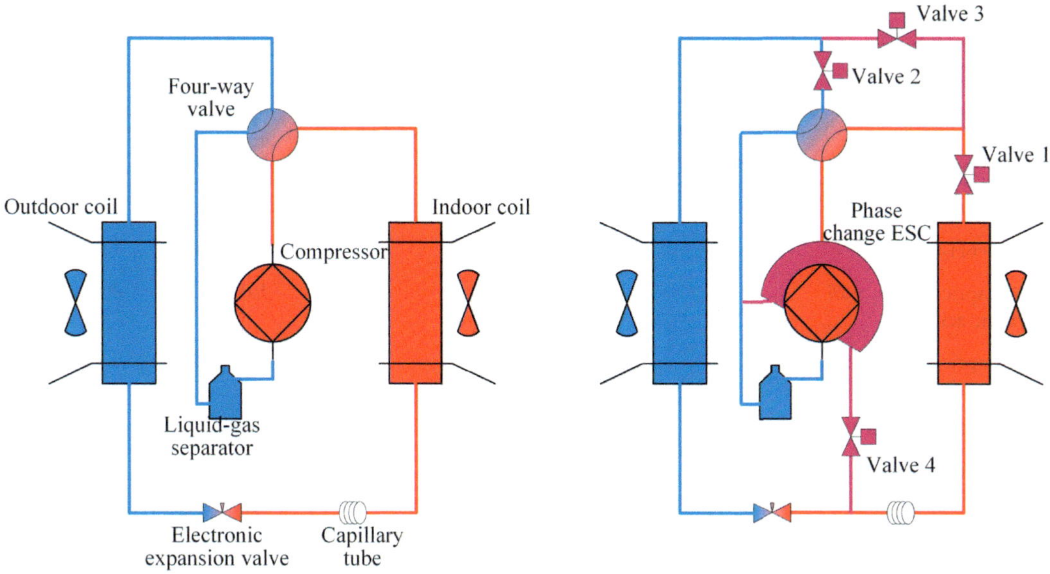

(A) Prototype of ASHP system (B) Sustainable defrosting ASHP system

FIGURE 10.2 Schematics of the (A) prototype; and (B) sustainable defrosting ASHP system. *ASHP*, Air source heat pump.

refrigerant evaporation temperature during the defrosting process. On the one hand, excessive phase change temperature will lead to excessive suction temperature of the compressor. Besides, it will increase the heat loss of the phase change ESC in the environment. On the other hand, excessively low phase change temperature will decrease the compressor surface temperature, which may affect the performance of the compressor. Accordingly, it is finally determined to use PCM with a phase change temperature of about 20°C. The physical property parameters of common suitable PCMs are shown in Table 10.1. Finally, the binary mixture of 65% Capric acid and 35% Lauric acid is selected in this experiment.

The next step is to calculate the amount of energy that can be stored in the phase change ESC. Dong et al. [61] indicated that the amount of energy used for melting the frost accounted for 70% of the total defrosting energy amount, and 60% of the total defrosting energy amount came from the low-level energy source. Therefore the amount of energy supplied by the phase change ESC for defrosting can be evaluated by

$$Q_1 = m_w \times L_w \div 0.6 \times 0.7 = 1.03 \times 334 \div 0.6 \times 0.7 = 401.4 \text{ kJ} \quad (10.1)$$

where m_w is the mass of the melting water, and L_w the latent heat of melting water.

Given that a part of the energy from the phase change ESC may be used for space heating, a margin of 35% is considered in calculating the amount of PCM. Besides, since the physical properties of PCMs change with their purity, it is necessary to check the phase change temperature and latent heat of the purchased PCM before designing the phase change ESC. In this work, a differential scanning calorimetry (DSC) test was conducted for the purchased PCM. The phase change temperature of the purchased PCM is 19.0°C, while

TABLE 10.1 The physical property parameters of common phase change materials.

PCM	Phase change temperature (°C)	Phase change latent heat (kJ/kg)	Density (kg/m^3)		Thermal conductivity (W/m °C)	
			Solid	Liquid	Solid	Liquid
$Mn(NO_3)_2 \cdot 6H_2O$	25.8	125.9	1795	1728	–	–
$CaCl_2 \cdot 6H_2O$	29	190	1800	1560	1.09	0.6
34% Capric acid + 66% myristic acid	24	147	1018	888	0.16	0.16
65% Capric acid + 35% lauric acid	19.5	132	900	895	0.14	0.14
Polyethylene glycol 400	8	99	1228	1125	–	0.19
Polyethylene glycol 600	20	127	1232	1126	–	0.19
Paraffin	22–24	189	900	760	0.21	–

its latent heat is 130.1 kJ/kg, which is a little different from the data provided in the literature. Therefore the volume of the PCM was evaluated by

$$V_{PCM} = \frac{(1+35\%) \times Q_1}{\rho_{PCM} L_{PCM}} = \frac{1.35 \times 401.4}{0.895 \times 130.1} = 4.7 \text{ L} \tag{10.2}$$

where ρ_{PCM} is the liquid density of PCM, and L_{PCM} the latent heat of PCM.

On the other hand, the height and diameter of the compressor are 210 and 114.8 mm, respectively. Considering that the liquid–gas separator is very close to the compressor, the effective arc length of the compressor that can be wrapped by the phase change ESC is 314 mm. As a result, the structure parameters of the phase change ESC are shown in Table 10.2, and its schematic is shown in Fig. 10.3. In general, the phase change ESC is composed of three parts, and they are the energy storage tank, PCM, and finned tube heat exchanger.

10.2.3 Design of defrosting modes

The sustainable defrosting modes of the ASHP system can be divided into the following two types: defrosting mode with indoor space heating and that without indoor space heating, as shown in Fig. 10.4A and B, respectively. For the defrosting mode with indoor space heating, Valve 2 is closed and the other valves are opened. The high-temperature and high-pressure refrigerant is discharged from the compressor through the four-way valve and divided into two parts. One part of the refrigerant enters the indoor coil through Valve 1 to condense and release heat and is then throttled by the capillary tube. The other part of the refrigerant enters the outdoor coil through Valve 3 to condense and defrost and then throttled by the electronic expansion valve. Thereafter, these two parts of refrigerant converge and enter the phase change ESC through Valve 4 to evaporate and absorb heat.

TABLE 10.2 Structure parameters of the phase change energy storage component.

Item	Value	Unit
Height	270.0	mm
Inner ring radius	58.0	mm
Inner ring arc length	314.0	mm
Outer ring radius	106.0	mm
Outer ring arc length	573.0	mm
Fin space	4.0	mm
Diameter of refrigerant tube	10.0	mm
Length of refrigerant tube	7.0	m
Volume of PCM	4.7	L

PCM, Phase change material.

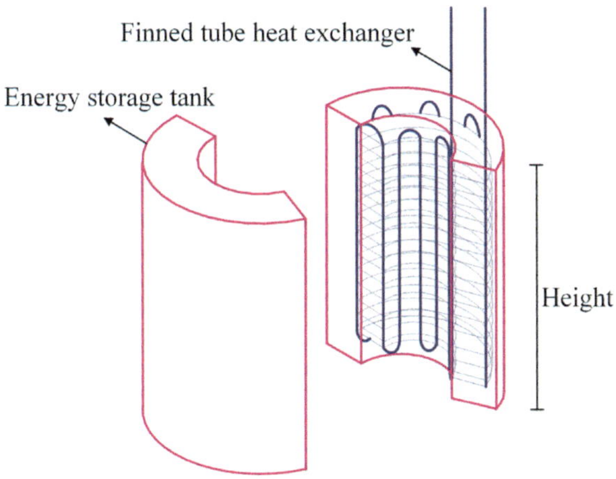

FIGURE 10.3 Schematic of the phase change energy storage component.

Finally, the refrigerant enters the liquid–gas separator and returns to the compressor to complete a cycle process.

For the defrosting mode without indoor space heating, Valves 1 and 2 are closed, and the other valves are opened. The high-temperature and high-pressure refrigerant is discharged from the compressor through the four-way valve and then enters the outdoor coil through Valve 3 to condense and defrost. Thereafter, the refrigerant is throttled by the electronic expansion valve and enters the phase change ESC through Valve 4 to evaporate and absorb heat. Finally, the refrigerant enters the liquid–gas separator and returns to the compressor to complete a cycle process.

Compared with the defrosting mode without indoor space heating, the defrosting mode with indoor space heating has the advantage of continuous heating of the room during the defrosting process. Accordingly, the defrosting mode with indoor space heating can

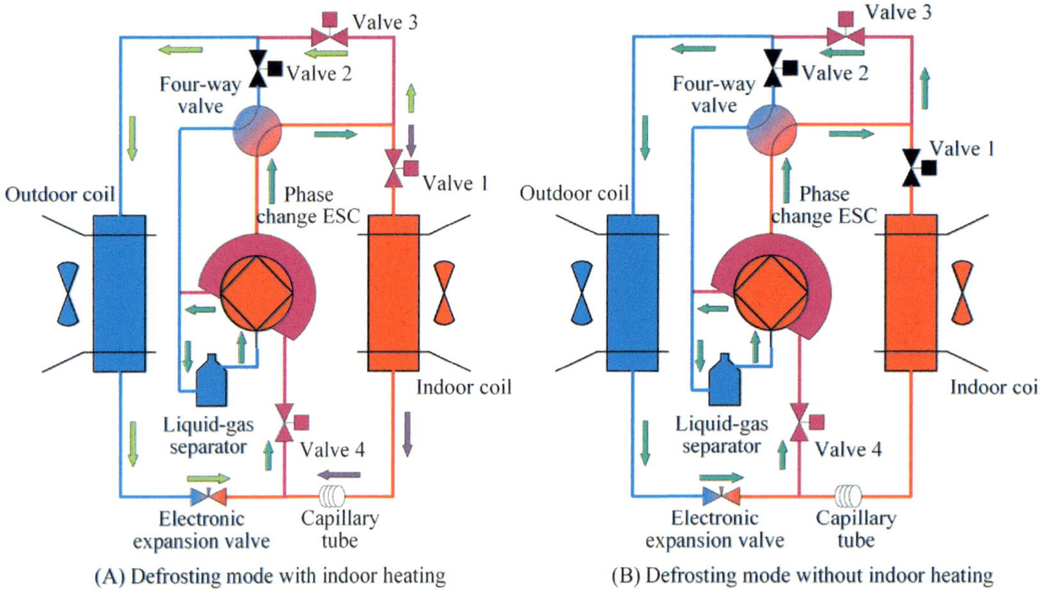

FIGURE 10.4 Schematics of defrosting mode (A) with indoor heating; (B) without indoor heating.

greatly improve indoor thermal comfort and ensure that there is no recovery heating stage after defrosting. However, in this defrosting mode, the refrigerant entering the indoor coil will absorb part of the heat of the phase change ESC, resulting in a reduction in defrosting energy, which may lead to an increase in defrosting time.

10.3 Thermal stability of the phase change material

Considering that the physical properties of PCM may change after a large number of phase change cycles, it is necessary to test the thermal stability of the selected PCM. In this work, the thermal stability of the PCM was mainly evaluated by comparing the variations of phase change temperature and latent heat after multiple cycles of energy charge and discharge.

10.3.1 Experimental setup

Fig. 10.5 shows the schematic of the accelerated thermal cycling experimental setup. It is composed of two parts: One is the cold and heat source unit, and the other is the automatic transposition unit. For the cold and heat source unit, it is composed of a compressor, condenser, evaporator, capillary, air fan, and automatic water refill device. The condenser is the high-temperature heat source with a temperature of 48.0°C ± 2.0°C, and the evaporator is the low-temperature cold source with a temperature of 6.0°C ± 2.0°C. The air fan is used as a supplementary cooling device. As the temperature of the heat source

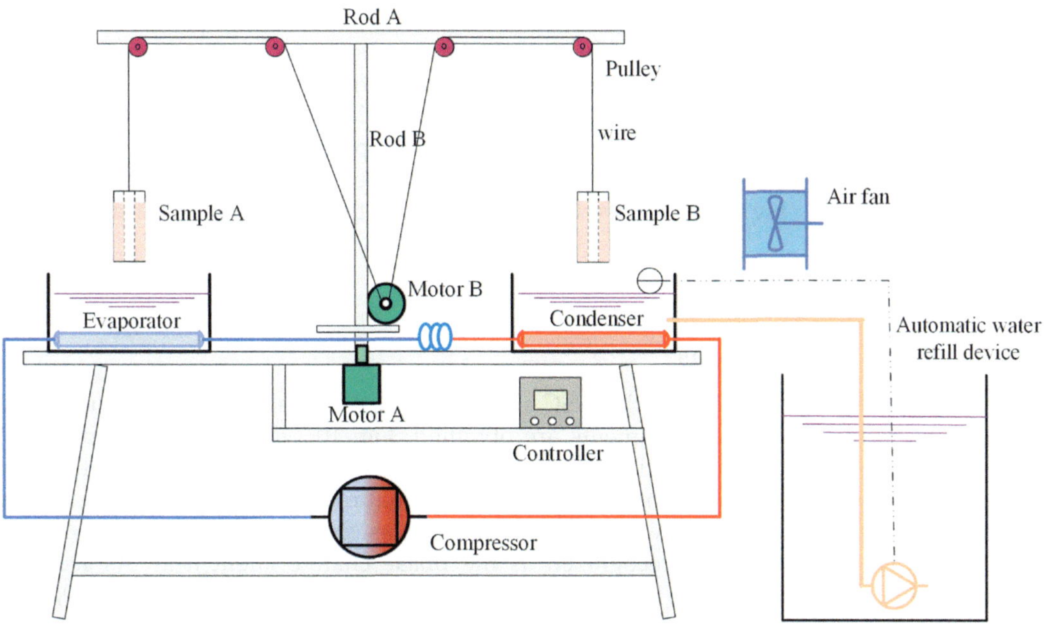

FIGURE 10.5 Schematic of the accelerated thermal cycling experimental setup.

increases, the temperature of the cold source increases. Therefore when the temperature of the cold source is higher than 8.0°C, the air fan is turned on to dissipate heat from the heat source, thereby reducing the temperature of the cold source. The automatic water refill device is turned on when the liquid level of the condenser drops to a certain height due to evaporation loss.

For the automatic transposition part, it is composed of Motor A, Motor B, Rod A, Rod B, Sample A, Sample B, pulleys, and a controller. Samples A, B, and Motor B are connected by pulleys and wires. Motor A drives Rods D and C to realize 180 degrees horizontal rotation to ensure the horizontal transposition of Samples A and B. Motor B realizes the vertical ascending and descending motion of Samples A and B through wires and pulleys, which ensures that Samples A and B are removed from the cold and heat sources before horizontal transposition and enter the cold and hot source after horizontal transposition. The controller realizes the programming control of Motors A and B to meet different action modes.

10.3.2 Experimental condition and procedure

The temperatures of the heat and cold sources of the accelerated thermal cycling setup are maintained at 48.0°C ± 2.0°C and 6.0°C ± 2.0°C, respectively, which meets the heat storage and release requirements of the selected PCM. Besides, the PCM is put into Samples A and B, which are made up of copper and have good thermal conductivity. A complete thermal cycle comprises three stages: heat storage, transposition, and heat release stages. The heat storage stage is immersing Sample A (B) into the heat source for about 10 mins to make sure all the housed PCM complete the energy storage process. Then the transposition stage is to transfer Sample A (B) from the heat source to the cold source. Finally, the heat release stage is immersing Sample A (B) into the cold source for about 10 minutes to make sure all the housed PCM complete the energy release process. The whole thermal cycle is accomplished by the accelerated thermal cycling experimental setup automatically. Fig. 10.6 shows the schematic of the accelerated thermal cycling procedure.

10.3.3 Results and analysis

According to the test scheme, the DSC results of the PCM after a certain number of accelerated thermal cycling experiments are obtained. Fig. 10.7 shows the DSC curves of the PCM after 0, 800, 1600, 2400, 3200, and 4000 times of accelerated thermal cycling experiments. The DSC tests are conducted for the thermal release process of PCM, and the initial phase change temperature point is recognized as the phase change temperature, while the area of the curve is recognized as the phase change latent heat. As seen, the phase change temperatures are 18.4°C, 18.5°C, 18.3°C, 18.0°C, and 18.2°C after 800, 1600, 2400, 3200, and 4000 times of accelerated thermal cycling experiments, while the corresponding phase change latent heats are 129.6, 123.5, 123.2, 120.1, and 119.8 kJ/kg, respectively.

The variation trends of phase change temperature and latent heat are shown in Fig. 10.8. As seen, the initial phase change temperature and latent heat are 19.0°C and 130.1 kJ/kg, respectively. In general, the phase change temperature of PCM decreases slightly with the thermal cycle, and the fluctuation range is 1.0°C. Besides, the phase change latent heat of PCM decreases continuously, but the magnitude of the decline is not large. After 4000 times of thermal cycles, the phase change temperature and latent heat decrease by 0.8°C and 7.9%, respectively. The experimental results show that the selected PCM has good thermal stability.

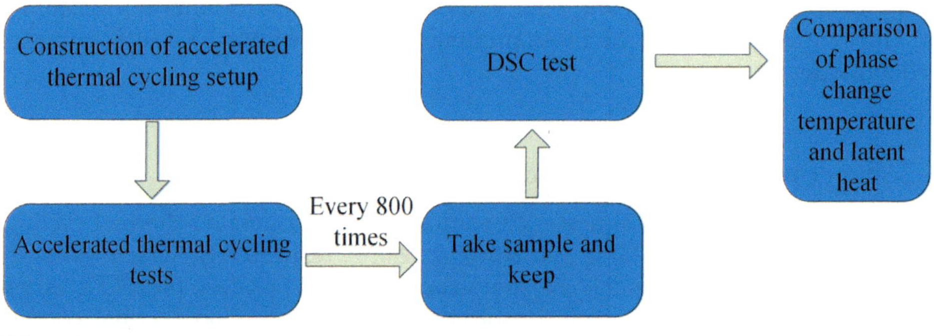

FIGURE 10.6 Schematic of the accelerated thermal cycling procedure.

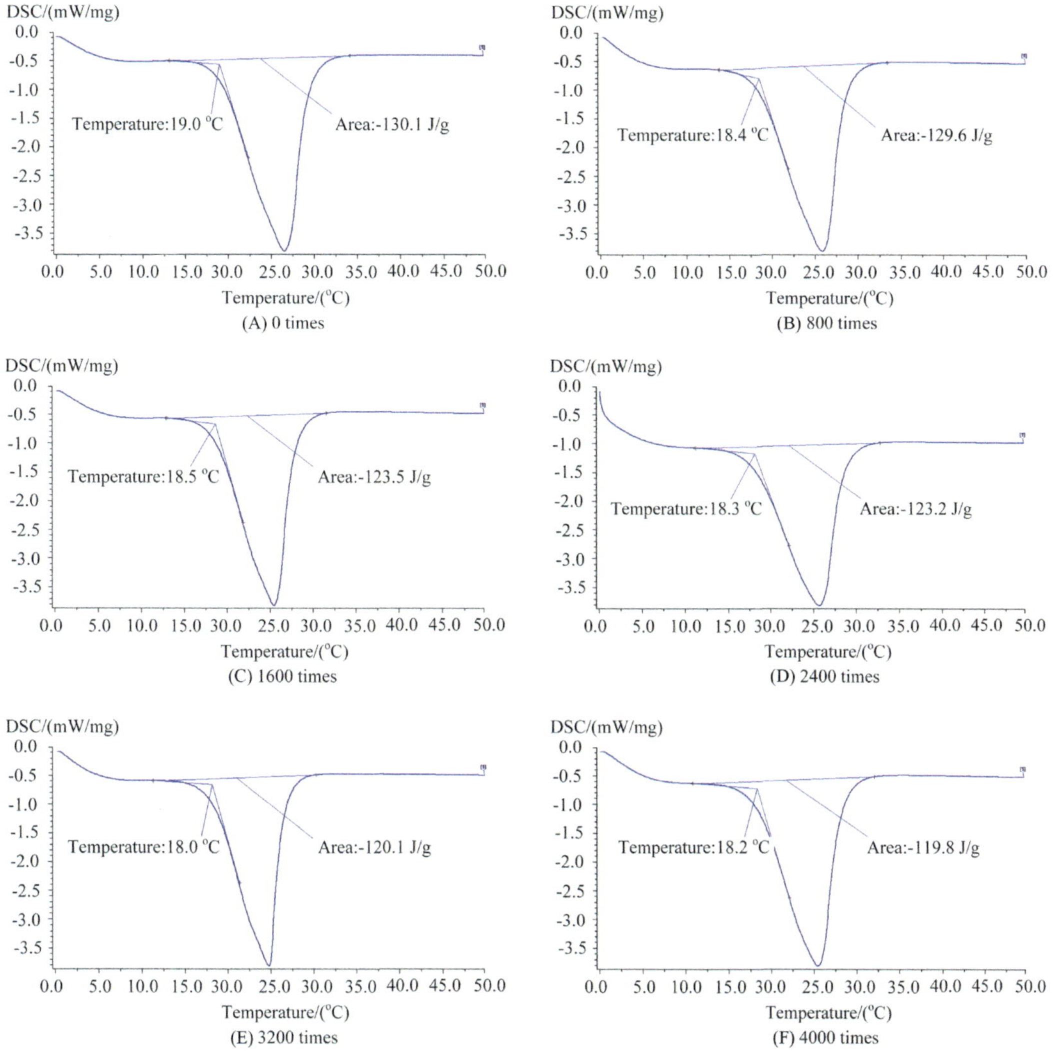

FIGURE 10.7 DSC curves of the PCM after (A) 0; (B) 800; (C) 1600; (D) 2400; (E) 3200; and (F) 4000 times of accelerated thermal cycling experiments. *DSC*, Differential scanning calorimetry; *PCM*, phase change material.

10.4 Optimization of the sustainable defrosting method

In this section, the traditional RCD and proposed sustainable defrosting performances are analyzed and discussed in detail. The specifications of the experimental ASHP are shown in Table 10.3. The evaluation indexes include the defrosting duration, defrosting energy consumption, compressor discharge and suction pressure, and so on. Based on the comparison results, the proposed sustainable defrosting method is further optimized, which can guide the control strategy of ASHP systems using the sustainable energy storage defrosting method.

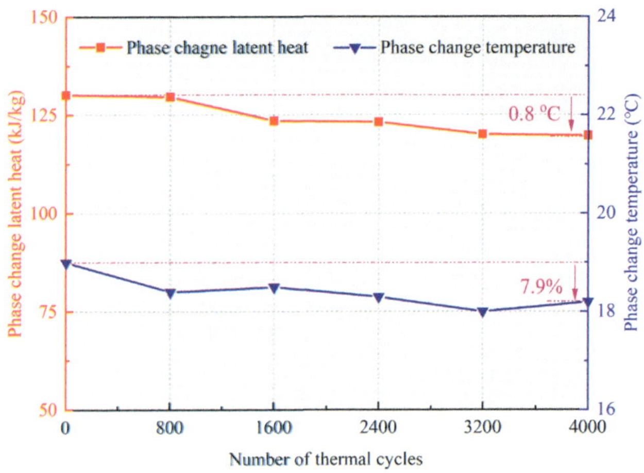

FIGURE 10.8 Variation trends of phase change temperature and latent heat.

TABLE 10.3 Specifications of the experimental air source heat pump.

No.	Item	Type/Value	Unit
1	Compressor type	Rolling rotor	–
2	Compressor input power for heating	2.13	kW
3	Rated heating capacity	8.9	kW
4	Indoor electric heating	2.0	kW
5	Indoor air flow rate	1200 (High)	m^3/h
		950	
		750	
6	Refrigerant	R410A	–

10.4.1 Traditional reverse cycle defrosting performances

Before investigating the sustainable defrosting performances of the improved ASHP system, its traditional RCD performances are first investigated. The simulated indoor dry bulb temperature is 20.0°C, and the outdoor dry and wet bulb temperatures are 2.0°C and 1.0°C, respectively. Besides, the frosting duration is fixed at 50 minutes, and then the system is turned into the traditional RCD operation. The whole traditional RCD operation can be divided into four stages, which are the predefrosting stage, RCD stage, postdefrosting stage, and resuming heating stage. The termination criteria for each stage are listed in Table 10.4. It should be noted that the outdoor air fan is closed in the predefrosting stage, RCD stage, and postdefrosting stage, while the indoor air fan is closed in all stages.

Fig. 10.9 shows the compressor suction and discharge pressures during the traditional RCD operation. As seen, the durations of the predefrosting stage, RCD stage, postdefrosting stage, and resuming heating stage are 75, 265, 60, and 75 seconds, respectively. In the

TABLE 10.4 Termination criteria of each stage of reverse cycle defrosting operation.

No.	Stage	Description	Termination criteria
1	Predefrosting	Compressor frequency step-down and shutdown	The difference between suction and discharge pressures is little
2	RCD	Four-way valve is reversed and compressor powered on	The outlet temperature of the bottom liquid refrigerant of the outdoor coil is greater than 15°C
3	Postdefrosting	Compressor frequency step-down and shutdown	The difference between suction and discharge pressures is little
4	Resuming heating	Four-way valve is reversed and compressor powered on	The outlet temperature of liquid refrigerant of the indoor coil is greater than 25°C

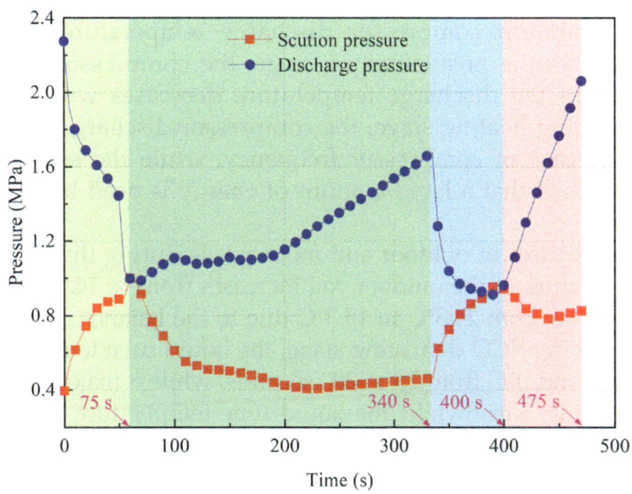

FIGURE 10.9 Compressor suction and discharge pressures during the RCD operation. *RCD*, Reverse cycle defrosting.

predefrosting stage, the compressor frequency gradually reduces and shuts down, which leads to the balance of compressor pressure. The suction pressure of the compressor increases from 0.39 to 1.01 MPa, and the discharge pressure decreases from 2.27 to 0.98 MPa. Then, the four-way valve reverses and the compressor powers on. In the RCD stage, the suction pressure first decreases and then slowly increases. The average and minimum suction pressures are 0.48 and 0.41 MPa, respectively. By contrast, the discharge pressure continues to rise and reaches the highest point of 1.68 MPa. The average discharge pressure in this stage is 1.23 MPa. In the postdefrosting stage, the compressor frequency gradually reduces and shuts down once again. At 400 seconds, the suction pressure rises to 0.94 MPa, and the discharge pressure drops to 0.92 MPa. In the resuming heating stage, the four-way valve reverses, and the compressor powers on. The suction pressure first decreases and then stabilizes at the value of 0.78 MPa, and the discharge pressure continues to rise. At 475 seconds, the system enters into a normal heating operation, while the indoor air fan is opened.

Fig. 10.10 shows the compressor suction and discharge temperatures during the traditional RCD operation. As seen, the compressor suction temperature increases from −18.2°C to 0°C due to the system pressure balance in the predefrosting stage. By contrast, the compressor discharge temperature gradually decreases from 45.2°C to 30.9°C. In the traditional RCD stage, the compressor suction and discharge temperatures gradually rise from 75 to 140 seconds due to the sufficient energy provided by the indoor coil in terms of the sensible energy stored in metal. From 140 to 200 seconds, the indoor coil cannot provide sufficient energy, which means that the refrigerant used for defrosting can only obtain heat through the compressor. As a result, the compressor suction and discharge temperatures show a declining trend and drop to 17.6°C and −15.2°C, respectively. From 200 to 340 seconds, the energy required for defrosting decreases as the frost layer of the outdoor coil reduces. Accordingly, the compressor discharge temperature continues to rise and reaches 27.4°C at 340 seconds. At this stage, the average, maximum, and minimum compressor suction temperatures are −11.4°C, 0.83°C, and −18.8°C, respectively. By contrast, the average, maximum, and minimum compressor discharge temperatures are 24.8°C, 33.8°C, and 17.6°C, respectively. In the postdefrosting stage, the compressor suction temperature gradually increases, but the discharge temperature decreases with the balance of system pressure. In the resuming heating stage, the compressor discharge temperature increases rapidly with the increase of compressor frequency, while the suction temperature increases slowly due to the fact that a large amount of energy is used to heat the coil and refrigerant.

Fig. 10.11 shows the liquid tube temperatures of outdoor and indoor coils during the RCD operation. As seen, the liquid tube temperature of the outdoor coil increases from −14.9°C to 0.7°C, while that of the indoor coil decreases from 24.3°C to 15.3°C due to the balance of system pressure in the predefrosting stage. In the RCD defrosting stage, the liquid tube temperature of the outdoor coil is stabilized at around 0°C from 75 to 200 seconds, while a majority of the frost layer on the outdoor coil is melting. Meanwhile, the liquid tube temperature of the indoor coil decreases from 15.0°C to −3.2°C, which means that the indoor coil provides

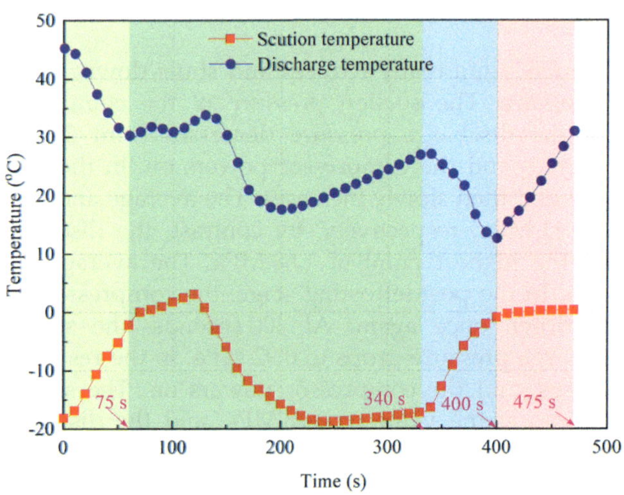

FIGURE 10.10 Compressor suction and discharge temperatures during the RCD operation. *RCD*, Reverse cycle defrosting.

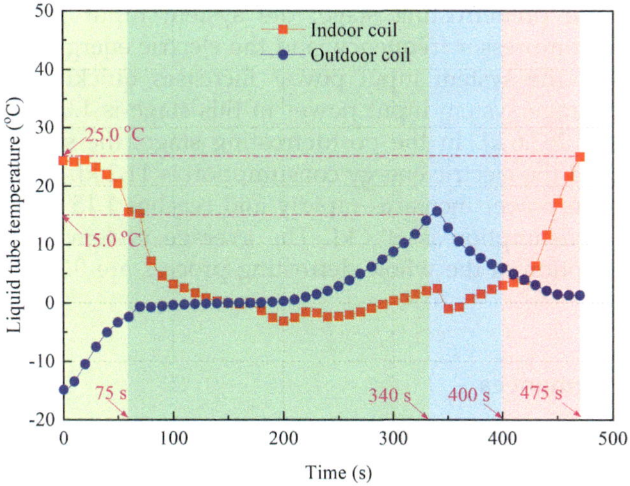

FIGURE 10.11 Liquid tube temperatures of indoor and outdoor coils during the RCD operation. *RCD*, Reverse cycle defrosting.

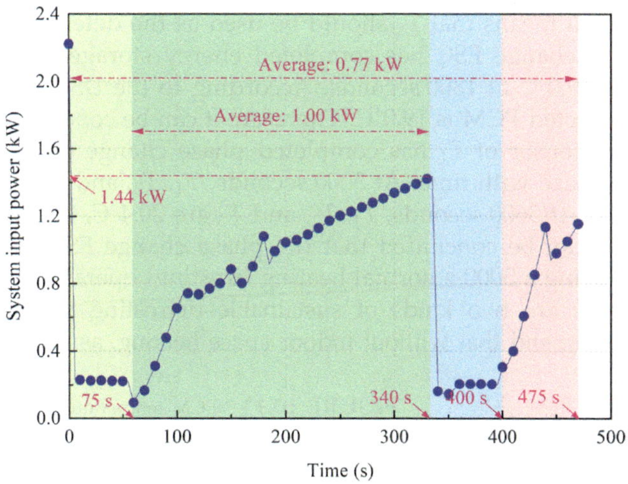

FIGURE 10.12 System input power during the RCD operation. *RCD*, Reverse cycle defrosting.

sufficient energy for defrosting. From 200 to 340 seconds, the liquid tube temperature of the outdoor coil consistently increases and reaches 15.4°C at 340 seconds, which means that all the frost layer has been removed. By contrast, the liquid tube temperature of the indoor coil gradually increases at the same time as the defrosting progresses. In the postdefrosting stage, the liquid tube temperature of the outdoor coil gradually decreases to 6.0°C, and that of the indoor coil decreases first and then increases to 2.9°C due to the balance of system pressure. In the resuming heating stage, the liquid tube temperature of the outdoor coil consistently decreases, and that of the indoor coil increases and reaches 25.0°C at 475 seconds when the indoor air fan is turned on for space heating.

Fig. 10.12 shows the system input power during the RCD operation. The system input power during the RCD operation mainly comprises the compressor power and other electronic components. As seen, in the predefrosting stage, the system input power decreases rapidly due to the decreased compressor frequency, and the electric energy consumption is 25.5 kJ. In the RCD stage, the system input power increases quickly and reaches 1.44 kW at 340 seconds. The average system input power in this stage is 1.00 kW, and the electric energy consumption is 267.6 kJ. In the postdefrosting stage, the system input power decreases rapidly again, and the electric energy consumption is 11.2 kJ. In the resuming heating stage, the system input power increases rapidly and reaches 1.15 kW at 475 seconds, and the electric energy consumption is 60.4 kJ. The average system input power and total electric energy consumption of the whole defrosting process are 0.77 kW and 364.7 kJ, respectively.

10.4.2 Sustainable defrosting performances

To better understand the sustainable defrosting performances of the improved ASHP system, the energy storage characteristics of the phase change ESC are investigated first. Fig. 10.13 shows the PCM temperatures during the normal heating operation. T_1, T_2, and T_3 are placed in the bottom, middle, and top of the phase change ESC. As seen, the PCM temperature increases with height, which means that T_1 should be used as the determination temperature of whether the phase change ESC has completed energy storage. As a result, it can be found that T_1 reaches 19.0°C at 1800 seconds. According to the DSC test, the phase change temperature of the selected PCM is 19.0°C. Therefore it can be concluded that the PCM around the temperature sensor of T_1 has completed phase change energy storage. Thereafter, T_1 continues to increase with time. At 3000 seconds, T_1, T_2, and T_3 are 20.0°C, 20.3°C, and 21.3°C, respectively. At 3600 seconds, T_1, T_2, and T_3 are 20.4°C, 20.6°C, and 21.7°C, respectively. In general, it can be concluded that the phase change ESC can complete the energy storage process during a 3000-s normal heating (frosting) operation.

As mentioned in Section 10.2.3, there are two kinds of sustainable defrosting modes: defrosting mode with indoor space heating and that without indoor space heating, as shown

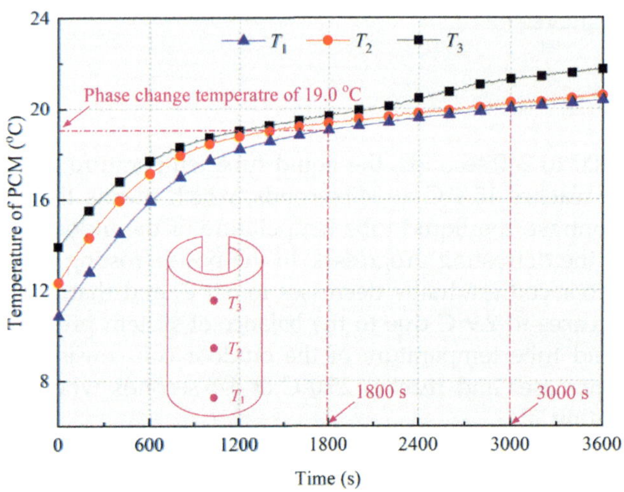

FIGURE 10.13 PCM temperatures during the normal heating operation. *PCM*, Phase change material.

in Fig. 10.4A and B, respectively. Considering that the phase change ESC cannot provide sufficient energy for space heating during the defrosting process, another sustainable defrosting mode, which is supplemented by indoor electric heating, is defined based on the defrosting mode with indoor space heating. The characteristics of these three kinds of sustainable defrosting modes are listed in Table 10.4. In theory, the defrosting rate of Mode 1 should be higher than those of Modes 2 and 3, because all the stored energy is used for defrosting in Mode 1, while that is also used for indoor space heating in Modes 2 and 3. Besides, the defrosting performances except for the indoor thermal comfort between Modes 2 and 3 should be similar, because the only difference between them is the indoor electric heating. In this work, the power of indoor electric heating is 2.0 kW, which can significantly improve indoor thermal comfort during the defrosting operation. Based on Table 10.5, the system characteristics during the defrosting process, including the compressor suction and discharge pressures and temperatures, the bottom liquid tube temperature of the outdoor coil, the liquid tube temperature of the indoor coil, the inlet and outlet air temperatures of indoor coil, and the system input power, are further analyzed in the following figures.

Fig. 10.14 shows the compressor suction pressures during the sustainable defrosting operation in Modes 1–3. As seen, the defrosting process in Mode 1 comprises the defrosting stage and resuming heating stage, while that in Modes 2 and 3 only comprises the defrosting stage. This is because a part of high temperature and pressure refrigerant pumping from the compressor passes through the indoor coil during the sustainable

TABLE 10.5 Sustainable defrosting modes.

Mode	System flow	Indoor space heating	Supplemented by indoor electric heating
1	Fig. 10.4B	No	No
2	Fig. 10.4A	Yes	No
3	Fig. 10.4A	Yes	Yes

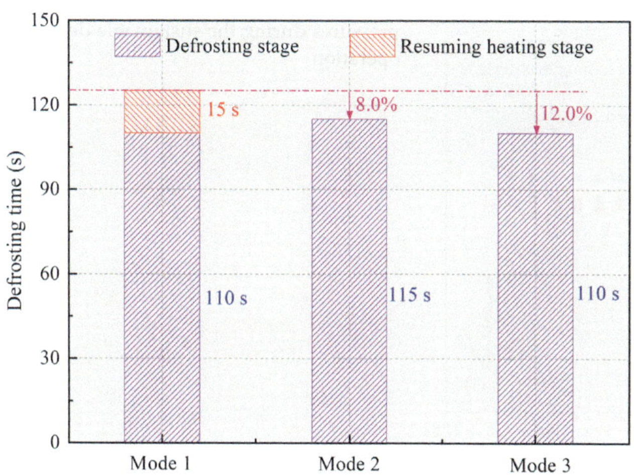

FIGURE 10.14 Defrosting time in sustainable defrosting modes.

defrosting operation in Modes 2 and 3, which effectively prevents the indoor coil temperature from being too low to result in a resuming heating stage. Besides, the durations of the defrosting stage in Modes 1–3 are 110, 115, and 110 seconds, which means that the difference in time spent on defrosting is little in these three modes. By contrast, given the entire defrosting time, which includes the defrosting stage and resuming heating stage, it reduces by 8.0% and 12.0% in Modes 2 and 3 when compared to Mode 1. Fig. 10.15 shows the compressor suction pressures during the sustainable defrosting operation in Modes 1–3. As seen, the overall variation trends of compressor suction pressure in Modes 1–3 are the same. That is, the compressor suction pressure rapidly increases in the first 10–15 seconds due to the closing and opening of valves and then fluctuates around 0.7 MPa until the end of the defrosting stage. The average compressor suction pressures during the defrosting stage in Modes 1–3 are 0.74, 0.71, and 0.71 MPa, and the corresponding maximum values in Modes 1–3 are 0.78, 0.79, and 0.79 MPa, respectively. For Mode 1, the compressor suction pressure increases rapidly at first and then decreases in the resuming heating stage. In general, it can be found that the compressor suction pressure during the sustainable defrosting operation is maintained at a high level due to the sufficient defrosting energy source from the phase change ESC.

Fig. 10.16 shows the compressor discharge pressures during the sustainable defrosting operation in Modes 1–3. Similar to the compressor suction pressure, the overall variation trends of compressor discharge pressure in Modes 1–3 are the same. That is, the compressor discharge pressure rapidly decreases in the first 5 seconds due to the closing and opening of valves and then consistently increases until the end of the defrosting stage. Compared to Modes 2 and 3, the variation trend of compressor discharge pressure in Mode 1 is more rapid due to that a part of the refrigerant remains in the indoor coil. The average compressor discharge pressures during the defrosting stage in Modes 1–3 are 1.80, 1.62, and 1.65 MPa, and the corresponding maximum values in Modes 1–3 are 2.59, 2.30, and 2.35 MPa, respectively. For Mode 1, the compressor discharge pressure decreases rapidly from 2.30 MPa at 110 seconds to 1.83 MPa at 115 seconds and then increases to 2.07 MPa at

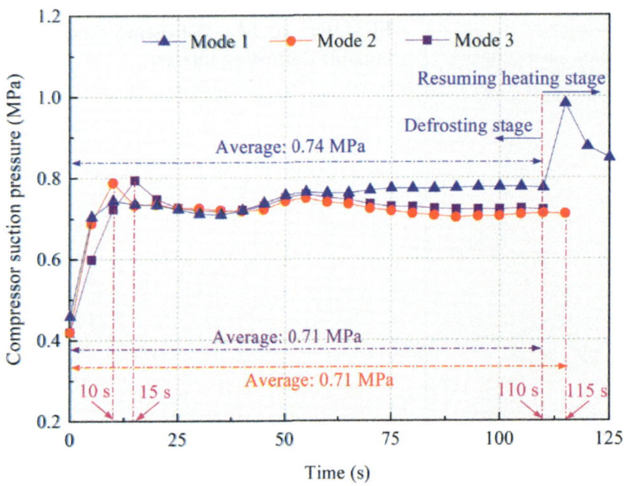

FIGURE 10.15 Compressor suction pressures during the sustainable defrosting operation.

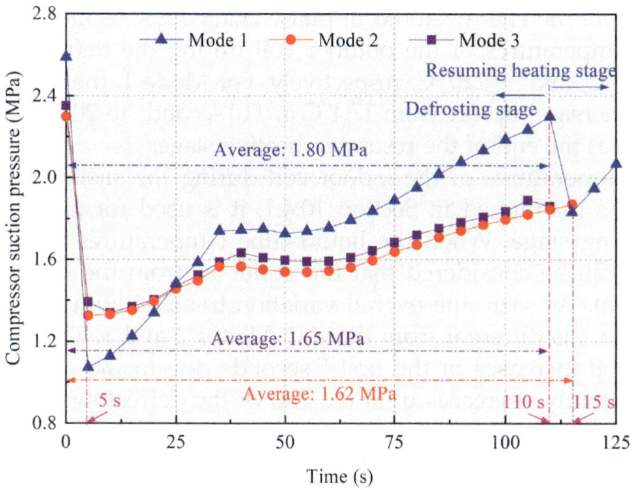

FIGURE 10.16 Compressor discharge pressures during the sustainable defrosting operation.

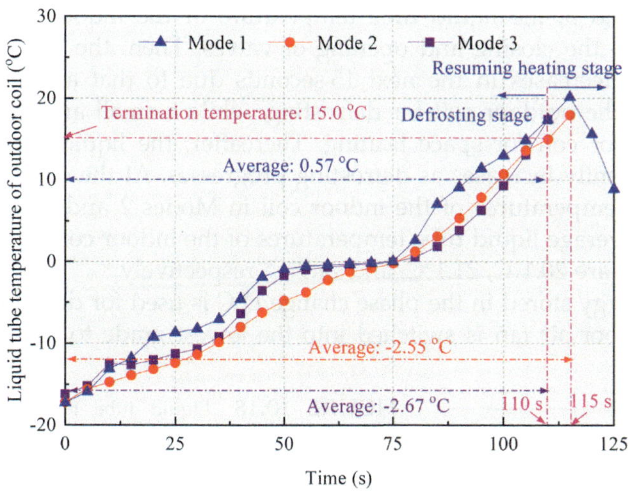

FIGURE 10.17 Liquid tube temperatures of the outdoor coil during the sustainable defrosting operation.

the end of the resuming heating stage. In general, it can be found that the compressor discharge pressure during the sustainable defrosting operation is also maintained at a high level due to the sufficient defrosting energy source from the phase change ESC.

Fig. 10.17 shows the liquid tube temperatures of the outdoor coil during the sustainable defrosting operation in Modes 1–3. As mentioned in Section 10.4.1, it is used for judging the termination of the defrosting stage. When the liquid tube temperatures of the outdoor coil are higher than 15.0°C, it can be considered that the entire frost layer on the outdoor coil has been melted and drained away. As seen, the overall variation trends of liquid tube temperatures of the outdoor coil in Modes 1–3 are the same. That is, the liquid tube temperatures of the outdoor coil consistently increase and fluctuate around 0°C for a while, and then continue to increase. Compared to Modes 2 and 3, the liquid tube temperatures of the outdoor coil in

Mode 1 are higher due to the fact that all the energy stored in phase change ESC is used for defrosting. The average liquid tube temperatures of the outdoor coil during the defrosting stage in Modes 1–3 are 0.57°C, −2.55°C, and −2.67°C, respectively. For Mode 1, the liquid tube temperatures of the outdoor coil increase rapidly from 17.4°C at 110 seconds to 20.2°C at 115 seconds and then decrease to 8.9°C at the end of the resuming heating stage.

Fig. 10.18 shows the liquid tube temperatures of the indoor coil during the sustainable defrosting operation in Modes 1–3. As mentioned in Section 10.4.1, it is used for judging the termination of resuming the heating stage. When the liquid tube temperatures of the indoor coil are higher than 25.0°C, it can be considered that the outlet air from the indoor coil will not cause a cold draft problem. As seen, the overall variation trend of liquid tube temperature of the indoor coil in Mode 1 is different from those in Modes 2 and 3. The liquid tube temperature of the indoor coil increases in the first 5 seconds due to the closing and opening of valves and then consistently decreases until the end of the defrosting stage. This is because the inlet to the indoor coil is closed by Valve 1, and there is no refrigerant circulation. At the end of the defrosting stage, the liquid tube temperature of the indoor coil in Mode 1 is 16.3°C, which is far below 25.0°C. Accordingly, a resuming heating stage is necessary for Mode 1, which increases the liquid tube temperature of the indoor coil to 25.4°C at 125 seconds. In Modes 2 and 3, the liquid tube temperature of the indoor coil increases in the first 5 seconds due to the closing and opening of valves. Then, the liquid tube temperature of the indoor coil decreases in the next 15 seconds due to that a large amount of refrigerant flows through the outdoor coil for defrosting, while a small amount of refrigerant flows through the indoor coil for space heating. Thereafter, the liquid tube temperature of the indoor coil consistently increases as defrosting progresses. At the end of the defrosting stage, the liquid tube temperatures of the indoor coil in Modes 2 and 3 are 25.2°C and 25.6°C, respectively. The average liquid tube temperatures of the indoor coil during the defrosting stage in Modes 1–3 are 20.1°C, 21.1°C, and 21.9°C, respectively.

Considering that the majority of energy stored in the phase change ESC is used for defrosting rather than space heating, the indoor air fan is switched into the lowest grade to avoid

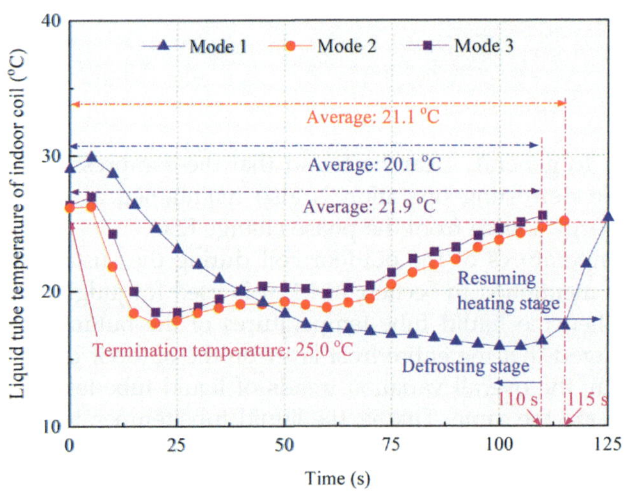

FIGURE 10.18 Liquid tube temperatures of the indoor coil during the sustainable defrosting operation.

insufficient defrosting energy during the sustainable defrosting operation in Modes 2 and 3. Fig. 10.19 shows the inlet and outlet air temperatures of the indoor coil during the sustainable defrosting operation in Modes 2 and 3. As seen, the inlet air temperatures of the indoor coil are both maintained at around 20.0 ± 0.3°C, which is the set value of the experimental conditions. The variation trends of the outlet air temperature of the indoor coil in Modes 2 and 3 are the same. That is, the outlet air temperature of the indoor coil decreases in the first 25 seconds due to that a large amount of refrigerant flows through the outdoor coil for defrosting, while a small amount of refrigerant flows through the indoor coil for space heating. Thereafter, the outlet air temperature of the indoor coil consistently increases as defrosting progresses. At the end of the defrosting stage, the outlet air temperatures of the indoor coil in Modes 2 and 3 are 27.2°C and 37.9 °C, respectively. The average outlet air temperatures of the indoor coil during the defrosting stage in Modes 2 and 3 are 23.9 °C and 33.5 °C, respectively. Actually, the only difference between Modes 2 and 3 is that Mode 3 is supplemented by indoor electric heating of 2.0 kW, which can significantly increase the outdoor air temperature of the indoor coil by 9.6 °C during the defrosting operation.

Fig. 10.20 shows the system input power during the sustainable defrosting operation. It should be noted that the system input power mainly comprises the compressor power, the indoor air fan power, and other electronic components except the indoor electric heating. As seen, the overall variation trends of system input power in Modes 1−3 are the same. That is, the system input power decreases in the first 25 seconds due to the redistribution of system refrigerant. Thereafter, the system input power consistently increases as defrosting progresses. At the end of the defrosting stage, the system input powers in Modes 1−3 are 2.77, 2.31, and 2.30 kW, respectively. The average system input powers during the defrosting stage in Modes 1−3 are 2.24, 2.09, and 2.11 kW, respectively. For Mode 1, the system input power decreases rapidly from 2.77 kW at 110 seconds to 2.06 kW at 120 seconds and then decreases to 2.32 kW at the end of the resuming heating stage. In general, compared to Modes 2 and 3, the system input power in Mode 1 is higher due to the fact that all the energy stored in phase change ESC is used for defrosting, which leads to an easier defrosting process as reflected by the compressor suction and discharge pressures shown in Figs. 10.15 and 10.16.

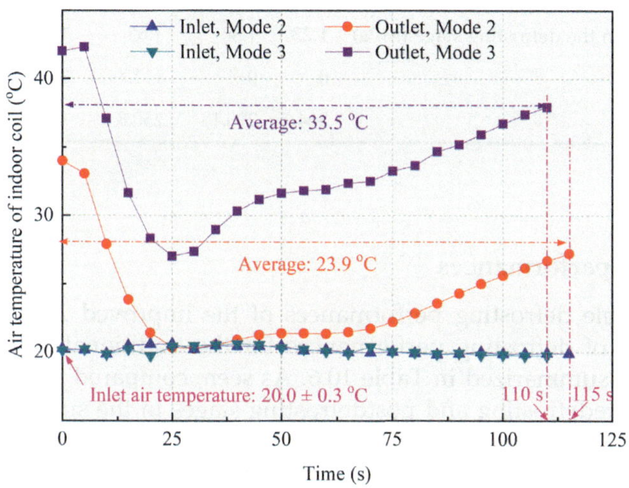

FIGURE 10.19 Inlet and outlet air temperatures of the indoor coil during the sustainable defrosting operation.

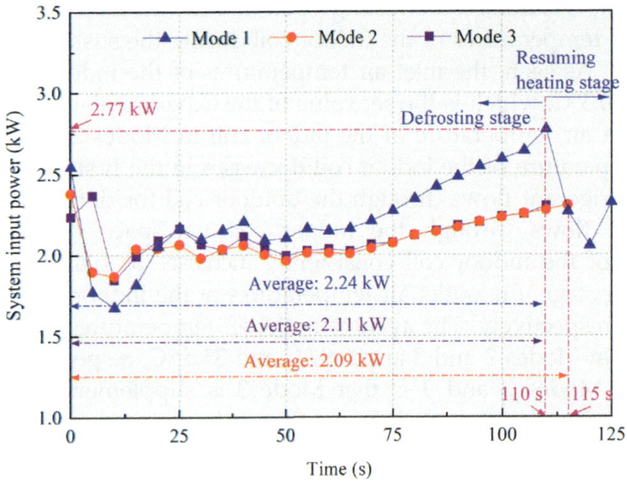

FIGURE 10.20 System input power during the sustainable defrosting operation.

TABLE 10.6 The main evaluation indexes of defrosting performances.

No.	Item	RCD	Sustainable defrosting		
			Mode 1	**Mode 2**	**Mode 3**
1	Duration of predefrosting stage(s)	75	0	0	0
2	Duration of defrosting stage(s)	265	110	115	110
3	Duration of postdefrosting stage(s)	60	0	0	0
4	Duration of resuming heating stage(s)	75	15	0	0
5	Total defrosting time(s)	475	125	115	110
6	Average compressor suction pressure in the defrosting stage (MPa)	0.48	0.74	0.71	0.71
7	Average compressor discharge pressure in the defrosting stage (MPa)	1.23	1.80	1.62	1.65
8	Total space heating energy amount (kJ)	0	0	111.4	336.6
9	Total defrosting energy consumption (kJ)	364.7	290.1	250.8	451.5

RCD, Reverse cycle defrosting.

10.4.3 Comparison of defrosting performances

To better understand the sustainable defrosting performances of the improved ASHP system, the main evaluation indexes of defrosting performances for the traditional RCD and sustainable defrosting modes are summarized in Table 10.6. As seen, compared to the traditional RCD mode, there are no predefrosting and postdefrosting stages in the sustainable defrosting modes because they do not need the reversing operation of the four-way valve and shutdown. Besides, the duration of the defrosting stage is significantly shortened in the sustainable defrosting modes, which is shortened by 58.5% when compared to the traditional RCD mode. The duration of resuming the heating stage is shortened by 80% in sustainable defrosting Mode 1 when compared to the traditional RCD mode. For the sustainable defrosting Modes 2 and 3, there is no resuming heating stage since they can realize continuous space heating during the defrosting process. Compared to the traditional RCD mode, the total defrosting times of sustainable defrosting Modes 1–3 are shortened by 73.7%, 75.8%, and 76.8%, respectively.

On the other hand, since the sustainable defrosting modes can provide sufficient defrosting energy, the defrosting performances such as average compressor suction and discharge pressure are improved. Compared to the traditional RCD mode, the average compressor suction pressures of sustainable defrosting Modes 1–3 are increased by 54.2%, 47.9%, and 47.9%, and the corresponding average compressor discharge pressures are increased by 46.3%, 31.7%, and 34.1%, respectively. For the traditional RCD mode and sustainable defrosting Mode 1, there is no space heating operation during the defrosting process. By contrast, the total space heating energy amounts in sustainable defrosting Modes 2 and 3 are 111.4 and 336.6 kJ, respectively. It can be found that the total space heating energy amount in sustainable defrosting Mode 3 is 3.02 times that in Mode 2 with the help of indoor electric heating. Accompanying, the total defrosting energy consumption in sustainable defrosting Mode 3 is 1.80 times that in Mode 2. Compared to the traditional RCD mode, the total defrosting energy consumptions of sustainable defrosting Modes 1 and 2 are reduced by 20.5% and 31.2%, respectively. In general, sustainable defrosting Mode 2 shows the best defrosting performance because it considers both energy consumption and indoor thermal comfort. If indoor thermal comfort is the main concern, the sustainable defrosting Mode 3 is a better choice.

10.5 Concluding remarks

In this chapter, a sustainable defrosting ASHP system, which adds a phase change ESC around the compressor to store the heat dissipated by the compressor, is proposed and experimentally investigated in detail, as well as the thermal stability of the PCM. Compared to the traditional RCD mode, there are no predefrosting and postdefrosting stages in the sustainable defrosting modes because they do not need the reversing operation of the four-way valve and shutdown. The total defrosting times of sustainable defrosting Modes 1–3 are shortened by 73.7%, 75.8%, and 76.8%, the corresponding average compressor suction pressures are increased by 54.2%, 47.9%, and 47.9%, and the corresponding average compressor discharge pressure is increased by 46.3%, 31.7%, and 34.1%, respectively, when compared to the traditional RCD mode.

Compared to the traditional RCD mode, the total defrosting energy consumptions of sustainable defrosting Modes 1 and 2 are reduced by 20.5% and 31.2%, respectively. With the help of indoor electric heating, the total space heating energy amount in sustainable defrosting Mode 3 is 3.02 times that in Mode 2. Accompanying, the total defrosting energy consumption in sustainable defrosting Mode 3 is 1.80 times that in Mode 2. In general, sustainable defrosting Mode 2 shows the best defrosting performance because it considers both energy consumption and indoor thermal comfort. If indoor thermal comfort is the main concern, sustainable defrosting Mode 3 is a better choice. The effects of the added phase change ESC on the system heating and cooling performances will be further investigated in future work. The results of this chapter may help to better understand the proposed sustainable defrosting ASHP system.

References

[1] P.M. Congedo, C. Baglivo, D. D'Agostino, D. Mazzeo, The impact of climate change on air source heat pumps, Energy Conversion and Management 276 (2023) 116554.

[2] G. Leonzio, P.S. Fennell, N. Shah, Air-source heat pumps for water heating at a high temperature: State of the art, Sustainable Energy Technologies and Assessments 54 (2022) 102866.

[3] L. Zhang, Y. Jiang, J. Dong, Y. Yao, Advances in vapor compression air source heat pump system in cold regions: a review, Renewable and Sustainable Energy Reviews 81 (2018) 353–365.

[4] M. Takayashiki, K. Nishimura, A. Sciazko, T. Okabe, J. Taniguchi, N. Shikazono, Three dimensional reconstruction of frost structure by replica method, International Journal of Heat and Mass Transfer 196 (2022) 123280.

[5] X. Bai, S. Liu, S. Deng, L. Zhang, M. Wei, An experimental study on achieving even-frosting for an air source heat pump using a novel dual-fan outdoor coil, Energy and Buildings 255 (2022) 111695.

[6] L. Zhang, M. Song, S. Deng, J. Shen, C. Dang, Frosting mechanism and behaviors on surfaces with simple geometries: a state-of-the-art literature review, Applied Thermal Engineering 215 (2022) 118984.

[7] R. Mukherjee, S.F. Ahmadi, H. Zhang, R. Qiao, J.B. Boreyko, Electrostatic jumping of frost, ACS Nano 15 (2021) 4669–4677.

[8] L. Zhang, M. Song, C.Y.H. Chao, J. Shen, An experimental study on the dynamic frosting characteristics on the edge zone of a horizontal copper plate under forced convection, International Journal of Heat and Mass Transfer 200 (2023) 123541.

[9] L. Zhang, M. Song, J. Shen, X. Zhang, Y. Xu, Y. Hu, Impacts of initial cooling rate on local frosting characteristics of horizontal cold plate surface with edge effect considered, International Communications in Heat and Mass Transfer 143 (2023) 106654.

[10] L. Zhang, M. Song, T. Zhan, J. Shen, Cold plate temperature effect on droplet and frost crystal behaviors at the early condensation frosting stage considering plate edge effect, Engineered Science (2022).

[11] L. Zhang, M. Song, C.Y.H. Chao, C. Dang, J. Shen, Localized characteristics of the first three typical condensation frosting stages in the edge region of a horizontal cold plate, Micromachines 13 (2022) 1906.

[12] X. Liu, J. Yu, G. Yan, An experimental study on the air side heat transfer performance of the perforated fin-tube heat exchangers under the frosting conditions, Applied Thermal Engineering 166 (2020) 114634.

[13] W. Lee, K.J. Bae, O.K. Kwon, Effect of hydrophobic surfaces on frost retardation in fin-tube heat exchangers with various fin pitches, Applied Thermal Engineering 176 (2020) 115424.

[14] D.L. Da Silva, C.J.L. Hermes, C. Melo, Experimental study of frost accumulation on fan-supplied tube-fin evaporators, Applied Thermal Engineering 31 (2011) 1013–1020.

[15] L. Zhang, Y.Q. Jiang, J.K. Dong, Y. Yao, S.M. Deng, An experimental study on the effects of frosting conditions on frost distribution and growth on finned tube heat exchangers, International Journal of Heat and Mass Transfer 128 (2019) 748–761.

[16] L. Zhang, Y.Q. Jiang, J.K. Dong, Y. Yao, S.M. Deng, A comparative study of frosting behavior on finned tube heat exchanger under different fan control modes, Applied Thermal Engineering 160 (2019) 114063.

[17] L. Zhang, Y.Q. Jiang, J.K. Dong, Y. Yao, S.M. Deng, An experimental study of frost distribution and growth on finned tube heat exchangers used in air source heat pump units, Applied Thermal Engineering 132 (2018) 38–51.

[18] L. Zhang, M. Song, S.H. Hosseini, J. Shen, Y. Jiang, A modeling study of spatial and temporal frost growth on the edge of windward fins for a tube-finned heat exchanger, International Journal of Heat and Mass Transfer 183 (2022) 122093.

[19] L. Zhang, M. Song, N. Mao, J. Dong, Temporal and spatial frost growth prediction of a tube-finned heat exchanger considering frost distribution characteristics, International Journal of Heat and Mass Transfer 183 (2022) 122192.

[20] W. Wu, J. Luo, D. Li, X. Feng, L. Tang, Z. Fang, et al., Experimental Investigation of Heat Transfer Performance of a Finned-Tube Heat Exchanger under Frosting Conditions, Sustainable Cities and Society 80 (2022) 103752.
[21] G.C. Gong, J.C. Tang, D.Y. Lv, H.J. Wang, Research on frost formation in air source heat pump at cold-moist conditions in central-south China, Applied Energy 102 (2013) 571–581.
[22] X.M. Guo, Y.G. Chen, W.H. Wang, C.Z. Chen, Experimental study on frost growth and dynamic performance of air source heat pump system, Applied Thermal Engineering 28 (2008) 2267–2278.
[23] M.J. Song, C.Y. Tso, C.Y. Chao, C.L. Wu, Techno-economic analysis on frosting/defrosting operations for an air source heat pump unit with an optimized multi-circuit outdoor coil, Energy and buildings 166 (2018) 165–177.
[24] K. Kim, D.R. Kim, K.S. Lee, Local frosting behavior of a plated-fin and tube heat exchanger according to the refrigerant flow direction and surface treatment, International Journal of Heat and Mass Transfer 64 (2013) 751–758.
[25] K. Kim, K.S. Lee, Frosting and defrosting characteristics of surface-treated louvered-fin heat exchangers: Effects of fin pitch and experimental conditions, International Journal of Heat and Mass Transfer 60 (2013) 505–511.
[26] A.J. Mahvi, K. Boyina, A. Musser, S. Elbel, N. Miljkovic, Superhydrophobic heat exchangers delay frost formation and enhance efficency of electric vehicle heat pumps, International Journal of Heat and Mass Transfer 172 (2021) 121162.
[27] D. Seo, S. Oh, B. Moon, H. Kim, J. Kim, C. Lee, et al., Influence of lubricant-mediated droplet coalescence on frosting delay on lubricant impregnated surfaces, International Journal of Heat and Mass Transfer 128 (2019) 217–228.
[28] F. Wang, C. Liang, X. Zhang, Research of anti-frosting technology in refrigeration and air conditioning fields: a review, Renewable and Sustainable Energy Reviews 81 (2018) 707–722.
[29] S. Yu, Y. Su, W. Cai, G. Qiu, Experimental investigation on an air source heat pump system with a novel anti-frosting evaporator, Applied Thermal Engineering 221 (2023) 119910.
[30] T. Yajima, H. Ohkubo, M. Seki, Research on defrost free air-source heat pump with surface stripe concavo-convex fins, Energy and Buildings 231 (2021) 110568.
[31] G.G. Heidinger, S.M. Nascimento, P.D. Gaspar, P.D. Silva, Experimental study of the fins arrangement pattern of refrigerated display cabinet evaporator towards thermal performance improvement, Applied Thermal Engineering 138 (2018) 246–253.
[32] M.H. Kim, H. Kim, D.R. Kim, K.S. Lee, A novel louvered fin design to enhance thermal and drainage performances during periodic frosting/defrosting conditions, Energy Conversion and Management 110 (2016) 494–500.
[33] K. Kim, M.H. Kim, D.R. Kim, K.S. Lee, Thermal performance of microchannel heat exchangers according to the design parameters under the frosting conditions, International Journal of Heat and Mass Transfer 71 (2014) 626–632.
[34] S. Liu, X. Bai, L. Zhang, Y. Lin, S. Deng, W. Wang, et al., Developing condensing-frosting performance maps for a variable speed air source heat pump (ASHP) for frosting suppression, Applied Thermal Engineering 211 (2022) 118397.
[35] Z.S. Fang, L.F. Li, Y. Cheng, H.H. Wang, M.O. Oladokun, Investigation on thermal-hydraulic performance of outdoor heat exchanger in air sources heat pump, Experimental Thermal and Fluid Science 84 (2017) 28–38.
[36] S. Liu, X. Bai, S. Deng, L. Zhang, M. Wei, Developing a novel control strategy for frosting suppression based on condensing-frosting performance maps for variable speed air source heat pumps, Energy and Buildings 289 (2023) 113049.
[37] S. Liu, X. Bai, S. Deng, L. Zhang, M. Wei, A modeling study on developing the condensing-frosting performance maps for a variable speed air source heat pump, Journal of Building Engineering 58 (2022) 104990.
[38] W. Su, W. Li, J. Zhou, X. Zhang, Experimental investigation on a novel frost-free air source heat pump system combined with liquid desiccant dehumidification and closed-circuit regeneration, Energy Conversion and Management 178 (2018) 13–25.
[39] Z. Wang, F. Wang, Z. Ma, M. Song, W. Fan, Experimental performance analysis and evaluation of a novel frost-free air source heat pump system, Energy and Buildings 175 (2018) 69–77.
[40] Z. Wang, G. Li, F. Wang, K. Li, Y. Lou, Techno-economic evaluation of a frost-free air source heat pump water heater, Sustainable Cities and Society 57 (2020) 102102.

[41] X. Bai, S. Liu, S. Deng, L. Zhang, M. Wei, A modelling study on the frosting characteristics of a novel dual-fan outdoor coil in an Air Source Heat Pump unit, Applied Thermal Engineering 222 (2023) 119933.
[42] M. Amer, C. Wang, Review of defrosting methods, Renewable and Sustainable Energy Reviews 73 (2017) 53–74.
[43] P. Carroll, M. Chesser, P. Lyons, Air source heat pumps field studies: a systematic literature review, Renewable and Sustainable Energy Reviews 134 (2020) 110275.
[44] M. Bezovsky, M. Stricik, M. Prascakova, Energy demand and comparison of current defrosting technologies of frozen raw materials in defrosting tunnels, Applied Energy 87 (2010) 2447–2454.
[45] O. Gustafsson, C. Teuillieres, H. Hellgren, M. Axell, J. Dalenbäck, Reversing air-source heat pumps – Noise at defrost initiation and a noise reducing strategy, International Journal of Refrigeration 62 (2016) 137–144.
[46] M. Song, Y. Tian, L. Zhang, C. Dang, Y. Hu, Y. Xu, et al., Review on the optimization studies of reverse cycle defrosting for air source heat pump units with multi-circuit outdoor coils, International Journal of Green Energy 20 (2023) 1401–1422.
[47] S. Xu, J. Niu, G. Ma, Experimental investigation of an air source heat pump with multigroup heat pipe radiators, Solar Energy 260 (2023) 137–146.
[48] L. Zhang, J. Dong, S. Deng, S. Yan, An experimental study on the starting characteristics of an improved radiant-convective air source heat pump system, Energy and Buildings 226 (2020) 110384.
[49] L. Zhang, J. Dong, Y. Jiang, S. Deng, S. Huang, An experimental study on frosting and defrosting performances of a novel air source heat pump unit with a radiant-convective heating terminal, Energy and Buildings 163 (2018) 10–21.
[50] L. Ma, F. Wang, Z. Wang, S. Zhang, Z. Liu, Y. Lou, Experimental investigation on an air source heat pump system with coupled liquid-storage gas-liquid separator regarding heating and defrosting performance, International Journal of Refrigeration 134 (2022) 176–188.
[51] M.L. Qu, L. Xia, S.M. Deng, Y.Q. Jiang, An experimental investigation on reverse-cycle defrosting performance for an air source heat pump using an electronic expansion valve, Applied Energy 97 (2012) 327–333.
[52] D. Huang, X. Yuan, X. Zhang, Effects of fan-starting methods on the reverse-cycle defrost performance of an air-to-water heat pump, International Journal of Refrigeration 27 (2004) 869–875.
[53] X. Rong, W. Long, J. Jia, L. Liu, P. Si, L. Shi, et al., Experimental study on a multi-evaporator mutual defrosting system for air source heat pumps, Applied Energy 332 (2023) 120528.
[54] J.Y. Jang, H.H. Bae, S.J. Lee, M.Y. Ha, Continuous heating of an air-source heat pump during defrosting and improvement of energy efficiency, Applied Energy 110 (2013) 9–16.
[55] G. Mader, C. Thybo, A new method of defrosting evaporator coils, Applied Thermal Engineering 39 (2012) 78–85.
[56] S. Li, J. Lu, W. Li, Y. Zhang, S. Huang, L. Tian, et al., Thermodynamic analyses of a novel ejector enhanced dual-temperature air source heat pump cycle with self-defrosting, Applied Thermal Engineering 215 (2022) 118944.
[57] J. Kim, H. Choi, K.C. Kim, A combined dual hot-gas bypass defrosting method with accumulator heater for an air-to-air heat pump in cold region, Applied Energy 147 (2015) 344–352.
[58] J. Dong, L. Zhang, Y. Jiang, A comparative study on system performances of multi-split air source heat pump with different energy accumulators and storage methods, Energy and Buildings 231 (2021) 110588.
[59] Q. Minglu, Z. Rao, C. Jianbo, C. Yuanda, Z. Xudong, Z. Tongyao, et al., Experimental analysis of heat coupling during TES based reverse cycle defrosting method for cascade air source heat pumps, Renewable Energy 147 (2020) 35–42.
[60] B. Yang, J. Dong, L. Zhang, M. Song, Y. Jiang, S. Deng, Heating and energy storage characteristics of multi-split air source heat pump based on energy storage defrosting, Applied Energy 238 (2019) 303–310.
[61] J.K. Dong, S.M. Deng, Y.Q. Jiang, L. Xia, Y. Yao, An experimental study on defrosting heat supplies and energy consumptions during a reverse cycle defrost operation for an air source heat pump, Applied Thermal Engineering 37 (2012) 380–387.

CHAPTER

11

High-efficiency frost-free air source heat pumps

Pengfei Zhang, Zhihua Wang, Longxia Ma and Fenghao Wang

School of Human Settlement and Civil Engineering, Xi'an Jiaotong University, Xi'an, Shaanxi, P.R. China

Nomenclatures

Symbol	Description	Unit
A	Area	m^2
C	Perimeter of air flow passage	m
c	The specific heat capacity	kJ/(kg · °C)
Co	Cost	CNY
D	The pipe diameter	m
d	Moisture content	g/kg
H_m	The PCM heat of fusion	kJ/m^3
h	The specific enthalpies	kJ/kg
K_y	Coefficient of mass convection	$kg/(m^2 \cdot s)$
M	The compressor displacement	kg/s
m	The mass flow rate	kg/s
Po	Power	kW
p	Pressure	Pa
Q	Heat capacity	kW
q_{st}	Heat of sorption	kJ/kg
R	Thermal resistance	$(m^2 \cdot K)/W$

Frosting and Icing for Efficient Energy Use in Engineering Applications
DOI: https://doi.org/10.1016/B978-0-443-15495-9.00017-4

r_w	Latent heat of vaporization of water	kJ/kg
T	Temperature	°C
v	Velocity	m/s
W	Extent of adsorption	
x	Length of heat exchanger in x direction	m
Y	Absolute humidity ratio of the air	kg/kg
z	Length of heat exchanger in the z direction	m
Greek symbol	**Description**	**Unit**
α	Heat transfer coefficient	$W/(m^2 \cdot K)$
ε	The latent heating mass transfer coefficient	$W/(m^2 \cdot K)$
η	Efficiency	%
ρ	Density	kg/m^3
τ	The system working time	min

Abbreviation	**Description**
ASHP	Air source heat pump
COP	Coefficient of performance
DA	Dry air
EEV	Electronic expansion valve
EHECSD	Extra heat exchanger coated with a solid desiccant
ESD	Energy storage device
FFASHP	Frost-free air source heat pump
FS	Full scale
HES	Hybrid energy system
HGBD	Hot-gas bypass defrosting
MA	Moisture air
OA	Outdoor air
PCM	Phase change material
RCD	Reverse cycle defrosting
RH	Relative humidity

Subscripts	**Description**
a	Air
Con	Conventional
com	Compressor
cond	Condenser

cs	Cross-section
DM	Defrosting mode
d	Desiccant layer
e	Mechanical
eva	Evaporator
f	Fluid
HM	Heating mode
i	Inner
il	Initial
in	Inlet
l	Liquid water
m	Matrix
mo	Motor
N	Novel
o	Outer
out	Outlet
PCM	Phase change material
RM	Regeneration mode
r	Refrigerant
rec	Reconstruction
rm	Refrigerant side mean value
ru	Running
SV	Solenoid valve
sat	Saturated
V	Volume
w	Water
wm	Waterside mean value
a	Air

11.1 Introduction

Under the background of increasingly severe energy shortage and environmental pollution [1], seeking and expanding more abundant and stable low-grade renewable energy as well as high-quality and efficient low-carbon heating technology have become the goal of many scholars. Air source heat pump (ASHP) is considered one of the best technical solutions to replace the low-temperature heat production of fossil energy because of its remarkable energy-saving and carbon reduction characteristics [2,3]. Developing ASHP technology and improving its performance can effectively solve the problems of low energy utilization efficiency and serious environmental pollution in the existing heat energy production methods, and it is also an important task to ensure human settlement environment and promote sustainable social development.

However, when operating in heating mode under low temperature and high humidity conditions (at ambient temperatures between −7°C and 5°C and relative humidity (RH) exceeding 65% [4]), the outdoor heat exchanger may experience frost formation, causing exorbitant compression ratio, discharge temperature and thus low coefficient of performance (COP) [5–8], which impede the further development of ASHP technology [9]. To address the issue of ASHP frosting while heating in low temperature and high humidity conditions, extensive studies have been carried out to investigate the mechanism of frost formation, performance under frosting, and defrosting technologies [10]. Also, active and passive antifrosting techniques have been developed [11]. Various antifrosting technologies such as electric and magnetic field assistance, ultrasonic and mechanical oscillation, and hydrophilic and hydrophobic surface treatments, as well as defrosting technologies such as reverse cycle, hot gas bypass, and electrical heating have been developed, and corresponding control strategies have been proposed. These technologies have shown good performance in preventing frosting and improving defrosting, but it is still challenging to completely avoid ASHPs running into frosting conditions [12]. There is a growing interest in the development of frost-free air-source heat pumps (FFASHP) as an alternative to conventional ASHPs.

There are two physical requirements for the development of frost on the exterior of the ASHP's outdoor heat exchanger: first, the heat exchanger's surface temperature must be below freezing point, and second, it must be below the air's dew point temperature [13,14]. The study of the influence of environmental parameters on frost formation characteristics of cold surfaces shows that the effect of air moisture on the growth of the frost layer is significant [15,16]. It is possible to avoid frosting by raising the heat exchanger's temperature or reducing the air's dew point temperature. As mentioned above, the FFASHP is mainly based on the destruction of the above frosting conditions, that is, adjusting the temperature or the humidity of air flowing through the heat exchanger. Therefore, there are two approaches: raising the air temperature (or exchanger temperature) and reducing the air humidity.

Kwak et al. [17] applied an electric heater that will turn on in frosting conditions in front of an outdoor unit. Experimental tests in accordance with ANSI/ASHRAE Standard [18,19] were conducted, and the findings revealed the following: when the outdoor temperature was 2°C/1°C (dry bulb/wet bulb), the frost was generated immediately on the heat exchanger and the conventional heat pump stopped operation while the proposed heat pump was still functioning normally. The proposed system outperformed the conventional one in terms of both heating capacity and COP, coming up at 38.0% and 57.0% greater, respectively. However, the high-quality power energy is directly applied to heat the outdoor air, it is a certain amount of waste. Solar energy is used in FFASHP because of its renewable and versatile characteristics [20]. Xu et al. [21] optimized the common solar-assisted ASHP system and proposed a new hybrid energy system (HES) (Fig. 11.1).

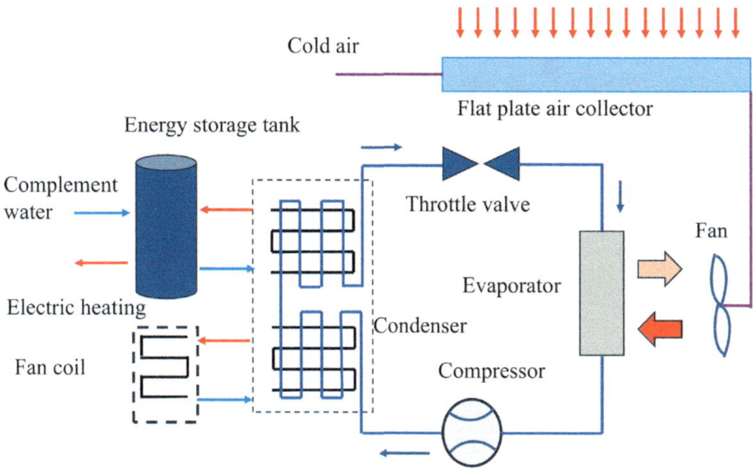

FIGURE 11.1 Schematic diagram of the HES [21]. *HES*, Hybrid energy system.

The water tank would storage parts of energy on clear days and supply it in the absence of sunlight. Air is applied as the heat collecting system's working fluid as it has good antifrosting properties. The proposed system COP is anticipated to be higher than 3.0 when applied in an environment where the outdoor heating calculation temperature is below $-30°C$ and the solar energy resource is not very abundant.

It is possible to raise the temperature of the exchanger by improving the structure of ASHP without extra energy input. Ma et al. [22,23] proposed a novel ASHP with a coupled liquid-storage gas–liquid separator, in which the refrigerant is heated before entering the outdoor heat exchanger, and thus the temperature of the exchanger is raised. The novel ASHP system represents frost-free characteristics under some conditions (for instance, an ambient temperature of 5°C and RH of 90%) in heating mode and no performance reduction in refrigerant mode [22–24].

Using solid or liquid desiccant to reduce the air's humidity is also an important approach to constructing an FFASHP. Su et al. [25,26] proposed a novel FFASHP combined with liquid desiccant (LiCl) dehumidification. The total system performance of the proposed FFASHP system is higher under low temperature and high humidity conditions compared with the reverse-cycle defrosting (RCD) system. Su et al. [27] also improved the FFASHP system, which combined with self-regeneration membrane-based liquid desiccant dehumidification, and performance is increased. Before entering the evaporator in the study of Wang et al. [28], the outside air was desiccated using solid desiccant. There was a unique adsorbent bed constructed. The air parameters through the adsorbent bed were outside the frost area, according to the simulation findings. Frost can be prevented when the exchanger is passed by the dehumidified air. Zhang et al. [29] added an extra heat exchanger coated with a solid desiccant in a conventional ASHP system to prevent the system from frosting, using regeneration air to regenerate the desiccant. But while in regeneration mode, when the refrigerant's temperature dropped below 0°C, the moisture from the heat exchanger coated with desiccant would continue to cause frosting in the evaporator [30].

In this chapter, a novel FFASHP system, coupled with an extra heat exchanger coated with a solid desiccant (EHECSD) and an energy storage device (ESD), is introduced.

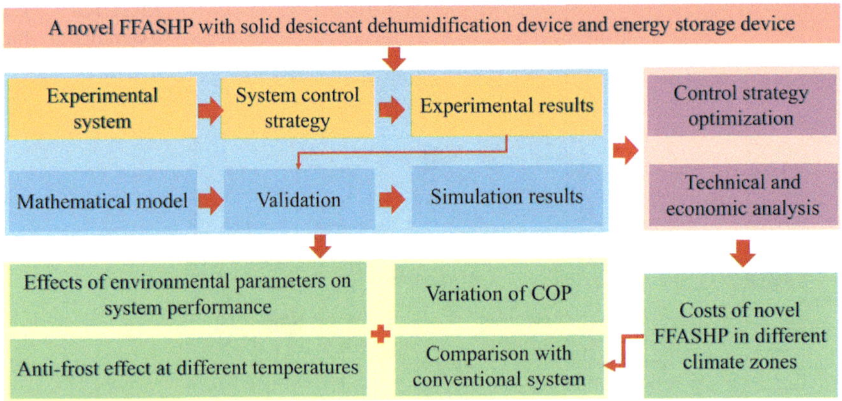

FIGURE 11.2 The outline of this chapter.

This system not only prevents frosting but also avoids frost formation and continuous heating in regeneration mode [30]. Fig. 11.2 is the frame of this chapter. First, the working principle of the FFASHP water heater is introduced, and experiments are carried out to investigate the system's performance. Second, a detailed mathematical model of the novel system is presented and validated by experimental investigations. Then, the effects of three environmental parameters (temperature, RH, and air velocity) on the dynamic performance of the system are studied further. Third, an optimal control strategy for the novel system is proposed. Finally, a technical and economic performance analysis of the system is conducted in detail.

11.2 Experimental system and procedure

11.2.1 System description

A schematic diagram of the novel FFASHP water heater system is illustrated in Fig. 11.3, with the process shown on a p-h diagram of the refrigerant in Fig. 11.4. The main components are a rotary compressor (1), high/low-pressure protection devices (2, 14), a four-way valve (3), a condenser that wraps around outer of the water tank (4), solenoid valves (5, 13, 15, 18), an ESD (6), filter driers (7, 10, 16), electronic expansion valves (EEVs) (8, 11, 17), an EHECSD (9) and an outdoor heat exchanger (12). There are two modes for this system, that is, heating mode and regeneration mode.

The system control methods during heating mode and regeneration mode are shown in Table 11.1. During heating mode, the solenoid valves (5, 13) are opened while the solenoid valves (15, 18) and EEV (17) are closed. The refrigerant from the compressor unit (1) flows into the condenser. The condenser wraps around the water tank (4) and releases heat to the cold water in the tank. Then, the refrigerant flows into the ESD (6), and the condensation waste heat will be absorbed by phase change material (PCM) in the ESD. After flowing out ESD, the refrigerant is one-stage throttled by the EEV (8) and then enters the EHECSD (9). After being two-stage throttled by the EEV (11), the refrigerant absorbs the latent heat of

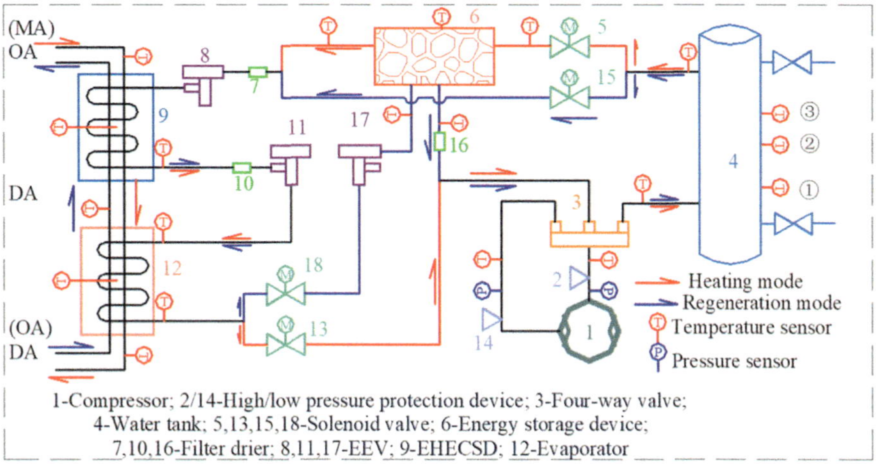

FIGURE 11.3 Schematic diagram of the novel FFASHP system [30–32]. *FFASHP*, Frost-free air source heat pump.

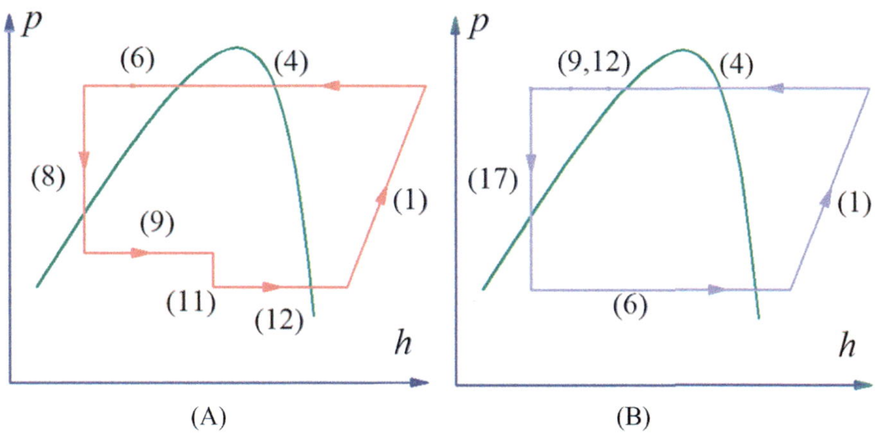

FIGURE 11.4 p–h diagram of refrigerant. (A) Heating mode and (B) regeneration mode.

TABLE 11.1 The system control method during heating mode and regeneration mode.

Heating mode				Regeneration mode			
EEVs		Solenoid valves		EEVs		Solenoid valves	
Number	Opening	Number	On/off	Number	Opening	Number	On/off
8	280 plus	5	On	8	Full	5	Off
10	40 plus	13	On	10	Full	13	Off
17	0	15	Off	17	40 plus	15	On
–	–	18	Off	–	–	18	On

EEV, Electronic expansion valve.

water vapor partly in EHECSD. Then absorbs heat from outdoor air through an outdoor heat exchanger (12). Finally, the refrigerant, which is superheated, goes back to the compressor, which marks the end of one cycle. Outdoor air (OA) passes through the EHECSD, absorbing water vapor and releasing sensible heat in the process. The frost-free operation can be accomplished because the dew point temperature of dry air (DA) that has been dehumidified by the solid desiccant is lower than the evaporation temperature. At state EA, the air is finally discharged off from (12). The system will switch to regeneration mode when the dehumidifying ability of solid desiccant drops as the moisture content of the solid desiccant rises to a point where its vapor partial pressure exceeds that of the surrounding air. What should be mentioned is that to guarantee that the desiccant will fully regenerate in regeneration mode, there needs to be adequate heat stored in the ESD (6) during heating mode.

During regeneration mode, the solenoid valves (15, 18) and EEVs (8, 11) are fully opened, while the solenoid valves (5, 13) are closed. The condenser receives the refrigerant that has been released from the compressor and rejects heat into the tank's cool water. After passing through the EHECSD (9), the refrigerant releases the condensation waste heat. Desiccant regeneration is realized as a result of heating the solid desiccant and raising the vapor partial pressure above that of the surrounding air. The refrigerant passes through the ESD (6), which serves as a low-temperature heat source, after entering the outside heat exchanger (12) and being throttled by the EEV (17). The superheated refrigerant finally enters the compressor. Compared to the heating mode, the flow of OA is in the opposite direction. It passes through the EHECSD (9) after being preheated by the outside heat exchanger (12), and the water vapor is removed from the solid desiccant. At last, the air in the state of MA is eliminated from EHECSD (9). It should be mentioned that the ESD (6) provides the majority of the heat needed to regenerate the solid desiccant rather than the regeneration air.

The energy balance of the system in a single period is given by Eq. (11.1), and the COP of the system is calculated by Eq. (11.2).

$$Q_{HM,eva} + W_{com} = Q_{cond} + Q_{EHECSD} \tag{11.1}$$

$$COP = \frac{Q_{cond}}{W_{com}} = \frac{\int_{T_1}^{T_2} c_{water} m_{water} dT}{\int_0^{\tau} P o d\tau} \tag{11.2}$$

11.2.2 Experimental setup

The EHECSD is one of the key components of the system, which integrates the solid desiccant and the fin-and-tube heat exchanger. Silica gel is an excellent type of adsorbent and has been widely used in dehumidification processes for its great pore surface area and good moisture adsorption capacity [33] and is used as the material for the experiment in this chapter. $CaCl_2 \cdot 6H_2O$ is employed as the PCM because its phase transformation temperature is between the condensing temperature and the evaporating temperature and has a wide temperature difference from the coil in the proposed system, according to the study of Hu et al. [34]. The main components used in the experiment system are shown in Table 11.2. The experiment system and the arrangements of measurement instruments are shown in Fig. 11.3.

TABLE 11.2 Main components used in the experiment system.

Component	Type	Parameters
Compressor	Mitsubishi RB174GHAC type rotary compressor	Swept volume:17.4cm^3/r; input power: 600 W
Evaporator	Finned-tube heat exchanger	Copper tubes with aluminum fins
Condenser	Tubular heat exchanger	Copper tube
Expansion valve	Electronic	–
Tank	–	60 L
Refrigerant	R134a	Density: 4.25 kg/m^3 (gas), boiling point: − 26.5°C

TABLE 11.3 Structural parameters of energy storage device (ESD) and Extra heat exchanger coated with a solid desiccant (EHECSD).

ESD		EHECSD	
Inner diameter	250 mm	Size	350 × 350 × 125 mm^3
Outer diameter	300 mm	Fin thickness	0.2 mm
Height	450 mm	Fins distance	3 mm
Cobber tube radius	9.52 × 0.5 mm	Cobber tube radius	9.52 × 0.5 mm
Cobber tube length	11 m	Silica gel quantity	2.8 kg

TABLE 11.4 Main characteristics of the plant instrumentation.

Measured variable	Measurement device	Calibration range	Calibrated accuracy
Temperature (°C)	Platinum resistance	− 100 to 150	± 0.1
Temperature (°C)	T-type thermocouple	− 50 to 150	± 0.2
Temperature (°C)	Temperature and humidity	− 10 to 85	± 0.5
RH (%)	Acquisition instrument	10–95	± 3
Pressure (MPa)	Pressure transducer	0–6	± 1% FS
Power (kW)	Power meter	0.01–24.00	± 0.01% FS
Weight (kg)	High-precision electronic scale	0–20	± 0.1 g

RH, Relative humidity.

The structural parameters of ESD and EHECSD are shown in Table 11.3, and the main characteristics of the plant instrumentation are in Table 11.4. The experiments were conducted in an enthalpy difference laboratory; the experiment conditions were ambient temperatures of 3°C, 0°C, and 3°C, RH of 85%. Operating conditions for the experimental FFASHP are shown in Table 11.5.

TABLE 11.5 Operating conditions.

Ambient temperature (°C)	Relative humidity (%)	Air volume (m^3/h)		Initial water temperature (°C)	Mass of water (kg)
		Heating mode	Regeneration mode		
−3					
0	85	1400	500	15	60
3					

11.2.3 System control strategy

Programmable CNC Stepper Motor Controllers were used to control EEVs (8, 11, 17), which would help the refrigerant absorb the latent heat from the solid desiccant and reduce heat losses in the EEV (8) while dehumidifying. The system would be switched to regeneration mode when the RH reached 77%, 75%, and 74% at the temperature of 3°C, 0°C, and 3°C, respectively. The air volume was lowered from 1400 m^3/h in heating mode to 500 m^3/h during the regeneration phase to make sure there was enough energy to regenerate the solid desiccant. If there was no longer a noticeable variation in the RH of the regeneration air, it is considered that the regeneration process was complete.

11.2.4 Results and discussions

The variation of air humidity ratio after dehumidification, air temperature after dehumidification, temperature profile of EHECSD, and temperature of PCM with time in a cycle are shown in Fig. 11.5A–D, respectively. As shown in Fig. 11.5A and B, while air temperature showed a reverse trend from the initial moment to 32 minutes, the humidity ratio increased as heating time increased. The humidity ratio increased by 0.54, 0.56, and 0.47 g/kg while the temperature decreased by 1.8°C, 2.18°C, and 3.22°C at the ambient temperature of −3°C, 0°C, and 3°C, respectively. When the outside heat exchanger started to frost, and the system was put into regeneration mode (in the first 32, 34, and 36 minutes), the temperature and air humidity ratio both increased significantly. EHECSD's temperature remained relatively constant from minute 0 to minute 32 (Fig. 11.5C). The temperature reached its highest points at 50.4°C, 58.4°C, and 60.2°C. According to the research [35,36] about silica gel desiccant used in this system, the ideal temperature for desiccant regeneration can be effectively produced by this novel system. The temperature of the PCM increased quickly during the first 10 minutes of heating mode, as shown in Fig. 11.5D. After that, the temperature's slope decreased as the time increased. After the heating mode was turned off, the temperature was 35°C, 32°C, and 27°C. In the regeneration phase, the PCM's temperature dropped quickly before starting to drop more gradually. This was due to the fact that, during regeneration, the helically smooth copper tube in the ESD served as the ASHPWH system's evaporator, absorbing a significant quantity of heat for heating.

Fig. 11.6 illustrates how the system COP changes over time in a cycle at an 85% RH. The system COP at different ambient temperatures increased to a point and the system's COP reached its maximum at around the 44th minute, then tended to decrease. When the

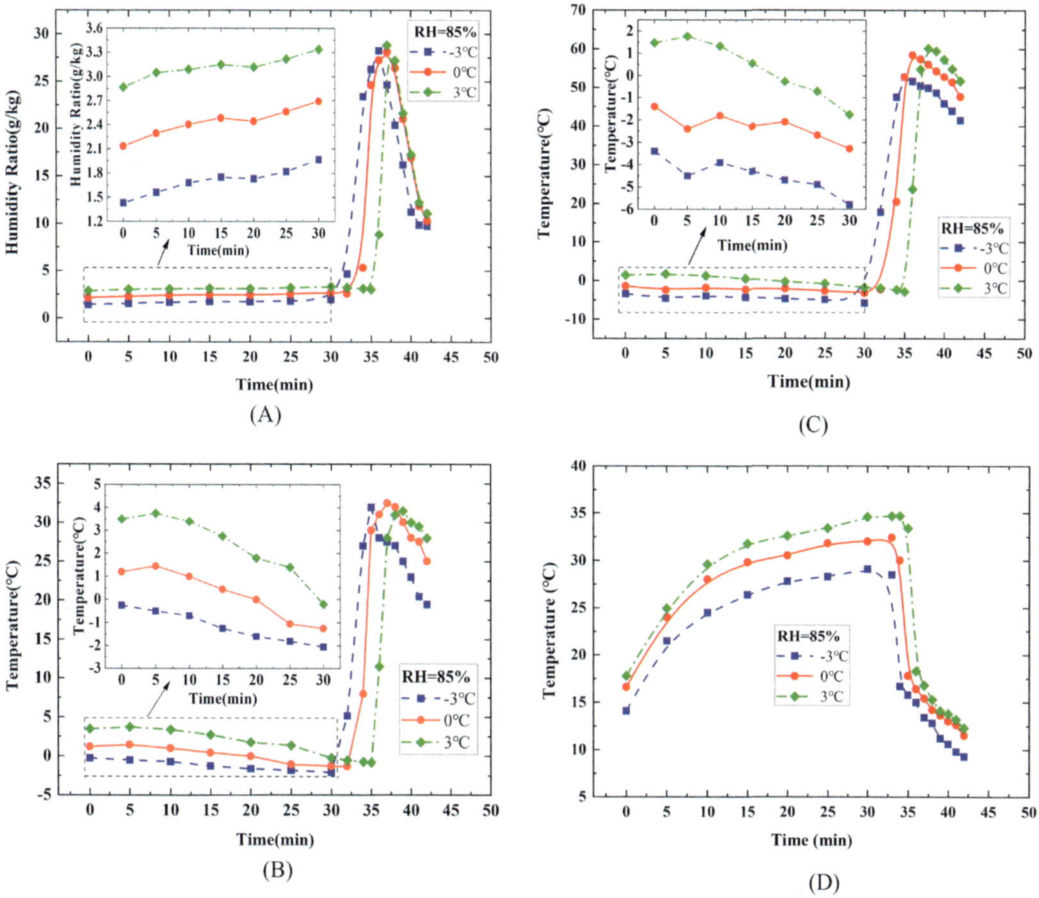

FIGURE 11.5 Variations of temperature/humidity with time in a cycle. (A) Air humidity ratio after dehumidification. (B) Air temperature after dehumidification. (C) Temperature of EHECSD. (D) Temperature of PCM. *EHECSD*, Extra heat exchanger coated with a solid desiccant; *PCM*, phase change material.

system achieved an operating time of about 44 minutes, its COP started to rise quickly before abruptly declining. This resulted from a rapid decline in heat storage and from a gradual heat loss of the system with time, hence the heat exchangers' performance decreased. In a cycle, the system mean COP was 2.47, 2.80, and 3.11 at the following ambient temperatures: 3°C, 0°C, and −3°C. This demonstrated how the system's COP was significantly impacted by the ambient temperature.

With the aim of comparing the performance of the FFASHP system with the conventional RCD system, some pipelines of the prototype ASHP were little modified. The components of EHECSD (9) and ESD (6) were eliminated. Since the heating capacity and COP are two important metrics that could reveal information about the system's performance, they were measured and computed under identical settings. In Fig. 11.7, the comparative findings are displayed. It can be seen that the capacity of RCD was 0.14 and 0.06 kW lower than that of the FFASHP at temperatures of −3°C and 0°C. At an ambient temperature of

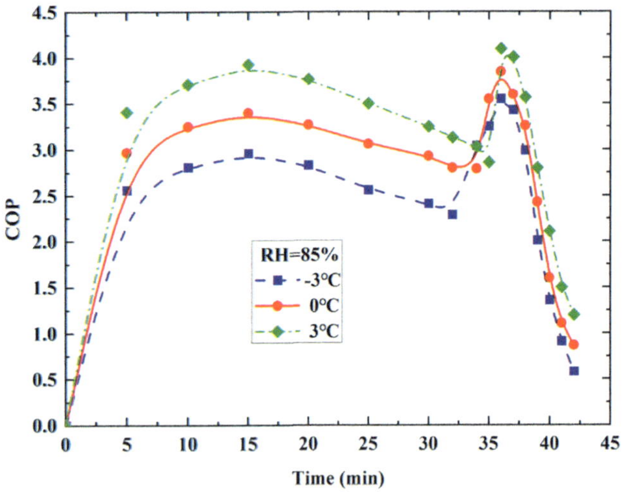

FIGURE 11.6 Variation of COP with time in a cycle. *COP*, Coefficient of performance.

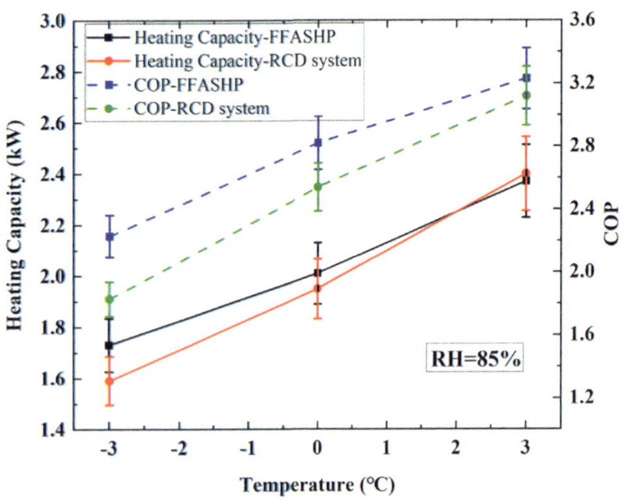

FIGURE 11.7 Performance comparison between the RCD system and FFASHP at different ambient temperatures. *FFASHP*, Frost-free air source heat pump; *RCD*, reverse-cycle defrosting.

−3°C and 3°C, respectively, the COP of the FFASHP was 17.9% and 3.4% greater than that of the RCD system, demonstrating that the proposed novel system performs better and is more significant at low temperatures.

11.3 Mathematical model and simulation process

In this section, to conduct a thorough analysis into the novel FFASHP system, a comprehensive mathematical model is developed and validated by comparison with experimental findings. Subsequently, more research is done on how environmental factors like ambient temperature, RH, and air velocity affect the system's dynamic performance.

11.3.1 Mathematical model

The model is based on the following assumptions: (1) Presuming that the compression process of refrigerant vapor is polytropic. (2) The refrigerant pressure drop in the pipes and condenser is insignificant. (3) It is isenthalpic for the expansion valve throttling mechanism. (4) The EHECSD surface has a uniform distribution of the desiccant substance. (5) The desiccant layer receives all the adsorption heat. (6) Ignoring axial conduction, the impact of natural convection in the PCM melts and the movement of refrigerant as the system transitions from heating to regeneration mode. (7) To expedite computation, it is assumed that the compressor, evaporator, and EEV are in a steady state.

The compressor's actual displacement is calculated by Eq. (11.3).

$$M_r = M_{r0}\eta_V \tag{11.3}$$

where M_{r0} is the theory of quality displacement compressors, kg/s; η_V is the volume efficiency of the compressor.

The shaft power is calculated by Eq. (11.4).

$$P_{el} = P_e/\eta_{mo} \tag{11.4}$$

where P_e is the mechanical power, kW; η_{mo} is the motor efficiency, usually 0.8–0.95.

The refrigerant side heat transfer equation and the water side heat transfer equation in the condenser are given by Eqs. (11.5) and (11.6), respectively. The refrigerant side heat transfer equation and the refrigerant side heat transfer equation in the evaporator are given by Eqs. (11.7) and (11.8), respectively.

$$Q_r = m_r\left(h_{r,in} - h_{r,out}\right) = \alpha_r A_{cond,i}(T_{rm} - T_{wall}) \tag{11.5}$$

$$Q_w = m_w c_{p,w}\frac{dT_w}{d\tau} = \alpha_w A_{cond,o}(T_{wall} - T_{wm}) \tag{11.6}$$

$$Q_r = m_r\left(h_{r,in} - h_{r,out}\right) = \alpha_r A_{eva}(T_{wall} - T_{rm}) \tag{11.7}$$

$$Q_a = \alpha_a(T_a - T_{wall})\left(A_{eva}c_p + A_f\eta_1\right) + \varepsilon r_w\left(d_a - d_{wall,sat}\right)\left(A_{eva} + A_f\eta_2\right) = \left(c_{p,a}\Delta T_a + r_w\Delta d_a\right) \tag{11.8}$$

where η_1 and η_2 are the efficiency of heat transfer.

One of the key components of the novel system is the EHECSD, whose model may be estimated using the technique recommended by Ge et al. [37]. The moisture conservation in the system is given by Eq. (11.9).

$$\rho_a A_{cs}\left(\frac{\partial Y_a}{\partial \tau} + v_a\frac{\partial Y_a}{\partial z}\right) + F_a\frac{\partial W}{\partial \tau} = 0 \tag{11.9}$$

The moisture conservation in a desiccant layer can be expressed as Eq. (11.10).

$$F_d\frac{\partial W}{\partial \tau} = K_y C(Y_a - Y_d) \tag{11.10}$$

For the EHECSD model, the energy conservation in the system is given by Eq. (11.11).

$$\rho_a A_{cs}c_{p,a}\left(\frac{\partial T_a}{\partial \tau} + v_a\frac{\partial T_a}{\partial z}\right) + \rho_w A_w c_{p,w}\left(\frac{\partial T_w}{\partial \tau} + v_f\frac{\partial T_w}{\partial z}\right) + \left[F_d\left(c_{p,d} + Wc_{p,l}\right) + F_m c_{p,m}\right]\frac{\partial T_d}{\partial \tau} = K_y C(Y_a - Y_d)q_{st} \tag{11.11}$$

The PCM in the ESD was calcium chloride hexahydrate (made up of mass fractions of 2% additive $SrCl_2 \cdot 6H_2O$ and 98% $CaCl_2 \cdot 6H_2O$). The refrigerant side heat transfer equation and the PCM side heat transfer equation in ESD are given by Eqs. (11.12) and (11.13), respectively.

$$c_{p,f} m_f(\tau) \frac{\partial T_f(x,\tau)}{\partial x} = T_f c D_i \left[T_{wall}(x,\tau) - T_f(x,\tau) \right] \tag{11.12}$$

$$H_m 2 R_{PCM}(x,\tau) \frac{\partial R_{PCM}(x,\tau)}{\partial \tau} = T_f(x,\tau) D_i \left[T_{wall}(x,\tau) - T_f(x,\tau) \right] \tag{11.13}$$

11.3.2 Model validation

The main equations of the mathematical model are presented above, and more details can be found in the literature [32]. The system model was established using these equations to assess the suggested system's thermal performance. The experimental studies then confirmed the validity of the mathematical model.

Fig. 11.8 displays the simulated results of the humidity ratio (1), air temperature (2), and system COP (3) exit from EHECSD in comparison with the experimental data. These statistics demonstrate how the experiment's results match the theory in general, particularly when the heating mode is engaged. The largest error value of the humidity ratio is 1.17 g/kg, and its lowest value is 0.61 g/kg. The inaccuracy value for the air temperature exit from the EHECSD ranges from 0.73°C to 2.89°C. Furthermore, COP has a maximum error value of just 0.4. Therefore, during the heating mode, the simulated results and the experimental data coincide fairly well. However, the mistakes of the results during the regeneration mode are a little far from the experiments. This could be explained by the system's actual pressure drop and heat loss during the transition from heating mode to regeneration mode. All of the numerical results, however, point to the acceptability and adoption of the established mathematical model for further analysis of the dynamic performance of the proposed FFASHP system.

11.3.3 Results and discussions

The effects of environmental parameters, including ambient temperature, RH, and air velocity, on the characteristics of the proposed system have been analyzed. The results are shown in Fig. 11.9. The regeneration efficiency changed in the opposite direction as the ambient temperature rose, but the dehumidification efficiency grew in tandem with the temperature rise. What should be mentioned is that the regeneration effectiveness of solid desiccant was 73% at 0°C but only 48% at 10°C, which went against the suggested system's safety operation. Consequently, more research is required to increase regeneration efficiency at lower temperatures, and so far, it has not been conducted. As can be seen in Fig. 11.9B, during the heating mode, the dehumidification efficiency increased by 0.85% and 0.45%, respectively, when the RH increased from 75% to 80% and from 80% to 85%. This meant that the dehumidification efficiency increased with the RH increased, but the slop of dehumidification efficiency marginally dropped. The regeneration efficiency decreased by 3.5% as the RH increased from 75% to 85% at a given ambient temperature of 0°C.

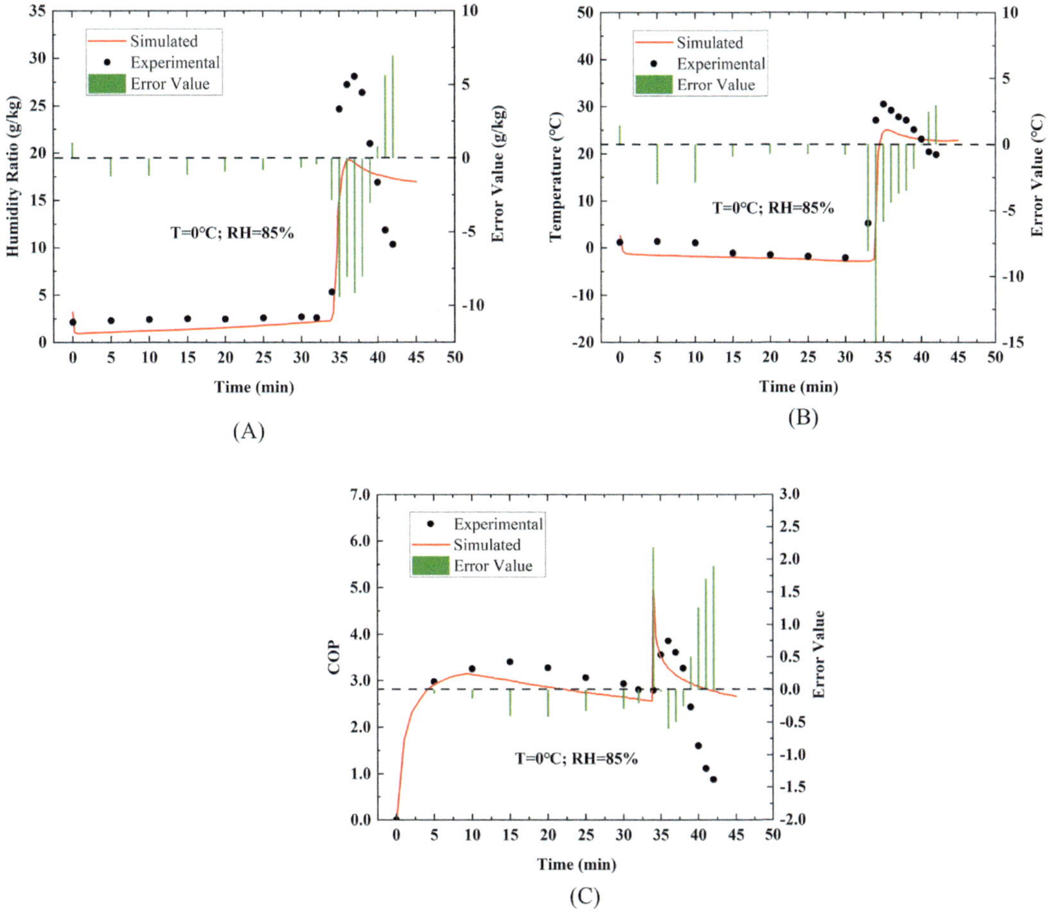

FIGURE 11.8 Comparison between simulated and experimental data. (A) Humidity ratio comparison, (B) temperature comparison, and (C) COP comparison. *COP*, Coefficient of performance.

Fig. 11.9C made it evident that the dehumidification increased from 32.5% to a maximum of 33% and then decreased to 31.6%. On the other hand, when air velocity increased from 2.5 to 3.5 m/s, the regeneration efficiency increased as well, rising from 71.9 to 75.5%. Fig. 11.9D–F are the effects of three ambient parameters on system COP. When the ambient temperatures were 0, −5, and −10°C for approximately 5 minutes, the system COP grew quickly for the first time and reached its maximum values of 3.27, 2.74, and 2.07, respectively. During the heating mode, the system COP then fell slightly. Furthermore, after operating for 34, 52, and 67 minutes, respectively, at an ambient temperature of 0°C, −5°C, and −10°C, the system transitioned from heating mode to regeneration mode. The maximum COPs were 4.9, 4.5, and 4.02, respectively, before falling to 2.85, 3.0, and 2.87. With the change of RH, the system COPs had the same trends at different RH, especially in heating mode, and increased rapidly and subsequently fell gradually during regeneration mode. The system average COPs were 3.03, 2.87, and 2.84 at the RH

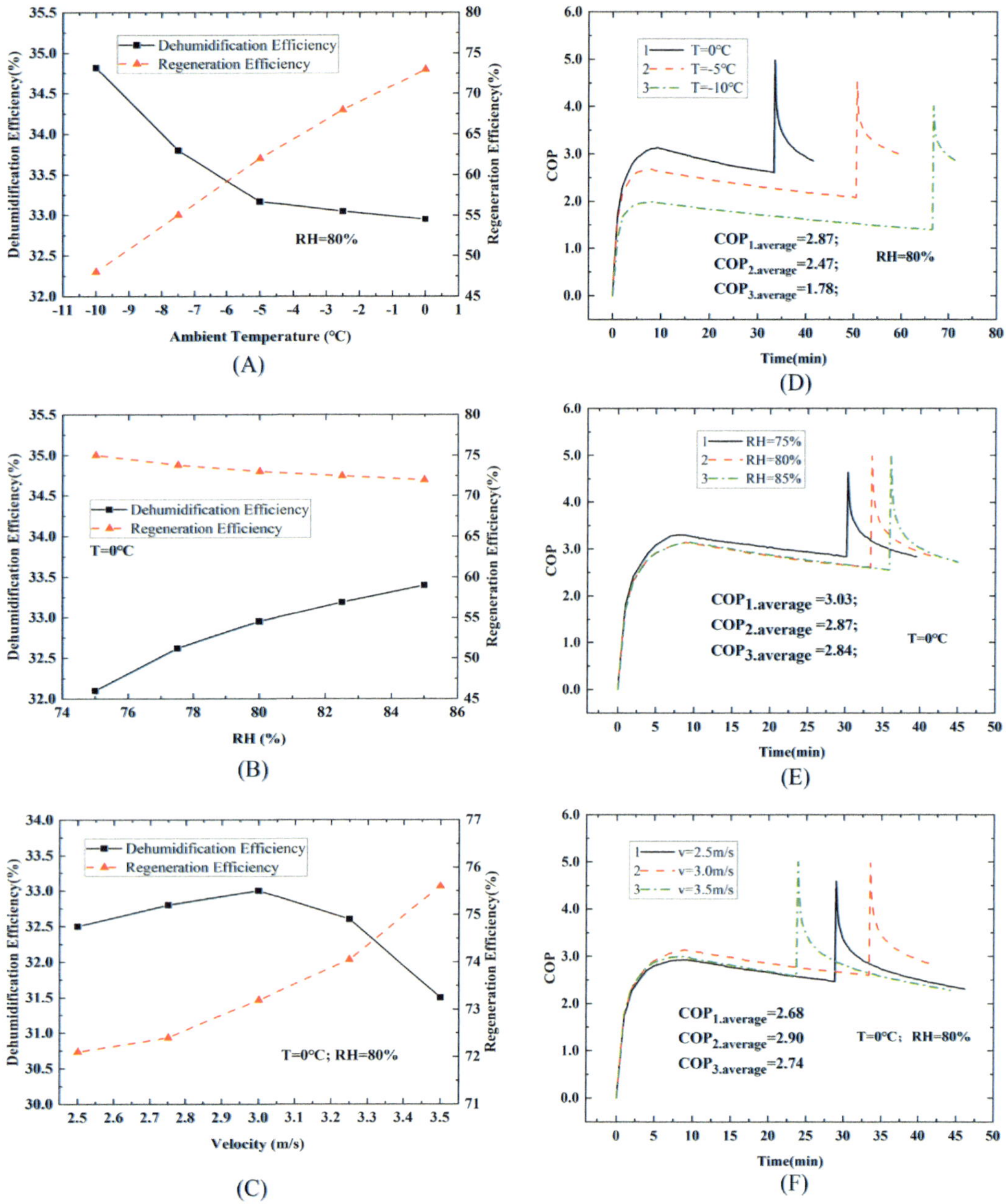

FIGURE 11.9 Effect of ambient temperature, RH, and air velocity on system performance. (A) Effect of ambient temperature on dehumidification efficiency and regeneration efficiency. (B) Effect of RH on dehumidification efficiency and regeneration efficiency. (C) Effect of air velocity on dehumidification efficiency and regeneration efficiency. (D) Effect of ambient temperature on system COP. (E) Effect of RH on system COP. (F) Effect of air velocity on system COP. *COP*, Coefficient of performance; *RH*, relative humidity.

of 75%, 80%, and 85%, respectively. As a result, the system's COP declined marginally with an increase in RH, a factor that had a smaller impact on COP than ambient temperature. From Fig. 11.9F, the changes of the system COP at the influence of air velocity were the same as the effect of ambient temperature and relative humidity. At an air velocity of 3.0 m/s, the system average COP was 0.22 and 0.16, greater than that at 2.5 and 3.5 m/s. The results showed that, in comparison to ambient temperature, air velocity had a far smaller effect on the system's performance.

11.4 Control strategy optimization for the novel system

In this section, to improve the system performance at different ambient temperature and RH and establish the optimal control strategy for the novel system, a three-dimensional curved surface (Fig. 11.10A) of the time for maintaining the evaporator frost-free (heating time) operation with ambient temperature and RH was obtained through the regression analysis of numerical results by Table Curve 3D software.

As can be seen in Fig. 11.10, to ensure the system's highly efficient operation, if the system operated time is under the three-dimensional curved surface for a given ambient temperature and RH, the outdoor heat exchanger will be kept frost-free status. Otherwise, the system should be switched from heating mode to regeneration mode. Aimed to calculate the system performance at a given ambient temperature and RH as well as compare the novel system performance with RCD and HGBD, the surface plot of the system average COP with ambient temperature and RH is given (shown in Fig. 11.10(b)) and a correlation is presented in the literature [32]. These results are convenient and helpful in the process of system optimization.

11.5 Technical and economic performance analysis of the system

Although the proposed FFASHP system demonstrates superior operational performance compared to conventional systems, it incorporates additional components such as an EHECSD, an ESD, three solenoid valves, and two EEVs, leading to an increase in initial costs. These higher initial costs and longer payback periods present challenges for the widespread adoption of the novel system. This section aims to provide essential information for users making decisions and to assess the economic viability before a wide range of promotions and applications. A detailed economic analysis of the proposed system compared to conventional ones is conducted, with more information available in the literature [38].

11.5.1 Economic analysis

From an economic point of view, the main components can be divided into initial cost and running cost. The initial cost can significantly impact the system's economic feasibility, encompassing the ASHP unit cost and installation costs. The proposed system, compared to the conventional ASHP system, includes additional components such as an

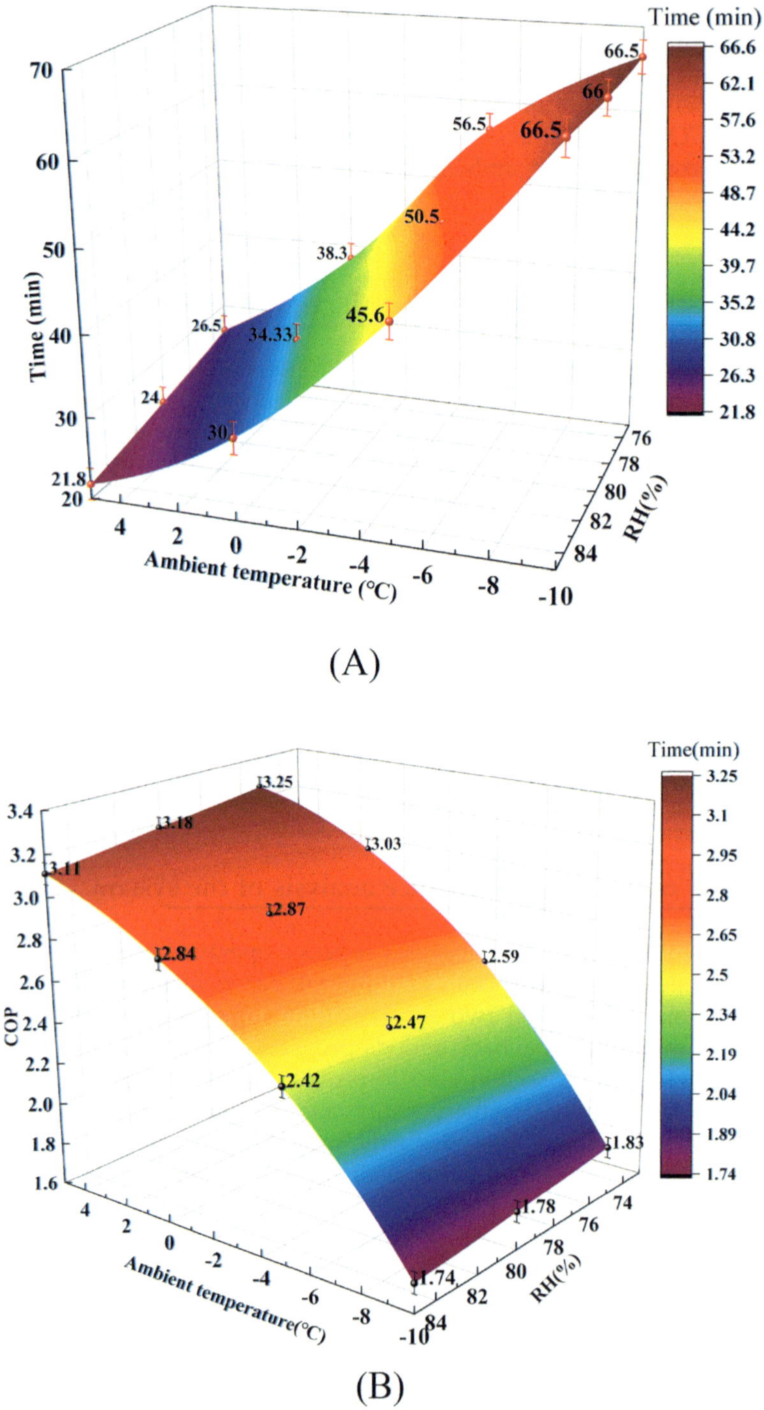

FIGURE 11.10 Effect of ambient temperature coupled with RH on the system performance. (a) Effect of ambient temperature and RH on the time of keeping the evaporator frost-free. (b) Effect of ambient temperature and RH on the system average COP. *COP*, Coefficient of performance; *RH*, relative humidity.

EHECSD, two EEVs, three solenoid valves, and an ESD, thereby increasing the total initial cost. This total cost comprises the ASHP unit cost, additional components cost, and reconstruction cost. The total cost can be expressed by the following:

$$Co_{il,N} = Co_{ASHP} + Co_{il,EHECSD} + Co_{il,EEV} + Co_{il,ESD} + Co_{il,SV} + Co_{il,rec} \tag{11.14}$$

The initial cost of the conventional system is denoted as Co_{ASHP}. Before conducting the techno-economic analysis, over 20 manufacturers were surveyed to determine the average price as a reference. According to the market survey, the price of a conventional ASHP system with the same heating capacity as the experimental prototype was approximately 4000 CNY. The additional auxiliary components in the experimental prototype, including an EHECSD, two EEVs, three solenoid valves, and an ESD, accounted for about 43% of the conventional ASHP system's initial cost. Therefore, the initial cost of the experimental prototype was about 5720 CNY. It should be noted that the prices of the ESD and EHECSD were higher because they were custom-made for this study in this chapter. It is believed that the initial cost of the proposed FFASHP system will decrease with batch production.

The running costs, including those in heating and regeneration modes, arise from electricity consumption by the compressor, fan, and water pump. According to previous experimental investigations presented above, the system operates for 35 minutes in heating mode and requires 5 minutes for desiccant regeneration in regeneration mode. So, the running cost of the system can be expressed as follows:

$$Co_{ru,N} = (Co_{ru,N,HM,com} + Co_{ru,N,HM,pump} + Co_{ru,N,HM,fan}) + (Co_{ru,N,RM,com} + Co_{ru,N,RM,pump} + Co_{ru,N,RM,fan}) \tag{11.15}$$

There is a cost in heating mode and defrosting mode for a conventional ASHP. For the former, the primary running cost is the same as the FFASHP; for the latter, the running cost would not include water pump power consumption as it would close. The electricity cost per kWh is set as 0.5 CNY (based on electricity prices in Xi'an, China). According to the previous studies, the conventional ASHP unit was operated in 60 minutes and needed 10 minutes for defrosting [39]. So, the running cost of the conventional ASHP unit can be calculated by the following:

$$Co_{ru,Con} = (Co_{ru,Con,HM,com} + Co_{ru,Con,HM,pump} + Co_{ru,Con,HM,fan}) + (Co_{ru,Con,DM,com} + Co_{ru,Con,DM,fan}) \tag{11.16}$$

11.5.2 Results and discussions

According to the previous work, the ambient temperature and RH are two main factors that affect the system performance and frost quantity of an ASHP. For the novel FFASHP system, these factors also impact the system's COP, frost-free duration, and regeneration duration. So, the system performance was investigated under the different climate zones for space heating in the heating season. To compare the economic feasibility of both conventional and novel systems, three typical cities in different climate zones in China were selected as operation regions: Xi'an (cold zone), Hefei (hot summer and cold winter zone), and Shenyang (severe cold zone). During the entire heating period, the average ambient temperatures and RH were 2.0°C and 63.6% in Xi'an, 5.1°C and 73.1% in Hefei, and −6.8°C and 86.9% in Shenyang (the data from typical meteorological years using TRNSYS software).

Fig. 11.11 shows the system operating total running costs of the FFASHP and conventional ASHP system in the three cities in the 20 operating years (N: novel system, Con: conventional system, the same below). Hefei had the shortest system running period

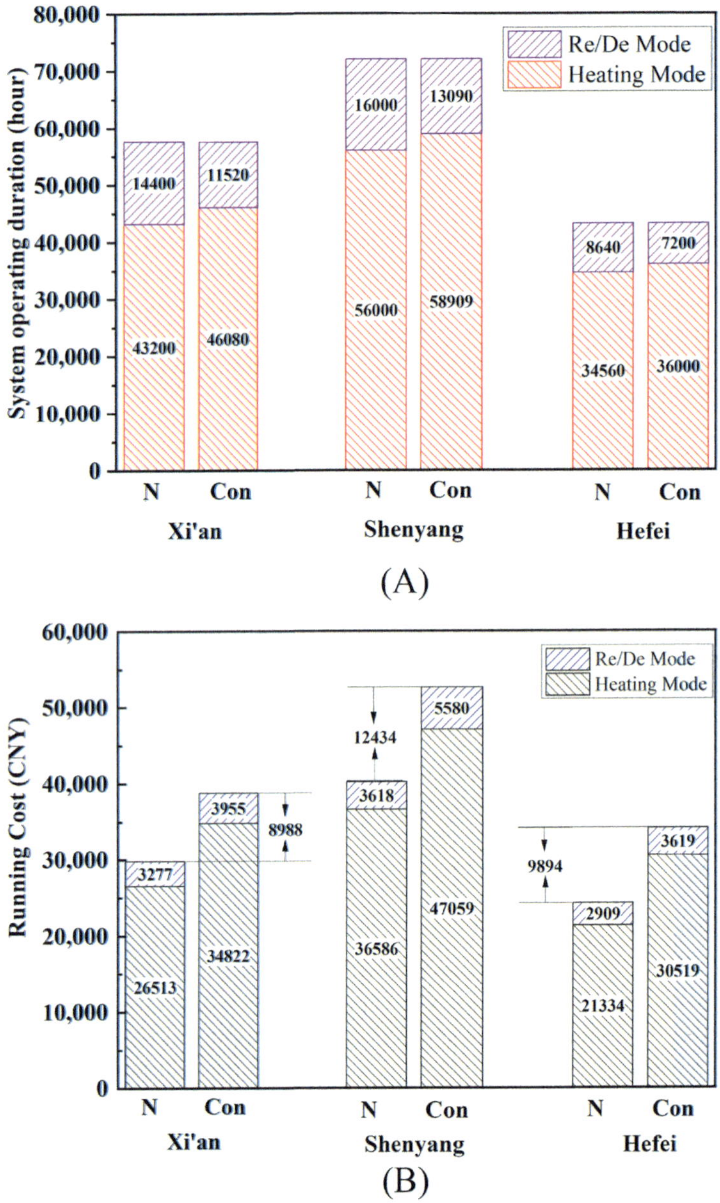

FIGURE 11.11 System operating duration and total running costs in three cities in the 20 operating years. (A) System operating duration and (B) total running costs.

during the heating season (43,200 hours), followed by Xi'an (57,600 hours) and Shenyang (72,000 hours). Different operating length times result that, in the cities of Xi'an, Shenyang, and Hefei, respectively, the total expenses of FFASHP are 23.2%, 23.6%, and 29.0% less than those of the conventional ASHP system. As demonstrated in Fig. 11.11B, the conventional system's running costs in the defrosting mode were 10.2%, 10.6%, and 10.6% of the total cost, whereas the FFASHP system's running costs in the regeneration mode account for approximately 11.0%, 9.0%, and 12.0% of the total cost, respectively.

Compared to the traditional system, the FFASHP system's heating mode operating duration time accounted for a smaller portion of the total operating duration time (Fig. 11.11A). Nonetheless, because the DES served as a low-temperature heat source, the FFASHP system performed well in regeneration mode. Additionally, in the case of the conversional ASHP system, the energy of water was absorbed during the defrosting process, resulting in a negative energy for water heating. These reasons suggested that the FFASHP system's operating costs were less than those of the conventional ones.

Fig. 11.12A shows the total cost differences between the FFASHP system and the conventional system in the three cities in the 20 operating years. As can be seen, in Xi'an, Shenyang, and Hefei, the overall cost disparities drop to zero after operating for around 4.2, 3.3, and 4.03 years, respectively. This indicates that the initial cost payback periods in these three cities were 4.2, 3.3, and 4.03 years. After 10 years of system operation, the total operational costs of the two systems in the cities of Xi'an, Shenyang, and Hefei were 2363, 3928, and 2755 CNY, respectively. After 20 years of operation, the overall cost savings were 55,839, 88,877, and 64,419 CNY, respectively. The difference in the total running expenses between the two systems increased as the operating duration increased, reaching 7268, 10,714, and 8174 CNY/year in Xi'an, Shenyang, and Hefei cities, respectively.

Fig. 11.12B shows the variation of the proportion of additional initial cost for the FFASHP system in total cost in the 20 operating years. In Xi'an, Shenyang, and Hefei, the proportions of the increased initial cost were initially 24.7%, 23.3%, and 25.6%, respectively. Following 20 years of operation, it decreased to 4.8%, 3.8%, and 5.7% in the respective cities. This indicates that the FFASHP system's higher beginning cost had less of an impact on the system's total cost over its lifetime. The system in Shenyang had the longest heating period while the system in Hefei had the shortest; hence, Shenyang had the lowest system initial cost proportion, and Hefei had the highest total cost proportion. This illustrates that the longer the heating time, the less of an influence the system's higher starting cost will have on the total cost.

11.6 Concluding remarks

In this chapter, a novel FFASHP system, coupled with an EHECSD and an ESD, is proposed. The chapter presents experimental and simulation results to demonstrate the system's performance. Based on these findings, the chapter discusses the optimization of control strategies for the proposed system alongside a technical and economic analysis. The conclusions drawn are as follows:

1. A novel FFASHP system, incorporating an ESD with PCM and an EHECSD, is proposed to prevent frosting. The system utilizes regeneration air to regenerate the desiccant,

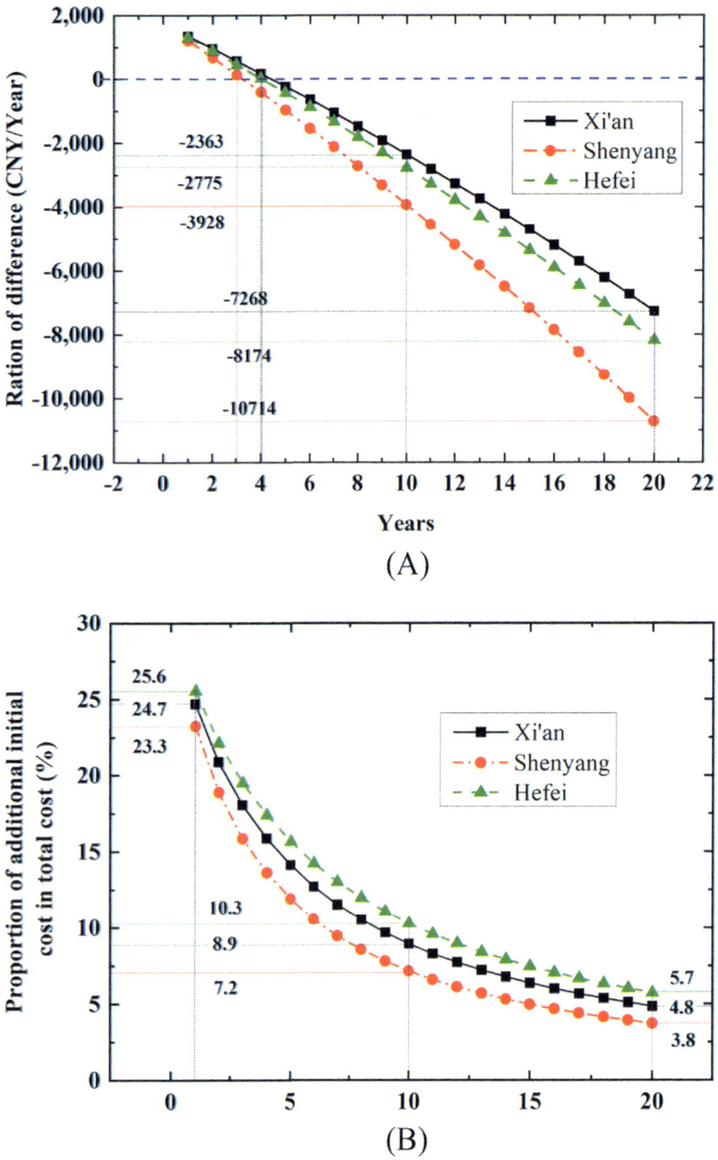

FIGURE 11.12 Ratio of total cost differences in the three cities and proportion of the additional first cost for the FFASHP system in total cost in the 20 operating years. (A) The ratio of total cost differences in the three cities. (B) Proportion of additional first cost for the FFASHP system in total cost. *FFASHP*, Frost-free air source heat pump.

which can balance aims and energy savings. This design dehumidifies the air with a solid adsorbent before it enters the outdoor heat exchanger, thus preventing frosting. Additionally, in regeneration mode, the system avoids frost formation and continuously heats as the ESD releases energy during the solid desiccant regeneration process.

2. An experiment was conducted to verify the feasibility of the proposed FFASHP system, and the results showed that experimental validation of the proposed FFASHP system indicated that the system kept the evaporator frost-free for 32, 34, and 36 minutes at

ambient temperatures of $-3°C$, $0°C$, and $3°C$ with an RH of 85%. The maximum temperatures of the EHECSD during these tests were 50.4°C, 58.4°C, and 60.2°C, respectively, which were sufficient for desiccant regeneration. Compared to the RCD system, the COP of the FFASHP was 17.9% and 3.4% higher at ambient temperatures of $-3°C$ and 3°C, respectively, with the system's advantages being more pronounced at lower temperatures.

3. A mathematical model and dynamic simulation program were developed to analyze the FFASHP system's dynamic characteristics. Results indicated that at ambient temperatures of 0°C, $-5°C$, and $-10°C$, the average COPs were 2.87, 2.47, and 1.78, respectively. The dehumidification efficiency increased by 0.85% and 0.45% as the RH increased from 75% to 80% and from 80% to 85%, respectively, while the regeneration efficiency decreased by 3.5% from 75% to 85% RH at an ambient temperature of 0°C. Additionally, the regeneration efficiency increased from 71.9% to 75.5% as the air velocity increased from 2.5 m/s to 3.5 m/s. Overall, ambient temperature and RH have the greatest impact on system performance.
4. To enhance system performance under various ambient temperatures and RH levels, an optimal control strategy was developed. This included a three-dimensional curved surface and a correlation formula (as found in the literature [32]) for maintaining the evaporator frost-free (heating time) relative to ambient temperature and RH. A surface plot of the system's average COP against ambient temperature and RH was also provided, offering insights for system optimization.
5. The initial cost, running cost, total life span cost, and payback period of the proposed FFASHP system were thoroughly analyzed in three representative cities in China (Xi'an, Hefei, and Shenyang), each in a different climate zone (cold zone, hot summer, and cold winter zone, and severe cold zone respectively). Compared to a conventional ASHP system, the running cost of the FFASHP system was reduced by 23.2%, 23.6%, and 29.0% in these cities, respectively. The payback periods for the initial cost were 4.2, 3.3, and 4.03 years in Xi'an, Shenyang, and Hefei, respectively. Considering both system performance and payback period, the proposed system demonstrated significant technical and economic feasibility compared to the conventional system.

The contents of this chapter may help better understand the proposed FFASHP and it is an honor if this chapter can inspire others. To sum up, using dehumidification devices to reduce the moisture of the air before entering the ASHP outdoor heat exchanger is one of the reliable approaches to deal with the frosting issue. Integrating with ESD, the system performance is guaranteed in regeneration mode. Through appropriately arranging the structure of the system, additional components, and control strategy, the performance in both technical and economic layers will possibly be better than the conventional ASHP system.

References

[1] F. Hao, W. Shao, What really drives the deployment of renewable energy? A global assessment of 118 countries, Energy Research & Social Science 72 (2021) 101880.

[2] A.A. Aganda, J.E.R. Coney, P.E. Farrant, C.G.W. Sheppard, T. Wongwuttanasatian, A comparison of the predicted and experimental heat transfer performance of a finned tube evaporator, Applied Thermal Engineering 20 (2000) 499–513.

[3] F. Wang, C.H. Liang, M.T. Yang, X.S. Zhang, Preliminary study of a novel defrosting method for air source heat pumps based on superhydrophobic fin, Applied Thermal Engineering 90 (2015) 136–144.
[4] Y. Yao, Y. Jiang, S. Deng, Z. Ma, A study on the performance of the airside heat exchanger under frosting in an ASHP water heater/chiller unit, International Journal of Heat and Mass Transfer 47 (2004) 3745–3756.
[5] L. Zhang, Y. Jiang, J. Dong, Y. Yao, Advances in vapor compression air source heat pump system in cold regions: a review, Renewable Sustainable Energy Reviews 81 (2018) 353–365.
[6] Y. Chung, J. Yoo, G. Kim, M. Kim, Prediction of the frost growth and performance change of air source heat pump system under various frosting conditions, Applied Thermal Engineering 147 (2018) 410–420.
[7] Z. Li, W. Wang, Y. Sun, S. Liang, S. Deng, Y. Lin, et al., A novel defrosting initiating method for air source heat pumps based on the optimal defrosting initiating time point, Energy and Buildings 222 (2020) 110064.
[8] S. Liang, W. Wang, Y. Sun, Z. Li, J. Zhao, Y. Lin, et al., A novel characteristic index for frosting suppression based on the configuration and operation of air source heat pumps, International Journal of Refrigerant 109 (2020) 161–171.
[9] Z.J. Xu, H. Lu, An experimental study on energy-storage based defrosting performance of an air source heat pump system with a micro-channel heat exchanger as outdoor coil, Applied Thermal Engineering 240 (2024) 122067.
[10] S.F. Huang, H.F. Yu, M.X. Zhang, H.S. Qu, L.B. Wang, C.Y. Zhang, et al., Advances, challenges and outlooks in frost-free air-source heat pumps: A comprehensive review from materials, components to systems, Applied Thermal Engineering 234 (2023) 121163.
[11] W. Huang, J.X. Huang, Z.G. Guo, W. Liu, Icephobic/anti-icing properties of superhydrophobic surfaces, Advances in Colloid and Interface Science 304 (2022) 102658.
[12] Y. Zhang, G.M. Zhang, A.Q. Zhang, Y.H. Jin, R.R. Ru, M.C. Tian, Frosting phenomenon and frost-free technology of outdoor air heat exchanger for an air-source heat pump system in China: an analysis and review, Energies 11 (2018) 2642.
[13] M.L. Qu, L. Xia, S.M. Deng, Y.Q. Jiang, A study of the reverse cycle defrosting performance on a multi-circuit outdoor coil unit in an air source heat pump – Part I: Experiments, Applied Energy 91 (2012) 122–129.
[14] W. Wang, Q.C. Guo, Y.C. Feng, W.P. Lu, X.G. Dong, J.H. Zhu, Theoretical study on the critical heat and mass transfer characteristics of a frosting tube, Applied Thermal Engineering 54 (2013) 153–160.
[15] Y.B. Lee, S.T. Ro, An experimental study of frost formation on a horizontal cylinder under cross flow, International Journal of Refrigeration 24 (2001) 468–474.
[16] C.H. Cheng, C.C. Shiu, Frost formation and frost crystal growth on a cold plate in atmospheric airflow, International Journal of Heat and Mass Transfer 45 (2002) 4289–4303.
[17] K. Kwak, C. Bai, A study on the performance enhancement of heat pump using electric heater under the frosting condition: Heat pump under frosting condition, Applied Thermal Engineering 30 (2009) 539–543.
[18] ANSI/ASHRAE 37-1978, Method of Testing for Rating Unitary Air Conditioning and Heat Pump Equipment, The American Society of Heating Refrigerating, and Air-Conditioning Engineers, 1978.
[19] ARI-310/380-2004, Standard for Packaged Terminal Air-conditioners and Heat Pumps, Air-conditioning and Refrigeration Institute (ARI) and the Canadian Standards Association, 2004.
[20] Y.F. Liu, Y. Zhou, Y.W. Chen, D.J. Wang, Y.Y. Wang, Y. Zhu, Comparison of support vector machine and copula-based nonlinear quantile regression for estimating the daily diffuse solar radiation: a case study in China, Renewable Energy 46 (2020) 1101–1112.
[21] W. Xu, C.P. Liu, A.G.B. Qiao, Feasibility and performance study on hybrid air source heat pump system for ultra-low energy building in severe cold region of China, Renewable Energy 146 (2020) 2124–2133.
[22] L.X. Ma, F.H. Wang, Z.H. Wang, S. Zhang, Z.X. Liu, Y.C. Lou, Experimental investigation on an air source heat pump system with coupled liquid-storage gas–liquid separator regarding heating and defrosting performance, International Journal of Refrigeration 134 (2022) 176. -168.
[23] L.X. Ma, F.H. Wang, Z.Y. Wang, Z.H. Wang, S. Zhang, Y.J. Sun, Thermodynamic mechanism of high energy performance of air source heat pump with coupled liquid-storage to gas–liquid separator, Solar Energy 255 (2023) 497–506.
[24] Z.H. Wang, P.F. Zhang, F.H. Wang, L.X. Ma, Z.J. Ma, The study of dynamic characteristics on a novel air source heat pump coupled with liquid-storage gas–liquid separator under non frosting and refrigeration conditions, Journal of Building Engineering 84 (2024) 108538.

[25] W. Su, W.H. Li, J.M. Zhou, X.S. Zhang, Experimental investigation on a novel frost-free air source heat pump system combined with liquid desiccant dehumidification and closed-circuit regeneration, Energy Conversion and Management 178 (2018) 13–15.
[26] W. Su, H. Li, B. Sun, S.H. Li, X.S. Zhang, Performance investigation on a frost-free air source heat pump system employing liquid desiccant dehumidification and compressor-assisted regeneration based on exergy and exergoeconomic analysis, Energy Conversion and Management 183 (2019) 167–181.
[27] W. Su, X.S. Zhang, Performance analysis of a novel frost-free air-source heat pump with integrated membrane-based liquid desiccant dehumidification and humidification, Energy and Buildings 145 (2017) 293–303.
[28] S.W. Wang, Z.Y. Liu, A new method for preventing HP from frosting, Renewable Energy 30 (2003) 753–761.
[29] L. Zhang, T. Fujinawa, M. Saikawa, A new method for preventing air-source heat pump water heaters from frosting, International Journal Refrigeration 35 (2012) 1327–1334.
[30] Z.H. Wang, Y.X. Zheng, F.H. Wang, X.K. Wang, Z. Lin, J.C. Li, et al., Experimental analysis on a novel frost-free air-source heat pump water heater system, Applied Thermal Engineering 70 (2014) 808–816.
[31] F.H. Wang, Z.H. Wang, Y.X. Zheng, Z. Lin, P.F. Hao, C. Huan, et al., Performance investigation of a novel frost-free air-source heat pump water heater combined with energy storage and dehumidification, Applied Energy 139 (2015) 212–219.
[32] Z.H. Wang, F.H. Wang, X.K. Wang, Z.J. Ma, X.Z. Wu, M.J. Song, Dynamic character investigation and optimization of a novel air-source heat pump system, Applied Thermal Engineering 111 (2017) 122–133.
[33] K.S. Chang, H.C. Wang, T.W. Chung, Effect of regeneration conditions on the adsorption dehumidification process in packed silica gel beds, Applied Thermal Engineering 24 (2004) 735–742.
[34] W.J. Hu, Y.Q. Jiang, M.L. Qu, et al., An experimental study on the operating performance of a novel reverse-cycle hot gas defrosting method for air source heat pumps, Applied Thermal Engineering 31 (2011) 363–369.
[35] S. Pramuang, R.H.B. Exell, The regeneration of silica gel desiccant by air from a solar heater with a compound parabolic concentrator, Renewable Energy 32 (2007) 173–182.
[36] R. Tu, X.H. Liu, Y. Jiang, Performance analysis of a new kind of heat pump-driven outdoor air processor using solid desiccant, Renewable Energy 57 (2013) 101–110.
[37] T.S. Ge, Y.J. Dai, R.Z. Wang, Performance study of silica gel coated fin-tube heat exchanger cooling system based on a developed mathematical model, Energy Conversion and Management 52 (2011) 2329–2338.
[38] Z.H. Wang, G.C. Li, F.H. Wanga, K.H. Li, Y.C. Lou, Techno-economic evaluation of a frost-free air source heat pump water heater, Sustainable Cities and Society 57 (2020) 102102.
[39] H.H. Tan, G.H. Xu, T.F. Tao, X.Q. Sun, W.D. Yao, Experimental investigation on the defrosting performance of a finned-tube evaporator using intermittent ultrasonic vibration, Applied Energy 158 (2015) 220–232.

CHAPTER

12

Atomization equipment for simulating icing environment

Huanyu Deng[1], *Mengjie Song*[2,3], *Xiaohong Liu*[1] *and Xiaofeng Li*[1]

[1]Beijing Institute of Radio Measurement, Beijing, P.R. China [2]Department of Energy and Power Engineering, School of Mechanical Engineering, Beijing Institute of Technology, Beijing, P.R. China [3]School of Mechanical Engineering, Hanyang University, Seoul, South Korea

12.1 Introduction

The aircraft brings excellent convenience to people's travel, while it is impossible to eliminate the impact of the weather on the flight safety of the aircraft. For winter flights, the crucial natural enemy that affects flight safety is aircraft icing [1,2]. Ice on the outside of the aircraft, especially on the surface of the wings, seriously affects flight safety [3,4]. Ice roughness equivalent to that of medium-grain sandpaper reduces maneuverability to dangerous levels, so even slight pollution caused by frost can damage the wing. As a kind of crystal, if the engine sucks ice in, the efficiency of the engine will decrease. In severe cases, ice could cause the engine to stop in the air [5]. Besides, aircraft speedometers and pressure sensors may also freeze. Accidents caused by aircraft icing have been numerous and widely reported in recent decades [6].

Given the considerable damage caused by ice, to remove ice accumulation on the parts of the aircraft or even eliminate ice accumulation on the surface of the aircraft becomes necessary. In the open literatures, there are two types of aircraft antiicing methods reported, electrothermal antiicing and antiicing coating [7]. Researches on electrothermal antiicing are very comprehensive. A novel electrothermal antiicing system for fiber-reinforced polymer composite airfoils had been developed by Mohseni and Amirfazli [8]. Two composite airfoil samples were made to study the effect of wires' spacing on the airfoils' surface temperature distribution and the effectiveness of the thermal elements' pattern for icing mitigation. To reveal the physical mechanism of coupled aerodynamic and thermal effects on antiicing, three types of actuators are designed and mounted on a

Frosting and Icing for Efficient Energy Use in Engineering Applications
DOI: https://doi.org/10.1016/B978-0-443-15495-9.00014-9

NACA 0012 airfoil by Meng et al. [9]. The effects of anisotropic composite skin on the electrothermal antiicing system were studied by Shen et al. [10]. There are also many pieces of research on antiicing coatings. By controlling the characteristics of surface features such as wettability, topography, and lubricity, researchers have developed various antiicing or icephobic coatings [11,12]. Typically, the formation of zwitterionic coatings with an aqueous lubricating layer for antifogging/antiicing applications was carried out by Tao et al. [13]. However, for some ice accumulation phenomena caused by lack of antiicing measures, such as aircraft used for flight training without functioning antiicing equipment fly under icing conditions, it is necessary to remove ice from the icing surfaces. Therefore, aircraft active deicing studies become necessary [14,15].

To understand the icing mechanism of aircraft and control the icing process, it is necessary to simulate the icing environment at high altitudes. An icing cloud simulation system in the climatic environmental test facility (CETF) provides a ground test simulation method by creating a ground icing condition comparable to high-altitude icing. However, it is challenging to simulate icing in flight via experiments. The first problem to be solved is the large relative velocity between the aircraft and the clouds. Supercooled water droplets in clouds, whose temperature can reach as low as $-40^{\circ}C$ [16,17], bring another difficulty in simulating aircraft icing through clouds [18,19]. The formation process of the water droplets from the atomization nozzles is so different from that of clouds, and the distribution of water droplets used in the icing test is far from the uniformity of the clouds. So, it is impossible to achieve complete consistency with both the particle size and distribution of water droplets in clouds [20,21].

The wind tunnel is one of the practical tools for aerodynamic experiments. It is widely used to simulate the flow field around the aircraft or other objects. In a wind tunnel, a modular model was built by Nicolosi et al. [22], and both complete aircraft aerodynamic characteristics and mutual effects among components were analyzed. To predict the hinge moments required for the initial design of the missile, a study on aerodynamic modeling was performed by Yim et al. [23]. Also, based on a wind tunnel, the dependence of the aerodynamic noise from the bogie region on different factors was studied by Latorre et al. [24]. A wind tunnel test program was implemented to measure wind-induced vibration of a yawed bridge tower at different wind speeds and directions by Marra et al. [25]. Wind tunnel tests on aerodynamic forces of road vehicles under unsteady wind conditions were conducted by Sumida et al. [26]. However, it is not enough to achieve the speed required by the flight, and the distribution and particle size of supercooled water droplets should conform to the real cloud conditions. Therefore, icing wind tunnels are widely used. For example, in the simulated environment of the icing wind tunnel, an integrated experiment method was proposed by Zhu et al. [27], with the icing tested and shear stress measured. In an icing wind tunnel, a superhydrophobic surface with increased solar radiation absorption was tested by Gonzales et al. [28].

The icing wind tunnels are always used as tools, while few studies take the optimization of the icing wind tunnels as the research object. Compared with an ordinary wind tunnel, it is challenging to control the formation size and distribution of supercooled water droplets in an icing wind tunnel. For example, the use of saline water in a spray-assisted natural draft dry cooling tower was performed by Sadafi et al. [29], and the effects of a supersonic nozzle arrangement angle on the performance of an impulse axial turbine were studied by Aghaei-Togh et al. [30]. These studies are both in ordinary wind tunnels. Around the icing wind tunnels, only two studies are found. The local collection coefficient along the wingspan was calculated under four different kinds of nozzle arrangements by Li et al. [31]. As reported, the uniformity of the icing cloud could only be improved slightly when the distribution of nozzles was changed from aligned to the staggered arrangement in certain numbers of nozzles. The other one was conducted by Deng et al. [32,33], with the antifreezing design verification of atomization equipment in an icing cloud simulation system reported.

Clearly, to accurate control of a simulated icing environment, it becomes necessary to investigate the nozzle arrangement. In this chapter, the optimization of the simulated icing environment by adjusting the arrangement of nozzles in an atomization equipment for the antiicing and deicing of aircraft are carried out. First, a model was developed and validated with ice shape and thickness. Then, the velocity field of water droplets and the distribution of liquid water content (LWC) are analyzed according to different arrangements of the 10 nozzles in the experimental system. Finally, after the limitations of this study are discussed, the most important results are summarized.

12.2 Methodology

The establishment of the icing cloud simulation systems required the combination of experiment and numerical study, and the geometric model was built according to the dimensional data determined by the test-bed. At the same time, the test conditions used in the experiment were taken as the initial conditions of the numerical study. Finally, the conclusions were drawn by comparing the two methods. The method of numerical study validated by the experimental results can deduce more reliable results. This paper mainly introduces the work of a numerical study. The design and numerical study process of the atomization equipment is illustrated in Fig. 12.1.

12.2.1 Experimental study

An icing cloud simulation system provides an icing ground test simulation method by creating a ground icing condition comparable to high-altitude icing. A relatively famous icing cloud simulation system is installed in the NASA Lewis Research Center's Icing Research Tunnel [34,35]. As an effective assessment for antiicing ability of helicopter systems, icing cloud simulation systems are mainly applied to the antiicing ground test of helicopters, using spray tower test facility to generate icing clouds at low temperature, and thus to simulate the high-altitude cloud environment under different freezing conditions and conduct icing and antiicing tests on helicopters with ground mooring and hovering conditions [36,37], restoring the situation of high altitude icing as much as possible. To create artificial icing clouds, the water spray nozzles of the atomization equipment inject microscopic droplets into the airflow field. The nozzles are always fixed in the front of the CETF, just at upstream of the test section where the test specimens are placed in the ice-forming test [38,39], as shown in Fig. 12.2.

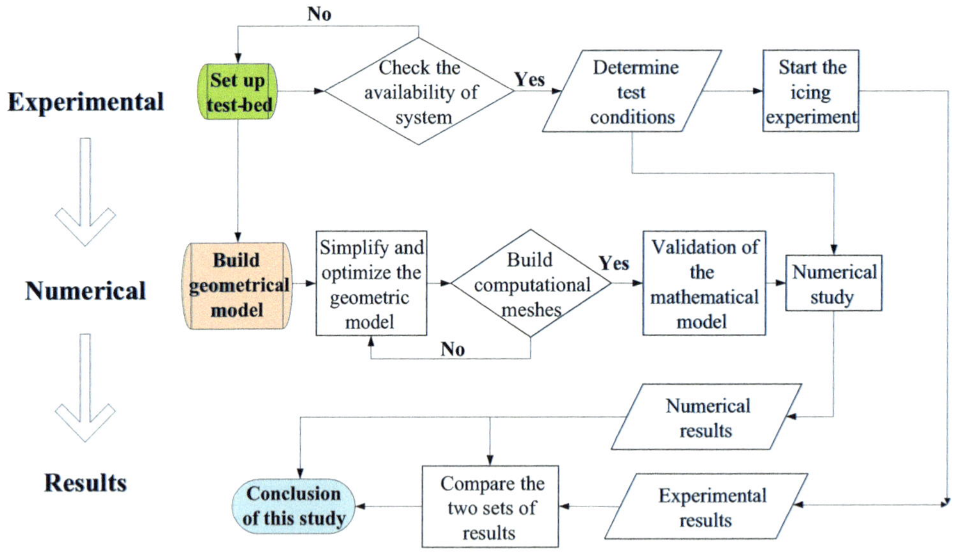

FIGURE 12.1 The design and numerical study process of the atomization equipment.

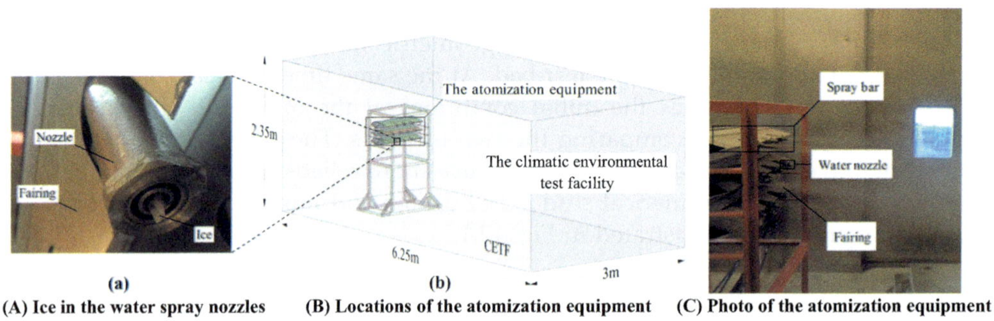

FIGURE 12.2 Locations of the atomization equipment and water spray nozzles in the CETF. (A) Ice in the water spray nozzles. (B) Locations of the atomization equipment. (C) Photo of the atomization equipment. *CETF*, Climatic environmental test facility.

All the nozzles are arranged in accordance with certain rules inside the aerodynamic fairings. Since the nozzle outlet may be blocked by the water freeze due to a lower temperature, the antifreezing design should be further added in the performance test of an icing cloud simulation system. This icing cloud simulation system used nozzles to atomize water into small droplets and then inject them into the low-temperature airflow to simulate icing clouds. A prerequisite for this scenario is that the water flows through the main water pipe into the branch pipes, and it can be atomized by mixing with the air. The mixture is heated to above ambient temperature before the water droplets leaves the nozzles, and air in the pipe also bears an auxiliary effect of antifreezing. The calculation of the temperature field not only predicts whether the droplets are frozen before exiting the nozzles but also provides a visual representation of the temperature distribution at different locations inside the spray bar. The entire spray system was placed in the CETF with a temperature range of 233K to 273K. Since the position of the air supply system was upstream of the spray system, the shape of the spray bars was designed as airfoils to minimize their influence on the flow of air in the flow field. As shown in Fig. 12.2, there were three spray bars, two of which had three spray nozzles and the other had four spray nozzles. Three spray bars were arranged vertically at a distance of 118 mm, with a four-nozzle bar in the middle, and nozzle holders were spaced every 150 mm on each spray bar.

12.2.2 Numerical method

In this chapter, optimization of the simulated icing environment was conducted by adjusting the arrangement of nozzles in an atomization equipment. First, the geometric model of the CETF was established and reasonably simplified. Multiple sets of computational domain grids were then generated, and the grid best suited for this study was selected after validating the grid independence. The numerical model was validated by the selected grid, and the numerical study of nozzle arrangement was carried out after validating the model successfully.

To show the effect of nozzle arrangement on parameters such as velocity and distribution uniformity of LWC in CETF, a variety of nozzle arrangements were designed. The nozzle arrangements involved in this study are shown in Fig. 12.3. Among them, the arrangement in Fig. 12.3A was adopted in the experiment, and the other arrangement patterns were derived from Case 1. Fig. 12.3F can be obtained by swapping the second and third rows in Fig. 12.3C. Similarly, Fig. 12.3E can be obtained by swapping the second and third rows in Fig. 12.3D. The middle spray bars in Fig. 12.3A–D are all four nozzles, of which Fig. 12.3A and B are symmetrical about the horizontal lines across the center. The influence of whether the nozzles are distributed symmetrically about the horizontal lines on the parameters of droplets when 4-nozzle spray bars are arranged in the middle can be obtained by comparing the arrangements shown in Fig. 12.3A and B and the comparison results from arrangements of Fig. 12.3E and F will present the influence when 2-nozzle spray bars are arranged in the middle. From the comparison of the nozzle arrangements shown in Fig. 12.3C–F, the effect of the positional change of the 2-nozzle spray bar on the droplet parameters can be further obtained. Meanwhile, Fig. 12.3A, B, E, and F could be used to study how the number of nozzles on the middle spray bar affects the water droplet parameters.

The computational domain is divided by the structural grid, and a finer grid scheme in the boundary layer is adopted. The overall grid of the computational domain is shown in Fig. 12.4, **as well as** the distribution of the grid in the boundary layer from the side view. The growth rate of the grid in the boundary layer is fixed at 1.1, with the number of layers of 30. To arrange the value of $y+$ within a reasonable range, take the height of the first layer of the grid as 5×10^{-5} m. In this way, better results can be obtained at the wall where the flow parameters vary greatly.

To comprehensively consider the relationship between computing accuracy and computing resource utilization efficiency, it is necessary to validate the grid independence

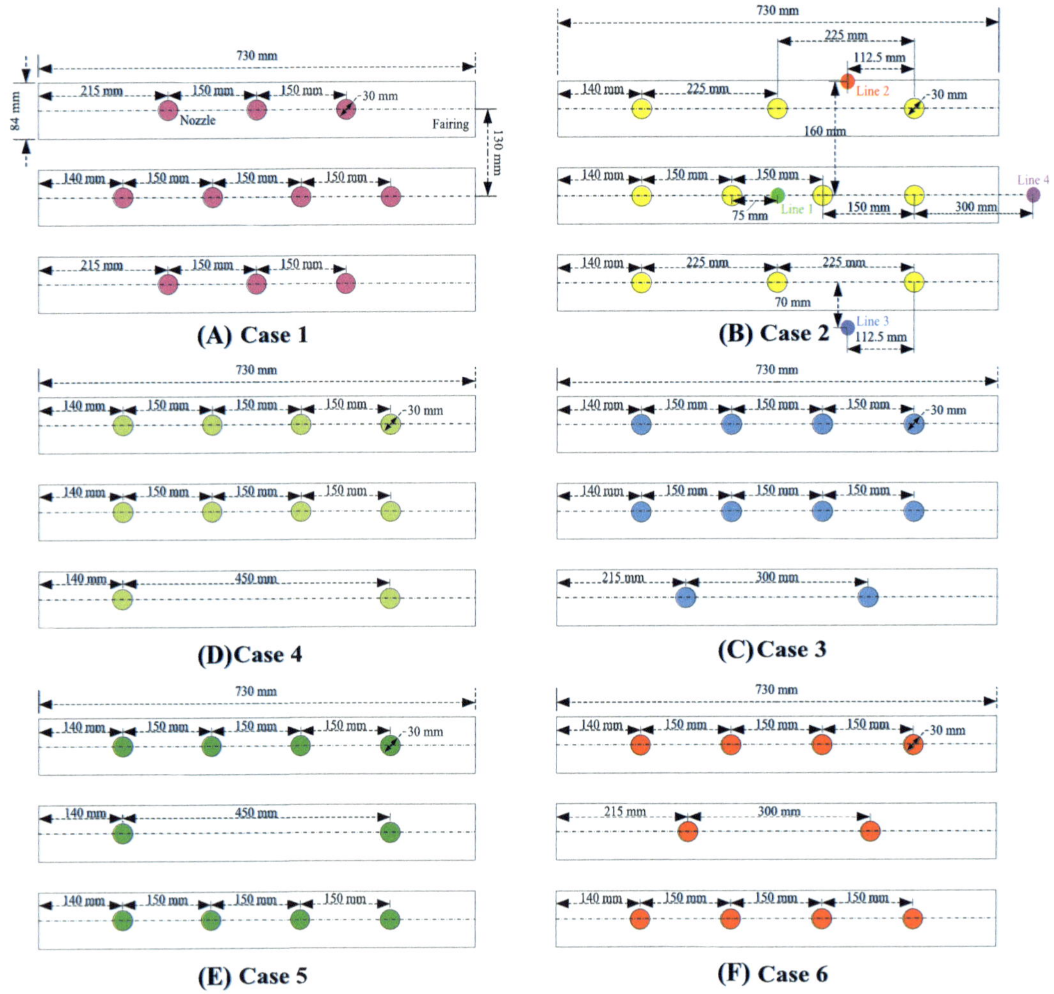

FIGURE 12.3 Position of nozzles and monitoring lines on the front view. (A)–(F) Show different nozzle arrangements.

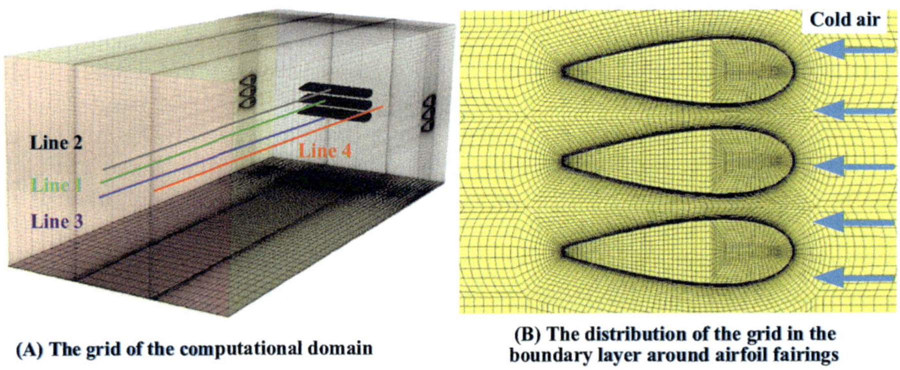

FIGURE 12.4 Computational domain and the grid in the boundary layer. (A) The grid of the computational domain. (B) The distribution of the grid in the boundary layer around airfoil fairings.

until the generated grid finds an optimal balance. As the local momentum coefficient of the middle airfoil fairing is very sensitive to the change in the number of grids in the vicinity of the airfoil fairings, it is selected as the primary basis for validating the independence of the grid. Another reason is that the flow field parameters near the middle airfoil fairing change sharply, and the accuracy of these fluid parameters directly affect the final results of the simulation. The grid independence study was performed on the configuration of Case 1 in Fig. 12.3. The velocity used throughout this validation was 15 m/s, which was also the velocity value used in the experiment, and the direction of the incoming flow was increasing x. Fig. 12.5 shows the local momentum coefficient of the middle airfoil fairings corresponding to different grids. The abscissa is the number of iteration steps, and the ordinate represents the value. As shown, the local momentum coefficient calculated with the 1.34 million-cells grid oscillates continuously and posts a significant deviation compared with the others. It can be seen that the use of the too-sparse grid in this study may lead to inaccurate or even wrong results. The local momentum coefficient calculated with the 2.94 million-cells grid varies very little from the result calculated with the 3.12 million-cells grid, which is 0.01002 and 0.00959, respectively. The error of the coefficient calculated by these two grids is 4.48%, which is within acceptable limits. It is demonstrated that the grid with 3.12 million elements can satisfy the requirement of grid independence.

12.2.3 Variation simulation of droplet parameters

It takes several steps to simulate the variation of droplet parameters in CETF. First, the airflow field in the entire computational domain was calculated using the reprogramming software FLUENT. The flow velocity of the external flow field used in the numerical

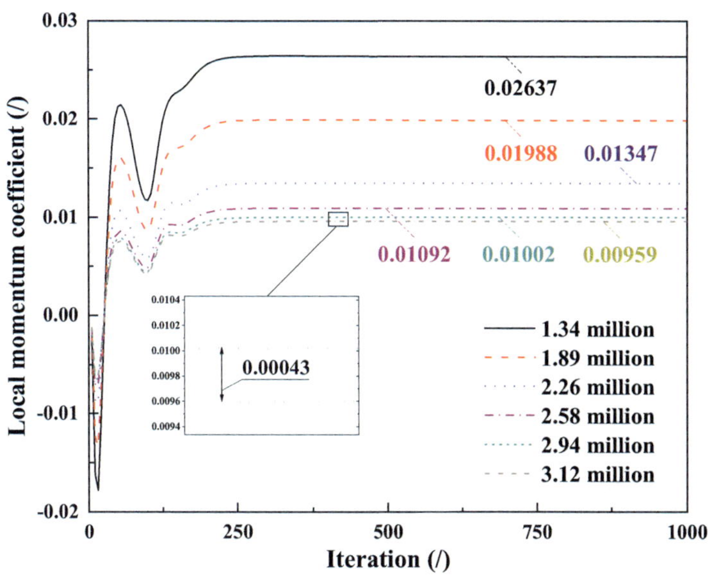

FIGURE 12.5 The local momentum coefficient of the airfoil fairings at different grids.

calculation was consistent with the velocity in the experiment, which was 15 m/s. The Reynolds number corresponding to this velocity and characteristic length, which was 0.76 m, was 1,396,500. The velocity inlet boundary condition was adopted at the entrance of the calculation domain, and the pressure outlet boundary condition was adopted at the exit. The reference pressure in the calculation domain is set to 101,325 Pa. Standard k-epsilon viscous model and spatial numerical scheme were adopted in this numerical study. Both the numerical calculation and the experiment were conducted at zero angle of attack, and then, the user-defined secondary development functions were loaded on the basis of the correct flow field results. At last, the water droplets were added to simulate the real spray system using the discrete phase model (DPM) of FLUENT so that the movement of water droplets could be captured. The hybrid grid was used to calculate the heat distribution of the spray bar, and close the wall at the inner side of the pipes was densified to capture the information of the boundary layer. Turn on the DPM model in FLUENT to simulate the movement of water droplets. The diameter of the water droplets was given according to the actual particle size emitted by the nozzle. The force balance equation [40] of droplets can be written as follows:

$$m_d \frac{d\vec{u}_d}{dt} = m_d \frac{\vec{u} - \vec{u}_d}{\tau_r} + m_d \frac{\vec{g}(\rho_d - \rho)}{\rho_d} + \vec{F} \tag{12.1}$$

The following two heat and mass transfer equations [34] of water droplets were also be taken into account in the model development.

$$\frac{dm_d}{dx} = \beta A_d(\rho_\infty - \rho_s) \tag{12.2}$$

$$\frac{d(Cp_d \cdot m_d T_d)}{dt} = hA_d(T_a - T_d) + L_d \cdot \frac{dm_d}{dt} \tag{12.3}$$

To estimate the mass and heat transfer coefficient β and h, for dual mass and heat transfer from a sphere, Nu and Sh correlations could be written as follows:

$$Nu = 2.0 + 0.6Pr^{1/3}Re^{1/2} \tag{12.4}$$

$$Sh = 2.0 + 0.6Sc^{1/3}Re^{1/2} \tag{12.5}$$

12.2.4 Validation of the model

The calculation model was validated by comparing the experimental results with the numerical results. In the experiment, a test rod with a height of 1.4 m and a diameter of 22 mm was erected at a distance of 3.3 m from the nozzles, and the icing cloud simulation system was started to conduct the icing test. The recorded ice shape and maximum ice thickness were taken as the final results of the test. The numerical method calculates the ice accretion on the surface of the test rod by simulation. The calculation conditions adopted by the numerical method were consistent with the test conditions of the icing test, and the specific data are shown in Table 12.1.

TABLE 12.1 Experimental conditions for the icing test in this study.

Item	Parameter	Value	Unit
1	Airflow velocity	15	m/s
2	Air pressure in the nozzles	0.38	MPa
3	Ambient temperature	− 20	°C
4	Initial spray velocity of water droplets	20	m/s
5	Spray angle	6.5	°
6	Test duration	0.5	H
7	The initial diameter of water droplets	18–22	μm
8	The initial temperature of water droplets	5	°C
9	Water pressure in the nozzles	0.19	MPa

TABLE 12.2 The measuring devices used in this experimental study.

Items	Parameters	Measuring devices	Measurement errors
1	Air temperature	Platinum resistance	± 0.2 (°C)
2	Air pressure	Pressure gauge	2%
3	Ice shape	High-speed camera	/
4	Ice thickness	Vernier caliper	± 0.02 (mm)
5	Time	Stopwatch	± 0.01 (s)

Platinum resistance was used to measure the temperature in the experiment. The temperature measurement range was − 70°C to + 150°C, and the measurement accuracy was ± 0.2°C. The ice thickness was measured by vernier caliper, and the ice shape was recorded by a high-speed camera with a maximum frame rate of 2070 fps. The measuring devices used in the experiments and the corresponding measurement errors of each device are shown in Table 12.2.

Different flow conditions correspond to different changes in water droplet parameters. By designing 12 cases and intercepting 20 planes, whose interval is about 0.2 m, along with the direction of airflow, we can get the statistical law of each parameter. Tables 12.3 and 12.4 give the calculation conditions and the position of the planes that are created to obtain flow parameters, respectively. The ambient air temperature and three airflow velocities in the external flow field were the real temperatures and velocity values used in the CETF experiment, while the pressure and mass flow rate at the nozzles' outlet adopted the four real working states of the nozzles calibrated by measurement. Given these initial conditions, we wished to get the effects of these parameters on the temperature, velocity, diameter, and LWC of water droplets along the path. The selection of the 20 planes in Table 12.4 contains the following criteria. First, the outflow field results of all cases were compared, and the

TABLE 12.3 Calculation conditions designed in this numerical study.

Case no.	Airflow velocity (m/s)	Water pressure (bar)	Air pressure (bar)	Medium volume diameter (MVD) (μm)	Mass flow rate (kg/s)	Ambient air temperature (K)
1	10	0.7	0.7	47.1	0.694	263
2	15					
3	20					
4	10	2	1.7	40.7	1.861	263
5	15					
6	20					
7	10	1.5	2.1	29.7	0.944	253
8	15					
9	20					
10	10	4	4.5	19.8	1.889	253
11	15					
12	20					

TABLE 12.4 The distance from each plane to the nozzles.

Plane No.	1	2	3	4	5	6	7	8	9	10
Distance (m)	0	0.022	0.122	0.222	0.322	0.522	0.722	0.922	1.122	1.322
Plane No.	**11**	**12**	**13**	**14**	**15**	**16**	**17**	**18**	**19**	**20**
Distance (m)	1.522	1.722	2.022	2.322	2.622	2.922	3.222	3.522	3.722	3.922

regions where the flow parameters changed significantly and smoothly were divided. The setting of plane spacing was smaller in the region where the flow parameters changed significantly and larger in the region where the flow parameters changed smoothly.

The ice thickness at different heights of the test rod is shown in Fig. 12.6, where the abscissa represents the ice thickness, the ordinate represents the height of the test rod, the solid line is the test result, and the dashed line is the numerical result. As shown in the figure, the maximum ice thickness measured in the test is 11.5 mm at 0.44 m from the ground, while 11.2 mm at 0.31 m from the ground in the simulation. The difference between the maximum ice thickness and its location is 0.3 mm and 0.13 m, respectively, and the error of the maximum ice thickness accounts for only 2.6% of the test results. The maximum and mean deviations between experimental and numerical results are 1.13 and 0.68 mm, respectively. The small and acceptable errors show that the numerical results agreed well with the experimental results. The water droplets moving in the flow field shift toward the bottom of the test rod under the action of gravity, which leads to a more

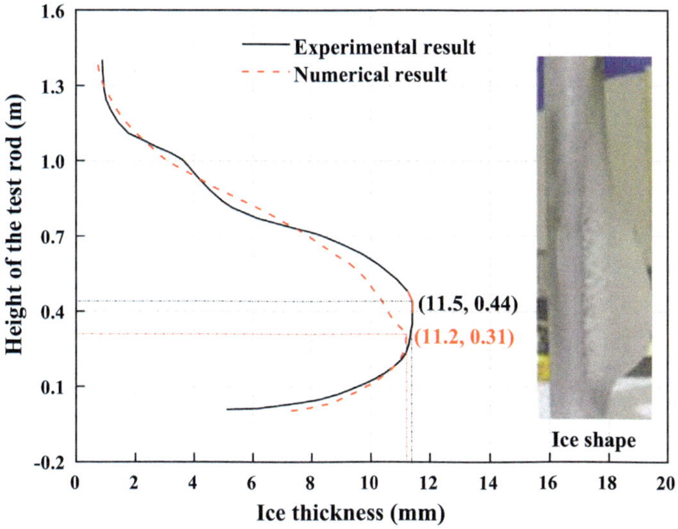

FIGURE 12.6 Ice thickness and shape on the test rod obtained and the validation.

significant local collection coefficient at the bottom of the test rod than that at the top, which results in a small amount of ice accumulation at the top of the test rod. A three-dimensional icing model had been validated by Cao et al. [41] based on the ice shape with a mean deviation of 18.56%. The mean deviation of the model established in this study was only 9.71%. The rationality of the model is fully explained.

12.3 Results and analysis

Four straight lines were selected in the calculation domain to show the calculation results at their positions. Fig. 12.4A shows the spatial positions of the four lines. Since the computational domain was utterly symmetric about the XOZ plane, so the monitoring lines were only selected on one side. The positions of four lines relative to the airfoil fairings on the front view are shown in Fig. 12.4B. By intercepting several planes, whose interval is about 0.2 m, along with the direction of flow, we can get the statistical law of each parameter. The position at the exit of the nozzles is $x = 0.478$ m in the global coordinates. Four self-written user-defined functions were added to the FLUENT to modify the vaporization law by reprogramming the solver so as to get more accurate results of the velocity field, the distribution of LWC, and the distribution of droplet size.

12.3.1 Velocity of the humidity airflow

The initial calculation conditions involved in the simulation are consistent with the experimental conditions in Table 12.1. The diameter distribution of the water droplets conforms to the Rosin–Rammler–Logarithmic distribution function, and the initial mean volume diameter is 20 μm. Contours of velocity in different sections corresponding to Case 1 are shown in Fig. 12.7, and the x values in the contours belong to the local coordinates whose

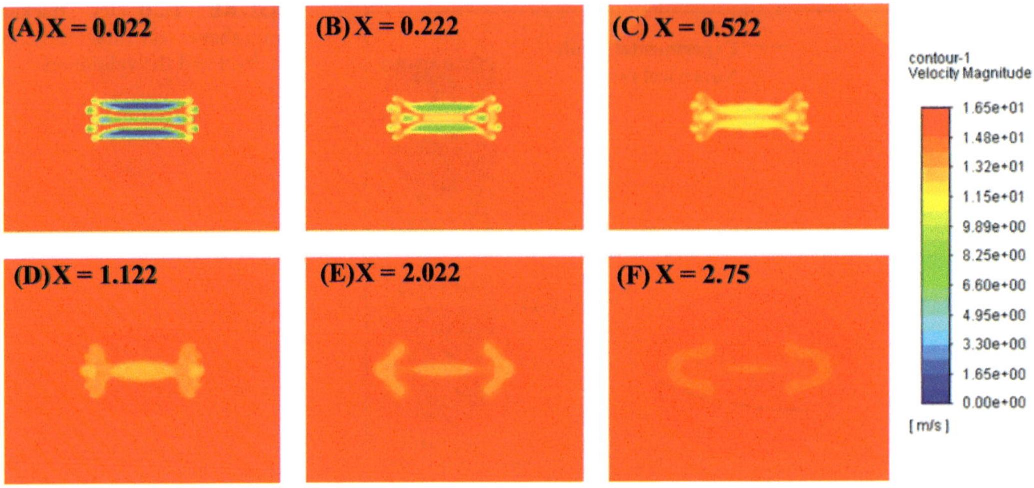

FIGURE 12.7 Contours of velocity in different sections corresponding to Case 1.

origin is the nozzle exit. In Fig. 12.7A, it can be seen that there are several low-velocity regions, and high-velocity airflow is mainly distributed around the airfoil fairings. The effect of the initial velocity of water droplets on the flow field can be clearly seen in Fig. 12.7B. Combined with the distribution position of the nozzles in Fig. 12.3A, it can be seen that the velocity at the positions of the nozzles in Fig. 12.7B is close to 16.5 m/s. The low-velocity region in Fig. 12.7C shows a vertical compression trend, which is caused by the upstream airflow gathering toward the center after bypassing the airfoil fairings. Fig. 12.7D–F shows that the velocity field gradually tends to be uniform from the nonuniform state.

Fig. 12.8 shows the velocity of the mixture on the four monitoring lines corresponding to the nozzle arrangement of Case 1. The relative position of the nozzles in Case 1 can be known from the nozzle arrangement in Fig. 12.3A and the positions of the four monitoring lines in Figs. 12.3B and 12.4A. In the development stage of the flow field, the velocity of mixtures on the four monitoring lines varies greatly. The velocity values at the positions of the four monitoring lines gradually tend to be consistent as the flow field enters the mature stage, which reflects the development process of the flow field from the perspective of data. In the range of $0.5\ \mathrm{m} < x < 0.625\ \mathrm{m}$, the velocity value of Line 1 jumps from 5.125 m/s to 11.5 m/s.

This variation is reflected in the color change in the velocity contour of the external flow field, as shown in Fig. 12.9. The initial velocities of Lines 1 and 2 in the region blocked by airfoil fairings are very low, especially Line 1, whose initial velocity is only 5.2 m/s. However, Lines 1 and 2 experience a considerable increase in velocity after leaving the wake zone of airfoil fairings. The velocity of Line 1 increases by 6.375 m/s and then enters a short period of stability, while the velocity of Line 2 increases by 6.19 m/s and begins to coincide with Line 3 at 15.72 m/s. In the stage before $x = 2$ m, no water droplets arrive at the position of Line 4, so the velocity of Line 4 in the first half of the journey is very stable. The velocity of water droplets has dropped below the airflow velocity by the time arriving at Line 4, which makes the velocity of Line 4 after $x = 2$ m decreases under the action of water droplets.

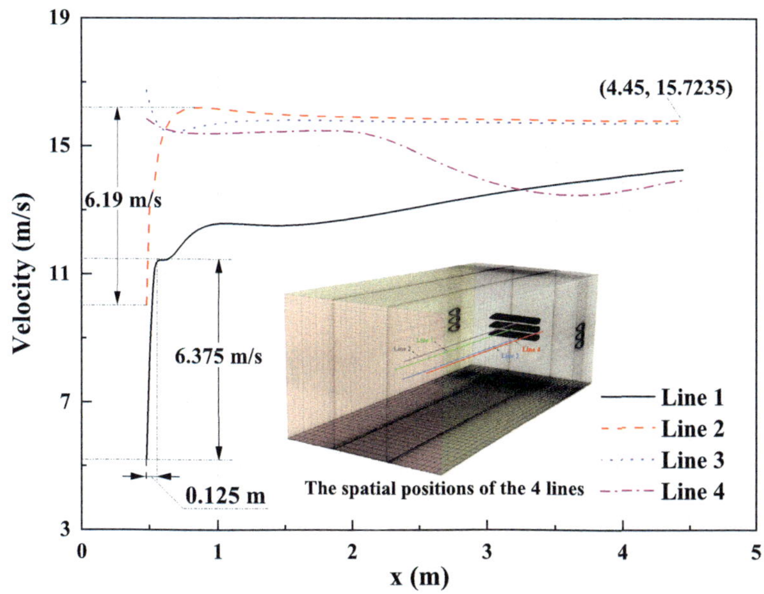

FIGURE 12.8 Velocity of the mixture on the monitoring lines of Case 1.

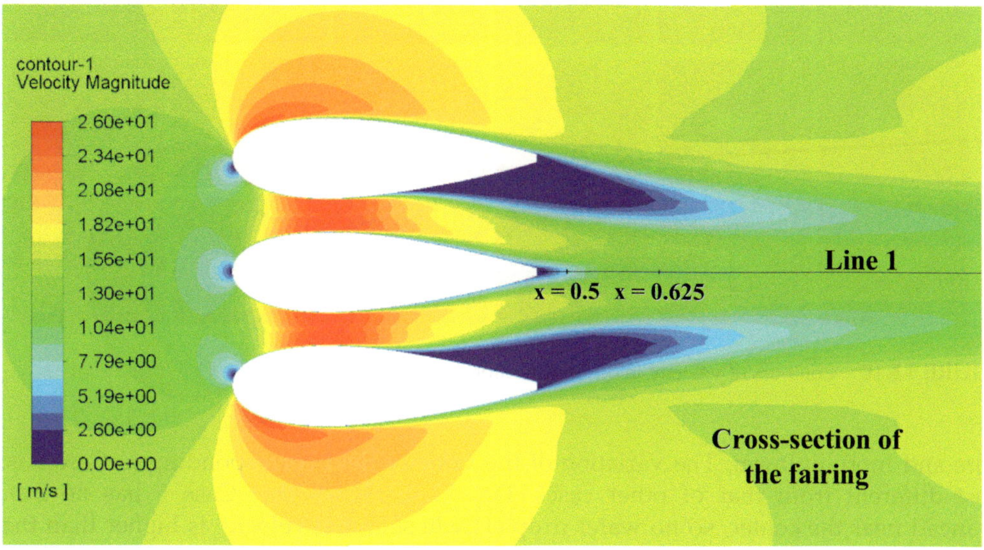

FIGURE 12.9 Velocity contour of the section where Line 1 is located in Case 1.

Fig. 12.10(1A–1F) are velocity contours of Case 1 at different sections along the *x*-direction, similar to the other cases. It can be seen from Fig. 12.10 that the velocity of water droplets does not have much influence on the subsequent development of the velocity field in a flow field dominated by airflow velocity. So, the velocity variation trend in Fig. 12.10 is

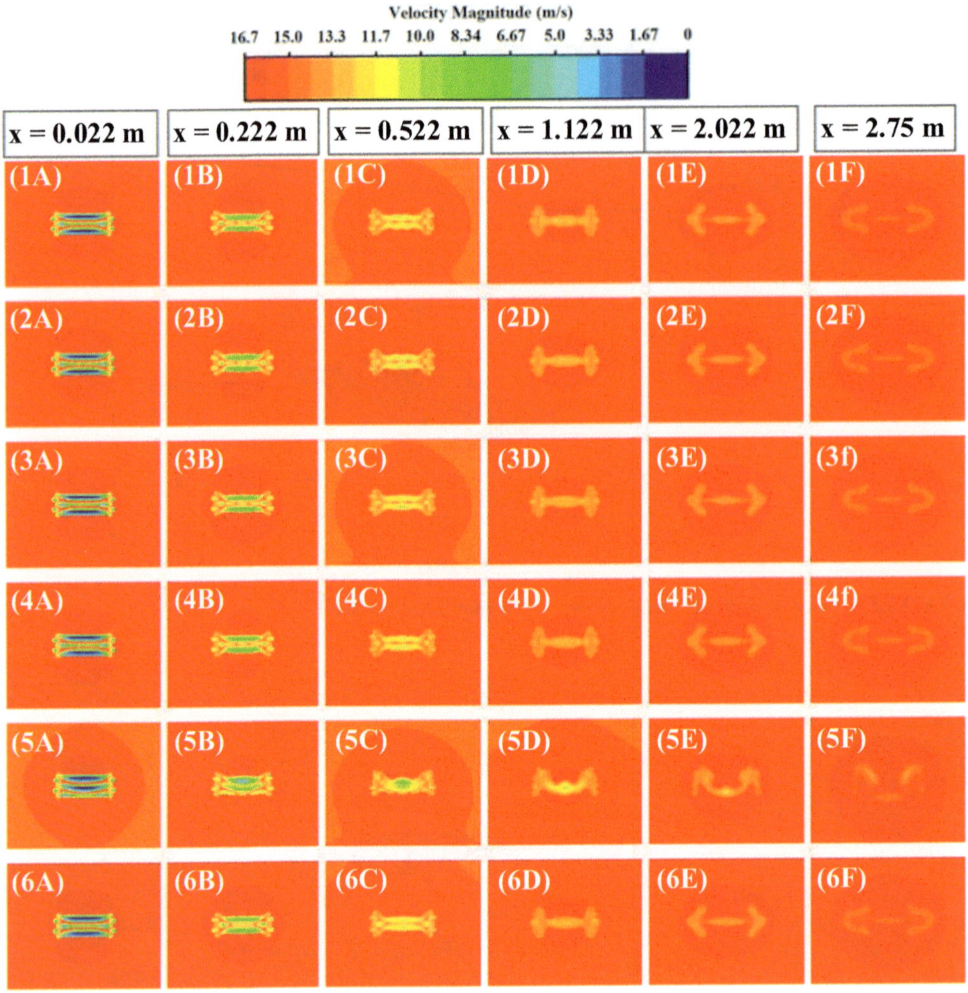

FIGURE 12.10 Contours of velocity in different sections of six cases.

more and more uniform. The variation of the velocity field corresponding to Case 5 is the most different from that of other cases. Fig. 12.3E shows that Case 5 has no nozzles arranged near the center, so no water droplets whose initial velocity is higher than that of airflow in this region. A low-velocity region then forms at the center position of Case 5. The water droplets ejected from the remote nozzles decelerate greatly once they enter this low-velocity region. The water droplets, after deceleration, continue to slow down the subsequent water droplets entering this zone and move downward under the action of gravity, forming an asymmetric low-velocity region. This low-velocity region is the root cause of the severe deformation of the following mature flow field, which leads to the most unreasonable velocity field distribution in Fig. 12.10(5F) compared with other cases.

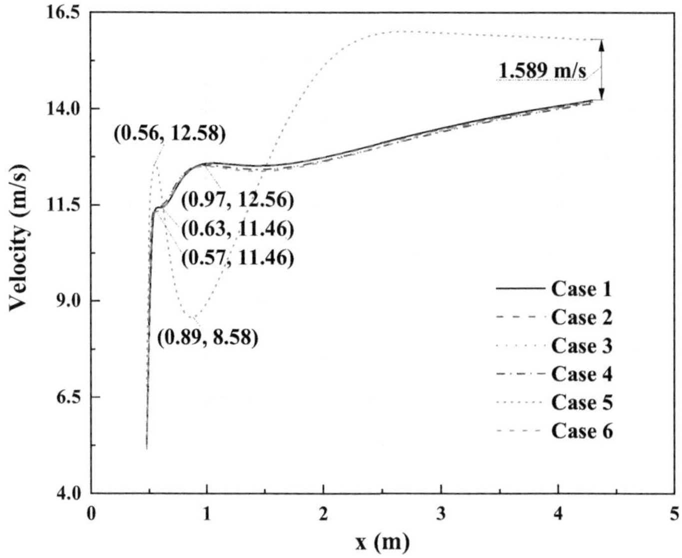

FIGURE 12.11 The velocity of the mixture at Line 1 corresponding to different cases.

Different arrangements of nozzles result in the change of the initial injection positions of water droplets, which inevitably affects the velocity of the flow field. Because the central position of the test specimen needs to be placed on Line 1 during the experiment, the velocity of Line 1 and its surroundings is particularly important. Fig. 12.11 shows the velocity of the mixture at Line 1, corresponding to different nozzle arrangements. Referring to Figs. 12.3B, E and 12.4A, there are no nozzles placed near Line 1 of Case 5, which results in a significant difference between Case 5 and other cases in Fig. 12.11 of the entire flow field. The variations of Cases 1–4 and 6 in the figure are basically the same. The velocity value at $x = 0.478$ m is the smallest, and then the velocity rapidly increases to 11.46 m/s. The velocity increases again between $x = 0.63$ m and $x = 0.97$ m after a short stable stage. When the speed reaches 12.56 m/s, the trend begins to slow down. The velocity at Line 1 of Case 5 is directly accelerated from the lowest value to 12.58 m/s and then enters the low-velocity region mentioned above. At $x = 0.89$ m, the velocity has decreased to 8.58 m/s, and finally, the velocity at $x = 4.3$ m is 1.589 m/s higher than other cases after a steady rise between $x = 0.63$ m and $x = 0.97$ m. The reason why the velocity of Case 5 is higher than other cases in the mature flow field stage is that the velocity of water droplets is lower than that of airflow in the later stage, and there are fewer water droplets in Line 1 of Case 5.

12.3.2 Liquid water content of the humidity airflow

The contours of the LWC in Case 1 are listed below. The contours in Fig. 12.12 give the most intuitive way to observe the distribution of water droplets. As shown in Fig. 12.12A and B, the spreading area of water droplets is tiny when leaving the nozzle exit, and the distribution of LWC still follows the arrangement of nozzles. With the continuous change

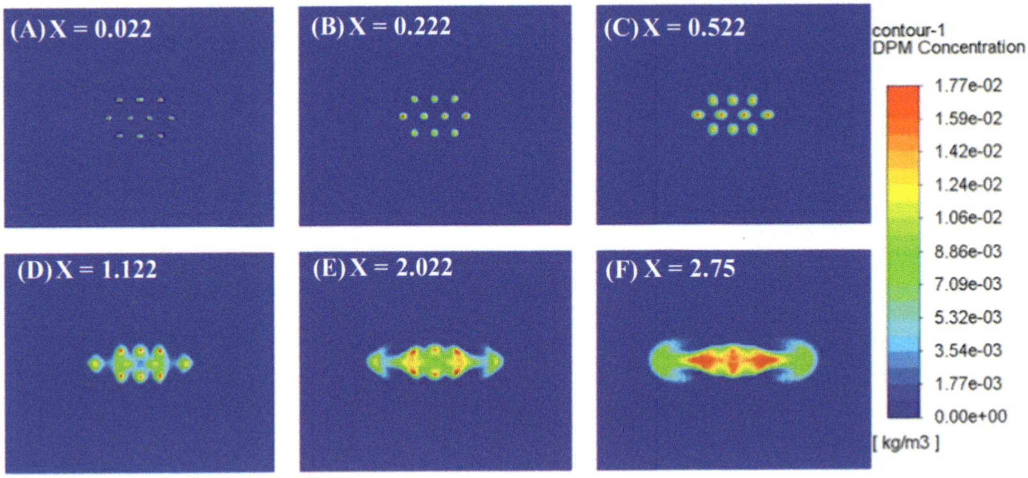

FIGURE 12.12 Contours of LWC in different sections of Case 1. *LWC*, Liquid water content.

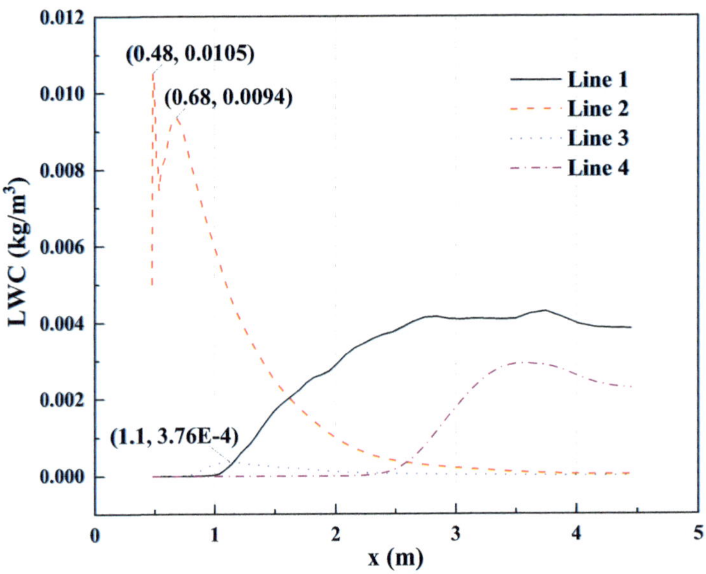

FIGURE 12.13 The LWC on the monitoring lines of Case 1. *LWC*, Liquid water content.

of the flow field, the spreading area of water droplets is increasing, and the position of the maximum value of LWC is also changing. By observing Fig. 12.12C–F, it is known that the distribution of water droplets continues to expand in the horizontal direction. Fig. 12.13 shows the variation of the LWC values at the location of the monitoring lines in Case 1. As shown in the figure, the change of Line 2 is the most dramatic, reaching the maximum value of 0.0105 kg/m^3 at $x = 0.48$ m and then reaches 0.094 kg/m^3 at $x = 0.68$ m

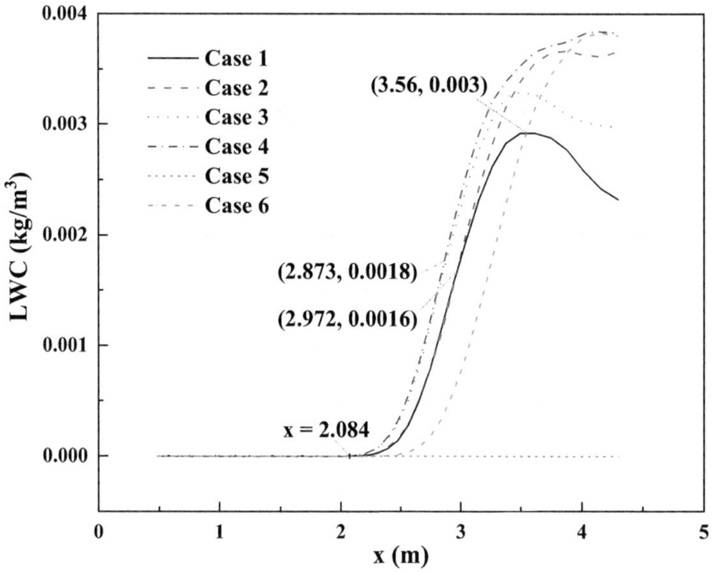

FIGURE 12.14 The LWC at Line 4 in six cases in this study. *LWC*, Liquid water content.

after a short sharp drop, followed by a continuous drop to 0 kg/m^3. There are no water droplets at Line 1 before $x = 1$ m, and then the LWC on Line 1 begins to have a value because of the arrival of water droplets from the surrounding nozzles. Line 3 reaches its maximum value of 3.76×10^{-4} kg/m^3 at $x = 1.1$ m and then slowly decreases to 0 kg/m^3. The LWC on Line 4 increases from 0 kg/m^3 after $x = 2.4$ m and decreases when approaching 0.003 kg/m^3. It can be known from the nozzle arrangement in Fig. 12.3A and the positions of the four monitoring lines in Figs. 12.3B and 12.4A that the reason for this change in LWC on Line 4 is that no water drops reach Line 4 before $x = 2.4$ m.

The value and spreading area of LWC at different distances in the x-direction directly affect the size and position of the test specimens. Fig. 12.14 shows the LWC at Line 4, corresponding to different cases. There are no water droplets at Line 4 of Case 5, so the value of LWC is always 0 kg/m^3. The LWC values of Cases 1–4 and 6 increase almost simultaneously at $x = 2.084$ m, in which Cases 3 and 4 are almost consistent between $x = 2.084$ m and $x = 2.873$ m, and the two curves separate after the values of LWC reach 0.0018 kg/m^3. Similar to Cases 3 and 4, Cases 1 and 2 are also consistent with each other between $x = 2.084$ m and $x = 2.972$ m, and the LWC value at the separation point of the two curves is 0.0016 kg/m^3. After separation from Case 2, the rising speed of Case 1 gradually slows down and finally reaches the maximum value of 0.003 kg/m^3 at $x = 3.56$ m.

12.4 Discussion on the limitations

The developed model was validated with experimental results, and the measured and predicted data agreed well. The enlightening conclusions obtained in this study can be directly used to guide the design and implementation of the experimental setup. Although the model was specially established for the CETF, the research methods introduced in this study could be referred for other related topics. Although a series of predicted data under different conditions were collected, there are many limitations for this model.

First, the model development was limited by the target object. The fairing part of the spray bar used in the experiment was drawn from a two-dimensional NACA 0024 airfoil. Only 10 nozzles were used, and all the designed arrangements of 10 nozzles were in three rows. In future work, the shape of the fairing, as well as the total number and row number of nozzles, could be adjusted. As expected, the model could be more flexible.

Second, a few icing models for the antiicing and deicing of aircraft are reported, and limited physical parameters are mentioned for the model validation. The ice thickness is always one of the validation parameters. For example, it was used as the sole validation parameter by Pena et al. [42] and Farag et al. [43] for NACA 23012 and NACA 0012 airfoils and windshield deicing pattern analysis, respectively. Also, the icing shape and icing velocity were used as validation and evaluation parameters for a three-dimensional icing model and antiicing effect for an aircraft by Xiao et al. [44]. In this study, the ice thickness on the test rod and the ice shape outlined by the thickness were taken as the validation factors.

In addition, not all the influence factors on the performance of the simulated icing environment are mentioned in this study. For example, the spray cone angle also has a significant influence on the parameters of water droplets [45,46], and changes in the velocity field also cause changes in the distribution of water droplets. That means an in-depth analysis of the spray cone angle and the change in the velocity field should be provided in our future work. The diameter of the spray nozzle, the mass ratio of liquid and vapor in the two-phase flow, and the dynamic control strategy of the spray nozzles should be further discussed.

12.5 Concluding remarks

A numerical investigation on the effect of the nozzle arrangement of atomization equipment in the icing cloud simulation system was carried out. The following conclusions could be summarized.

1. A model of the atomization equipment in an ice cloud simulation system was developed with the secondary development of FLUENT. The inhomogeneous sizes, gravity, and evaporation effects of water droplets were considered. The model was validated with the parameters of ice shape and maximum ice thickness on the test rod. The error of maximum ice thickness, maximum, and mean deviations are 2.6%, 1.13 mm, and 0.68 mm, respectively.
2. Six different nozzle arrangements were simulated, and the performances of six simulated icing environments were evaluated with the spreading area of water droplets and the uniformity of LWC distribution. The uniformity of LWC was determined by the area-weighted average value on the section.
3. For the best nozzle arrangement, the spreading area is 0.24009 m^2, and the area-weighted average of LWC is 0.168 g/m^3. For the worst nozzle arrangement, the spreading area and the area-weighted average of LWC are 0.22511 m^2 and 0.17056 g/m^3, respectively.

In Case 1, the nozzle arrangement is the best for the experimental system, with the minimum area-weighted average of LWC and a spreading area of 0.24009 m^2.

4. For the arrangements with two nozzles in the middle spray bar, the distance between the two nozzles determines the size of the spreading area and the uniformity of the LWC distribution. For the arrangements of two nozzles on the lower spray bar, the distance of two nozzles has little effect on the LWC distribution and the spreading area. Increasing the horizontal distance between two nozzles in the middle spray bar adversely affects the performance of the simulated icing environment.

References

[1] G. Thompson, M.K. Politovich, R.M. Rasmussen, A numerical weather model's ability to predict characteristics of aircraft icing environments, Weather and Forecasting 32 (1) (2017) 207–221.

[2] Y. Cao, W. Tan, Z. Wu, Aircraft icing: an ongoing threat to aviation safety, Aerospace Science and Technology 75 (2018) 353–385.

[3] S.R. Yu, M.J. Song, R.M. Gao, A review of icing prediction techniques for four typical surfaces in low-temperature natural environments, Applied Thermal Engineering 241 (2024) 122418.

[4] Q. Su, S.N. Chang, M.J. Song, Y.Y. Zhao, C.B. Dang, An experimental study on the heat transfer performance of a loop heat pipe system with ethanol-water mixture as working fluid for aircraft anti-icing, International Journal of Heat and Mass Transfer 139 (2019) 280–292.

[5] J. Haggerty, E. Defer, A. De Laat, K. Bedka, J.M. Moisselin, R. Potts, et al., Detecting clouds associated with jet engine ice crystal icing, Bulletin of the American Meteorological Society 100 (1) (2019) 31–40.

[6] X. Zhang, Y. Gao, X. Liu*, J. Min*, K. Li, M. Song, Modelling of the icing processes of sessile supercooled water droplets on cold spheres, Applied Thermal Engineering (IF = 6.4). 243 (2024) 122596.

[7] L. Ding, S.N. Chang, X. Yi, M.J. Song, Coupled thermo-mechanical analysis of stresses generated in impact ice during in-flight de-icing, Applied Thermal Engineering 181 (2020) 115681.

[8] M. Mohseni, A. Amirfazli, A novel electro-thermal anti-icing system for fiber-reinforced polymer composite airfoils, Cold Regions Science and Technology 87 (2013) 47–58.

[9] X. Meng, H. Hu, C. Li, A.A. Abbasi, J. Cai, H. Hu, Mechanism study of coupled aerodynamic and thermal effects using plasma actuation for anti-icing, Physics of Fluids 31 (3) (2019) 037103.

[10] X. Shen, X. Liu, G. Lin, X. Bu, D. Wens, Effects of anisotropic composite skin on electrothermal anti-icing system, Proceedings of the Institution of Mechanical Engineers Part G: Journal of Aerospace Engineering (2019) 0954410019845980.

[11] S. Gam-Derouich, J. Pinson, A. Lamouri, P. Decorse, S. Bellynck, R. Herbaut, et al., Micro-patterned anti-icing coatings with dual hydrophobic/hydrophilic properties, Journal of Materials Chemistry A 6 (40) (2018) 19353–19357.

[12] R.M. Gao, M.J. Song, C.Y.H. Chao, Review on condensation frosting and defrosting experiments for superhydrophobic surfaces, Applied Thermal Engineering 236 (2024) 121691.

[13] C. Tao, S. Bai, X. Li, C. Li, L. Ren, Y. Zhao, et al., Formation of zwitterionic coatings with an aqueous lubricating layer for antifogging/anti-icing applications, Progress in Organic Coatings 115 (2018) 56–64.

[14] M. Most, L. Berentsen, C. Rodriguez, B. Cheek, Developing a system for the prediction and management of aircraft deicing/anti-icing fluids concentrations in airport effluents, The Collegiate Aviation Review International 26 (2) (2018).

[15] T. Xie, J. Dong, H. Chen, Y. Jiang, Y. Yao, Experiment investigation on deicing characteristics and energy efficiency using infrared ray as heat source, Energy 116 (2016) 998–1005.

[16] T.F. Whale, B.J. Murray, D. O'Sullivan, T.W. Wilson, N.S. Umo, K.J. Baustian, et al., A technique for quantifying heterogeneous ice nucleation in microlitre supercooled water droplets, Atmospheric Measurement Techniques 8 (6) (2015) 2437–2447.

[17] S. China, G. Kulkarni, B.V. Scarnato, N. Sharma, M. Pekour, J.E. Shilling, et al., Morphology of diesel soot residuals from supercooled water droplets and ice crystals: implications for optical properties, Environmental Research Letters 10 (11) (2015) 114010.

[18] C. Wang, S. Chang, M. Leng, H. Wu, B. Yang, A two-dimensional splashing model for investigating impingement characteristics of supercooled large droplets, International Journal of Multiphase Flow 80 (2016) 131–149.
[19] C. Wang, S. Chang, H. Wu, Lagrangian approach for simulating supercooled large droplets' impingement effect, Journal of Aircraft 52 (2) (2015) 524–537.
[20] Y.S. Djikaev, E. Ruckenstein, Dependence of homogeneous crystal nucleation in water droplets on their radii and its implication for modeling the formation of ice particles in cirrus clouds, Physical Chemistry Chemical Physics 19 (30) (2017) 20075–20081.
[21] Z. Wang, H. Letu, H. Shang, C. Zhao, J. Li, R. Ma, A supercooled water cloud detection algorithm using himawari-8 satellite measurements, Journal of Geophysical Research: Atmospheres 124 (5) (2019) 2724–2738.
[22] F. Nicolosi, S. Corcione, P. Della Vecchia, Commuter aircraft aerodynamic characteristics through wind tunnel tests, Aircraft Engineering and Aerospace Technology: An International Journal 88 (4) (2016) 523–534.
[23] K.J. Yim, A study on the modeling method of missile fin aerodynamic coefficient using wind tunnel test and CFD, Journal of the Korea Institute of Military Science and Technology 22 (3) (2019) 360–368.
[24] E. Latorre Iglesias, D.J. Thompson, M. Smith, T. Kitagawa, N. Yamazaki, Anechoic wind tunnel tests on high-speed train bogie aerodynamic noise, International Journal of Rail Transportation 5 (2) (2017) 87–109.
[25] A.M. Marra, C. Mannini, G. Bartoli, Wind tunnel modeling for the vortex-induced vibrations of a yawed bridge tower, Journal of Bridge Engineering 22 (5) (2017) 04017006.
[26] M. Sumida, S. Morita, Wind tunnel tests on aerodynamic forces of road vehicles under unsteady wind conditions, International Journal of Automotive and Mechanical Engineering 15 (4) (2018) 6064–6077.
[27] C.X. Zhu, C.L. Zhu, W.W. Zhao, M.J. Tao, Experimental study on the shear adhesion strength between the ice and substrate in icing wind tunnel, Journal of Mechanics 34 (2) (2018) 209–216.
[28] J. Gonzales, D. Kurihara, T. Maeda, M. Yamazaki, T. Saruhashi, S. Kimura, et al., Novel Superhydrophobic Surface with Solar-Absorptive Material for Improved De-Icing Performance, Materials 12 (17) (2019) 2758.
[29] M.H. Sadafi, I. Jahn, K. Hooman, Nozzle arrangement effect on cooling performance of saline water spray cooling, Applied Thermal Engineering 105 (2016) 1061–1066.
[30] R. Aghaei-Togh, A.M. Tousi, Effects of nozzle arrangement angle on the performance of partially admitted turbines, Journal of Mechanical Science and Technology 32 (1) (2018) 455–464.
[31] X. Li, S. Chang, M. Leng, W. Yu, Numerical calculation for water droplet parameters change and its effect on impingement property in icing wind tunnel, 2018 CSAA/IET International Conference on Aircraft Utility Systems 214(6) (2018).
[32] H. Deng, S. Chang, M. Song, Numerical investigation on the performance and anti-freezing design verification of atomization equipment in an icing cloud simulation system, Journal of Thermal Analysis and Calorimetry (2019) 1–13.
[33] H.Y. Deng, S.N. Chang, M.J. Song, The optimization of simulated icing environment by adjusting the arrangement of nozzles in an atomization equipment for the anti-icing and deicing of aircrafts, International Journal of Heat and Mass Transfer 155 (2020) 119720.
[34] M.B. Bragg, A.P. Broeren, L.A. Blumenthal, Iced-airfoil aerodynamics, Progress in Aerospace Sciences 41 (5) (2005) 323–362.
[35] A.P. Broeren, E.A. Whalen, G.T. Busch, M.B. Bragg, Aerodynamic simulation of runback ice accretion, Journal of Aircraft 47 (3) (2010) 924–939.
[36] G.W. Wilson, Helicopter icing-testing and certification, Journal of the American Helicopter Society 27 (2) (1982) 66–72.
[37] J.L. Palacios, Y. Han, E.W. Brouwers, E.C. Smith, Icing environment rotor test stand liquid water content measurement procedures and ice shape correlation, Journal of the American Helicopter Society 57 (2) (2012) 29–40.
[38] T. Irvine, J. Oldenburg, D. Sheldon, The new icing cloud simulation system at NASA Lewis' icing research tunnel, 36th AIAA Aerospace Sciences Meeting and Exhibit, Reno, Nevada, USA, (1998) 143–164.
[39] C. Wang, S. Chang, H. Wu, L. Ding, J.M. Thompson, Theoretical modeling of spray drop deformation and breakup in the multimode breakup regime, Atomization and Sprays 25 (10) (2015) 857–869.
[40] Y. Hou, Y. Tao, X. Huai, Y. Zou, D. Sun, Numerical simulation of multi-nozzle spray cooling heat transfer, International Journal of Thermal Sciences 125 (2018) 81–88.

[41] Y. Cao, J. Huang, J. Yin, Numerical simulation of three-dimensional ice accretion on an aircraft wing, International Journal of Heat and Mass Transfer 92 (2016) 34–54.
[42] D. Pena, Y. Hoarau, É. Laurendeau, Development of a three-dimensional icing simulation code in the NSMB flow solver, International Journal of Engineering Systems Modelling and Simulation 8 (2) (2016) 86–98.
[43] A. Farag, L.J. Huang, CFD analysis and validation of automotive windshield de-icing simulation, SAE Technical Paper 2003-01-1079 (2003).
[44] C. Xiao, Z. Hu, Y. Gui, G. Lin, H. Zhang, Test study on anti-icing effects of hydrophobic coating in icing wind tunnel, Journal of Experiments in Fluid Mechanics 27 (2) (2013) 41–45.
[45] A. Marchewicz, A.T. Sobczyk, A. Krupa, A. Jaworek, Induction charging of water spray produced by pressure atomizer, International Journal of Heat and Mass Transfer 135 (2019) 631–648.
[46] A. Panda, A.R. Pati, S. Pradhan, B. Saha, A. Kumar, S.S. Mohapatra, High mass flux air atomized spray with subcooled water: A novel methodology to enhance the heat transfer rate in film boiling regime, International Journal of Heat and Mass Transfer 120 (2018) 1287–1304.

CHAPTER

13

The thermodynamic evolution of landfast ice in the polar regions

Jiechen Zhao [1,2]

[1]Qingdao Innovation and Development Base (Centre) of Harbin Engineering University, Qingdao, Shandong Province, P.R. China [2]Key Laboratory for Polar Acoustics and Application of Ministry of Education (Harbin Engineering University), Harbin, Heilongjiang Province, P.R. China

13.1 Introduction

Along the Antarctic shoreline and shelf edges, landfast ice frequently forms in small bands with varying widths up to 150 km from the Antarctic continent. Since landfast ice normally makes up between 5% and 35% of the continent's sea ice area in the summer and winter, it is essential to understanding fluctuations in Antarctic Sea ice cover [1,2]. Furthermore, nearshore landfast ice can be used to study variations in Antarctic Sea ice thickness, allowing for regular access to sea ice observation [3,4].

Since the late 1970s, the total Antarctic Sea ice area has expanded by around 0.4 million km^2, in contrast to the rapid reduction of sea ice extent and volume in the Arctic [5,6]. A significant portion of our knowledge of Antarctic Sea ice variability comes from in-situ surveys and space-borne sensor measurements of sea ice. However, measuring sea ice thickness on a large scale from space using remote sensing is still very difficult [7–9]. The conventional method for mapping Antarctic landfast ice thickness is field observations [10].

To coordinate landfast ice observations from research stations run by worldwide contributors, the Antarctic Fast-Ice Network (AFIN) was formed [3,11]. Studying landfast sea ice in Lützow–Holm Bay, researchers discovered that heavy snow cover on multiyear ice (MYI) caused significant thickness growth in the summer because it formed snow-ice on top of the ice, but minimal growth in the winter [12]. The evolution of multiyear snow-ice in McMurdo Sound was examined, and the ice temperature was discovered to remain at $\sim -2.5°C$ until the first week of March [13]. Additionally, they observed that $\sim$200 cm thick ice exhibited bottom melting from mid-March to the end of April at a steady rate of

Frosting and Icing for Efficient Energy Use in Engineering Applications
DOI: https://doi.org/10.1016/B978-0-443-15495-9.00015-0

0.2 cm/day. Strong easterly winds result in thicker sea ice and snow cover in the west of Atka Bay, according to landfast ice observations, which also revealed that supercooling water, including platelet ice, contributes significantly to sea-ice formation [14]. The maximum ice thickness in Prydz Bay's Davis Station varied between years, rising from 130 cm to 180 cm over a decadal period, according to long-term field observations [15,16].

Fieldwork on first-year ice (FYI) at Zhongshan Station (ZS) in the winter of 1992 revealed that FYI was primarily made up of congelation ice [17]. An examination of landfast ice in Nella Fjord during the summer of 2003 revealed that the average ice thickness was 170.0 cm, which indicates the existence of MYI. Snow and ice thickness also increased in the direction of the predominant wind [18]. The annual mass balance and heat flux were measured and observed from mid-February to late November of 2006, offering a crucial material comparison with the current work [19].

This chapter provides an examination of the landfast ice's temporal and spatial evolution close to ZS using field measurements made between December 2011 and December 2012. Landfast ice breakup and refreezing were essentially captured on camera. A sonar and a stainless-steel ruler were used to measure the thickness of the snow. Borehole measurements were used to personally observe the thickness of the ice. The snow and ice thickness distribution within Nella Fjord was assessed. The oceanic heat flux was estimated using the residual approach. These data were analyzed to determine the major parameters of the annual FYI and SYI cycle. The objectives of this study were a scientific study of climate change and expedition services for the Chinese icebreakers Xuelong and Xuelong 2.

13.2 Data and methods

ZS (76°22′W, 69°22′S) is situated near the Russian Progress Station and the Australian Davis Station in the vicinity of the Larseman Hill in East Antarctica (Fig. 13.1). These stations are all run throughout the year. Annual observations at ZS during the operational year 2011–12 began in December 2011 following the arrival of the Chinese polar icebreaker R/V Xuelong with observation equipment. The observations continued until December 2012, when a significant amount of snow and ice melted.

13.2.1 Meteorological observations

At a height of 15 m, the ZS Automatic Weather Station (WMO No. 89573) measured meteorological parameters, such as surface pressure, air temperature, wind speed and direction, and humidity (Fig. 13.2). A meteorologist manually recorded visibility and cloudiness every day at UTC 00:00, 06:00, 12:00, and 18:00. Additionally, from March 25 to November 30, 2012, a sensor T109 (Campbell Scientific, Inc., Logan, UT, USA) was placed on a tripod close to the ice observation location P1 to record the air temperature on the ice every minute. The sensor T109 had a height of 2 m.

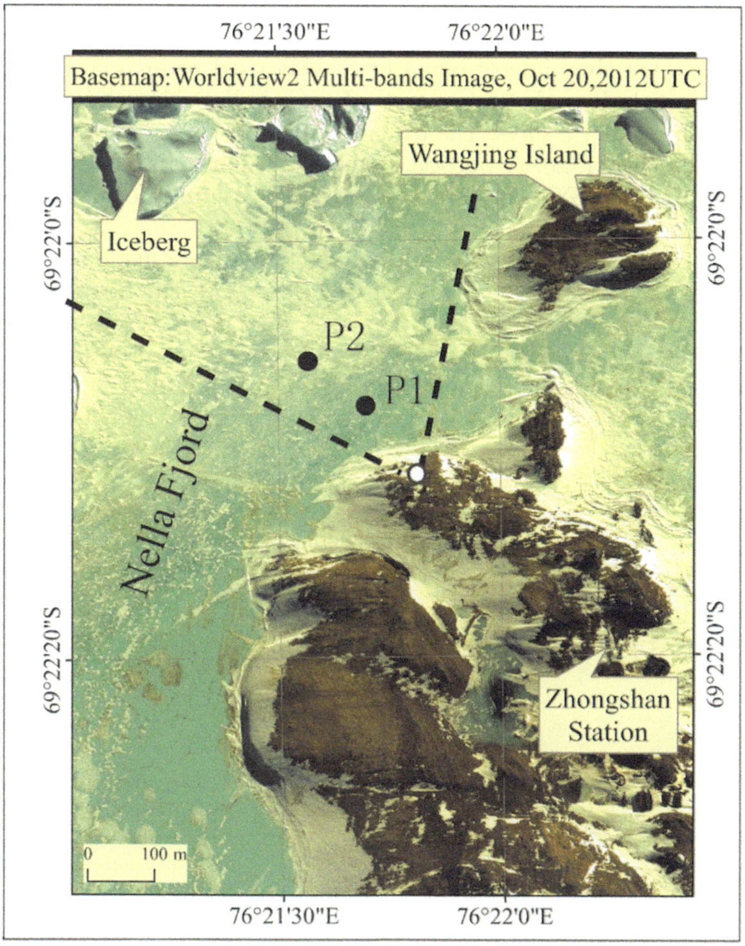

FIGURE 13.1 A satellite photo taken in mid-October 2012 of the regular landfast ice observation site close to Zhongshan Station. P1 and P2 are FYI and SYI sites, respectively. The hollow circle and two dotted lines represented the location and directions where the sea ice photos shown in Fig. 13.8 were taken. *FYI*, First-year ice; *SYI*, second-year ice [20]. *Source: Adapted from Zhao JC, Yang QH, Cheng B, Leppäranta M, Hui FM, Xie SR, et al. Spatial and temporal evolution of landfast ice near Zhongshan Station, East Antarctica, over an annual cycle in 2011/2012. Acta Oceanologica Sinica 2019;38(5):51-61.*

FIGURE 13.2 Automated weather station at Zhongshan Station.

13.2.2 Snow and ice observations

A digital camera installed on the shoreline of Nella Fjord was used to manually track the progression of landfast ice, including shattering and refreezing. From December 2011 to March 2012, pictures were shot every 5 days. The different surfaces were obvious in P1 and P2, as shown in the photos (Fig. 13.3). A portion of the near-shore landfast ice may last until the next winter and turn into SYI. These images provided useful information for identifying the distinction between FYI and SYI.

At P1 and P2, manual measurements of the snow and ice thickness were made at a distance of approximately 100 and 200 m off the coast, respectively (Fig. 13.4). The frequency of the measurements was 2–3 days from March 10, 2012. However, they paused from the end of May to the end of July during the 2 months of polar night and then resumed. A stainless-steel ruler was used to measure the thickness of snow by pushing it into the snow and down to the ice surface. There was 0.1 cm of accuracy. A Kovacs ice auger (5 cm in diameter) was

FIGURE 13.3 Photos of the regular ice monitoring station in winter 2012. The closer smooth surface is FYI and the farther rough surface is SYI. *FYI*, First-year ice; *SYI*, second-year ice.

FIGURE 13.4 On-site manual measurement of sea ice and snow thickness.

used to drill boreholes into which ice thickness was measured using a gauge with an accuracy of 0.1 cm. Boreholes were drilled carefully at each observation to prevent repeating the earlier ones. Every observation day, one borehole was typically drilled at P1; however, on rare occasions, three boreholes were drilled to evaluate the changes in thickness from several boreholes. However, because there were noticeable, significant differences in the ice thickness between several boreholes at P2 in the initial measurements, more than three boreholes were dug for each observation day.

A sonar sensor SR50A (Campbell Scientific, Inc., Logan, UT, USA) was fixed to a tripod that was set up close to P1 in order to automatically record the thickness of the snow. It was also equipped with an infrared sensor SI-111 (Campbell Scientific, Inc., Logan, UT, USA) to measure the temperature of the snow or ice surface. Both sensors had a 1-min sample frequency, and the observation period ran from March 25 to November 30, 2012.

From September 29 to October 2, 2012, five sections were planned to measure the thickness of the snow and ice inside the Nella Fjord. In total, 34 measurement sites spread over five sections were included in winter spatial observations in this region (Fig. 13.5). The measurements of snow and ice thickness were done manually, the same as for the regular observation sites. Every site's location was recorded using a portable GPS receiver. A Kriging interpolation technique was used to produce spatial distributions throughout the entire fjord.

Every December, the R/V Xuelong, a Chinese icebreaker, sails to ZS to replace crew and transport scientific and logistical supplies. Usually, the edge of the landfast ice zone is where the Xuelong anchors. A team of three to four scientists from ZS used snowmobiles to investigate a logistic transport route survey (TRS) between ZS and the R/V Xuelong. Between 2011 and 2016, the route surveys covered over 200 km^2 of ice field. Measurements of the ice thickness and snow depth were made along the routes (Fig. 13.6). On flat, deformed, ridged, and rafted ice, a total of 511 measurements were made. In this chapter, we included 137 measurements taken on level ice floes.

13.3 Temporal evolution of landfast ice

13.3.1 Meteorological conditions

The mean air temperature for the annual cycle from December 2011 to December 2012 was −10.4°C (Fig. 13.7b). January 2012 had the highest monthly mean air temperature (−0.6°C), while July 2012 had the lowest (−19.6°C). On January 18 and May 24, respectively, the warmest and lowest daily mean air temperatures were recorded at 6.1°C and −34.5°C. The average relative humidity was 54% (Fig. 13.7C). For 48% of the annual cycle period, the predominant wind direction was east (Fig. 13.7E). The maximum hourly mean wind speed of 32.8 m/s was observed on June 29. The annual mean wind speed was 7.1 m/s (Fig. 13.7f). Although precipitation was not recorded at ZS, it could be obtained from the Progress Station. June through August saw the highest monthly totals of precipitation, ranging from 18 to 30 mm, whereas the remaining months saw monthly totals of less than 10 mm (Fig. 13.7g).

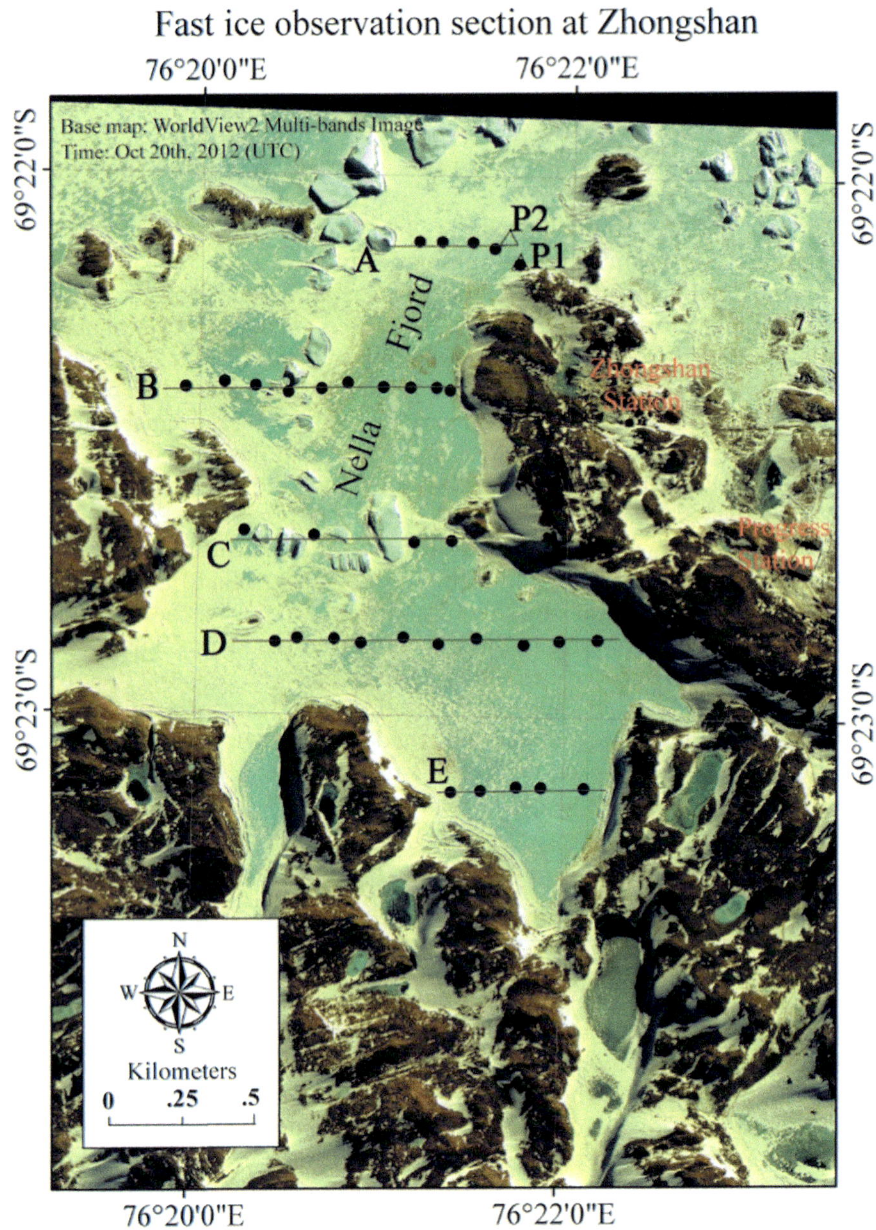

FIGURE 13.5 The five sections of snow and ice observation at Nella Fjord, Zhongshan Station, with P1 and P2 serving as the FYI and SYI sites, respectively. *FYI*, First-year ice; *SYI*, second-year ice.

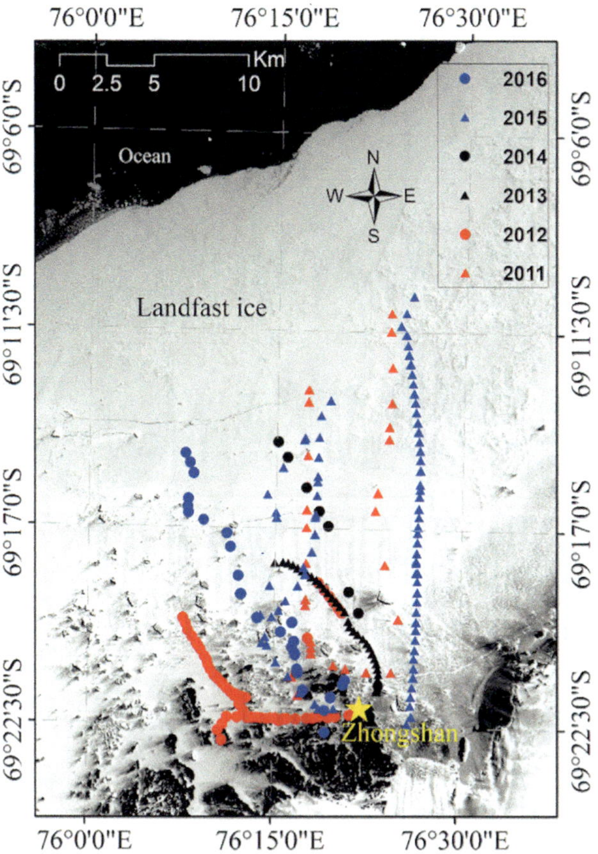

FIGURE 13.6 Locations for measuring snow and ice along the logistic transportation routes of R/V Xuelong (TRS). Source: *Adapted from J.C. Zhao, B. Cheng, T. Vihma, et al., Observation and thermodynamic modeling of the influence of snow cover on landfast sea ice thickness in Prydz Bay, East Antarctica, Cold Regions Science and Technology 168 (2019) 1–11.*

The frequent snowfall at Larseman Hill caused by Prydz Bay cyclones was dispersed over the area by strong winds that carried the snow away. As a result, the recorded snow accumulation was not accurate enough to reflect the actual amount of snowfall. The annual mean air temperature was 0.6°C lower, the relative humidity was 4% lower, and the wind speed was 0.1 m/s higher in comparison to the climatology from 1989 to 2014.

13.3.2 Ice melting and breakup

High-resolution satellite imagery, field data, and photos taken along the shore were used to map the landfast ice properties. Strong and predominately easterly winds carried ice floes of varying sizes, which gathered northeast of ZS and typically survived the summer to become MYI. At the end of October, the landfast ice near the Sound of Nella Fjord began to melt; melt ponds became visible in early December 2011 (Fig. 13.8). After one powerful cyclone visit, the ice began to break up in parts on January 23, 2012. After then, there was open water for roughly 20 days until the day–night melting–freezing cycles started in the middle of February. At the end of February, new FYI begins to develop.

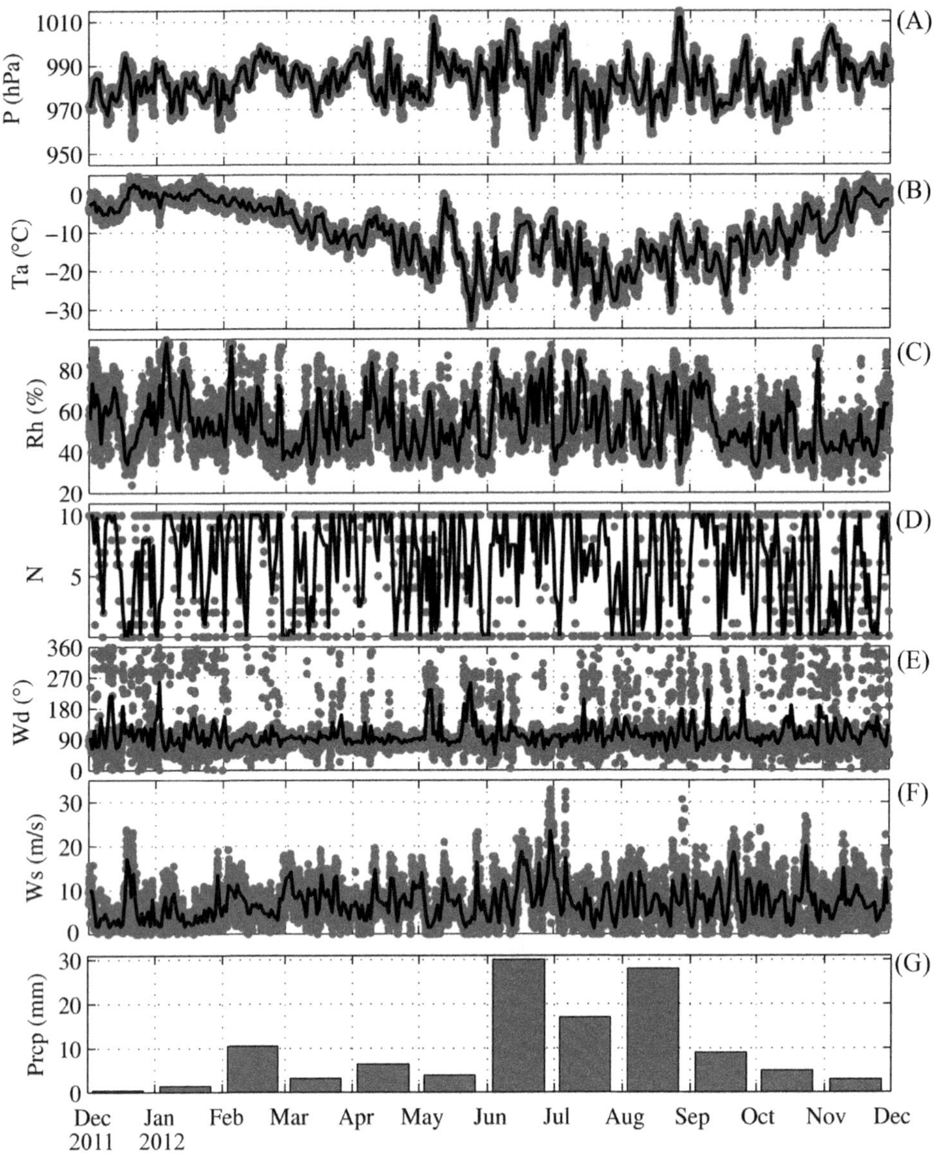

FIGURE 13.7 Surface air pressure (A), temperature (B), relative humidity (C), cloudiness (D), surface wind direction (E), wind speed (F), and precipitation (G). Gray dots are hourly data in (A), (B), (C), (E), and (F); nevertheless, for 6-hourly data in (D). The daily mean values are shown by black lines. The monthly accumulated precipitation at Progress Station is represented by the gray bar in (G).

FIGURE 13.8 The progression of the ice in the Nella Fjord sound.

During summer, melt ponds were abundant, although the ice was maintained intact by fast to nearby islands and tiny icebergs. Thick layers of snow frequently melted, but the ice underneath the snow held up until the next winter. Field recordings indicate that the ice at P2 was SYI, which evolved differently from MYI within the fjord.

The late-January and mid-February, which are the ice breakup and refreezing times, matched observations made at Davis Station since 2000 [16] and at ZS in 2006 [19]. The timing and power of the cyclones have a significant impact on the timing and level of breakage. The landfast ice may persist until the next winter, suggesting that dynamic mechanisms rather than melting cause the ice in the Nella Fjord to vanish [22]. On the Arctic Siberian coast, a similar phenomenon has also been noted [23].

13.3.3 Ice refreezing and growth

Fig. 13.9 shows the snow thickness measured manually at the filed, as well as automatically at the sonar site. With a Pearson's correlation coefficient of 0.71, the automatic and manual observations at the tripod site showed a strong association. This suggests that the sonar sensor has a high degree of reliability when recording changes in snow thickness. All measurement sites reported significant snow accumulation between July and August, although the quantities varied greatly. Strong cyclones promoted moisture transport, which resulted in snowfall around ZS. Strong winds would typically swiftly redistribute the snow cover. After June, the sonar detected five significant snowfall episodes. After snowfall events S1 and S2, a strong wind blew away half of the accumulated snow in 24 hours. Due to the severe wind, the snow thickness at S5 decreased in 72 hours from 31.0 cm to 15.0 cm.

The ice surface was essentially snow-free prior to June. On October 29, the tripod site's maximum thickness record was 26.0 cm by ruler; however, on October 31, during the snowstorm event S5, sonar measurements revealed a maximum thickness of 31.0 cm. This significant greatest value and date were missed by manual observations, but sonar detected it.

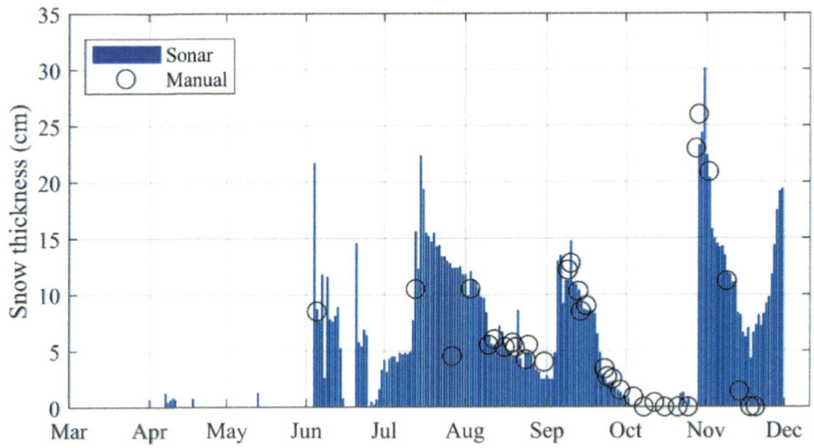

FIGURE 13.9 Snow thickness was measured using the sonar sensor, and by the observer manually. S1, S2, S3, S4, and S5 designate the five peak snowfall occurrences.

Fig. 13.10 shows the evolution of SYI at P2 and FYI at P1. The ice thickness at P2, where the initial field measurement was conducted on March 10, was 116.0 cm. About 1 day later, on March 11, the initial measurement at P1 was carried out. The results showed that the ice thickness there was 22.0 cm, which is around 100 cm less than SYI at P2 (Fig. 13.10). The FYI increased steadily from March and October. The average growth rate of the ice was 0.7 cm/day from March to May (Fig. 13.10), and then it decreased to 0.4 cm/day on average from August to October. The greatest growth rates resembled the landfast ice observations off ZS in 2006 [19], at about 1.0 cm/day. On October 25, 2012, a maximum thickness of 160.5 centimeters was recorded for FYI. According to ZS wintering personnel, this was less than 174.0 cm in 2006 [19], 176.0 cm in 2010, and 186.0 cm in 2011. Beginning around the end of October, bottom melting occurred at a pace of − 0.4 cm/day. November had a little decrease in melting, to − 0.3 cm/day, and December saw an increase to − 1.0 cm/day. The mean growth rate during the polar night was calculated to be 0.8 cm/day for the no-data period.

In March 2012, the SYI displayed a very distinct trajectory, melting at a monthly mean rate of − 2.7 cm/day (Fig. 13.11), resulting in a fall in ice thickness from 116.0 cm (March 10) to 57.0 cm (March 31). An easy-to-drill-through layer was found between 60.0 and 80.0 cm below the ice surface on March 10 and 11. SYI thickness varied minimally over the first 2 weeks of April, maintaining a steady value of roughly 60.0 cm. April 15 was when SYI really started to grow. Then FYI and SYI had comparable growth trends, while SYI was typically 10.0–20.0 cm thicker. August saw a slowdown in SYI growth, and SYI started to melt around the end of October. On October 29, the highest thickness of SYI measured 167.0 cm.

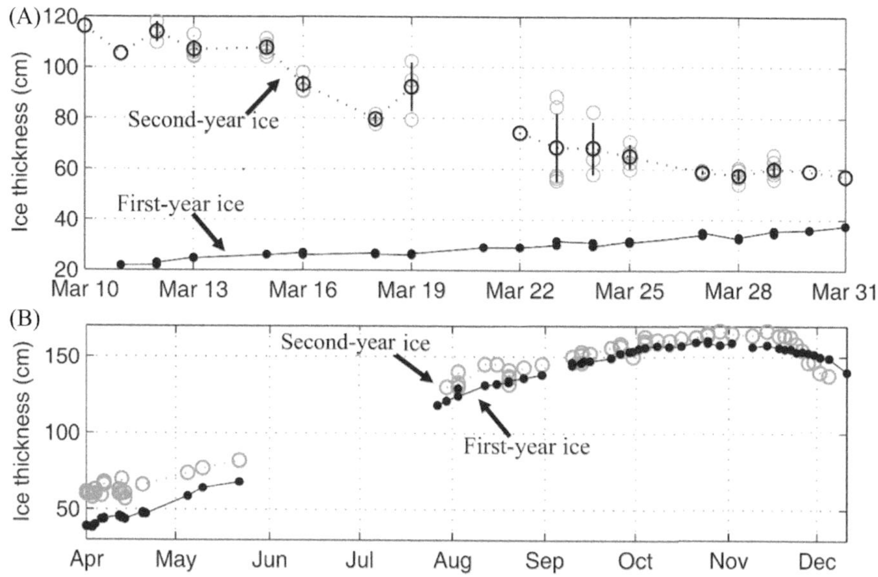

FIGURE 13.10 FYI and SYI ice thickness in March (A) and April–December 2012 (B). Black circles show averaged data; bars show standard deviation, and gray circles show each drilling result. For FYI, each drilling result is represented by a black dot. *FYI*, First-year ice; *SYI*, second-year ice.

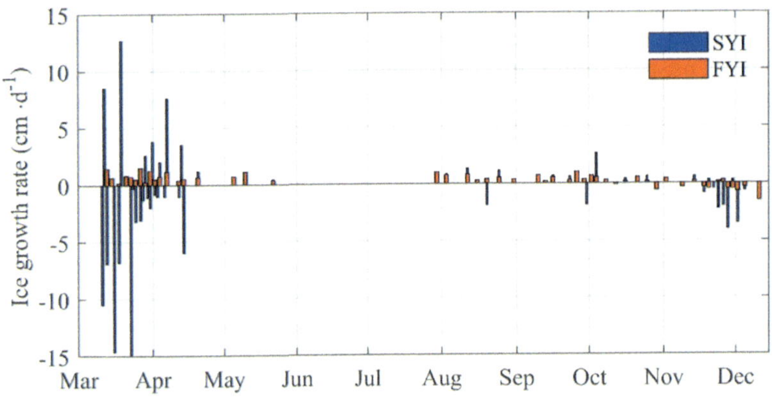

FIGURE 13.11 FYI and SYI ice growth rates from March to December 2012. *FYI*, First-year ice; *SYI*, second-year ice.

TABLE 13.1 Variations in second-year ice thickness at P2 over the course of several months.

	Mar.	Apr.	Aug.	Sep.	Oct.
Number of measurements	17	10	6	7	7
Max drillings in a single measurement	6	4	5	3	3
Mean ice thickness (cm)	81.4	61.8	141.5	153.0	162.6
Largest variation (cm)	33.0	10.0	10.0	7.0	4.5

The monthly mean melting rate was only -0.5 cm/day in December, half of the FYI melt rate. It was hypothesized that SYI started to exhibit greater spatial variation throughout the melting season during November and December. According to a prior in-depth analysis, the concurrent SYI ablation and FYI growth in March were caused by internal melting and a nonlinear temperature profile.

For every observation day at P1, the differences in ice thickness across the several boreholes were less than 1.0 cm. This suggested that the FYI grew synchronously and that the bottom was rather smooth, especially when given the level surface. Table 13.1 illustrates the variance in ice thickness at P2. In March, P2 was visited 17 times, and during each visit, at least three drilling operations were carried out. The highest number of boreholes dug this month was six on March 23. The highest thickness disparity among the drilling sets (33.0 cm) varied from 55.5 cm to 88.5 cm. In March, the average ice thickness was 81.4 cm. According to Table 13.1 results, SYI thickness showed a significant spatial deviation at a short spatial scale ($\sim$10 m), and that variation decreased over time, most likely as a result of thinner ice growing faster. During the winter, the ice surface unevenness was measured to be between 5.0 and 10.0 cm. This suggests that an uneven bottom accounted for 70% of the changes and an uneven surface for 30%.

This chapter assessed the ice thickness and freeboard height multiple times along the direction of the prevailing wind in an attempt to investigate the link between the ice thickness and the direction of the wind. The Nella Bay region at the end of September likewise showed a minor increase in sea ice thickness along the direction of the prevailing wind, as seen in

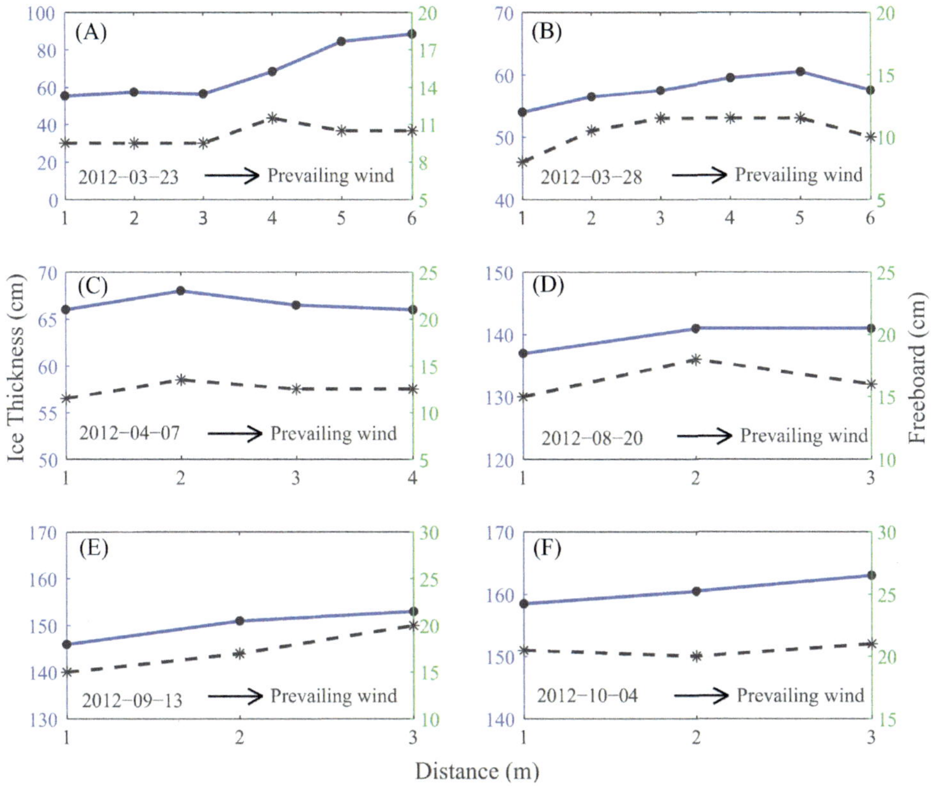

FIGURE 13.12 The ice thickness among different measuring holes along the predominant wind direction on the different dates. 23 March 2012 (A), 28 March 2012 (B), 7 April 2012 (C), 20 August 2012 (D), 13 September 2012 (E) and 4 October 2012 (F).

Fig. 13.12, over numerous measurements at different periods. It is possible that the prevailing wind direction affects the distribution of snow, resulting in thicker snow along the wind's direction. The thicker snow then forms more snow ice and superimposed ice, resulting in thicker sea ice. Snow ice and superimposed ice both occupy a certain proportion in the thickness composition of MYI, both of which are the result of the transformation of snow to ice.

Field observations of ice cores have not yet verified this theory. The ice thickness differential across boreholes can be greater than 0.1 m within a maximum distance of 10 m, suggesting that the MYI's lower surface is undulating. On the other hand, the difference in ice thickness measured by simultaneous porosity observation at the 1-year ice observation point is often less than 0.01 m, suggesting that the 1-year ice's lower surface is rather flat.

13.3.4 Ocean-to-ice heat flux

The traditional Stefan's Law estimates the thickness of ice during the growth season. This rule has been applied inversely to estimate the oceanic heat flux. However, surface air temperature is typically employed because ice surface temperature is rarely accessible.

The present investigation yielded simultaneous measurements of the snow/ice surface temperature (T_o), field surface air temperature (T_a), and surface air temperature at ZS (T_{zs}). As can be seen from the daily mean values in Fig. 13.13, T_o was 1.3 ± 1.5°C colder than T_a. Over the time series, 83% of the cases were $T_o < T_a$. T_a was larger than T_{zs} by 0.1 ± 3.4°C. In general, there was an oscillation between T_a and T_{zs} within a range of ±5.0°C.

Using the three different observed temperatures mentioned above, the residual approach [24] was used to determine the oceanic heat flux for FYI at P1. Prior research [25–27] has previously made extensive use of this methodology. The residual method was also applied by [20] with a numerical model of ice thickness.

Assuming a quasi-steady heat flux at the sea ice base, the thermal energy balance is made simpler by the following:

$$\rho_i L_f dH/dt = k_i(T_f - T_o)/H - F_w \tag{13.1}$$

where ρ_i is the sea ice density at the basal layer, L_f is the sea ice latent heat of freezing, dH/dt is the ice growth rate, k_i is the thermal conductivity of sea ice, T_f is the seawater freezing point, T_o is the ice surface temperature, H is the ice thickness, and F_w is the oceanic heat flu13.

Latent heat flux (F_l) from ice growth is represented by the term on the left in Eq. (13.1), and the conductive heat flux (F_c) at the base ($F_c \geq 0$) is represented by the first term on the right. The quasi-steady technique ignores thermal inertia for simplicity.

The original measurements were subjected to a three-record moving average in the computations, and the measured ice thickness was then linearly interpolated into daily values. The interpolated daily numbers served as the basis for calculating the growth rate. Thermal conductivity

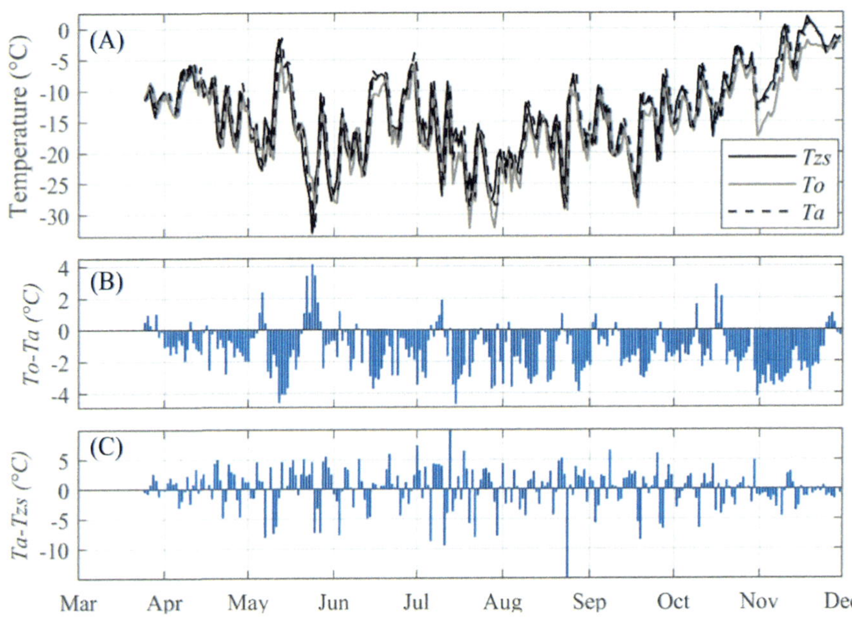

FIGURE 13.13 Time series of daily mean air temperature observed in ZS (T_{zs}), snow/ice surface temperature (T_o), and field air temperature (T_a) (A); The defferences between T_o and T_a (B); The defferences between T_{zs} and T_a (C).

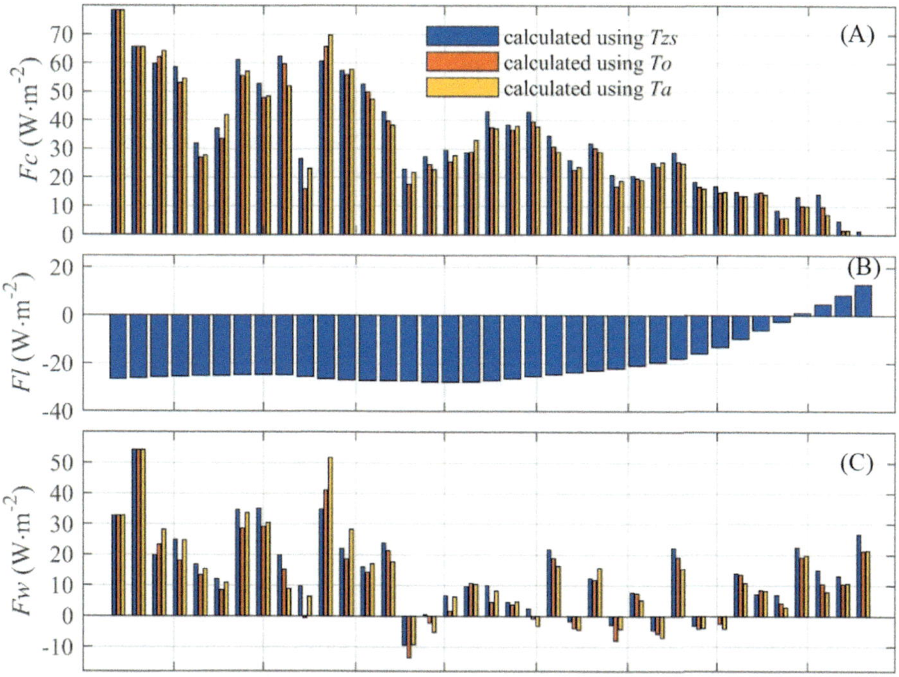

FIGURE 13.14 Conductive heat flux at ice base (F_c) (A), latent heat flux (F_l) (B), and oceanic heat flux (F_w) (C) of FYI. *FYI*, First-year ice.

was adjusted at 2.2 W/m°C, sea ice latent heat of freezing to 333.4 kJ/kg, and ice density to 910 kg/m^3. T_o, T_a, and T_{zs} were utilized independently to investigate the temperature sensitivities of F_c from various sources. When using T_o instead of T_a, the computed F_c was 2.7 ± 3.9 W/m^2 higher, and when using T_{zs}, it was 2.2 ± 9.1 W/m^2 higher.

Fig. 13.14 displays the heat fluxes following a 7-day running mean. As the ice thickened in mid-March, F_c dropped, and by the end of November, when the ice surface temperature was almost 0°C, it was almost zero (Fig. 13.10A). F_l remained consistent at approximately −20.0 W/m^2 until the end of September, at which point it started to decline and turned positive by the end of October. This suggested that the direction of the latent heat transport was reversed with the commencement of basal melting. During the first phase of ice growth, F_w stayed between 20.0 and 40.0 W/m^2, reaching a maximum monthly mean value of 33.8 ± 7.5 W/m^2 in March. F_w dropped to about 10.0 W/m^2 after July. The monthly mean minimum was 4.7 ± 8.0 W/m^2 in September.

13.4 Spatial evolution of landfast ice

13.4.1 Landfast ice classification

The majority of the sea ice outside ZS will melt and lose intensity in the summer due to rising temperatures [28,29]. It will break up and then melt after being subjected to strong winds and other weather phenomena. Sea ice will only exist until the following winter in the nearshore region and in a few fjords. The sea ice characteristics in February 2012 around ZS and Nella Bay are depicted in Fig. 13.15. There is a disorganized ice area called Area I in front of ZS. The predominant east wind causes broken floating ice to gather here, blocking the shoreline and resulting in an uneven, chaotic ice area when the ice refreezes in the winter. This area is constantly covered by sea ice and will not melt over the summer, even if there are melt pools on the sea ice surface. When combined with the predominant east wind, the floating ice that has accumulated here is also difficult to float away.

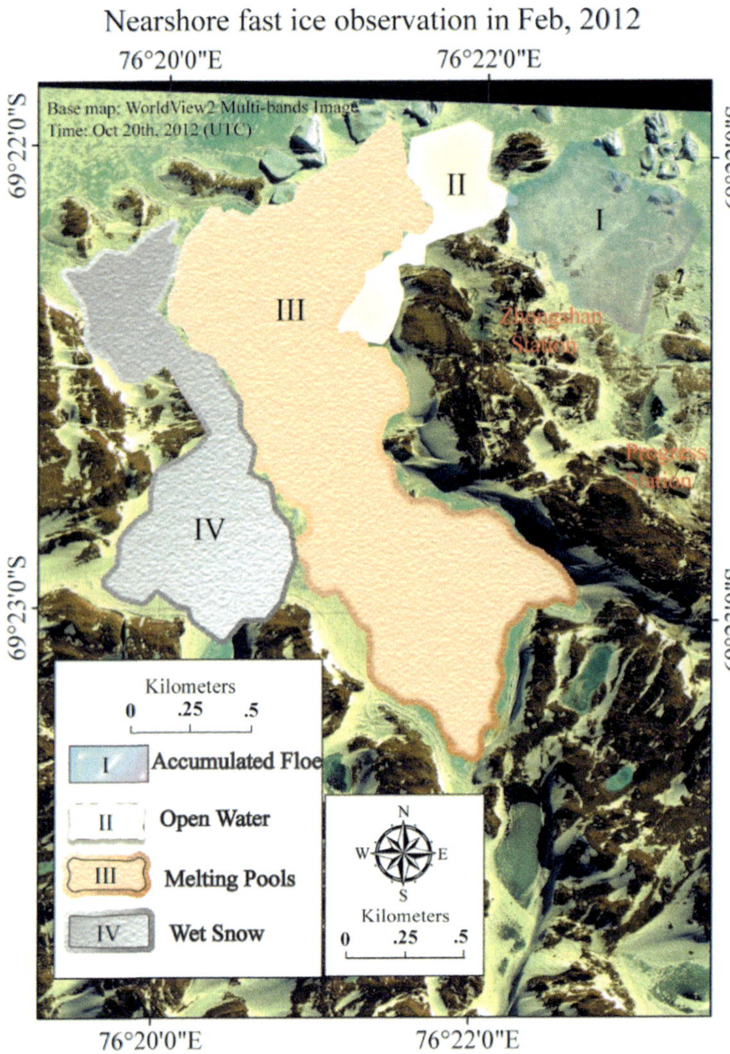

FIGURE 13.15 Sea ice characteristics in the vicinity of Zhongshan Station and Nella Bay during the Antarctic summer in February 2012.

Sea ice crevasses form along the shore in winter in Area II, as a result of the tidal sea level rise and fall. Strong winds cause the sea ice here to break and melt in the summer, creating a tiny stretch of open water as the temperature rises. Area II refreezes and turns into FYI in mid-to-late March.

Mid-December saw the emergence of sizable melt pools in the center of Nella Bay (Area III), which caused the sea ice strength to decline. Typically, this meant that ice unloading had to be halted. Nevertheless, the sea ice in this region essentially did not break and melt, and the melt pool refroze in the next winter because of the fixed function of the nearby hills and small icebergs. It was discovered that there was a "gap layer" phenomenon—that is, larger water layers between the two layers of sea ice—during the field observations of sea ice in March and April.

Area IV in the figure, which is the vicinity of Nella Bay's eastern coast, experiences year-round, heavy snowfall, with the deepest patches reaching depths of more than one meter. The snow on the surface will melt in the summer's high temperatures, creating a layer of moist snow. Area III is a MYI area, but the thicker snow successfully blocks out the heat of the atmosphere and shields the sea ice beneath it, preventing it from melting in the summer.

In winter, a large portion of Nella Bay is MYI, as Fig. 13.16 illustrates. This is mostly due to the unique fjord terrain that shields it from wind processes, as well as the small icebergs that have become stranded, which also prevent the sea ice in the area from breaking up. Here, the sea ice usually lasts for several summers before fully melting, with large surface melting pools only formed during the summer. A tiny area near the mouth of Nella Bay that is unhindered by hills is vulnerable to wind processes that melt and fracture sea ice in the summer, creating a small stretch of open water that freezes again in the winter when the temperature drops. With ZS serving as the platform, it is convenient for us to monitor the growth process of both multiyear and FYI due to their simultaneous presence in Nella Bay.

13.4.2 Nella Fjord section

Within the Nella Fjord, the spatial mean snow thickness was 11.3 cm. The fjord's eastern and western sections had different amounts of snow cover. The easternmost portion of Section D had the deepest snowfall, beyond the 100-cm ruler's range. The distribution of snow was affected by small icebergs, with the snow surrounding them being 5.0–10.0 cm thicker than the surrounding area (Fig. 13.17). The terrain and direction of the surface winds influenced the overall distribution of snow in this fjord. The snow was driven westward by the strong easterly winds that predominated, and the snow was forced to pile in the east by the hills that faced east to west. This resulted in a pattern that can be seen in Fig. 13.17, which is described as "west thicker, east thinner."

The ice thickness in the east was often greater than in the west, following the distribution of snow thickness, and there was a 0.52 spatial correlation coefficient between the two. The greatest thickness of the ice in Section D was 307.0 cm, and in 55% of the 34 sites, the ice was thicker than 200.0 cm. The spatial mean thickness of the ice was 202.7 cm (Fig. 13.18). In January 2003, the Nella Fjord showed a similar "west thicker, east thinner" trend, with a maximum ice thickness of 201.0 cm [30].

FIGURE 13.16 Sea ice characteristics in the vicinity of Zhongshan Station and Nella Bay during the Antarctic winter, March–December 2012.

In Nella Bay, the distribution of sea ice thickness and freeboard height is essentially the same, with smaller values in the east and bigger values in the west (Fig. 13.19). The freeboard height in Section B is similarly impacted by the little iceberg, making it approximately 0.1 m higher than the flat ice. The influence of small icebergs is primarily responsible for the increased freeboard height in Nella Bay, whereas the higher ice thickness and sea ice lifting caused by the coastal terrain in the west are the main causes.

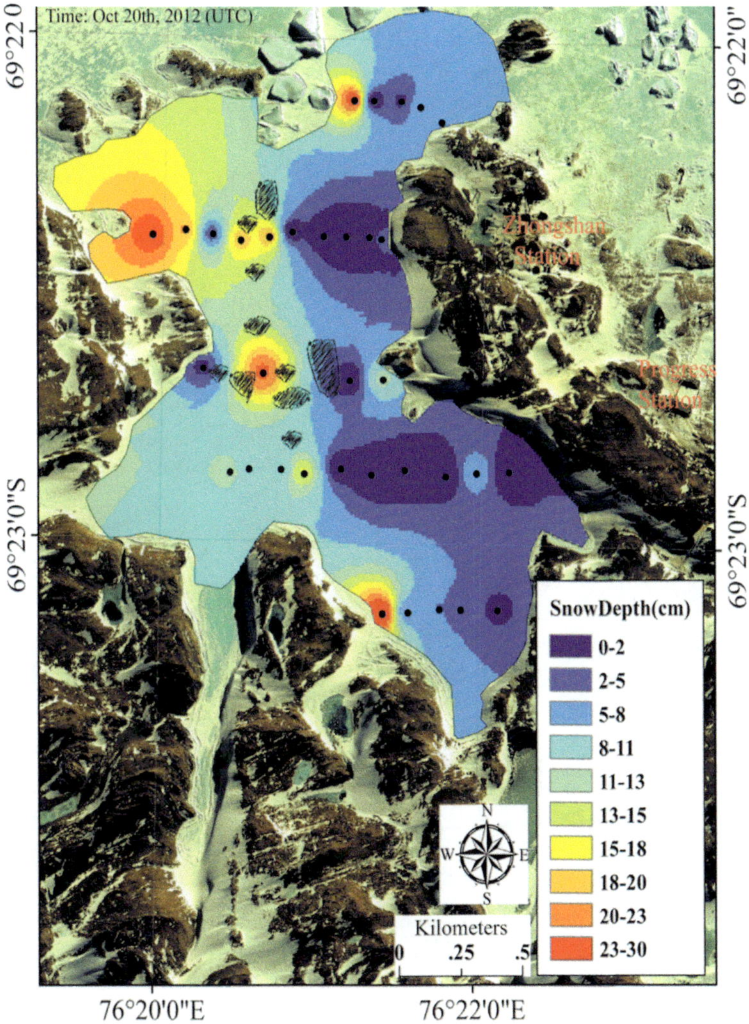

FIGURE 13.17 Spatial pattern of snow thickness in the Nella Fjord, measured between September 29 and October 2, 2012. Measurement sites are represented by solid black circles. The positions and shapes of minor icebergs are shown by the black slashes.

13.4.3 Transport route survey section

Fig. 13.20 illustrates the link between snow depth and ice thickness as determined during TRS. Since measurements were conducted on level ice floes, thermodynamic growth and, most likely snow–ice interaction—that is, snow ice or superimposed ice formation—are the main drivers of ice thickness. The thickness of the undeformed ice ranged from 1.0 to 1.9 m, whereas the snow depth varied from 0.06 to 1.0 m. There was a very modest negative connection (-0.2, $P < .02$) between snow depth and ice thickness. This happened as

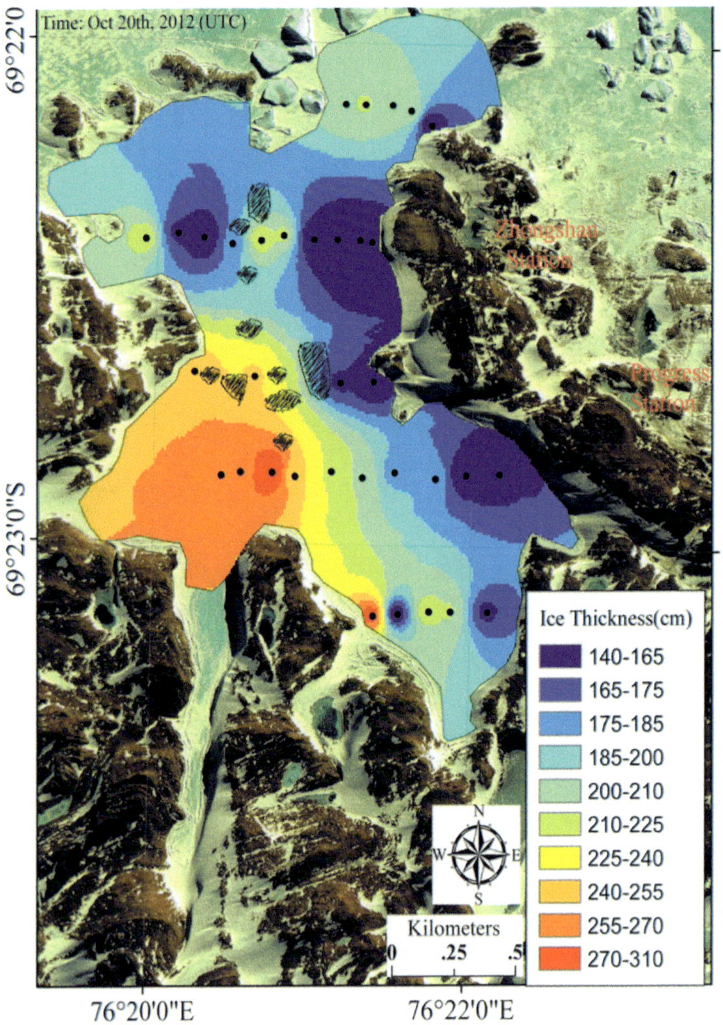

FIGURE 13.18 The Nella Fjord's regional distribution of ice thickness was recorded between September 29 and October 2, 2012. Measurement sites are represented by solid black circles. The positions and shapes of minor icebergs are shown by the black slashes.

a result of the measurements being dispersed over a wide area and taken over a comparatively short period of time in November and December. The relationship between snow/ice thickness and spatial variability is primarily dominated by it. A strong negative association between snow depth and ice thickness would most likely have been seen if the data had spanned the seasonal scale on level ice with a high spatial sampling precision.

The meridional direction was used to average the snow depth and ice thickness data gathered from TRS (Fig. 13.21). A negative association was found between the snow depth and ice thickness at a distance of ~6.75 km from the shore. These results are consistent

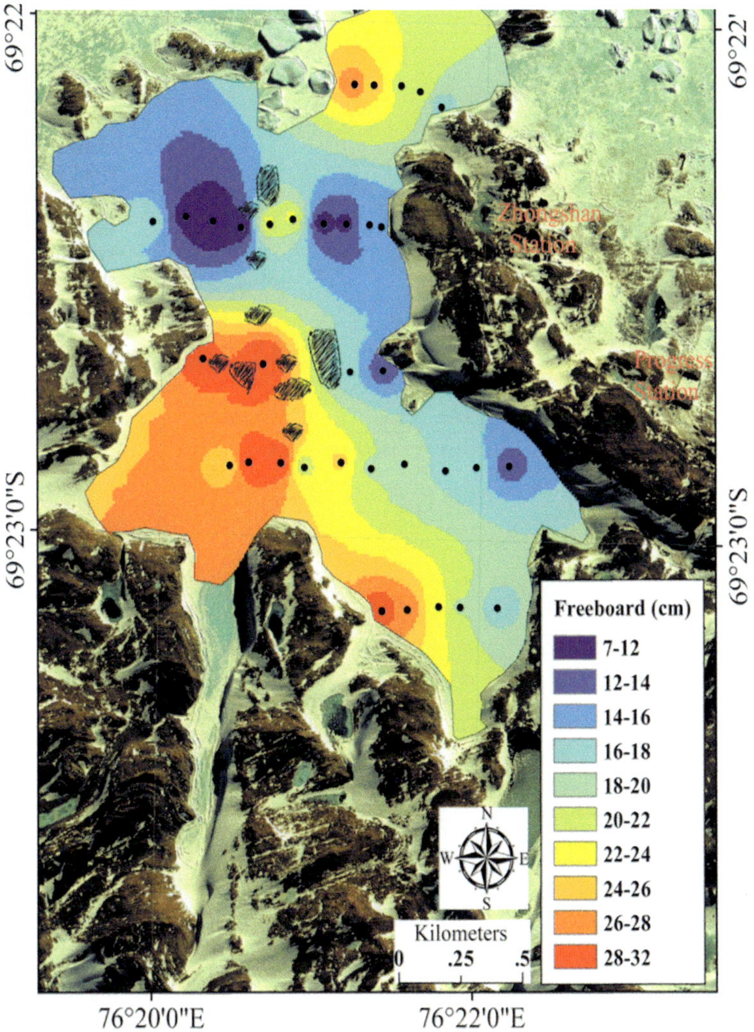

FIGURE 13.19 The Nella Fjord's freeboard spatial distribution, measured between September 29 and October 2, 2012. Measurement sites are represented by solid black circles. The positions and shapes of minor icebergs are shown by the black slashes.

with observations made at the SIP site from October to December, where the average snow depth was 0.05 ± 0.07 m and the average ice thickness was 1.53 ± 0.15 m. The ice thickness was 1.35 ± 0.13 m, and the snowpack was even heavier (0.52 ± 0.11 m) at locations 9 km further north. According to Archimedes' Law, an average of 0.52 m of snow on top of 1.35 m of sea ice might result in a negative freeboard of 0.06 m, assuming average densities of 330 kg/m^3 for snow and 910 kg/m^3 for sea ice. In the cold season, the flooded snow slush is probably going to turn into snow ice. In late spring and early summer, superimposed ice may form as a result of surface snowmelt.

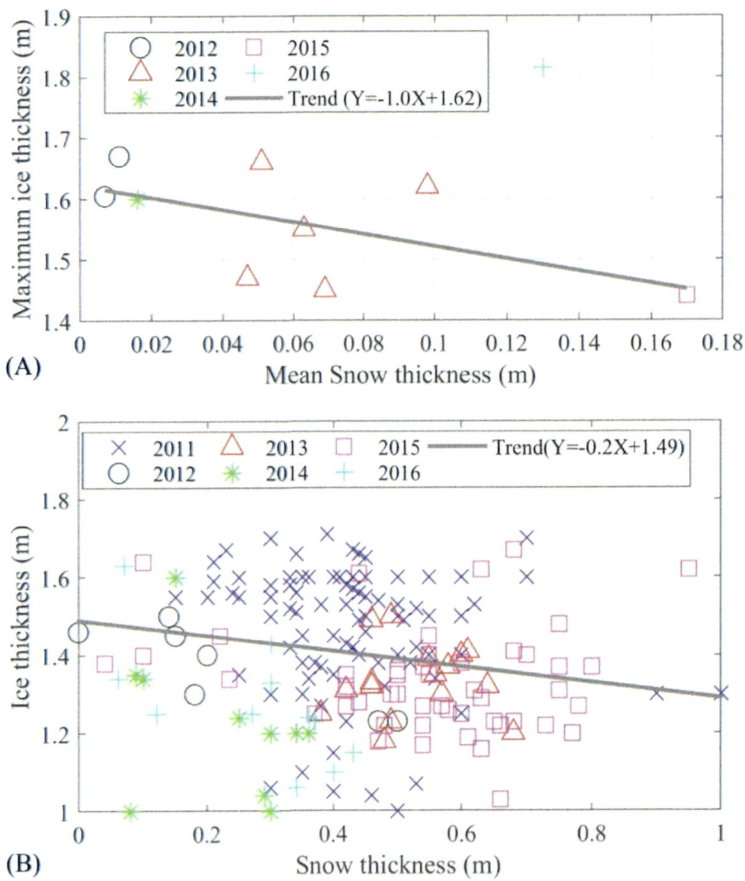

FIGURE 13.20 The relationship between (A) annual maximum ice thickness and mean snow depth at SIP and (B) snow and ice thickness measured on an undeformed level ice floe during TRS. The observations from different years are shown by different symbols, while the trends are shown by gray lines. *TRS*, Transport route survey.

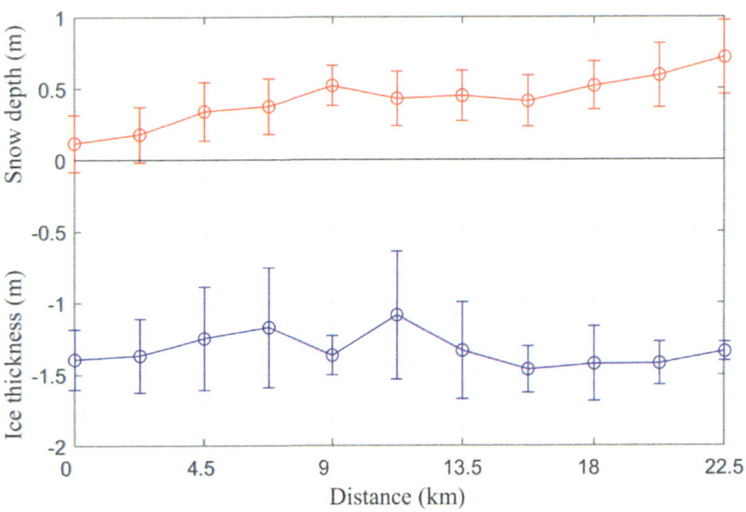

FIGURE 13.21 The average snow depth and ice thickness measurements on undamaged ice floes from TRS between 2011 and 2016 along the meridional fetch. *TRS*, Transport route survey.

At SIP, the annual maximum ice thickness was significantly impacted by snow insulation. A negative connection was found between the highest ice thickness and the mean snow depth ($r = -0.58$, $P < .05$). The modeling studies conducted with prescribed snow depth indicated a significantly larger negative relationship ($r = -0.83$, $P < .01$) between maximum ice thickness and snow depth.

The relationship between snow depth and ice thickness was, however, substantially less for the TRS domain farther off the coast due to (1) deeper snowpack and (2) temporal and geographical fluctuations in wind-driven snow redistribution in the area [21]. The insulating effect there prevented sea ice from growing. However, the modeling findings showed that this effect was somewhat offset by ice growth when snow ice and superimposed ice formed.

13.5 Discussion

The dense layer of snow on FYI typically slowed down the ice growth in the winter, resulting in thinner ice when compared to the area that is free of snow. But for MYI in the Nella Fjord, summer melting was restricted to the snow layer because of the heavy snow cover that prevented solar radiation from penetrating into ice in the summer [31]. The ice thickness was increased by the overlaid ice that was created by the melting and refreezing of snow layers. These mechanisms explain why substantial snow cover followed the dense MYI. Previous observations of this occurrence in McMurdo Sound on MYI have been made [13].

Based on field records, the surface of SYI was nearly level, indicating that the variation in the bottom geometry happened. Archimedes' Principle states that when the horizontal scale is greater than the characteristic length of floating ice, which is tens of meters for 1 m of ice, the scale of bottom topography variations can be 10 times that of surface topographical variations. Compared to MYI in the Arctic (~100 cm) [32] and SYI in McMurdo Sound (~50 cm) [13], the highest spatial fluctuations of SYI thickness in this chapter were 33.0 cm.

The dry and cold weather of winter can result in significant sublimation. The sonar sensor saw a total of 1 cm of surface sublimation between March 24 and April 3, when there was no snowfall. According to the previous estimate, the mean sublimation rate was 0.09 cm/day [22]. Ponds developed during the melting season as the surface grew wet, weakening sublimation.

Using a quasi-steady technique, the scaling and heat balance of the FYI and MYI annual cycle were investigated [33]. Due to the significant temperature difference between the ice surface and the bottom, FYI began to expand rapidly when the open water refrozen in mid-February, whereas SYI continued to melt until mid-April. Eq. (13.1) accurately represented the quasi-steady heat balance. It was evident that depending on thickness, ice grows or melts because $F_w > 0$. The thickness for equilibrium status (no melting and increasing) is $H \approx 70$ cm if $T_f - T_o \approx 10°C$ and $F_w \approx 30\ W/m^2$. This is consistent with field observations in that ice will melt at thicknesses greater than 70 cm. In McMurdo Sound, the bottom melting of SYI was also detected; however, the melting rate (0.23 ± 0.02 cm/day) was significantly lower than that reported in this chapter [13].

Using Eq. (13.1) to estimate the F_w of SYI in March, the result was approximately 100 W/m^2, far higher than the FYI range. A mushy layer that was seen between 60.0 and 80.0 cm below the ice surface in March indicates that internal melting may have taken place. Modeling research revealed that SYI was isothermal in the low section and had a nonlinear temperature profile in March [10]. Because of the consistent temperature structure, there was significant interior and bottom melting because conductive heat transport was hindered.

At least in the following characteristics, the results obtained for Prydz Bay characterize circumstances over the majority of the Antarctic land-fast sea ice zone. The snowpack is thinner close to the shore than it is farther out over the sea ice because of katabatic winds and summer snowmelt thickens the ice by forming superimposed ice. The latter impact counteracts the snow-insulating effect, which slows the formation of sea ice. Nonetheless, several regional traits also have an impact on the snow and sea ice conditions in Prydz Bay. First, Prydz Bay has stronger wind speeds than the majority of regions along the Antarctic coastline [34]. This tends to decrease nearshore snow depth and increase spatial variability in snow depth. Second, the bulk of the yearly precipitation in Prydz Bay comes from extremely uncommon days of intense precipitation, with the majority of precipitation occurring on days with little to no precipitation [35]. Large temporal changes in snow depth are favored by this. The ice thickness is also influenced by the spatiotemporal fluctuations in snow depth brought on by high winds and intense precipitation episodes. Our findings are consistent with [16], which were obtained at the Australian Davis Station, ~100 km from ZS, and which showed a correlation between a rise in cumulative solid precipitation and a reduction in the annual maximum landfast sea ice thickness, given the conditions present in Prydz Bay.

Most manual measurements of ice thickness and snow depth have been made at SIP. To improve our understanding of the regional and temporal variation of snow depth and ice thickness, autonomous measurements across a significant area off the ZS—for example, in the TRS domain—are required. It would be beneficial if ice core samples were gathered during upcoming field surveys in order to determine the areas of snow ice and superimposed ice in the overall ice thickness. For the land-fast ice region in Prydz Bay, better snow and ice services are necessary to operate ZS logistically.

13.6 Concluding remarks

Using field measurements from the wintering team, this chapter examined the regional and temporal evolution of the landfast ice, including FYI and SYI near the Antarctic ZS, from December 2011 to December 2012. Regular site measurements were made during the whole year, and winter surveys were carried out at various locations throughout the Nella Fjord.

1. In the Nella Fjord, the landfast ice began to exhibit melt ponds in early December 2011. Following a powerful cyclone in late January, the ice cover partially broke, and in mid-February 2012, as the air temperature dropped, the open water froze again. Small islands and icebergs around other patches of ice withstood the winter till the following year. Due to internal ice melting and the nonlinear temperature profile, FYI increased rapidly in March, while SYI underwent rapid melting. Midway through April, the SYI began to expand swiftly and attained a maximum thickness, which was greater than

the maximum thickness of FYI. In contrast, the SYI thickness fluctuated significantly, while the FYI thickness only displayed slight spatial variations. The notably uneven SYI bottom was linked to this significant thickness variance.

2. Snow and ice in the Nella Fjord exhibited a similar trend of "east thicker, west thinner," according to sectional surveys conducted in the winter. The topography and dominant wind both had an impact on the snow thickness spatial distribution, while summertime snowmelt in the form of overlying ice significantly increased MYI thickness. The identical snow and ice pattern for MYI in the Nella Fjord was explained by this process. The measured oceanic heat flux differed from the two-peak pattern at Mawson [36] but exhibited an annual cycle similar to that reported at ZS [19] and Davis [16]. The yearly average oceanic heat flux was similar to the findings using a 1-D sea ice thermodynamics model [22] but greater than the results at ZS in 2006 [19].

It should be mentioned that this inadequate examination of SYI melting and the superimposed ice contribution to MYI was brought on by the absence of ice core data. The regional scale modeling of snow depth and ice thickness is an ongoing project. A model experiment like this might benefit the Year of Polar Prediction (YOPP) Project and be useful for the area's snow and ice services [37]. In the future, to further improve our understanding of the mass balance of FYI and SYI near ZS, simultaneous oceanographic measurements beneath the ice and ice temperature/salinity profiles should be conducted [38]. The study will improve our knowledge of landfast ice for efficient energy use in engineering applications.

References

[1] A.D. Fraser, R.A. Massom, K.J. Michael, et al., East Antarctic fast sea ice distribution and variability, 2000-08, Journal of Climate 25 (4) (2012) 1137–1156.

[2] K.A. Giles, S.W. Laxon, A.L. Ridout, Circumpolar thinning of Arctic sea ice following the 2007 record ice extent minimum, Geophysical Research Letter 5 (2008) L22502.

[3] P. Heil, S. Gerland, M.A. Granskog, An Antarctic monitoring initiative for fast ice and comparison with the Arctic, The Cryosphere Discussion 5 (2011) 2437–2463.

[4] J.C. Zhao, G.H. Hao, J. Li, et al., Construction of a comprehensive sea-ice observation system at Zhongshan Station, Antarctica, Marine Forecast 35 (5) (2018) 41–52 (in Chinese).

[5] J.P. Liu, J.A. Curry, Accelerated warming of the Southern Ocean and its impact on the hydrological cycle and sea ice, PNAS 107 (34) (2010) 14987–14992.

[6] C.L. Parkinson, D.J. Cavalieri, Antarctic sea ice variability and trends, 1979-2010, The Cryosphere 6 (2) (2012) 871–880.

[7] R. Kwok, D. Sulsky, Arctic Ocean sea ice thickness and kinematics: Satellite retrievals and modeling, Oceanography 23 (4) (2010) 134–143.

[8] A.P. Worby, C.A. Geiger, M.J. Paget, et al., Thickness distribution of Antarctic sea ice, Journal of Geophysical Research 113 (2008) C05S92.

[9] P. Lu, Z.J. Li, Uncertainties in retrieved ice thickness from freeboard measurements due to surface melting, Annals of Glaciology 55 (66) (2014) 205–212.

[10] J.C. Zhao, B. Cheng, Q.H. Yang, et al., Observations and modelling of first-year ice growth and simultaneous second-year ice ablation in the Prydz Bay, East Antarctica, Annals of Glaciology 58 (2017) 59–67.

[11] Zhao J.C. Experimental study on key process mechanism and numerical prediction of polar sea ice, PhD Graduation thesis, Ocean University of China, 2020.

[12] T. Kawamura, K.I. Oshima, T. Takizawa, et al., Physical, structural and isotopic characteristics and growth processes of fast sea ice in Lützow-Holm Bay, Antarctica, Journal of Geophysical Research 102 (1997) 3345–3355.
[13] A.J. Gough, A.R. Mahoney, P.J. Langhorne, T.G. Haskell, Salinity evolution and mechanical properties of snow-loaded multiyear sea ice near an ice shelf, Antarctic Science 25 (6) (2013) 821–831.
[14] M. Hoppmann, M. Nicolau, The influence of platelet ice and snow on Antarctic fast sea ice, American Geophysical Union (2011). C23A-0483.
[15] P. Heil, I. Allison, V.I. Lytle, Seasonal and interannual variations of the oceanic heat flux under a landfast Antarctic sea ice cover, Journal of Geophysical Research 101 (C11) (1996) 25741–25752.
[16] P. Heil, Atmospheric conditions and fast ice at Davis, East Antarctica: A case study, Journal of Geophysical Research 111 (2006) C05009.
[17] J.F. He, B. Chen, K. Wu, Developing and structural characteristics of first-year sea ice and with effects on ice algal biomass of Zhongshan Station, East Antarctica, Journal of Glaciology and Geocryology 4 (1998) 358–367.
[18] S.L. Tang, D.H. Qin, J.W. Ren, Sea ice characteristics between Middle Weddell Sea and Prydz Bay, Antarctic during the 2003 Australian summer, Earth Science Frontiers 13 (3) (2006) 213–218.
[19] R.B. Lei, Z.J. Li, B. Cheng, et al., Annual cycle of landfast sea ice in Prydz Bay, East Antarctica, Journal of Geophysical Research 115 (2010) C02006.
[20] J.C. Zhao, Q.H. Yang, B. Cheng, M. Leppäranta, F.M. Hui, S.R. Xie, et al., Spatial and temporal evolution of landfast ice near Zhongshan Station, East Antarctica, over an annual cycle in 2011/2012, Acta Oceanologica Sinica 38 (5) (2019) 51–61.
[21] J.C. Zhao, B. Cheng, T. Vihma, et al., Observation and thermodynamic modeling of the influence of snow cover on landfast sea ice thickness in Prydz Bay, East Antarctica, Cold Regions Science and Technology 168 (2019) 1–11.
[22] Y. Yang, Z.J. Li, M. Leppäranta, B. Cheng, et al., Modelling the thickness of fast sea ice in Prydz Bay, East Antarctica, Antarctic Science 28 (1) (2015) 59–70.
[23] Y. Yang, M. Leppäranta, Z.J. Li, et al., Model simulations of the annual cycle of the fast ice thickness in the East Siberian Sea, Advances in Polar Sciences 26 (2) (2015) 168–178.
[24] M.G. McPhee, N. Untersteiner, Using sea ice to measure vertical heat flux in the ocean, Journal of Geophysical Research 87 (1982) 2071–2074.
[25] D.K. Perovich, B. Elder, Estimates of ocean heat flux at SHEBA, Geophysical Research Letter 29 (9) (2002) 1344.
[26] V.I. Lytle, R. Massom, N. Bindoff, et al., Wintertime heat flux to the underside of East Antarctic pack ice, Journal of Geophysical Research 105 (28) (2000) 759–769.
[27] C.R. Purdie, P.J. Langhorne, G.H. Leonard, et al., Growth of first-year fast Antarctic sea ice determined from winter temperature measurements, Annals of Glaciology 44 (2006) 170–176.
[28] J.C. Zhao, B. Cheng, T. Vihma, P. Heil, F.M. Hui, Q. Shu, et al., Fast ice prediction system (FIPS) for land-fast sea ice at Prydz Bay, East Antarctica: an operational service for CHINARE, Annals of Glaciology 61 (83) (2020) 271–283.
[29] J.C. Zhao, T. Yang, Q. Shu, H. Shen, Z.X. Tian, G.H. Hao, et al., Modelling the annual cycle of landfast ice near Zhongshan Station, East Antarctica, Acta Oceanologica Sinica 40 (7) (2021) 129–141.
[30] S.L. Tang, D.H. Qin, J.W. Ren, et al., Structure, salinity and isotopic composition of multi-year fast sea ice in Nella Fjord, Antarctica, Cold Regions Science and Technology 49 (2007) 170–177.
[31] G.A. Maykut, N. Untersteiner, Some results from a time dependent thermodynamic model of sea ice, Journal of Geophysical Research 76 (1971) 2–4.
[32] S.F. Ackley, W.D. Hibler, F.K. Kugzruk, et al., Thickness and roughness variations of Arctic multi-year sea ice, Proceedings of Ocean 74-IEEE International Conference Engineering in the Ocean Environment (1974) 109–117.
[33] M. Leppäranta, A review of analytical models of sea-ice growth, Atmosphere-Ocean 31 (1) (1993) 123–138.
[34] Z. Zhang, T. Vihma, A. Stössel, P. Uotila, The role of wind forcing from operational analyses for the model representation of Antarctic coastal sea ice, Ocean Modelling 94 (2015) 95–111.
[35] L. Yu, Q. Yang, T. Vihma, S. Jagovkina, J. Liu, Q. Sun, et al., Features of extreme precipitation at Progress Station, Antarctica, Journal of Climate 31 (22) (2018) 9087–9105.

[36] I. Allison, Antarctic ice growth and oceanic heat flux, Proceedings of the Canberra Symposium, 131, IAHS Publications, 1981, pp. 161–170.
[37] J.C. Zhao, J.J. Cheng, Z.X. Tian, X.P. Han, H. Shen, G.H. Hao, et al., Snow and ice thicknesses derived from Fast Ice Prediction System Version 2.0 (FIPS V2.0) in Prydz Bay, East Antarctica: Comparison with in-situ observations, Big Earth Data 6 (4) (2022) 492–503.
[38] J.C. Zhao, B. Cheng, T. Vihma, P. Lu, H.W. Han, Q. Shu, The internal melting of landfast sea ice in Prydz Bay, East Antarctica, Environmental Research Letters 17 (7) (2022).

CHAPTER

14

Ice prediction technology and applications for typical stationary and moving surfaces

Sirui Yu[1], *Mengjie Song*[1,2], *Long Zhang*[1], *Xuan Zhang*[1], *Zekang Zhen*[1], *Keke Shao*[1,2], *Mohasan Muhammad*[1] *and Lizhen Huang*[1]

[1]Department of Energy and Power Engineering, School of Mechanical Engineering, Beijing Institute of Technology, Beijing, P.R. China [2]School of Mechanical Engineering, Hanyang University, Seoul, South Korea

Nomenclatures

Symbols	Description	Unit
T_a	Air temperature	°C
T_e	Ambient temperature	°C
T_s	Surface temperature	°C
T_g	Generator temperature	°C
P_a	Atmospheric pressure	Pa
RH	Relative humidity	%
H_s	Surface humidity	g/m^3
v	Wind speed	m/s
v_b	Blade speed	r/s
v_f	Flight speed	Km/h

Frosting and Icing for Efficient Energy Use in Engineering Applications
DOI: https://doi.org/10.1016/B978-0-443-15495-9.00004-6

v_d	Raindrops fall speed	m/s
θ	Wind direction	°
θ_i	Ice angle	°
H	Precipitation per unit of time	mm/h
P	Precipitation	mm
Q	Solar radiation	W/m^2
S_i	Ice shape	\
t_i	Icing time	min
Δt	Prediction interval	min
t	Icing duration	h
h	Icing thickness	mm
Δh	Icing thickness increments	mm
h_{max}	Maximum icing thickness	mm
p	Icing probability	\
λ	Surface thermal conductivity	W/(m · K)
$LWC(w)$	Liquid water content	g/mm^3
mvd	Median volume diameter	μm
S_f	Airfoil	\
M	Icing quality of transmission line	g/m
M_0	Icing mass of transmission lines at the initial moment	g/m
ρ_0	Density of liquid water	g/m^3
ρ_i	Icing density	g/m^3
R	Radius of the transmission line after icing	mm
R_0	Traverse radius	mm
ΔR	Icing thickness of transmission line	mm
I	Icing quality of transmission lines per unit longitudinal sectional area	g/m^2
N	Number of icing hours	\
E	Collision coefficients	\
f	Freezing factor	\
C	Constant	\
k	Constant	\

14.1 Introduction

Icing is widely prevalent in both natural and industrial environments, often causing negative impacts on production and daily life [1–4]. To address this challenging issue and minimize its adverse effects, researchers have been conducting observations and studies on icing since the last century [5]. As shown in Fig. 14.1, research on ice development can include ice recognition, ice prediction, ice formation, ice detection, ice control, ice storage, ice prevention, ice breaking, and ice removal. Antiicing and deicing technologies, as effective means to tackle existing icing problems, can directly provide measures to resolve icing issues. Over the years, antiicing and deicing technologies have garnered significant attention and undergone in-depth research, with various methods such as mechanical deicing [6], thermodynamic deicing [7], surface coatings [8], and liquid flow deicing [9] emerging successively. Although surface antiicing treatments coupled with system optimization control can achieve a certain degree of ice disaster control, ice prediction, occurring prior to antiicing operations, can effectively reduce or even eliminate ice disasters by taking necessary measures in advance. Therefore, research on ice prediction technology has gradually gained widespread attention.

Considering the widespread occurrence of icing in transportation, energy and power, aviation, and food refrigeration fields, ice prediction technology holds significant importance as a critical applied technology. In terms of transportation, icing can decrease road capacity, leading to a significant increase in the occurrence of traffic accidents by over tenfold [10], thereby adversely affecting transportation and economic activities [11]. Accurate prediction and monitoring of road icing conditions facilitate the timely implementation of deicing measures by transportation authorities and provide necessary safety guidance for road users through proactive scheduling. In the energy and power sector, icing can cause reduced production or failure of energy facilities such as wind turbines, solar panels, and transmission lines [12,13], potentially leading to significant disruption of the entire energy system, resulting in substantial economic losses or casualties [14]. Through ice prediction, energy companies and relevant management departments can take proactive measures to ensure the stability and reliability of the energy supply. In the aerospace field, icing can disrupt aircraft navigation and communication systems, affecting the safe operation of aircraft and also have adverse effects on critical components and equipment [15], endangering flight safety [16]. Ice prediction technology can help optimize flight decisions, reduce the risks of aircraft operation in icing conditions, and ensure flight safety.

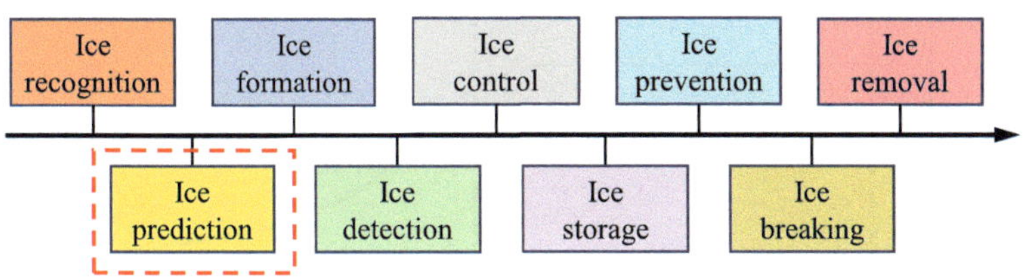

FIGURE 14.1 Diagram of the research process of ice.

14.2 Introduction to ice prediction techniques

14.2.1 Mechanism of icing

Ice formation is a complex phase transition process involving multiple fundamental domains such as heat and mass transfer, fluid dynamics, and surface chemistry. The essence of the icing process lies in the freezing of individual liquid droplets; hence, studying the solidification process of droplets contributes to understanding icing mechanisms, thereby advancing the development of ice prediction and antiicing technologies.

As illustrated in Fig. 14.2, previous studies [17,18] have divided the freezing process of individual liquid droplets on stationary cold surfaces into four stages: liquid supercooling, nucleation and recalescence, solidification, and solid cooling. During the liquid supercooling stage, the droplet temperature continuously decreases due to contact with the cold surface until it falls below the phase transition temperature T_1, while the interior remains liquid, thus maintaining a supercooled state. When the droplet temperature reaches the nucleation temperature T_3, the droplet enters the nucleation and recalescence stage, where a homogeneous mixture of ice and water forms internally, and the temperature rises back to T_1 due to the release of latent heat. This stage is brief and marks the onset of droplet freezing. Subsequently, the freezing front of the droplet gradually grows towards the droplet apex at temperature T_1, ultimately forming a pointed protrusion, indicating the end of the freezing stage, with the droplet temperature decreasing to the completion freezing temperature T_2. The time taken for the droplet to decrease from T_1 to T_2 is related to the ratio of the height of the thermocouple measuring the droplet temperature to the droplet height. After the completion of the solidification process, the temperature of the frozen droplet continues to decrease until it matches the temperature of the flat surface, marking the final solid cooling process. Research on icing mechanisms mainly focuses on the nucleation and recalescence stage and the solidification stage.

Fig. 14.3A–D illustrates the schematic diagrams of the ice formation process of individual liquid droplets under the influence of flat, curved, vertical flat, and wind-affected surfaces [19–22]. Among them, Fig. 14.3A and B can be regarded as basic simplifications of

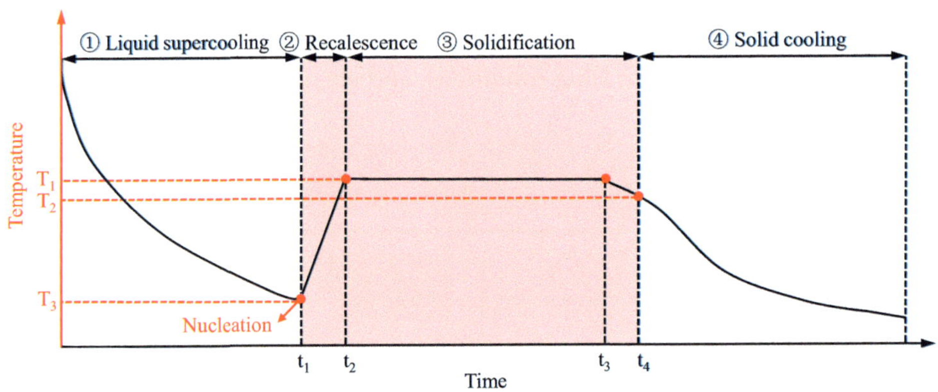

FIGURE 14.2 The solidification process of static droplets.

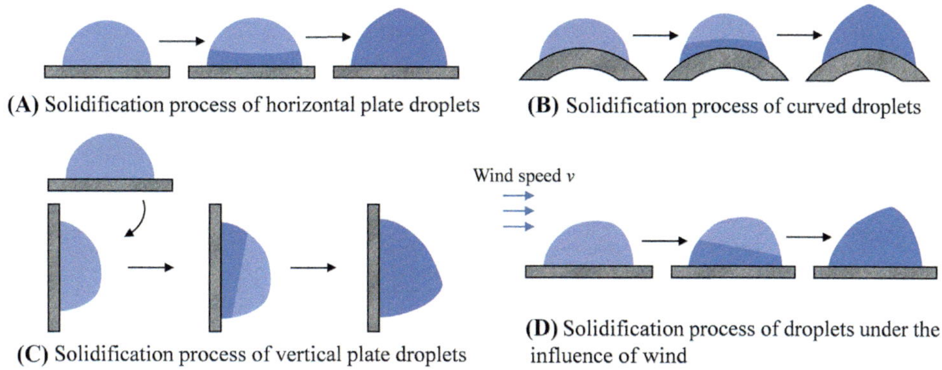

FIGURE 14.3 The solidification process of droplets under four different conditions: (A) Horizontal plate, (B) Curved surface, (C) Vertical plate, (D) Under the influence of wind.

the ice formation process on road surfaces and transmission line surfaces. The icing process on wind turbine blades and aircraft surfaces is more complex, involving the collision icing of supercooled water droplets. Fig. 14.3C and D demonstrate the freezing process of droplets under asymmetric conditions, including the influence of gravity in the vertical state and the influence of wind force in the horizontal state, indicating that the morphology of droplets and freezing fronts undergo more complex changes under asymmetric conditions. Currently, there is limited mechanistic analysis of collision icing of supercooled water droplets, with research primarily focusing on experimental and numerical simulation aspects [18].

14.2.2 Ice prediction techniques

Ice prediction technology is an application technology with vast development prospects. It achieves advanced prediction and dynamic monitoring of ice formation and evolution by coupling meteorological data with machine learning algorithms, ice sensors, and other software and hardware. According to the Google Patent Database, in 2019, there were 51 publicly disclosed invention patents in the field of ice prediction technology in China, accounting for approximately 0.11‰ of the national total of disclosed invention patents. By 2023, the number of publicly disclosed patents had increased to 163, accounting for 0.18‰ of the national total, indicating a significant upward trend and reflecting the promising development prospects of ice prediction technology applications.

As shown in Fig. 14.4, based on different principles, ice prediction techniques can be divided into two categories: model-driven and data-driven approaches. Model-driven ice prediction techniques are based on fundamental principles from disciplines such as fluid mechanics, phase-change heat transfer, and aerodynamics. They construct mathematical models based on numerical equations in the fields of heat and mass transfer and fluid dynamics, thereby calculating methods for predicting icing on specific low-temperature surfaces. They rely on known physical laws and rules to more accurately and reliably describe the icing process. Model-driven approaches typically require comprehensive

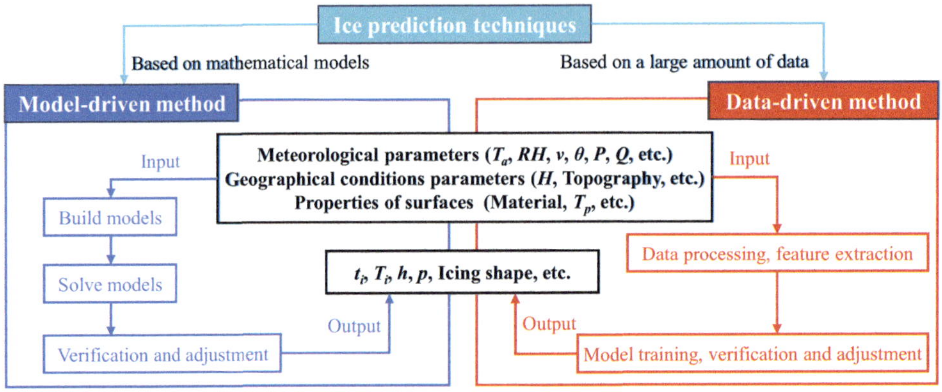

FIGURE 14.4 Classification of ice prediction technologies.

consideration of multiple icing influencing factors such as surface properties and meteorological parameters to establish mathematical models, although research on multifactor coupled models is currently limited. Additionally, the establishment and solving process of models in this method is often complex, requiring calibration and validation of multiple relevant parameters through comparison with experimental data under various conditions. Data-driven methods involve analyzing large amounts of actual data to uncover hidden patterns and numerical rules and, based on this, predicting icing conditions. When the characteristics and parameters of the data model are reasonably selected and adjusted under specific conditions, relatively high accuracy in ice prediction can be achieved. Therefore, this ice prediction technique exhibits high flexibility and adaptability. However, due to its reliance on extensive data for training and adjustment, it demands high quality and quantity of data.

Ice prediction for various types of surfaces mainly depends on parameters such as meteorological conditions, geographical information, and surface properties, including environmental temperature, humidity, wind speed, wind direction, precipitation, altitude, surface temperature, and surface hydrophobicity. Model-driven approaches use these parameters as inputs to solve the constructed mathematical equations and obtain prediction results. Data-driven approaches, on the other hand, analyze and model large amounts of historical and real-time data to achieve ice prediction. Various criteria can be adopted as the basis for ice prediction, such as icing thickness, icing time, icing temperature, icing shape, and icing probability. Due to different application scenarios of different surfaces, the predictive features of interest may vary. For instance, ice prediction on aircraft surfaces may focus on the shape of ice layers on wing surfaces, as different ice shapes have varying impacts on aircraft aerodynamic performance. In contrast, icing on road surfaces may be less concerned with ice shape and more focused on icing time and ice thickness.

As shown in Fig. 14.5A, various types of icing cold surfaces where ice prediction techniques are applied can be broadly categorized into stationary surfaces and moving surfaces. The ice prediction on stationary surfaces is relatively straightforward as these surfaces do not move, but they are also more susceptible to external factors. More attention is paid to the static accumulation of water droplets and the process of ice formation. On the other

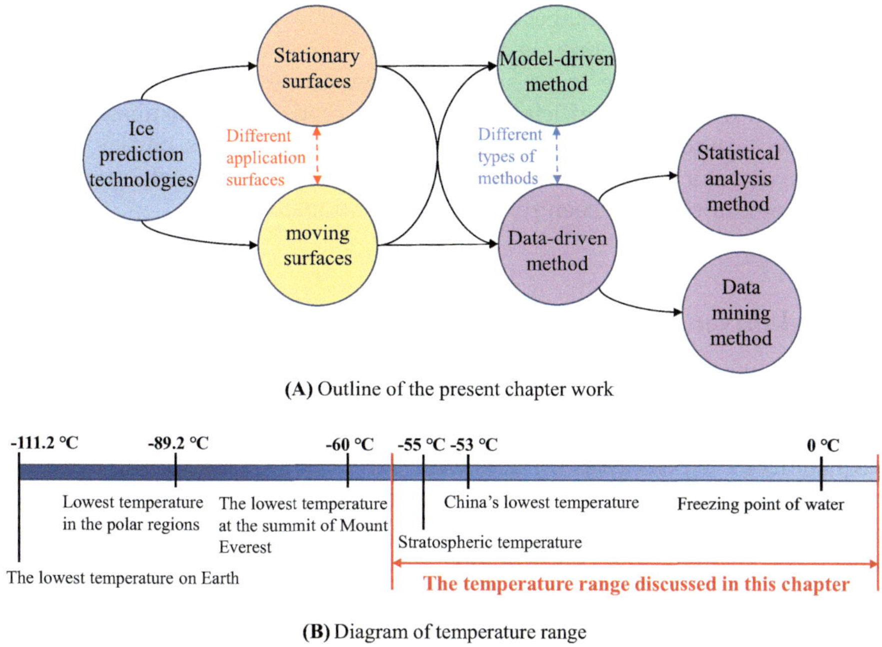

FIGURE 14.5 (A) Outline of this chapter, (B) Diagram of temperature range

hand, ice prediction on moving surfaces poses greater challenges. Due to their inherent dynamic nature, various dynamic factors make the mechanism of ice formation and the process of ice layer growth on moving surfaces more complex. In this chapter, two types of stationary surfaces (road and transmission line) and two types of moving surfaces (wind turbine blade and aircraft) are selected for investigation. Fig. 14.5B illustrates the temperature range of the two discussed types of cold surfaces, approximately from 0°C to around −55°C. To summarize and review the research progress and development trends of existing ice prediction technologies, this chapter analyzes and introduces existing ice prediction technologies for both stationary surfaces, such as roads and transmission lines, and moving surfaces, such as wind turbine blades and aircraft.

14.3 Ice prediction techniques for stationary surfaces

Road surface and transmission line surface are two representative stationary surfaces in the application of ice prediction technologies. The formation of ice on road surfaces reduces the coefficient of friction, significantly decreasing the frictional force between the vehicle tires and the road surface [23]. The vehicle's antiskid capability sharply declines, inevitably increasing braking distance and reducing the time to avoid accidents [24]. As a meteorological disaster, icing on transmission line surfaces is often caused by a combination of complex factors, such as meteorological conditions, geographical environment, and conductor conditions [25]. Many countries have experienced power system accidents, such as conductor dancing, line breakage, and support collapse due to ice accretion on transmission line surfaces. Ice accretion on transmission line surfaces remains a research challenge in the field of icing due to its significant negative impact and complex formation factors [26]. Given the severity of icing disasters on roads and transmission lines, the use of ice prediction technologies for early prevention and control is particularly important. For example, exploratory research in the field of highway traffic meteorology on road ice prediction and monitoring has been reported internationally [27].

14.3.1 Model-driven method

For road icing, researchers have established basic predictive models based on heat transfer and phase change theory. Among them, phase change theory is used to calculate the latent heat generated during the ice-water phase change process [28]. Specific phase change heat transfer equations for asphalt road surfaces were derived by Yang et al., and predictive models were established using finite element software [29]. Combining indoor icing simulation experimental results, the model's predicted road surface temperature error is within 10%, and the relative error in predicting icing duration is 5.04%, validating the model's effectiveness. Another study integrated climate factors and various pavement material characteristics to establish an ice prediction model based on the finite difference method [30]. The model inputs meteorological data and pavement material parameters and outputs road surface temperature, road surface status, and icing thickness. The study also designed indoor experiments to verify the results and analyzed the influence of relevant parameters. The results show that the prediction errors of ice thickness and road surface temperature under constant temperature are within 10%, and the prediction error of ice thickness under variable temperature is 7.83%. A short-term road surface ice prediction model for highways was designed by You based on basic principles of radiation heat transfer, using radiation heat balance to predict road surface temperature, with a prediction error of less than 15% [31]. The predictions and experimental results of Yang [29], Chen [30], and You et al. [31] are shown in Fig. 14.6A.

System dynamics is also a commonly used foundational theory in road surface model-driven modeling. Based on system dynamics, spatial data combined with meteorological data were employed by Hong et al. to design parameter methods and simulate the frost thickness and icing locations on the Honam highway in Korea for 14 hours [32]. Fig. 14.6B shows the proportion of different frost thickness sections along the entire length of the expressway. It is evident that the frost thickness exceeds 360 g for the entire expressway for 14 hours, with over 15% of the areas having frost thickness exceeding 1000 g, indicating severe frosting conditions. The study conducted hypothetical testing on the results but did not validate them. Building upon this research, a combination of deep neural networks and numerical methods was utilized by Hong et al. to compute the crucial intermediate factor, road temperature, in the predictive model, thereby further enhancing the model's reliability [33]. They validated the model using actual sensor data. The model yielded an average of black ice occurrences and locations over 3 days, with a root mean square error (RMSE) of 1.09, achieving a reliability of 94.5%, representing improvements of 148.6% and 11.43%, respectively, compared to previous studies.

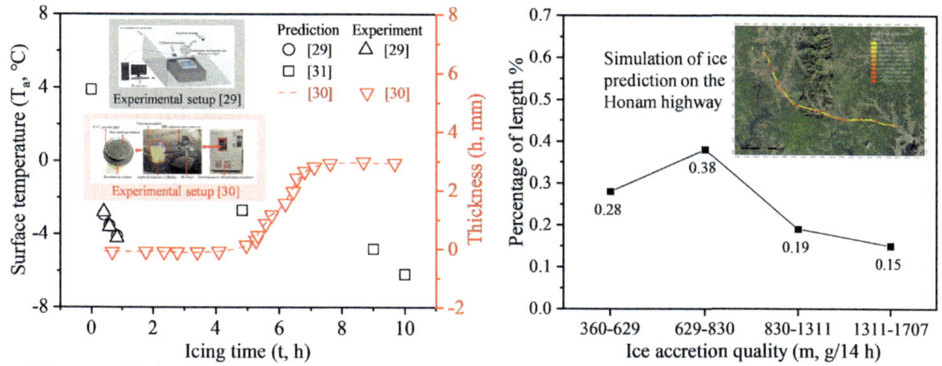

(A) Ice prediction based on phase transition theory **(B)** Ice prediction based on system dynamics

FIGURE 14.6 Road ice prediction based on model-driven approach: (A) Phase transition theory, (B) System dynamics.

TABLE 14.1 Classical transmission line ice prediction model.

Num.	Name	Expression	References
1	Lenhard model	$M = C_1 + C_2P$	[34]
2	Imai model	$R^{\frac{3}{2}} = C\sqrt{v}(-T_a)t$	[35]
3	McComber model	$M = M_0e^{kt}$	[36]
4	Goodwin model	$\Delta R = \frac{\rho_0}{\rho_i}\frac{P}{\pi}\sqrt{1+\left(\frac{v}{v_d}\right)^2}$	[37]
5	Chaine model	$\Delta R = \left[\frac{R_0k}{2}\sqrt{(Ht)^2+(wvt)^2}+R_0^2\right]^{\frac{1}{2}} - R_0$	[38]
6	Jones model	$\Delta R = \frac{1}{\rho_i\pi}\sum_{j=1}^{N}\left[(H\rho_0)^2+(3.6wv)^2\right]^{\frac{1}{2}}$	[39]
7	Makkonen model	$I = \frac{2}{\pi}Efvw$	[40]

Model-driven methods are also classic approaches for predicting surface icing and ice accretion on transmission lines, including classic surface ice prediction models, as shown in Table 14.1. Among them, the Lenhard model [34], Imai model [35], and McComber model [36] are empirical models derived early based on observational experimental results. The Goodwin model [37], the Chaine model [38], and the Jones model [39] are simple dry growth numerical models. Although they may not accurately reflect the icing process on transmission lines, they can provide relatively accurate prediction results under certain conditions due to the ease of obtaining their parameters. In 1984, a more rigorous ice accretion prediction model combining fluid mechanics and heat transfer principles was proposed by Makkonen [40]. Subsequent research on the icing mechanism of transmission lines has mainly focused on determining the collision coefficient and freezing coefficient [41].

Based on the fundamental principles of thermodynamics and aerodynamics, a local ice accretion prediction model for transmission lines was established by Zhu et al., which yielded relatively accurate ice mass and ice shape with an error of less than 6% [42]. An optimal time-step model for predicting ice accretion mass and shape on transmission lines was proposed by Jiang et al., with prediction deviations within 10% [43]. While physical models of transmission lines can predict surface icing conditions to some extent, the ice prediction models based on model-driven methods have limitations in accuracy due to the difficulty in obtaining precise meteorological parameters in the icing model and the inclusion of numerous simplifying assumptions in the model construction process.

14.3.2 Data-driven method

Data-driven methods for stationary surfaces can be divided into statistical analysis and data mining methods based on different ways of utilizing data. The statistical analysis method utilizes historical climate data and observed icing data to establish mathematical models through statistical analysis, thus predicting future icing situations. Statistical analysis methods can provide intuitive formulas with certain predictive abilities, but they lack theoretical explanations. Moreover, statistical analysis methods rely on historical databases, and their accuracy and reliability are influenced by data quality and model selection, resulting in a certain degree of false positives and difficulties in generalization. On the other hand, the data mining method is a new approach born with the development of computer technology and emerging artificial intelligence technology. It analyzes data and extracts features, combining with machine learning algorithms to achieve ice prediction. Compared to the statistical analysis method, the data mining method focuses more on mining hidden patterns and rules from large-scale datasets, and it can use more diverse data types, making it more suitable for handling complex nonlinear relationships. However, it has a stronger black-box nature and poor interpretability, and the performance of the model depends on the quality of the data and the selection of features.

14.3.2.1 *Statistical analysis method*

A probabilistic forecasting method was provided by Berrocal et al., who utilized deterministic numerical weather forecasts to establish two joint distribution probability models for temperature and precipitation, thereby obtaining the probability of road icing [44]. Both models yielded reliable results, with the spatial model demonstrating higher accuracy in spatial cluster prediction while employing this method reduced forecasting costs by nearly 50%. However, further refinement is needed for the theoretical model defined by this method and the evaluation of model parameters. A statistical model was developed by Liu et al., integrating the dynamic dependencies among multiple factors contributing to road icing incidents [45]. Copula theory was used to establish joint distributions of these factors and thereby derive the probability of icing occurrence. Based on this, they categorized road vulnerability levels and ice disaster risk zones. The partitioning results of ice disaster risk zones in Heilongjiang Province, as shown in Fig. 14.7A, reveal a higher risk in the southeastern region, consistent with actual road information. This approach holds significant importance for the prevention

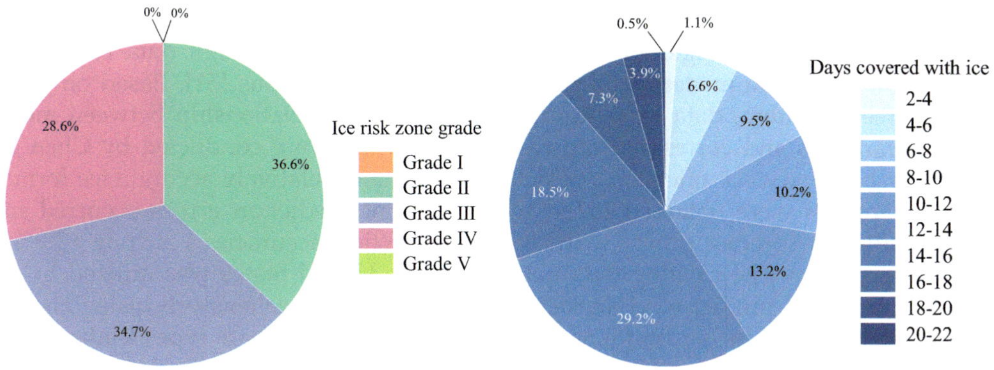

(A) Proportion of area with different grades of ice risk in Heilongjiang Province **(B)** Proportion of area with different icing days in Hunan Province

FIGURE 14.7 Ice prediction on stationary surfaces based on a statistical analysis method: (A) Liu's statistical model, (B) Lu's statistical model.

and rapid assessment of road icing disasters in cold regions, although the model does not consider the impact of wind speed and direction on icing.

The statistical analysis method commonly used for predicting icing on transmission lines is the classical model of extreme values, including Extreme Value Type I distribution, Extreme Value Type II distribution, and Extreme Value Type III distribution [46]. The extreme value model mainly focuses on the return period of disasters, that is, the frequency of disaster occurrence. The observed data of icing days at 97 meteorological stations in Hunan Province over 58 years (1951–2008) were statistically analyzed by Lu et al. using the Extreme Value Type I distribution [47]. They obtained a return period of 24.8 years for areas with particularly severe icing in Hunan Province and plotted a distribution chart of icing days once in a century, as shown in Fig. 14.7B, as a reference for antiicing and ice-control measures for power grids. It can be seen from the figure that most areas in Hunan Province belong to regions with particularly severe icing, with lighter icing severity in northern Hunan and more severe icing in southern Hunan. The prediction of the 50-year return period was conducted by Wang et al. using three different sampling methods and extreme value distribution models, combined with historical icing observation data from three different meteorological stations in Hunan Province [48]. They also analyzed the factors affecting the applicability of the model and the predictive models applicable to three different regional stations. The prediction of the return period can provide important reference for the prevention and control of icing disasters on transmission lines, and also can provide guidance value for the design of transmission lines. However, it lacks observational data on parameters such as icing thickness for prediction.

14.3.2.2 Data mining method

Data mining methods commonly used in existing research for predicting ice formation on stationary surfaces include decision tree algorithms, support vector machine (SVM) algorithms [49], neural network models [50], extreme learning machine algorithms, and so on. The decision tree algorithm is an important classification method in data mining, which has certain advantages in overcoming problems such as high data requirements, high complexity of parameters, and difficult solution calculations [51]. Based on the C4.5 decision tree algorithm, data learning and training on the relationship between meteorological parameters and corresponding road surface icing were conducted by Chen et al., constructing a binary classification prediction model, and relatively accurate ice formation prediction results were obtained [52]. Only 2 days of prediction errors occurred in the selected different forecast times, with overall prediction accuracies reaching 93.3% and 86.4%, respectively. The existing data network in the United States was utilized by Toms et al. to assess and predict road surface icing risks by comprehensively using risk parameters [53]. They conducted frequency analysis on different climate types leading to icing using a newly developed model, obtaining monthly and daily icing frequencies on roads under three types of icing climates: hoarfrost, freezing rain, and freezing fog. The data showed that the prediction accuracy of road surface icing frequencies under freezing fog conditions using this model could exceed 80%.

SVMs is a binary classification algorithm capable of handling high-dimensional data with good generalization performance. Models based on SVM have shown promising feasibility and effectiveness in predicting road surface icing. A prediction model for road surface icing duration and thickness was developed by Yang et al. using the support vector regression (SVR) algorithm [54]. They integrated a Bayesian optimization algorithm to automatically optimize and adjust relevant parameters, achieving prediction accuracies of 97.7% for icing duration and 84.4% for icing thickness. Compared to SVM models, relevance vector machine (RVM) models exhibit better generalization capability and prediction accuracy. An adaptive model for predicting the probability of icing faults on transmission line surfaces based on the RVM model was established by Zhao et al. [55]. They optimized the model using a quantum particle swarm optimization algorithm and a K-fold cross-validation method. Fig. 14.8A depicts the predicted results of icing thickness for road surfaces based on SVM and for transmission lines based on RVM. The SVM model adopts a Gaussian kernel function, while the RVM model adopts a radial basis

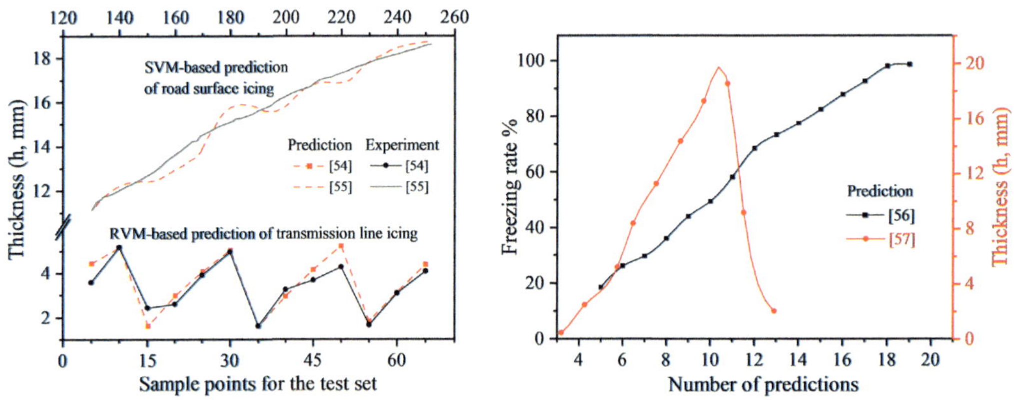

(A) Ice prediction based on SVM and RVM **(B)** Ice prediction based on neural network models

FIGURE 14.8 Ice prediction on stationary surfaces based on the data mining method: (A) SVM and RVM model, (B) Neural network model.

kernel function. Both models demonstrate a good fit with actual data, with a maximum relative error of 5.05% for the RVM model.

Neural network models, known for their capability to handle complex nonlinear relationships and strong learning abilities, are suitable for research on ice prediction techniques. Using a time-series-based backpropagation neural network, the formation process and status of road surface ice and frost coverage were estimated and alerted by Xiong et al. [56]. The study predicted the freezing rate of road surface water over time, as shown in Fig. 14.8B, with relative errors consistently within 2% compared to experimental results. Despite its high accuracy, this model has drawbacks such as slow learning speed and unstable training. A model for predicting the thickness of ice on transmission line surfaces was proposed by He et al. by analyzing data mapped onto parallel coordinates and using convolutional neural networks (CNN) [57]. As depicted in Fig. 14.8B, the model achieved an average prediction accuracy of ~91.18% within an average training time of 48.43 seconds. Extreme learning machine (ELM) is a type of machine learning method based on feedforward neural networks. The hybrid prediction model called WT-BA-ELM was introduced by Sun et al., which utilizes wavelet transform (WT) to handle nonlinear and nonstationary data and employs ELM optimized by bat algorithm (BA) for prediction [58]. An ELM algorithm model that combines ensemble empirical mode decomposition, random forest, and ELM optimized by chaotic gray wolf algorithm was proposed by Wang et al. [59]. They used it to predict the thickness of ice on two different transmission line surfaces, yielding high accuracy with RMSE values of 0.864 and 0.273 and R^2 values of 0.988 and 0.9976, respectively.

14.3.3 Summary

In the application of antiicing and deicing on stationary surfaces such as road surfaces and transmission lines, ice prediction technology plays a crucial role. Ice prediction technology mainly includes model-driven methods and data-driven methods, with the latter further divided into statistical analysis and data mining based on different ways of utilizing data. Table 14.2 summarizes various ice prediction technologies for stationary surfaces introduced in this paper, including input–output methods and prediction accuracy. It can be observed that ice prediction on stationary surfaces mainly relies on meteorological parameters such as air temperature, wind speed, and humidity, while the prediction results primarily focus on indicators such as ice thickness, surface temperature, and icing time. Currently, model-driven methods still play a key role in ice prediction on stationary surfaces, with prediction errors for icing time and ice thickness within 10%. Data-driven methods are also showing potential in the development of ice prediction, as they can adapt to the complexity of real-world situations through learning and training with large amounts of historical and observational data, achieving over 80% prediction accuracy for indicators such as icing time, ice thickness, and icing accuracy. Comparatively, the difficulty in improving ice prediction accuracy is greater for road surfaces than for other stationary surfaces, mainly due to their complex influencing factors such as geographical conditions, road surface conditions, traffic flow, and chemical emissions, all of which affect actual icing conditions on roads.

TABLE 14.2 Summary of predictive techniques for icing on stationary surfaces.

Num.	Input	Output	Evaluation indicators	Method	Surface
1	v, T_e, depth of stagnant water	t_i, T_s	Error < 10%	Model-driven	RS [29]
2	T_a, v, P, λ, surface material	T_s, h, road conditions	Error < 10% (constant temperature) Error = 7.83% (variable temperature)		RS [30]
3	Q, Δt, T_s	T_s	RMSE = 2.71		RS [31]
4	T_a, v, P, P_a, cloud cover, hillshades, etc.	T_s, H_s, m, icing location	RMSE = 1.05		RS [32]
5	T_a, v, P, P_a, Q, cloud cover, hillshades, etc.	T_s, H_s, m, icing location	Error < 6%		RS [33]
6	T_s, overflow water, etc.	m, S_i	Error < 10%		TS [42]
7	v, θ, LWC, mvd, T_a, etc.	m, S_i	Birer score < 0.2		TS [43]
8	T_a, P	p	\	Statistical analysis	RS [44]
9	T_a, P	p	\		RS [45]
10	Number of icing days	Number of icing days	\		TS [47]
11	h	h_{max}	Accuracy > 80%		TS [48]
12	T_a, P, etc.	Icing situation on the road	Accuracy > 80% (freezing fog)	Data mining	RS [52]
13	T_a, RH, v, θ, P, etc.	Frequency of icy weather	Accuracy > 80%		RS [53]
14	v, T_e, depth of stagnant water	t_i, h	Accuracy > 80%		RS [54]
15	T_a, RH, θ	Δh	Relative error < 10%		TS [55]
16	Freezing rate	t_i, freezing rate	Relative error < 2%		RS [56]
17	T_a, RH, maximum tension, etc.	h	Average accuracy = 10%		TS [57]
18	T_a, RH, v, θ, P_a, light intensity	h	MAPE < 3%		TS [58]
19	T_a, RH, v, θ, P_a, light intensity, t_i, etc.	h	RMSE < 0.9, R^2 > 0.98		TS [59]

RS, Road surface; *TS*, transmission line surface.

For stationary surfaces, the future development trends of model-driven approaches mainly involve refining models and conducting multiscale simulations to more accurately capture complex scenarios in real environments. Simultaneously, integrating advanced meteorological and geographical information system data aims to enhance the spatiotemporal resolution and predictive capabilities of the models. Moreover, strengthening field measurements and validation of model parameters, including exploring the icing mechanisms of various surfaces, will contribute to improving the reliability of the models. The optimization of data-driven methods involves more complex model structures, the integration of diverse datasets, and the application of real-time data. Enhancing the training of data-driven models under specific regional and meteorological conditions helps enhance their accuracy and generalization ability. Additionally, combining the strengths of both data-driven and model-driven approaches to develop integrated methods is also a future research direction for ice prediction technologies.

14.4 Ice prediction techniques for moving surfaces

Wind energy, as a mature, widely applicable, and renewable green energy source, holds a strategic position in the global energy, environmental, and economic fields [60]. Wind turbines, as crucial facilities for converting wind energy into electricity, directly determine energy acquisition efficiency. Due to the geographical distribution characteristics of wind resources, wind turbines are often installed in climatically cold, high-altitude, or high-latitude regions, or offshore areas [61]. In such high-humidity, low-temperature meteorological environments, icing on wind turbine blade surfaces becomes the greatest threat to the overall operation of the machinery. Icing can lead to a decrease in lift-to-drag ratio and power, shortened service life, and deteriorated aerodynamic performance of the blades, resulting in economic losses and even endangering the safety of wind farm personnel and equipment [62]. Although certain measures such as superhydrophobic coatings [63], chemical agent spraying [64], and surface preheating [65] can achieve some degree of antiicing and deicing effects, predictive technology for icing on wind turbine blade surfaces is of great research value due to its early detection potential.

During flight, an aircraft undergoes various maneuvers, including pitching, rolling, and yawing, making its surface a typical moving surface. As the aircraft traverses through clouds, supercooled water droplets collide with its surface. In high-altitude and low-temperature environments, these supercooled water droplets rapidly freeze upon contact with the aircraft surface [66]. This situation not only affects the aerodynamic characteristics of the aircraft, causing increased drag, reduced lift, stall angle, and flight speed [67–69] but also leads to roll and pilot control malfunction [70], thereby affecting the stability and controllability of the flight. The random shedding of irregular ice chunks poses a safety threat to the aircraft's tail components and flight status. All components on the aircraft's windward surface are susceptible to icing during flight, such as windshields, wings, nose cones, and propellers. Among them, the parts with the greatest impact on flight safety include wings, tail surfaces, rotors, and engines [71]. Icing primarily occurs at the leading edges of control surfaces, such as wings, tails, and rotors, which are the primary lifting surfaces of the aircraft.

The working state wind turbine blades rotating around the axis and aircraft surfaces traversing through clouds are typically moving low-temperature surfaces, whose icing principle is the formation of ice due to the collision of supercooled water droplets. Currently, research methods for predicting icing on wind turbine blade surfaces and aircraft surfaces can also be divided into two types: model-driven and data-driven approaches. Predicting icing on moving surfaces relies on understanding icing mechanisms, and related icing studies include flight tests [72], icing wind tunnel experiments [73,74], and natural wind field tests [75]. For moving low-temperature surfaces such as wind turbine blades and aircraft surfaces, there are already some mature icing calculation software available for simulating and predicting ice growth, such as LEWICE [66], FENSAP-ICE [76], ONERA [77], TRAJICE [78], etc.

14.4.1 Model-driven method

Using an extended Messinger icing model combined with blade element momentum theory (BEMT), a numerical method to predict the shape, thickness, and power loss of wind turbine blade icing under different atmospheric conditions was successfully developed by Yirtici et al. [79]. The accuracy range of the model-predicted ice volume and thickness is within 3%–20%. The Makkonen icing model combined with high-resolution weather forecasts was utilized by Strauss et al. to study ice prediction for two wind farm turbines, achieving relatively accurate icing forecasts within 6 hours [80]. A prediction model based on gray theory was established by Feng et al. [81], with outputs of temperature and humidity combined with LEWICE simulation software to further predict ice shape and thickness.

For simulating and predicting aircraft wing icing, classic icing models include the Messinger model [82], the Myers model [83], the SWIM model [84], and others. A method to predict the shape and accumulation rate of wing icing was proposed by Shad et al. based on the Euler model [85]. Ice and water were treated as a fluid by Dai et al., and an ice-shape prediction model was established based on the phase-field method [86]. Compared with the LEWICE model results, this model predicted an average ice thickness error of 10.88% and could well predict the main icing characteristics, such as ice shape, ice thickness, and icing rate under different icing conditions. Further, the unsteady icing model proposed based on the closely coupled method can predict the droplet shadow region [87], with a prediction accuracy 35% higher than that of LEWICE and an average ice thickness error of 9.05%. Considering the centrifugal force effect, a three-dimensional numerical prediction method for

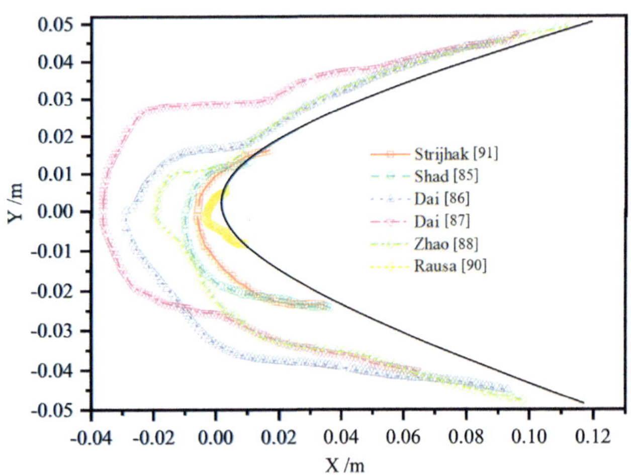

FIGURE 14.9 Results plot of six models that predict icing thickness of NACA0012.

helicopter rotor icing based on the Messenger model in hover was proposed by Zhao et al. [88]. Experimental validation showed a good correlation between the model's predictions and experimental data, especially at the blade tips. Building upon the BEMT model proposed by Brouwers et al. [89], Rausa et al. [90] further improved the blade icing simulation by creating a 3D volume grid analysis method. Fig. 14.9 shows the ice prediction results for the NACA0012 wing using the models by Shad [85], Dai [86,87], Zhao [88], and Rausa [90]. Due to the high-speed rotation of the rotor, there is a possibility of ice shedding from the blades.

14.4.2 Data-driven method

The data-driven approach collects a large amount of measured icing data and uses techniques such as data mining and machine learning to establish predictive models. It correlates input parameters with the degree of icing, thereby achieving predictions of icing on moving surfaces. Combining neural networks and numerical simulations, ice shapes for four different airfoil types were predicted by Strijhak et al. and validated using the LEWICE software [91]. The results show that fully connected networks and convolutional networks can obtain more accurate predictions of ice shapes, with prediction errors in ice thickness within 5% as indicated by the black dashed lines in Fig. 14.9. A deep neural network combined with a stacked autoencoder was used by Yi et al. to establish an ice shape prediction model, which demonstrated that the icing morphology of aircraft is accurately predicted by the model, with a consistency index of 94.4% [92]. Using a deep neural network prediction model, ice predictions for arbitrary airfoil shapes under low-speed incompressible flow conditions were achieved by Qu et al., with prediction errors in ice shapes within 15% [93].

In recent years, with the widespread application of supervisory control and data acquisition (SCADA) systems in the field of wind power generation and the vigorous development of computer technology, data-driven methods for predicting icing on wind turbines have gradually become a research hotspot in the field. A method for predicting blade icing

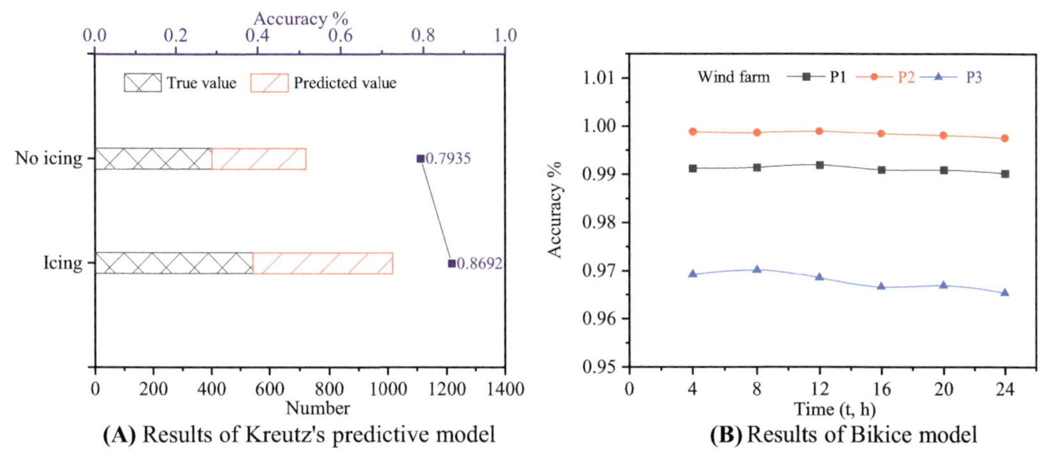

FIGURE 14.10 Ice prediction of wind turbine blades based on a data-driven approach.

using historical weather data from SCADA and forecasted ambient temperature data was proposed by Kreutz et al. based on supervised machine learning [26]. Fig. 14.10A shows the accurate quantity and precision of icing events predicted by this model. It can be seen that the model's accuracy in predicting icing situations is higher than that of nonicing situations, with an overall accuracy of 83.7% for predicting icing events. However, both the input and output data of this model are limited by time ranges. Based on this research, a dual-input one-dimensional convolutional neural network model called the Bikice model was further proposed by Kreutz et al., with input and output time ranges extended to 24 hours [25]. Fig. 14.10B demonstrates the 24-hour ice prediction for three different wind farms using the Bikice model. The prediction accuracy for wind farm P1 is the highest, while that for P3 is the lowest, yet all three have an average accuracy exceeding 96%. Subsequently, this model was evaluated by Kreutz et al. using the latest winter data from 2020 and 2021 [94]. After the essence of predicting icing on wind turbine blades based on data-driven time series analysis was analyzed, a prediction method combining recurrent neural networks and a t-distributed stochastic neighbor embedding dimensionality reduction algorithm was proposed by Zhang et al. [95]. This model takes historical data and labels of the blades as input and can predict the icing status of wind turbine blades at a future time. Compared to other algorithms, the predictive accuracy of this model has improved by over 100%. A combination of one-dimensional CNN with stacked bidirectional gated recurrent units was utilized by Li et al. to predict icing on wind turbine blades, and a set of innovative prediction accuracy evaluation metrics was proposed [96]. The accuracy of this model can reach 94.58%, with an improvement in accuracy of over 14.44% compared to other algorithms.

14.4.3 Summary

The ice prediction techniques for moving surfaces such as wind turbine blades and aircraft surfaces can also be categorized into model-driven and data-driven approaches. Table 14.3 summarizes their input parameters, output results, and prediction accuracy. Compared to stationary surfaces, the icing process and influencing factors on moving surfaces are more complex, and the relevant features of ice prediction techniques also differ. Regarding input parameters, in addition to meteorological data, the main input parameters for moving surfaces include microphysical parameters such as median volume diameter and liquid water content, as well as self-motion parameters such as flight speed and blade rotation speed. Ice prediction for wind turbine blades also considers cabin properties like generator temperature. As for output results, icing shape is the most concerning feature for moving surfaces. After simulating and predicting icing shape, additional features such as ice thickness and ice accumulation can be obtained. The model-driven approach remains an important means of prediction for moving surfaces. Compared to stationary surfaces, it further considers the dynamic characteristics of moving surfaces, such as simulating the icing process of aircraft or wind turbines in the air by combining principles of aerodynamics. However, models typically require complex calculations and precise input parameters, posing challenges in practical applications. On the other hand, data-driven approaches can better capture the nonlinear relationships of the icing process

TABLE 14.3 Summary of predictive techniques for icing on moving surfaces.

Num.	Input	Output	Evaluation indicators	Method	Surface
1	P_a, RH, v, LWC, mvd, T_e, etc.	S_i, h, Loss of power generation	3% < Error < 25%	Model-driven	WS [79]
2	T_a, RH, v, mvd, density of liquid water, etc.	T_s, v, Dew point temperature	\		WS [80]
3	T_a, P_a, v, LWC, RH, v_b, etc.	T_s, H_s, S_i, h	Average error < 15%		WS [81]
4	T_a, LWC, v_f, etc.	S_i, h, unfrozen water film thickness	\		AS [85]
5	LWC, mvd, T_e	S_i, h, ice mass fraction	Average error = 10.88% (for h)		AS [86]
6	T_a, P_a, LWC, mvd, etc.	S_i, h, etc.	Average error = 9.05% (for h)		AS [87]
7	LWC, mvd, T_s, S_f, centrifugal force, etc.	S_i	\		WS [88]
8	LWC, T_e, etc.	S_i	\		WS [90]
9	T_a, P_a, S_f, LWC, mvd, v_f, t	S_i	Error < 5% (for h)	Data-driven	AS [91]
10	T_a, v, LWC, mvd, t	S_i	Average ice shape consistency index = 94.4%		AS [92]
11	T_a, LWC, mvd, t, etc.	S_i, h, θ_i, etc.	Error < 15% (for S_i)		AS [93]
12	v, θ, v_b, T_a, T_g, cabin temperature, etc.	t, icing events	Accuracy = 83.4%		WS [26]
13	T_a, RH, v, P_a	t, icing events	Accuracy > 96%		WS [25]
14	T_a, RH, v, P_a	t, icing events	Accuracy < 83%		WS [94]
15	T_e, v, θ, v_b, T_g, etc.	Icing situation	Accuracy < 72%		WS [95]
16	v, θ, T_a, v_b, T_g, etc.	p	Accuracy < 95%		WS [96]

AS, Aircraft surface; *WS*, wind turbine blade surface.

and can be adjusted for different moving surfaces and meteorological conditions. In existing research on wind turbine blades and aircraft surfaces, both methods can achieve a prediction accuracy of over 80% for indicators such as ice thickness and icing profiles.

The future development directions of dynamic surfaces include improving the accuracy and real-time capability of icing models to adapt to rapidly changing environmental conditions. Further integration of advancements in aerodynamics and materials science to enhance the understanding of dynamic surface icing behavior. Simultaneously, leveraging advanced numerical computation and simulation techniques to enhance the computational efficiency of models. For data-driven approaches, further integration with real-time monitoring data is required to establish more refined and personalized data-driven models. Additionally, strengthening quality management and processing of data, as well as addressing the data heterogeneity associated with dynamic moving surfaces, to achieve more accurate predictions of icing behavior on dynamic surfaces.

14.5 Concluding remarks

This chapter provides a review and summary of existing ice prediction techniques for both stationary surfaces like roads and transmission lines, and moving surfaces like wind turbine blades and aircraft. Three conclusions and prospects are drawn from this review.

1. Ice prediction techniques for both stationary and moving surfaces can be classified into two categories: model-driven and data-driven methods. The temperature range for practical application of these techniques falls between -55°C and 0°C. Both types of surface ice prediction techniques rely on various meteorological data. Additionally, prediction for moving surfaces also considers parameters such as their own motion and microphysical parameters. In terms of prediction characteristics, stationary surfaces focus on metrics like ice thickness and freezing time, while moving surfaces are more concerned with the shape of the ice formation. Existing ice prediction techniques can achieve an accuracy of over 80% for predicting ice thickness on the four types of surfaces. Compared to stationary surfaces, moving surfaces exhibit more complex and variable ice formation behavior and mechanisms due to their motion characteristics, making their prediction more challenging.
2. Data-driven approaches for ice prediction techniques have great potential for future development. On the one hand, they can comprehensively consider various icing influencing factors, automatically learn, and construct complex models. On the other hand, they can combine real-time monitoring data to automatically adjust model parameters, thereby achieving self-optimization and adaptation of the model. However, this method lacks interpretability and cannot reflect the physical significance of the icing process. Therefore, if it can be combined with icing physical models, the prediction accuracy of this method can be further improved. Continuous breakthroughs in basic research during experimental studies of the icing process can lead to sustained improvement in the prediction accuracy of ice prediction techniques.
3. Existing ice prediction technologies still face difficulties in achieving high-precision real-time dynamic prediction. The challenges in achieving high-precision real-time prediction are partly due to limitations in data acquisition and transmission with current technologies. Real-time acquisition of meteorological data or sensor data relies on advancements in monitoring technology and the optimization of novel mathematical algorithms. Another aspect is the limited theoretical research on the icing mechanisms in complex environments. Future efforts need to delve deeper into fundamental icing research to unveil the underlying principles of icing processes, ice layer morphology, mass transfer flow, and other basic laws, thereby achieving a better understanding of the essence of icing and optimizing sensing technologies and ice prediction models. The resolution of the aforementioned issues relies on interdisciplinary collaboration. Collaborative efforts across meteorology, mathematics, computer science, materials science, and other fields will contribute to the development of a more comprehensive ice prediction framework, thereby enhancing the accuracy, practicality, and timeliness of predictions.

References

[1] L. Zhang, M.J. Song, S.M. Deng, et al., Frosting mechanism and behaviors on surfaces with simple geometries: a state-of-the-art literature review, Applied Thermal Engineering 215 (2022)118984.

[2] M.J. Song, S.M. Deng, C.B. Dang, et al., Review on improvement or air source heat pump units during frosting and defrosting, Applied Energy 211 (2018) 1150–1170.

[3] M.J. Song, S.W. Lei, S.H. Hosseini, et al., An experimental study on the effect of horizontal cold plate surface temperature on frosting characteristics under natural convection, Applied Thermal Engineering 211 (2022) 118416.

[4] S.W. Lei, M.J. Song, C.B. Dang, et al., Experimental study on the effect of surface temperature on the frost characteristics of an inverted cold plate under natural convection, Applied Thermal Engineering 211 (2022)118470.

[5] J.L. Laforte, M. Allaire, J. Laflamme, State-of-the-art on power line de-icing, Atmospheric Research 46 (1998) 143–158.

[6] C. Qiu, J.L. Yang, Y. Kou, Research progress of anti-icing and de-icing technology for aircraft, Journal of Xi'an Aeronautical Institute 41 (2023) 1–8. + 14. (In Chinese).

[7] Z.J. Zhang, H. Zhang, S. Yue, et al., A review of icing and anti-icing technology for transmission lines, Energies 16 (2023) 601.

[8] A. Baidya, M.A. Ganayee, S.J. Ravindran, et al., Organic solventfree fabrication of durable and multifunctional superhydrophobic paper from waterborne fluorinated cellulose nanofiber building blocks, ACSNano (2017) 11091–11099.

[9] H.J. Coffman, Helicopter rotor icing protection methods, Journal of the American Helicopter Society 32 (1987) 34–39.

[10] Y. Ou, Review on icing detection techniques of pavement, Highway 4 (2013) 191–196.

[11] R. Salerno, A. Ceppi, M. Mancini, et al., A study of an algorithm for the surface temperature forecast: from road ice risk to farmland application, Applied Sciences (2020) 4952.

[12] C. Wang, J.T. Huang, Y. Zhang, et al., Status and advance in research on blade icing of wind turbines, Power Construction 35 (2014) 70–75. In Chinese.

[13] Y. Hu, Analysis and countermeasures discussion for large area icing accident on power grid, High Volt (2008) 215–219. In Chinese.

[14] T.N. Ma, D.X. Niu, M. Fu, Icing forecasting for transmission lines based on a wavelet support vector machine optimized by a quantum fireworks algorithm, Applied Sciences 6 (2016) 54.

[15] Y.Q. Dong, An application of deep neural networks to the in-flight parameter identification for detection and characterization of aircraft icing, Aerospace Science and Technology (2018). 77.

[16] K.R. Petty, C.D.J. Floyd, A statistical review of aviation airframe icing accidents in the US, Conference on Aviation, Range, and Aerospace Meteorology 11 (2004) 623–628.

[17] F. Tavakoli, S.H. Davis, H.P. Kavehpour, Freezing of supercooled water drops on cold solid substrates: initiation and mechanism, Journal of Coatings Technology and Research 12 (2015) 869–875.

[18] X. Zhang, Research on freezing and impact processes of supercooled water droplet and their coupling characteristics, Tsinghua University (2019). (In Chinese).

[19] L.Z. Huang, Y.Y. Tian, M.J. Song, L. Zhang, C. Chao, Experimental study of relative humidity effect on the edge effect of frosting characteristics on a vertical cold plate surface, Applied Thermal Engineering 236 (2023)121537.

[20] X. Liu, X. Zhang, J.C. Min, Shape variation of sessile water droplet during freezing, Journal of Engineering Thermophysics (2020) 704–708. In Chinese.

[21] X. Zhang, Y.B. Gao, X. Liu, J.C. Min, K.L. Li, M.J. Song, Modelling of the icing processes of sessile supercooled water droplets on cold spheres, Applied Thermal Engineering 243 (2024)122596.

[22] M.F. Ismail, P.R. Waghmare, Universality in freezing of an asymmetric drop, Applied Physics Letters 109 (2016)234105.

[23] J. Ilkka, N. Pertti, H. Marjo, Statistical modelling of wintertime road surface friction, Meteorological Applications 20 (2013) 318–329.

[24] S. Sahar, N. Somayeh, B. Alireza, et al., Lateral coefficient of friction for characterizing winter road condition, Canadian Journal of Civil Engineering 43 (2015) 73–83.

[25] M. Kreutz, A.A. Alla, K. Varasteh, et al., Convolutional neural network with dual inputs for time series ice prediction on rotor blades of wind turbines, Procedia CIRP 104 (2021) 446–451.

[26] M. Kreutz, A.A. Alla, K. Varasteh, et al., Machine learning-based ice prediction on wind turbines, Procedia CIRP 81 (2019) 423–428.
[27] W. Iii, W. Mahoney, W. Myers, Predicting weather and road conditions: integrated decision-support tool for winter road-maintenance operations, Transportation Research Record 1824 (2003).
[28] A. Fujimoto, R.A. Tokunaga, M. Kiriishi, et al., A road surface freezing model using heat, water and salt balance and its validation by field experiments, Cold Regions Science and Technology 106 (2014) 1–10.
[29] E.H. Yang, J.H. Peng, L. Luo, et al., Analysis on influencing factors of asphalt pavement icing and establishment of ice prediction model, Road Materials and Pavement Design (2023).
[30] J.Q. Chen, C.Q. Sun, X. Sun, et al., Finite difference model for predicting road surface ice formation based on heat transfer and phase transition theory, Cold Regions Science and Technology 207 (2023)103772.
[31] J.Y. You, Research on ice warning and traffic flow scheduling strategy of expressway, Chang'an University (2016). (In Chinese).
[32] S.B. Hong, B.W. Lee, C.H. Kim, et al., System dynamics modeling for estimating the locations of road icing using GIS, Applied Sciences 11 (2021) 8537.
[33] S.B. Hong, H.S. Yun, S. Yum, et al., Development of black ice prediction model using GIS-based multi-sensor model validation, Natural Hazards and Earth System Sciences 22 (2022) 3435–3459.
[34] R.W. Lenhard, An indirect method for estimating the weight of glaze on wires, Bulletin of the American Meteorological Society 36 (1955) 1–5.
[35] L. Imai, Studies on ice accretion, Researches on Snow and Ice 3 (1953) 35–44.
[36] P.J.W. McComber, An analysis of selected ice accretion measurements on a wire at Mount Washington, Montreal: Proceedings of the Forty-second Annual Eastern Snow Conference, (1985).
[37] E.J.I. Goodwin, J.D. Mozer, A.M.J. Digioia, et al., Predicting ice and snow loads for transmission lines, Proceedings. First IWAIS. (1983) 267–273.
[38] P.M. Chaine, G. Castonguay, New approach to radial ice thickness concept applied to bundle-like conductors, Industrial Meteorology StudyIV Environment (1974).
[39] K.F. Jones, A simple model for freezing rain ice loads, Atmospheric Research 46 (1998) 87–97.
[40] L. Makkonen, Heat transfer and icing of a rough cylinder, Cold Regions Science and Technology 10 (1985) 105–116.
[41] Q. He, J.H. Li, M.Y. Deng, et al., Calculation and influencing factors of Icing freezing coefficient of overhead transmission line, Transactions of China Electrotechnical Society 34 (2019) 4162–4169. In Chinese.
[42] Y.C. Zhu, X.B. Huang, J.Y. Jia, et al., Numerical simulation for icing and influence on transmission line, Journal of Xi'an Jiaotong University 49 (2015) 120–125. In Chinese.
[43] X.L. Jiang, F.Y. Jiang, Q.L. Wang, et al., Prediction of rime accretion on transmission line based on optimal time step model, Transactions of China Electrotechnical Society 33 (2018) 4408–4418. In Chinese.
[44] V.J. Berrocal, A.E. Raftery, T. Gneiting, et al., Probabilistic weather forecasting for winter road maintenance, Journal of the American Statistical Association 105 (2010) 522–537.
[45] Q. Liu, A.P. Tang, Z.Y. Wang, et al., Exploring the road icing risk: considering the dependence of icing-inducing factors, Natural Hazards 115 (2023).
[46] C. Liu, Research of transmission lines icing thickness prediction model based on Markov process, North China Electric Power University, 2015. In Chinese.
[47] J.Z. Lu, H.X. Zhang, J.W. Peng, et al., Calculation of Hunan power grid icing recurrence interval based on extreme-value type I probability distribution model, High Voltage Engineering 38 (2012) 464–468. In Chinese.
[48] J.C. Wang, S. Su, X.Y. Sheng, et al., Comparative study of applicability of methods for estimating transmission line icing return period based on various extreme value distributions, Power System Technology 39 (2015) 2614–2620. In Chinese.
[49] Y. Chen, Y.J. Lu, R.H. Duan, et al., A forecast method of ice flashover fault early warning based on partial mutual information and support vector machine, Automation of Electric Power Systems 42 (2018) 92–98.
[50] D.Z. Li, J.R. Liu, H.Y. Zhang, Prediction method of fan blade icing based on deep fully connected neural network, Electric Power Science and Engineering 35 (2019) 39–44. In Chinese.
[51] W. Zhang, S. Gao, B. Chen, et al., The application of decision tree to intensity change classification of tropical cyclones in western North Pacific, Geophysical Research Letters 40 (2013) 1883–1887.

[52] K. Chen, G.Q. Xu, T. Xiong, et al., Construction and application of road icing forecast model based on C4.5 decision tree algorithm, Transportation Research 4 (2018) 57–62.

[53] B. Toms, J.B. Basara, Y. Hong, Usage of existing meteorological data networks for parameterized road ice formation modeling, Journal of Applied Meteorology and Climatology 56 (2017) 1959–1976.

[54] E.H. Yang, Q.L. Yang, J. Li, et al., Establishment of ice prediction model of asphalt pavement based on support vector regression algorithm and Bayesian optimization, Construction and Building Materials 351 (2022) 128955.

[55] J. Zhao, H.X. Zhang, H.L. Zou, et al., Probability prediction method of transmission line icing fault based on adaptive relevance vector machine, Energy Reports 8 (2022) 1568–1577.

[56] Z. Xiong, J.J. Qin, A study of road surface ice prediction based on simulate experiment, Highway Engineering 39 (2014) 216–220. In Chinese.

[57] L.F. He, J. Luo, X.D. Zhou, A novel deep learning model for transmission line icing thickness prediction (2021), 733–738.

[58] W. Sun, C.F. Wang, Staged icing forecasting of transmission lines based on icing cycle and improved extreme learning machine, Journal of Cleaner Production 208 (2019) 1384–1392.

[59] W.J. Wang, D. Zhao, L.G. Fan, et al., Study on ice prediction of transmission lines based on ensemble empirical mode decomposition and feature selection optimized extreme learning machine, Energies 12 (2019) 21–63.

[60] C.Q. Kang, L.Z. Yao, Key scientific issues and theoretical research framework for power systems with high proportion of renewable energy, Automation of Electric Power Systems 41 (2017) 2–11. In Chinese.

[61] P. Enevoldsen, M. Jacobson, Data investigation of installed and output power densities of onshore and offshore wind turbines worldwide, Energy for Sustainable Development 60 (2021) 40–51.

[62] J.Y. Jin, M.S. Virk, Study of ice accretion and icing effects on aerodynamic characteristics of DU96 wind turbine blade profile, Cold Regions Science and Technology 160 (2019) 119–127.

[63] B. Li, J. Guo, F. Wang, Prospect of application of superhydrophobic materials in the field of anti-icing of wind turbine blades, Henan Electric Power (2021) 71–73. + 82. (In Chinese).

[64] O. Parent, A. Ilinca, Anti-icing and de-icing techniques for wind turbines: critical review, Cold Regions Science and Technology 65 (2011) 88–96.

[65] L.C. Shu, J.H. Qi, H. Qin, et al., Anti-icing model and sectionalized anti-icing method by electrical heating for wind turbine blades, Proceedings of the CSEE 37 (2017) 1448–1455. In Chinese.

[66] G.M. Potapczuk, Aircraft icing research at NASA Glenn Research Center, Journal of Aerospace Engineering 26 (2013) 260–276.

[67] Y.H. Cao, K. Chen, J. Sheridan, Flowfield simulation and aerodynamic performance analysis of complex iced airfoils with hybrid multi-block grid, Proceedings of the Institution of Mechanical Engineers 222 (2008) 417–422.

[68] W.O. Valarezo, F.T. Lynch, R.J. Mcghee, Aerodynamic performance effects due to small leading-edge ice (roughness) on wings and tails, Journal of Aircraft 30 (1993) 807–812.

[69] T. Cebeci, Effect of ice on airfoil stall at high Reynolds numbers, AIAA Journal 33 (1995) 1351–1352.

[70] Q.F. Yu, Flying test research of the icing and its effects on flight performance for Y12-II aircraft, Flight Dynamics 2 (1995) 63–70. In Chinese.

[71] Z. Li, H.J. Xu, Y. Xue, et al., Research on flight safety effect mechanism and protection for aircraft icing, Flight mechanics 34 (2016) 10–14. In Chinese.

[72] J.L. Ding, L.L. Zhao, T. Yang, et al., Flight test technology of natural icing, Acta Aeronautica et Astronautica Sinica 44 (2023) 50–65. In Chinese.

[73] M.G. Potapczuk, Aircraft icing research at NASA Glenn Research Center, Journal of Aerospace Engineering 26 (2013) 260–276.

[74] S.N. Li, L. Caracoglia, Wind tunnel experimental variability of aerodynamic loads for wind turbine blades, Journal of Physics: Conference Series 1452 (2020) 012054.

[75] L.Y. Gao, T. Tao, Y.Q. Liu, A field study of ice accretion and its effects on the power production of utility-scale wind turbines, Renewable Energy 167 (2020).

[76] H. Beaugendre, F. Morency, W.G. Habashi, FENSAP-ICE's three-dimensional in-flight ice accretion module: ICE3D, Journal of Aircraft 40 (2003) 239–247.

[77] T. Hedde, D. Guffond, ONERA three-dimensional icing model, AIAA Journal 33 (1995) 1038–1045.

[78] R. Gent, A combined water droplet trajectory and ice accretion prediction program for aerofoil, Royal Aerospace Establishment, Technical Report (1990), 90054.

[79] O. Yirtici, S. Ozgen, I.H. Tuncer, Predictions of ice formations on wind turbine blades and power production losses due to icing, Wind Energy 22 (2019).
[80] L. Strauss, S. Serafin, M. Dorninger, Skill and potential economic value of forecasts of ice accretion on wind turbines, Journal of Applied Meteorology and Climatology 59 (2020) 1–58.
[81] C. Feng, C. Papachristou, Grey-model based ice prediction sensor system on wind turbine system, Aerospace and Electronics Conference (2012) 78.
[82] B.L. Messinger, Equilibrium temperature of an unheated icing surface as a function of air speed, Journal of the Aeronautical Sciences 20 (1953) 29–42.
[83] T.G. Myers, D.W. Hammond, Ice and water film growth from incoming supercooled droplets, International Journal of Heat and Mass Transfer 42 (1999) 2233–2242.
[84] S.P. Otta, A.P. Rothmayer, A simple boundary-layer water film model for aircraft Icing, In Proceedings of the 45th AIAA Aerospace Sciences Meeting & Exhibit 902 (2007) 8–11.
[85] A. Shad, S.A. Sherif, Numerical modeling of atmospheric rime ice accretion on an airfoil using an Eulerian approach, Journal of Fluids Engineering 144 (2022).
[86] H. Dai, C.L. Zhu, H.Y. Zhao, et al., A new ice accretion model for aircraft icing based on phase-field method, Applied Sciences 11 (2021) 5693.
[87] H. Dai, C.X. Zhu, N. Zhao, et al., An unsteady model for aircraft icing based on tightly-coupled method and phase-field method, Aerospace 8 (2021) 373.
[88] G.Q. Zhao, Q.J. Zhao, X. Chen, New 3-D ice accretion method of hovering rotor including effects of centrifugal force, Aerospace Science and Technology 48 (2016).
[89] E.W. Brouwers, L.P. Jose, C.S. Edward, A.P. Andrew, The experimental investigation of a rotor hover icing model with shedding, American Helicopter Society 66th Annual Forum (2010) 1863–1875.
[90] A. Rausa, M. Morelli, A. Guardone, A novel method for robust and efficient prediction of ice shedding from rotorcraft blades, Journal of Computational and Applied Mathematics 391 (2021)113452.
[91] S. Strijhak, D. Ryazanov, K. Koshelev, et al., Neural network prediction for ice shapes on airfoils using iceFoam simulations, Aerospace 9 (2022) 96.
[92] X. Yi, Q. Wang, C.C. Chai, et al., Prediction model of aircraft icing based on deep neural network, Transactions of Nanjing University of Aeronautics and Astronautics 38 (2021) 535–544.
[93] J.G. Qu, B. Peng, X. Yi, et al., Ice prediction method for arbitrary airfoil using deep neural networks, Acta Aerodynamica Sinica 41 (2023) 48–55. In Chinese.
[94] M. Kreutz, A.A. Alla, M. Lütjen, et al., Ice prediction for wind turbine rotor blades with time series data and a deep learning approach, Cold Regions Science and Technology 206 (2022)103741.
[95] Z.Q. Zhang, B. Fan, Y. Liu, et al., Rapid warning of wind turbine blade icing based on MIV-tSNE-RNN, Journal of Mechanical Science and Technology 35 (2021) 5453–5459.
[96] Y.Y. Li, L.Y. Hou, M. Tang, et al., Prediction of wind turbine blades icing based on feature Selection and 1D-CNN-SBiGRU, Multimedia Tools and Applications 81 (2022) 4365–4385.

CHAPTER

15

Ultrasonic application on ice detection and removal

Jifeng Zhang[1], *Yuan Tian*[1], *Zhonghua Shi*[2], *Wen Luo*[1], *Huanyu Zhao*[3] *and Yudong Guo*[4]

[1]College of Aerospace and Civil Engineering, Harbin Engineering University, Harbin, P.R. China [2]Xi'an Electronic Engineering Research Institute, Xi'an, P.R. China [3]Liaoning Provincial Key Laboratory of Aircraft Ice Protection, AVIC Aerodynamics Research Institute, Shenyang, P.R. China [4]Wuhan Aviation Instrument Co., Ltd., Wuhan, P.R. China

Nomenclatures

Symbol	Description	Unit
c	Velocity	m/s
d	Thickness	mm
f	the body load distribution	N/m
g	Gravity acceleration	m/s^2
p	Poynting vector	W/m^2
t	Time	s
u	Displacement distribution	mm
v	Velocity	m/s
x	Distance ahead of a interface	m
y	Position	m
z	Contact radius	m
A	Antisymmetric mode	/

Frosting and Icing for Efficient Energy Use in Engineering Applications
DOI: https://doi.org/10.1016/B978-0-443-15495-9.00005-8

E	Kinetic energy	N · m
L	Lamb wave	/
N	Cycle number	/
P	Poynting vector	W/m^2
S	Symmetric mode	/
T	Temperature	°C
U	Strain energy	/
V	Volume	mm^3
Greek symbol	**Description**	**Unit**
α	Incidence angle	°
ω	Angular frequency	rad/s
σ	Stress distribution	N/m
τ	In-plane shear stresses	N/m
Subscript	**Description**	**Unit**
0	Initial	/
k	kinetic	/
m	The mode numbers of UGW	/
n	The mode numbers of UGW	/
g	Group	/
i	Ice	/
c	Velocity	m/s
d	Thickness	mm
f	the body load distribution	N/m
g	Gravity acceleration	m/s^2
p	Poynting vector	W/m^2
t	Time	s
u	Displacement distribution	mm
v	Velocity	m/s
x	Distance ahead of a interface	m
y	Position	m
z	Contact radius	m

15.1 Introduction

For structures in cold environments, freezing is a critical issue that must be solved to guarantee the workability of structures. The accumulation of ice on structures would detrimentally affect performance and reliability. Ice accretions have serious impacts on the performance of offshore platforms, marine vessels [1,2], aircraft [3,4], wind turbine blades [5–8], power lines [9], etc. Particularly in aviation, modern aircraft are required to operate normally in all kinds of weather conditions to accomplish the task. However, aircraft accidents due to adverse weather conditions such as low-altitude wind shear, thunderstorms, atmospheric turbulence, frost, hail, lightning, rainfall, and ice accretion are frequently reported [10,11]. Over 50% of commercial aviation accidents are weather-related. Among these, the ice accretion is a major safety threat [3]. Therefore, timely detection and removal of icing at the early stage are significant in improving energy efficiency and safety.

In ice detection, the existing techniques for ice detection include direct and indirect detection. Indirect detection refers to the indirect prediction of the probability of ice accretion by collecting environmental data, such as ambient temperature, noise, and air humidity [12,13]. Direct detection includes impedance and capacitance methods [14,15], optical imaging identification methods [16], thermal infrared radiometry methods [17], intrinsic frequency detection methods [18], microwave reflection methods [19], ultrasonic echo methods [20], etc. In comparison, the ultrasonic-guided wave (UGW) based ice detection method bears the advantage of low energy consumption, high precision, and wide detection range,16 and thus has been widely used in the fields of nondestructive testing (NDT) [21–23].

In deicing, several thermal and mechanical methods associated with deicing have been developed, including pneumatic, electrothermal, electro-impulse, vibratory, etc. [24–26]. However, these technologies are limited due to energy and weight constraints [27]. Take the widely used thermal deicing methods for aircraft as an example; the electrothermal deicing system is heavy, up to 162 lbs. and requires power 25 kW. Therefore, the electrothermal deicing system can only operate for short periods. After the deicing system is turned off, the liquid on the aircraft can freeze again. An ultrasonic deicing system [28–30] is promising for achieving deicing by using lightweight and simple transducers, thus having the advantage of low power consumption.

In recent years, with the development and maturity of UGW-based NDT, the potential application of UGW in icing detection has attracted considerable attention [31,32]. Meanwhile, some studies have shown that the shear forces are generated at the interface due to the phase difference when the UGWs propagate in a laminate [33–35]. The deicing efficiency with different UGWs' mode-frequency pairs based on the cross-sectional stress concentration coefficient has been reported, and the effectiveness of the UGWs' technology for deicing was verified [36–38]. In this chapter, the in-depth research methods and results are presented on the above basis.

15.2 Ultrasonic ice detection

15.2.1 Basic theory

The presence of an ice layer on a surface can significantly impact the propagation of UGWs, resulting in notable changes to key wave characteristics such as amplitude

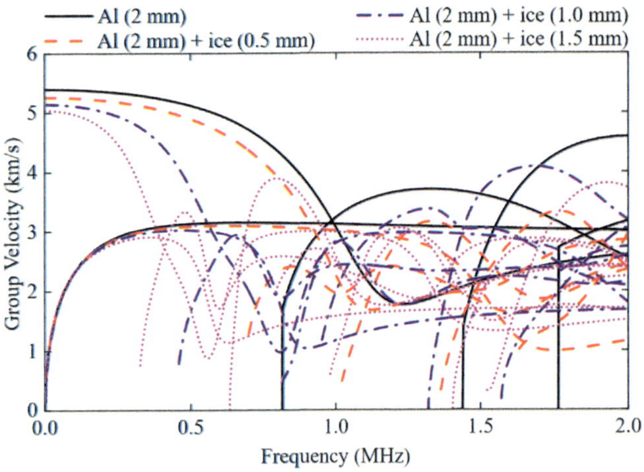

FIGURE 15.1 Dispersion curves of phase velocity with different ice thicknesses.

attenuation and signal arrival time delay. To illustrate this phenomenon, Fig. 15.1 presents a comparison of the group velocity dispersion curves for two distinct scenarios: an aluminum plate without an ice layer and an aluminum plate with an accumulated ice layer on its surface. Analysis of the dispersion curves shows that the introduction of the ice layer leads to a significant decrease in the group velocity of Mode 2 waves relative to the S0 mode. This observation highlights the importance of considering the presence of ice when assessing UGW propagation, as the ice layer can significantly alter the wave behavior and potentially influence the interpretation of the acquired data in practical applications.

Assuming the group velocity of ultrasonic-guided waves (UGW) in an aluminum plate is represented by $v_g^{Al}(f,m)$, and in an icing plate by $v_g^{ice}(f,n)$, where f denotes the frequency and m and n represent the mode numbers of UGW, respectively, the global matrix method (GMM) [39] can be employed to determine the group velocity at a specific frequency within the structure, provided that the thicknesses of the aluminum plate and ice layer are known. By comparing the difference in time-of-flight (ToF) Δt before and after icing, the length of the ice layer, denoted as l, can be obtained.

$$l = \frac{v_g^{Al} - v_g^{ice}}{v_g^{Al} v_g^{ice}} \Delta t \tag{15.1}$$

Similarly, the group velocity of UGW can be derived from the difference in ToF, enabling the determination of the ice thickness, represented by d_{ice}, as shown in Eq. 15.2.

$$v_g^{ice} = \frac{l}{\frac{l}{v_g^{Al}} + \Delta t} \Rightarrow d_{ice} \tag{15.2}$$

15.2.2 Effect of ice layers on the propagation of ultrasonic-guided wave

The UGW-based ice detection method, as illustrated in Fig. 15.2, leverages the unique properties of an aircraft wing's skin, which is a characteristic thin plate structure that also

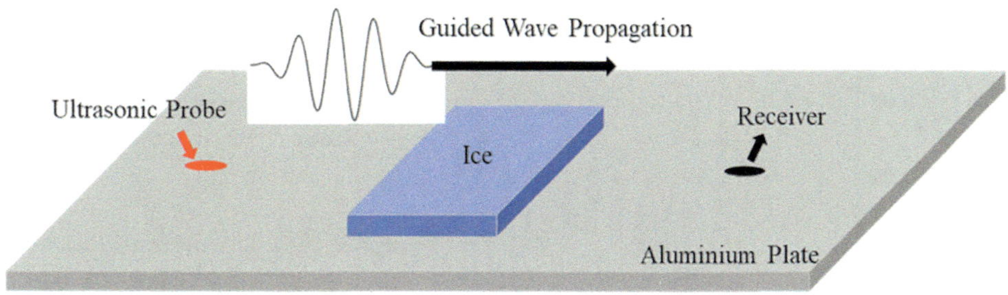

FIGURE 15.2 UGW-based ice detection method. *UGW*, Ultrasonic-guided wave.

functions as a waveguide. In this approach, two piezoelectric transducers are bonded with adhesive at opposite ends of the designated detection area, serving as the actuator and sensor, respectively. The actuator generates the UGW, which propagates through the region of interest, potentially affected by icing conditions. As the UGW traverses the surface, it interacts with any ice accumulation present, leading to changes in its propagation characteristics. These alterations are subsequently captured by the sensor upon the wave's arrival. By analyzing the received signal and comparing it to a baseline measurement taken under ice-free conditions, it becomes possible to accurately detect and quantify the presence of ice on the aircraft wing's surface. This method offers a noninvasive and efficient means of monitoring ice accumulation, contributing to enhanced aviation safety and operational efficiency.

The concept originates from an acousto-ultrasonic approach proposed by Vary in 1988 [40]. Initially, two normal beam probes were positioned in close proximity to each other, and an ultrasonic signature of the test piece was acquired, often revealing the good or bad character of the component under examination. Fig. 15.3 illustrates the influence of an ice layer on UGWs. When the length of the ice layer remains constant, an increase in its thickness results in a reduction in the displacement and a slight delay in the time of flight (ToF) of the received signal (the first wave packet). Conversely, when the thickness of the ice layer remains constant but its length increases, there is a minor decrease in displacement and a delay in the ToF of the received signal (the first wave packet).

15.2.3 Quantitative analysis of Ultrasonic-guided wave mode conversion

The preceding analysis illustrates that the formation of an ice layer alters the dispersion curves, leading to the conversion of UGW modes at the same frequency, which is typically observed as a reduction in group velocity. To gain a comprehensive understanding of the UGW modes and their corresponding group velocities in both icing and ice-free areas, it is imperative to introduce a frequency domain finite element (FDFE) method. This method enables a quantitative analysis of the mode conversion of the UGW, providing valuable insights into the complex wave propagation behavior [41]. The FDFE approach assumes the excitation to be a continuous sinusoidal signal with a single frequency, simplifying the analysis and allowing for a focused examination of specific frequency-mode pairs. Furthermore, by strategically selecting the spatial loading to match the desired UGW

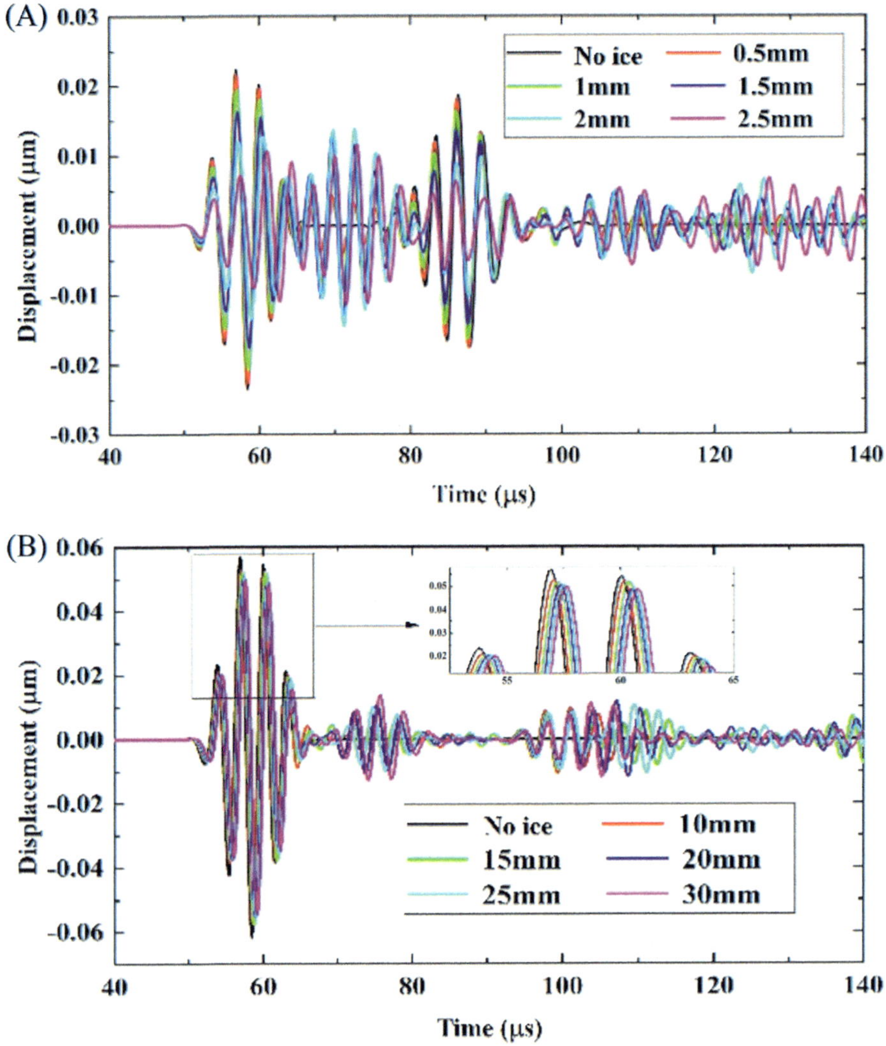

FIGURE 15.3 Effects of the variation in ice layer (A) thickness and (B) length on UGWs. *UGW*, Ultrasonic-guided wave.

mode, any single UGW with a specified frequency-mode pair can be dominantly excited, ensuring a targeted investigation of the wave propagation characteristics. Consequently, the FDFE method enables a detailed and quantitative analysis of UGW propagation and its interaction with the ice coating, effectively minimizing the wave signal complexity that arises from the dispersive nature and multimode features inherent to UGW. By employing this powerful numerical technique, we can gain a deeper understanding of the intricate wave behavior in the presence of ice, paving the way for the development of more accurate and reliable ice detection and monitoring systems.

The FDFE model is built in COMSOL® Multiphysics. The body load distribution in the loading area is related with the wave structure of the selected mode. Stress distribution $(\sigma_{zx}(z), \sigma_{zz}(z))$ and displacement distribution $(u_x(z), u_z(z))$ are obtained from wave structure according to constitutive equation of isotropic material. To dominantly excite the selected UGW mode-frequency pair, the body load distribution in the loading area can be expressed as the multiplication of the stress distribution by the sinusoidal spatial distribution, which is as follows:

$$f_{xx}(x,z) = \sigma_{xx}(z)\exp(i\omega x/c_p),$$
$$f_{zz}(x,z) = \sigma_{xz}(z)\exp(i\omega x/c_p), \tag{15.3}$$

where c_p is phase velocity. The wave structure and phase velocity are calculated based on the selected UGW mode-frequency pair, and are then substituted into the above equations to obtain the selected loading signal acted on the loading area in Fig. 15.4. In summary, a pure UGW with a single frequency and mode is excited and propagates until interacting with the first interface (between the incident area and icing area) and the second interface (between icing area and receiving area). In this process, the received UGW at the receiving area becomes the superposition of multiple modes.

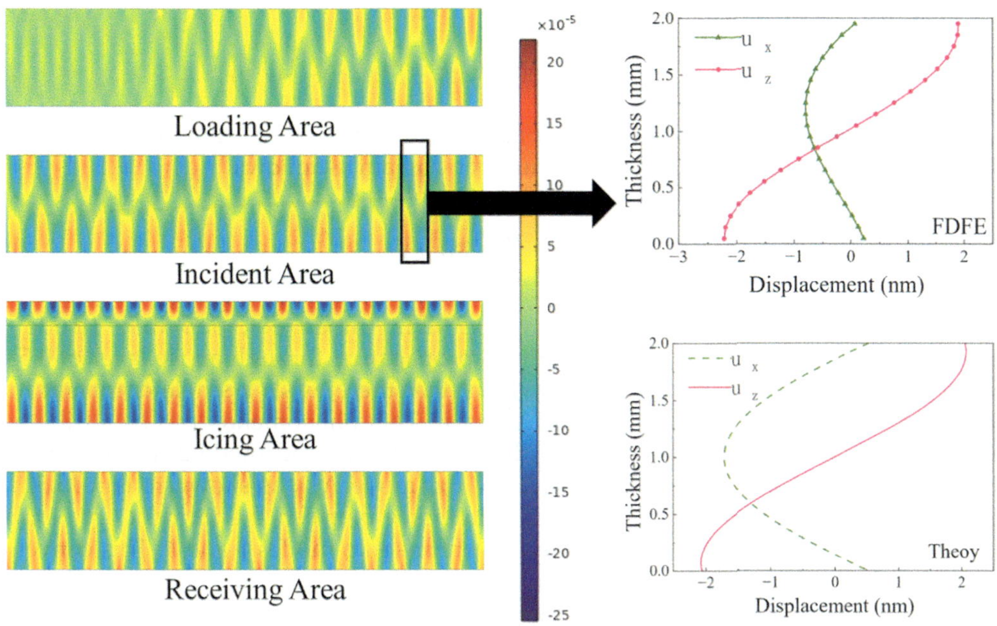

FIGURE 15.4 Displacement contour along the x direction for the UGW (1.2 MHz, mode S0) with ice and the comparison of the wave structure of simulation and theory in the incident area (the $x-z$ ratio in the zoom-in of the contour is adjusted as 1:10 for clear display). *UGW*, Ultrasonic-guided wave.

In the section, the Poynting vector, which represents the power flow density vector of acoustic waves in solids, is introduced to quantitatively analyze the mode conversion process in the icing skin. Poynting vector can be expressed as follows [39]:

$$\mathbf{P} = -\frac{1}{2}\boldsymbol{\nu}^{*}\boldsymbol{\sigma},$$
$$\boldsymbol{\nu} = \begin{bmatrix} v_x & v_y & v_z \end{bmatrix}, \boldsymbol{\sigma} = \begin{bmatrix} \sigma_{xx} & \sigma_{xy} & \sigma_{xz} \\ \sigma_{yx} & \sigma_{yy} & \sigma_{yz} \\ \sigma_{zx} & \sigma_{zy} & \sigma_{zz} \end{bmatrix}, \tag{15.4}$$

where the $\mathbf{v}^*$ is conjugate of the velocity vector, and σ is stress tensor. The calculation steps are as follows:

1. The discrete data of the velocity vector $\mathbf{v}(x, z)$ and stress tensor $\boldsymbol{\sigma}(x, z)$ on the element nodes are extracted from the results of FDFE. The length of extraction range along the x direction is 400 mm for all the incident, icing, and receiving areas to guarantee a consistent comparison of power flow density.
2. Velocity vector $\mathbf{F}_{\mathbf{v}}(k, z)$ and stress tensor matrix $\mathbf{F}_{\boldsymbol{\sigma}}(k, z)$ in the wave number domain are obtained with the fast Fourier transform (FFT) along the x direction.
3. The Poynting vector is calculated according to Eq. (15.4). For the developed 2-D FDFE model, the Y component of the velocity vector is zero, so the equation can be reduced as follows:

$$P(k, z) = -\frac{1}{2}\left(F^{*}_{v_x}(k, z)F_{\sigma_{xx}}(k, z) + F^{*}_{v_z}(k, z)F_{\sigma_{zx}}(k, z)\right). \tag{15.5}$$

4. The components of the Poynting vector along the z direction are summed to obtain the resultant Poynting vector in the wave number domain as follows:

$$P(k) = \sum_{z=0}^{z_{\max}} P(k, z). \tag{15.6}$$

Taking the UGWs with frequencies of 0.5 MHz (mode S0), 0.85 MHz (mode S0), 1.1 MHz (mode A1), and 1.24 MHz (mode A1) propagating through an ice layer of 0.5 mm thickness as illustrated in Fig. 15.5, it is evident that there exists a single dominant mode in all three areas, as depicted in Fig. 15.5A and B. This phenomenon can be attributed to the mode matching between mode S0 in the incident area and mode 2 in the icing area. However, when considering the UGW at 1.1 MHz (mode A1), as shown in Fig. 15.5C, the primary mode conversion occurs at the first interface between the incident area and the icing area, where mode A1 predominantly transforms into modes 3 and 4. This conversion can be ascribed to the absence of any mode in the icing area that matches mode A1, as illustrated in Fig. 15.6. In contrast, for the UGW at 1.24 MHz (mode A1), as presented in Fig. 15.6, the quasi-matching between mode A1 in the incident area and mode 4 in the icing area results in the principal conversion of mode A1 into mode 4 at the first interface, followed by a reconversion into mode A1 at the second interface. Notably, during this

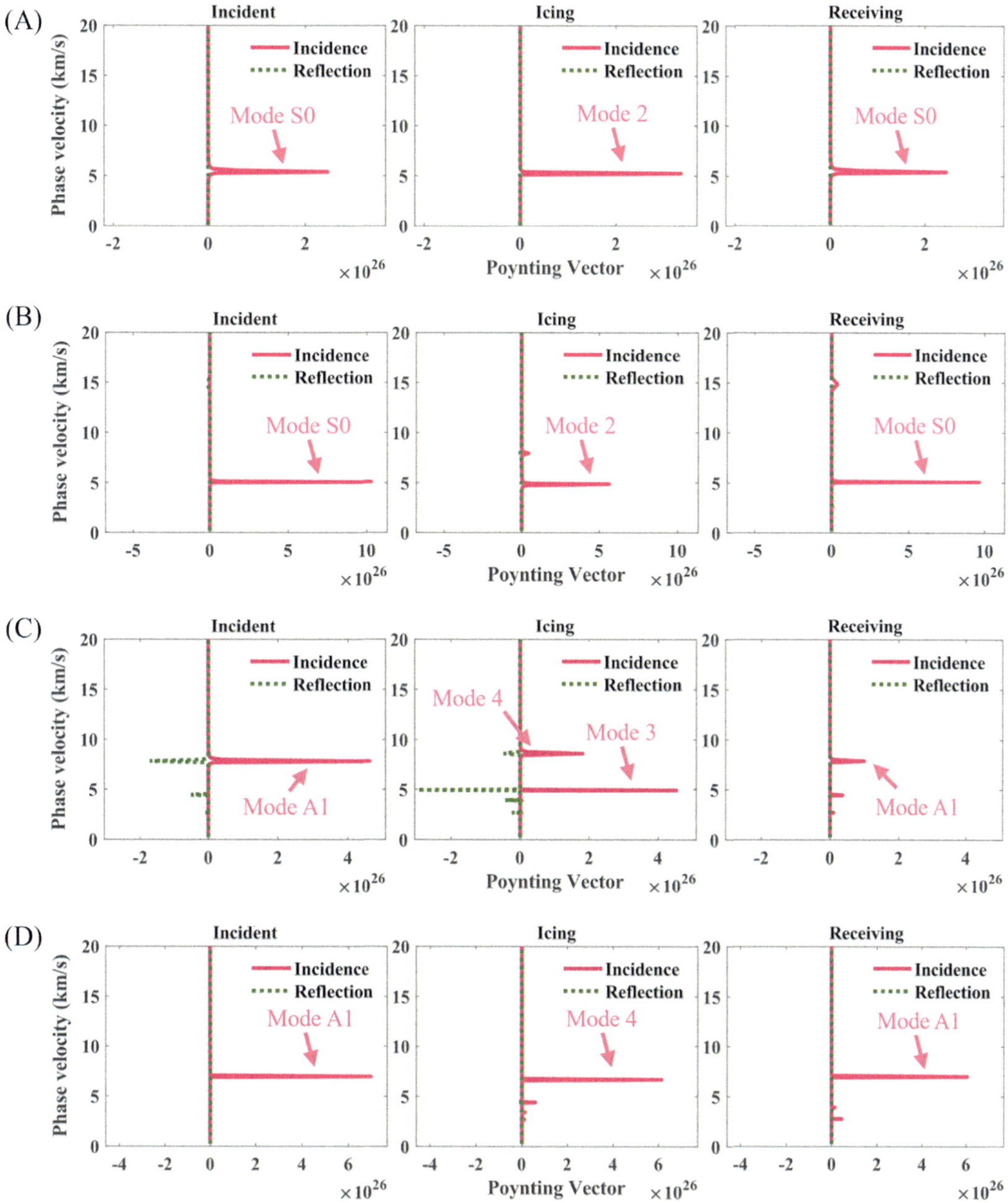

FIGURE 15.5 Poynting vector in three areas for the UGWs (A) (0.5 MHz, mode S0), (B) (0.85 MHz, mode S0), (C) (1.1 MHz, mode A1), and (D) (1.24 MHz, mode A1) with ice (0.5 mm thick). *UGW*, Ultrasonic-guided wave.

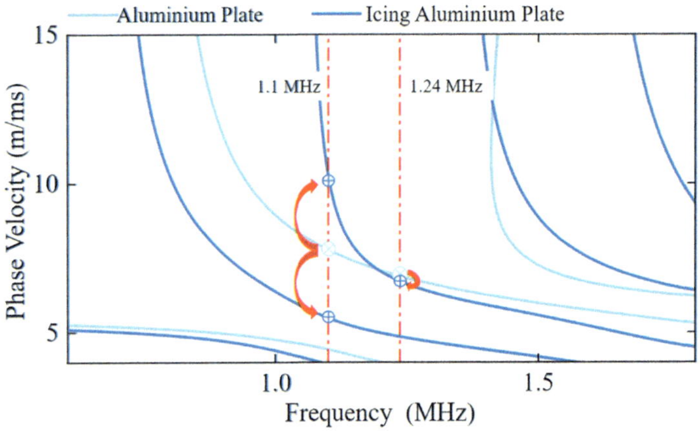

FIGURE 15.6 The phase velocities in skin and icing skin for the UGWs (1.1 MHz, mode A1) and (1.24 MHz, mode A1) with ice. *UGW*, Ultrasonic-guided wave.

process, there is a minimal generation of extra UGW modes apart from mode A1 in the receiving area, suggesting the feasibility of utilizing ToF measurements for the purpose of icing detection.

15.2.4 Ice monitoring on plates

Based on the comprehensive analysis conducted in the preceding section, the optimal combinations of UGW frequencies and modes have been identified, which include (0.5 MHz, mode S0), (0.85 MHz, mode S0), and (1.24 MHz, mode A1). These selected UGWs exhibit stable and pure mode characteristics during propagation, with the energy of the UGWs being relatively concentrated. It is anticipated that these optimal combinations will yield clearer and more reliable sampling signals, thereby enhancing the effectiveness of ice detection. The subsequent step in this research endeavor is to further validate the ice detection efficacy of these frequency and mode combinations through rigorous time domain finite element (TDFE) simulations or meticulously designed experiments.

Regarding the design of the excitation signal, a pulse signal consisting of N cycles modulated by a Hanning window is employed to concentrate the signal energy in the frequency domain, thus mitigating the likelihood of frequency dispersion. However, it is crucial to note that an excessive number of periods can lead to a wide time range for each individual wave packet, resulting in the superposition of wave packets corresponding to multiple modes, which may adversely affect signal analysis. Consequently, it is imperative to judiciously select an appropriate number of cycles based on the distance between the sensor and the receiver, as well as the excitation frequency, to ensure optimal signal quality and minimize potential interference.

In terms of the excitation method, the TDFE approach utilizes multipoint time delay loading applied to the loading area with a spatial loading interval of 1/10 of the UGW wavelength. This technique is employed to generate the selective combination of frequency and mode in the excitation signal, as illustrated in Fig. 15.7. By controlling the spatial and temporal distribution of the loading, the desired UGW characteristics can be efficiently generated, facilitating accurate and reliable ice detection.

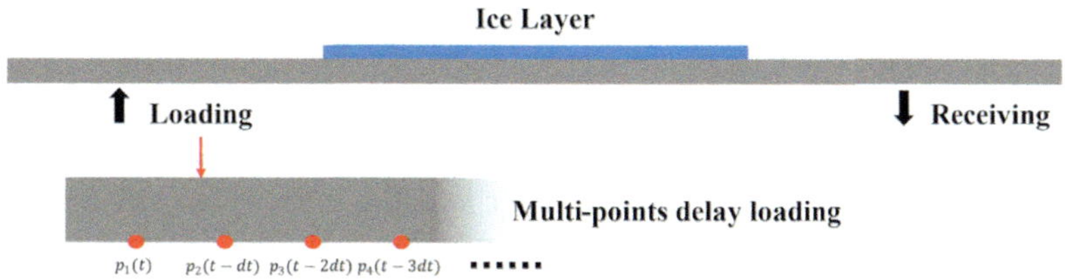

FIGURE 15.7 TDFE model with multipoints delay loading. *TDFE*, Time domain finite element.

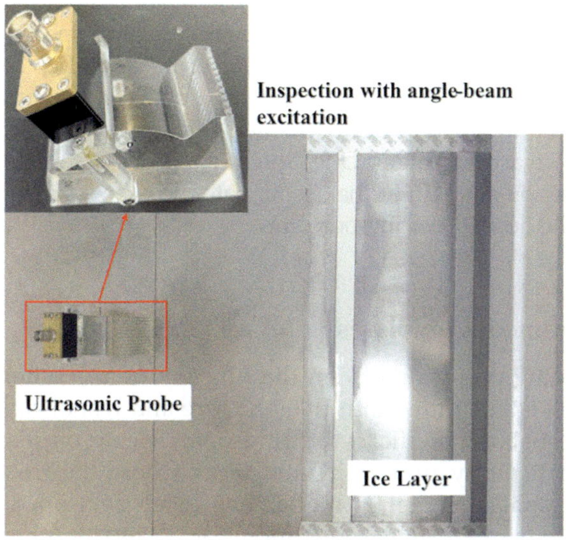

FIGURE 15.8 Experiment setup.

In the experiment, the oblique incidence method was employed to generate a dominant single-mode-guided wave. As illustrated in Fig. 15.8, the ultrasonic transducers were coupled to the flexible angle Plexiglas wedges, which were further coupled to both sides of the icing area using a coupling agent. By adjusting the angle of the incident wave, the phase of the selected UGW mode was aligned with the incident wave. Consequently, the amplitude of the selected UGW mode was enhanced through superposition, while the amplitudes of other modes were offset due to phase mismatch. The angle of incidence was determined using the following equation:

$$\alpha = \arcsin\left(\frac{c_p^{Plexiglass}}{c_p^{Aluminium}}\right), \tag{15.7}$$

where the α, $c_p^{Plexiglass}$ and $c_p^{Aluminium}$ are incidence angle, the velocity of ultrasonic wave in plexiglass, and the phrase velocity of UGW in aluminum, respectively. Similarly, the receiver can also employ the flexible angle Plexiglas wedges to achieve a similar effect to the signal superposition in TDFE.

15.3 Basic theory of ultrasonic ice removal

The fundamental concept behind ultrasonic deicing lies in leveraging the disparities in propagation characteristics, such as wave velocity and energy, between the Lamb waves and shear horizontal (SH) waves in the ice layer and the underlying substrate. These differences lead to distinct displacement and deformation patterns in the ice layer and the substrate. When the shear stress generated at the interface between the ice and the substrate surpasses the tangential adhesion strength of the ice layer, the ice layer detaches from the surface. To illustrate this phenomenon, consider a double-layer plate model consisting of the substrate and the ice layer. When the wave propagation direction aligns with direction 1, the in-plane shear stresses τ_{31} and τ_{32} play a dominant role in the peeling of the ice layer, while the shear stress τ_{21} primarily contributes to the fracturing of the ice layer. Notably, the shear stresses τ_{31} and τ_{32} exhibit zero values at the interface between the substrate and the accumulated ice, whereas the shear stress τ_{21} is distributed throughout the thickness direction of both the substrate and the ice accumulation. Once the in-plane shear stress at the ice layer interface exceeds the shear adhesion strength of the accumulated ice, the ice detaches from the substrate, effectively achieving the desired deicing outcome.

15.3.1 Ice removal model on plates

The ultrasonic deicing process relies on the function of an exciter within the ultrasonic deicing system, which is integrated into the structure and responsible for producing ultrasonic waves. To investigate the efficacy of ultrasonic deicing on plate-like structures with finite boundaries, particularly under circumstances where the ice detachment is incomplete, a three-dimensional geometric model has been developed and is presented in Fig. 15.9. This model serves as a foundation for analyzing the complex interactions between the ultrasonic waves, the structure, and the ice layer, enabling researchers to gain valuable insights into the factors that influence the effectiveness of the ultrasonic deicing process and to optimize the design of ultrasonic deicing systems for various applications.

The properties of the aluminum plate and ice materials used in the model are shown in Table 15.1.

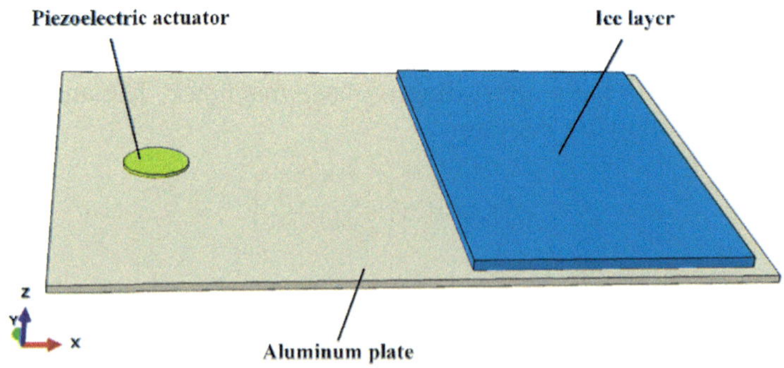

FIGURE 15.9 Three-dimensional geometric model [42].

TABLE 15.1 The material properties [42].

Material	Density (kg/m^3)	Young's modulus (GPa)	Poisson's ratio
Ice	920	9	0.28
Aluminum	2700	70	0.30

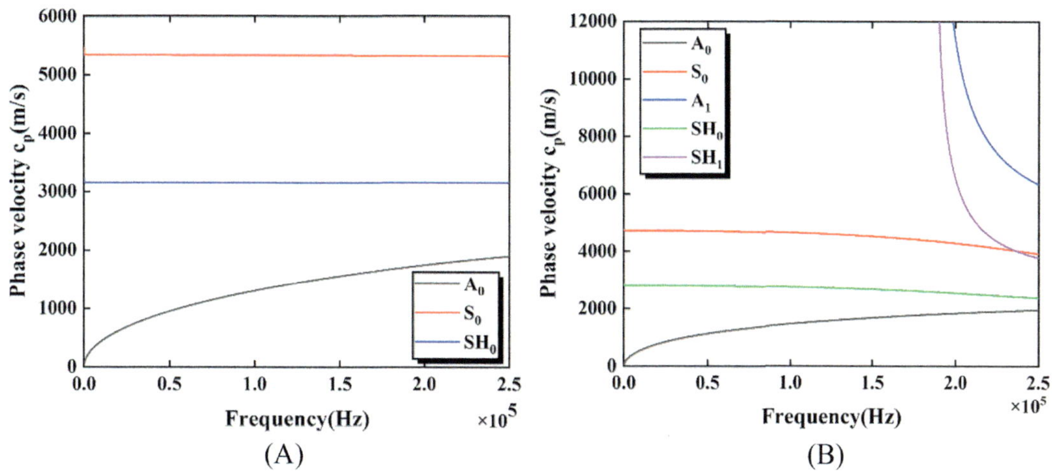

FIGURE 15.10 Phase velocity of the aluminum plate and ice-plate layered structure (A) 2 mm aluminum plate; (B) 2 mm aluminum plate with 3 mm ice [42].

The GMM [43] can be effectively employed to resolve the dispersion curves of the ice-plate layered structure that forms when ice accumulates on the surface of the plate. By solving the dispersion curve equations for both Lamb waves and SH waves, it is possible to accurately calculate the phase velocities of these wave types. The application of this method is demonstrated in Fig. 15.10, which presents the results obtained for two distinct cases: a 2 mm aluminum plate and a 2 mm aluminum plate with an additional 3 mm ice layer on its surface. The utilization of the GMM in this context provides a robust and reliable approach for determining the dispersive behavior of guided waves in such layered structures, thereby facilitating a deeper understanding of the impact of ice accretion on the propagation characteristics of these waves in plate-like structures.

An analysis was performed to evaluate the shear stress present at the ice-substrate interface across various icing scenarios, ensuring an equivalent interfacial area for direct comparison. The results of this analysis are depicted in Fig. 15.11, which elucidates the significant variations in interfacial shear stress magnitudes corresponding to distinct icing conditions. Notably, the first icing condition exhibited the highest magnitude of delamination shear stress at the ice interface, suggesting a heightened propensity for interfacial failure and ice detachment under these specific circumstances. Conversely, the second icing condition demonstrated the lowest interfacial shear stress values among the assessed scenarios. This observation implies an enhanced interfacial adhesion and a reduced likelihood

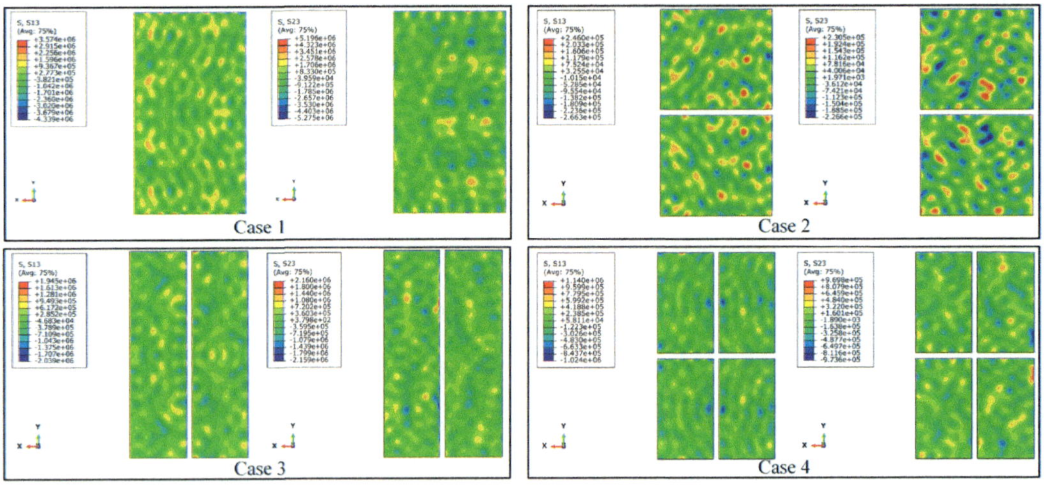

FIGURE 15.11 Ice-substrate interface shear stress nephogram for four different icing cases [42].

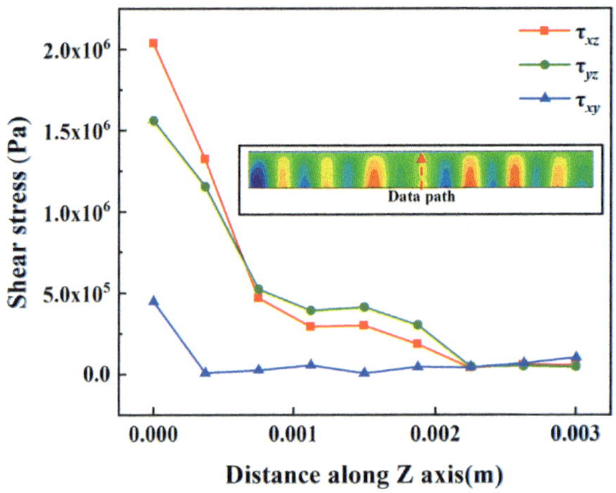

FIGURE 15.12 Shear stress amplitude varies along the Z axis [42].

of ice delamination when the second icing condition prevails. The divergent behaviors of the interfacial shear stress across different icing conditions underscore the complex interplay between the icing parameters and the resulting ice adhesion characteristics. These findings provide valuable insights for the development of targeted strategies to mitigate ice accretion and promote efficient ice removal in practical applications [42].

To further investigate the shear stress distribution within the ice, 15 paths were defined along the ice thickness direction. The resulting shear stress distributions along these paths are presented in Fig. 15.12. The analysis results exhibit good agreement with guided wave propagation theory, validating the approach used in this study. These findings provide valuable insights into the through-thickness variation of shear stress in ice, which is crucial for understanding the mechanical behavior and failure mechanisms of ice under dynamic loading conditions.

15.3.2 Ice removal model on coatings

Fig. 15.13 presents a schematic diagram of a generalized three-layer plate configuration consisting of a substrate or host plate, a coating layer, and an accreted ice layer. The present study focuses on investigating the interfacial shear stress developed between the ice layer and the coating, which plays a critical role in the deicing process. Understanding the magnitude and distribution of this interfacial shear stress is essential for designing effective ice protection systems and predicting the adhesion strength of ice to the surface. The three-layer model serves as a foundation for analyzing the complex interactions between the ice, coating, and substrate under various environmental and loading conditions typically encountered in icing scenarios.

To estimate the ability of inducing stress at the interface from a given amount of power, the concept of interfacial stress concentration coefficients (ISCC) is proposed to calculate the normalized interface shear stress for different UGW modes and frequencies [30,44,45]. The larger the ISCC, the larger the modal shear stress at the interface will be induced for a given power, and vice versa. The complex acoustic Poynting's vector is defined in Eq. 15.4.

To quantify the capability of inducing stress at the interface from a given input power, we cited the concept of ISCC. These coefficients are used to calculate the normalized interface shear stress for various UGW modes and frequencies [30]. The ISCC serves as an indicator of the modal shear stress magnitude at the interface for a fixed power input; a higher ISCC value corresponds to a larger induced shear stress, and conversely, a lower ISCC value indicates a smaller induced shear stress [45]. The complex acoustic Poynting vector, which is essential for determining the power flow, is defined in Eq. 15.4.

The total power flow can be obtained by integrating the x_1 component p_{x1} in **P** across the entire thickness of a multilayered plate and is expressed as follows:

$$\text{Power} = \int_{Thickness} p_{x1} dx_3 \tag{15.8}$$

The ISCC values for SH and Lamb waves are defined in Eqs. (15.9) and (15.10) [45],

$$ISCC_S = \frac{\sigma_{32}\Big|_{Layer\ interface}}{\sqrt{Power}} \tag{15.9}$$

$$ISCC_L = \frac{\sigma_{31}\Big|_{Layer\ interface}}{\sqrt{Power}} \tag{15.10}$$

where the subscripts "*S*" and "*L*" indicate SH and Lamb waves.

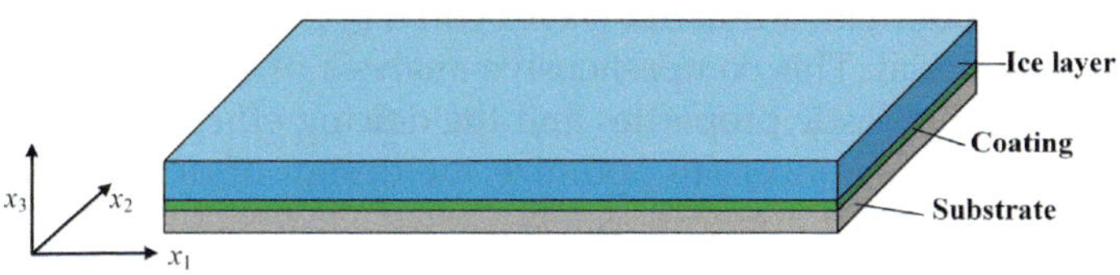

FIGURE 15.13 Schematic of a general three-layer plate model [44].

TABLE 15.2 Material properties [44].

Layer	Density (kg/m^3)	Young's modulus (GPa)	Poisson's ratio	Thickness (mm)
Ice	920	9.1	0.28	4
Coating	900	0.5	0.30	1
Aluminum	2700	70	0.27	2

To validate the accuracy of the GMM in computing dispersion curves and wave structures, a critical prerequisite for investigating ultrasonic deicing, a three-layered plate model was employed. This model consisted of a host plate, a coating layer, and an ice layer. The material properties of each layer were carefully selected and are presented in Table 15.2. By applying the GMM to this layered structure, the computed dispersion curves and wave structures were compared against well-established analytical or numerical methods to assess the model's precision and reliability.

The ISCC value at a given frequency for a particular mode represents the efficacy of ultrasonic deicing. To better understand the factors influencing deicing performance, this section presents a parametric study investigating the effects of coating properties and ice layer thickness on the ISCC. For ease of comparison and analysis, the ISCC values are superimposed on dispersion curves, as depicted in Fig. 15.14. This graphical representation allows for a clear visualization of the relationships between the ISCC, frequency, and the studied parameters, facilitating a comprehensive assessment of the ultrasonic deicing system's performance under various conditions.

Fig. 15.14 presents an illustrative example of the ISCC values superimposed on the dispersion curves, specifically the SH and Lamb waves, corresponding to three distinct Young's modulus of the coating material. The results demonstrate a clear trend: as Young's modulus of the coating increases, the number of modes for both SH and Lamb waves decreases within the considered frequency range. This effect is particularly pronounced for Lamb waves, where the mode count exhibits a dramatic increase with decreasing Young's modulus. Interestingly, while the maximum ISCC value for Lamb waves shows a consistent decrease with increasing Young's modulus, the SH waves do not display a regular pattern in this regard. Moreover, the optimal mode associated with the maximum ISCC shifts from a low-order mode to a high-order mode as Young's modulus of the coating decreases.

A key finding of this study is that, within the considered frequency range and for a given Young's modulus, the maximum ISCC value observed in Lamb waves is consistently higher than that in SH waves. This suggests that Lamb waves may be more effective in achieving deicing compared to SH waves. To further investigate this phenomenon, Fig. 15.15 summarizes the maximum ISCC values for both Lamb and SH waves corresponding to 18 different Young's modulus of the coating material. This comprehensive analysis provides insights into the relationship between the coating's elastic properties and the deicing efficiency of the two wave types, enabling researchers and engineers to optimize the design of deicing systems based on the specific requirements of their applications [44].

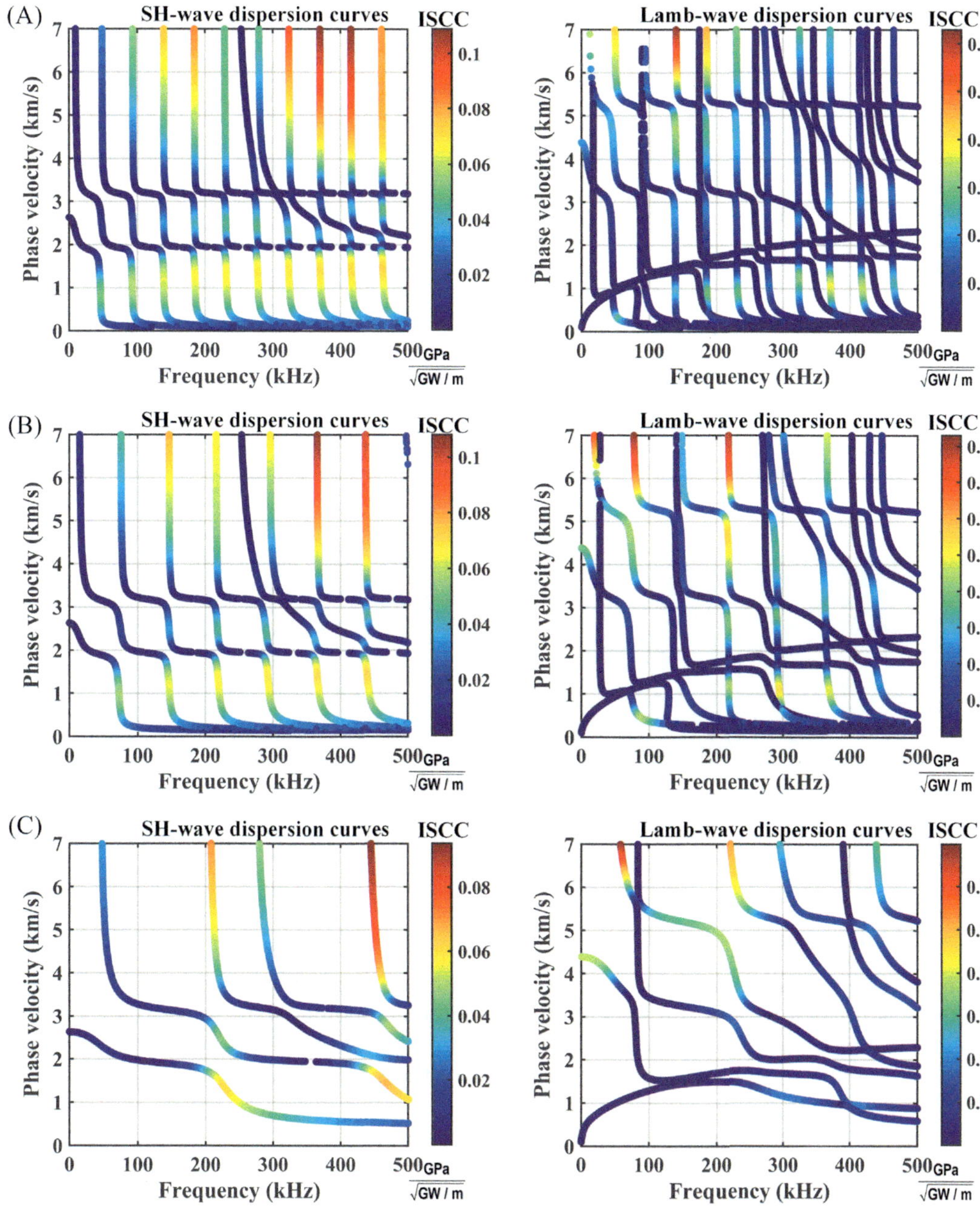

FIGURE 15.14 Dispersion curves and ISCC values of SH waves (left) and Lamb waves (right) when the young's modulus of the coating is different (A) 20 MPa; (B) 50 MPa; (C) 500 MPa [44]. *ISCC*, Interfacial stress concentration coefficients; *SH*, shear horizontal.

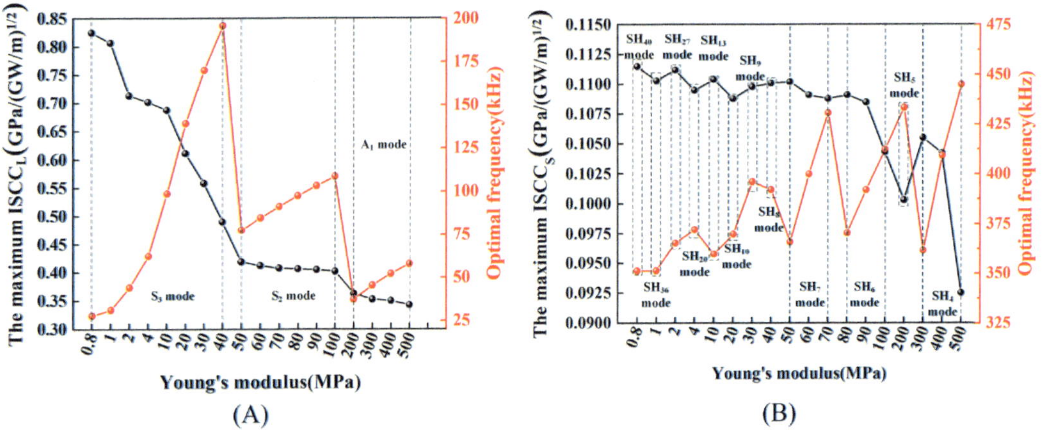

FIGURE 15.15 Effect of Young's modulus of coating on maximum ISCC value and optimal frequency (A) Lamb waves; (B) SH waves [44]. *ISCC*, Interfacial stress concentration coefficients; *SH*, shear horizontal.

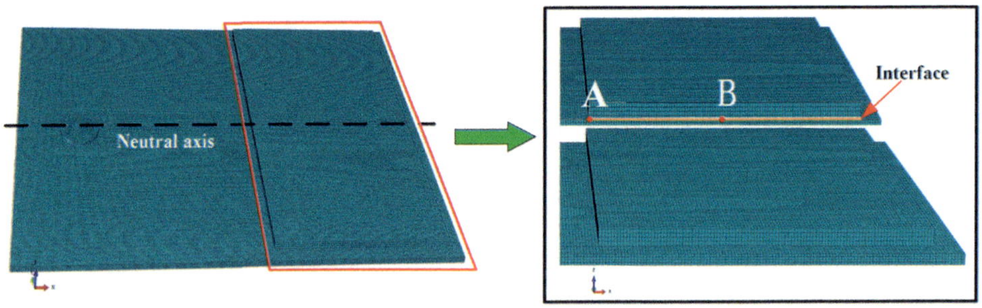

FIGURE 15.16 The finite element (FE) mesh and reference points (A and B) of the deicing model [42].

15.3.3 Deicing energy efficiency

To investigate the deicing energy efficiency on an aluminum substrate, a finite element model was developed and implemented. Two reference points were selected for detailed analysis in this study: point A, located at the lower edge of the ice layer, and point B, situated at the center of the ice layer. The positions of these reference points are illustrated in Fig. 15.16.

Harmonic analyses were conducted on point A and point B, located respectively at the edge and at the center of the interface between the substrate and the ice layer. This analysis aimed to identify the optimal frequency for deicing by examining the vibrational response of the system. Independent of their locations, whether at point A or point B, the process of ice detachment is predominantly governed by the in-plane shear stresses, τ_{13} and τ_{23}.

In the context of ultrasonic deicing technology, the mechanism by which ice is detached from surfaces predominantly involves the in-plane shear stresses, τ_{13} and τ_{23}.

These stresses play a critical role in the ice-peeling process. The cumulative in-plane shear stress responsible for peeling can be Eq. 15.11:

$$\omega = \sqrt{\tau_{13}{}^2 + \tau_{23}{}^2}, \tag{15.11}$$

Besides deicing shear stresses analysis, energy consumed for the deicing system is another important aspect needed to take into consideration. For simplicity, the energy dissipation and heat loss were ignored; only the kinetic energy and strain energy for the whole system were considered to estimate the energy consumed for each model at optimal frequency. Therefore, the energy density for deicing can be written as Eq. 15.12:

$$\omega = \frac{E_k + U}{V}, \tag{15.12}$$

where ω, E_k, U, and V represent energy density, kinetic energy, strain energy, and volume of the system, respectively, for deicing. From these data, the energy density needed to detach ice from the host plate for each model can be estimated. From the computation results, the conclusion can be drawn that the energy density needed for deicing varies from the radius of the piezoelectric wafer with the system at the optimal frequency for each configuration. Meanwhile, the deicing effect also differs from one to another. Therefore, it is difficult to distinguish which model is better. The ideal goal is to achieve deicing with minimum energy consumption. Therefore, a coefficient called the deicing efficiency factor (DEF) was proposed to describe the relationship between the deicing effect and deicing energy density. DEF was defined as Eq. (15.13) [42]:

$$\mathrm{DEF} = \frac{\tau_{total}}{\omega} = \frac{\sqrt{\tau_{13}{}^2 + \tau_{23}{}^2}}{\frac{E_k + U}{V}} = \frac{\sqrt{\tau_{13}{}^2 + \tau_{23}{}^2} \cdot V}{E_k + U}, \tag{15.13}$$

where τ_{total}, τ_{13}, τ_{23}, ω, E_k, U, and V have the same meaning as mentioned prior. This coefficient can describe the deicing efficiency for each model. The greater this coefficient is, the greater total shear stress obtained at the same energy density, indicating that more energy is applied to detach ice, which represents the system as more efficient. From the above data and Eq. (15.13), DEF for each configuration can be calculated.

15.4 Test and verification

15.4.1 Icing wind tunnel

The ice detection and removal experiments were conducted on the surface of the wing structure in the YBF-05 icing wind tunnel. The YBF-05 is a closed-circuit wind tunnel facility that consists of several key components, including the tunnel body, power system, refrigeration system, water spraying system, water treatment system, compressed air system for spray atomization, altitude simulation system, and integrated measurement and control system. These subsystems work together to create a controlled environment that simulates various icing conditions encountered during aircraft flight, enabling researchers to study ice accretion processes and evaluate the performance of ice detection and removal technologies on the wing surface under representative operating conditions (Fig. 15.17).

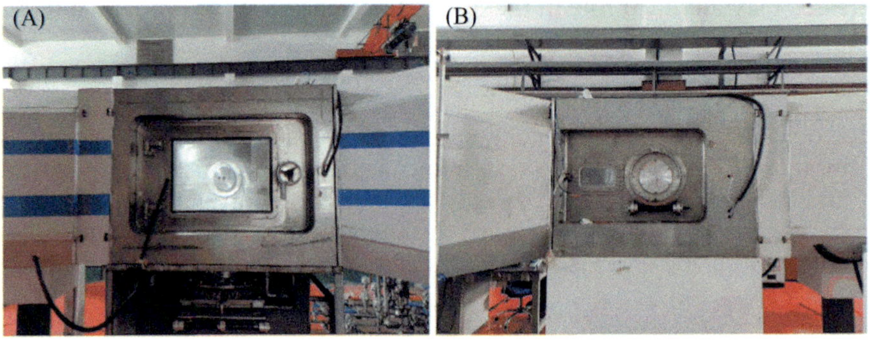

FIGURE 15.17 Test section of YBF-05 icing tunnel (A) the front side; (B) the back side.

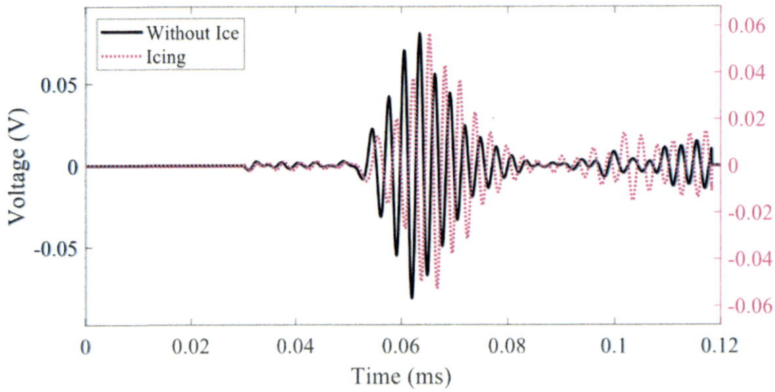

FIGURE 15.18 The sampling signal without ice and with ice.

15.4.2 Ice monitoring

For the purpose of experimental verification, a frequency-mode point exhibiting high sensitivity to the presence of an ice layer was selected to facilitate an accurate evaluation of the ice layer dimensions. Fig. 15.18 presents a comparison between the sampling signals obtained from a carbon fiber reinforced polymer (CFRP) laminate in two distinct scenarios: one without an ice layer and another with an ice layer measuring 50 mm in length and 1.17 mm in thickness. Fig. 15.19 illustrates the time-frequency domain distribution of the sampling signal, from which the frequency-ToF curves were derived. This approach enables a comprehensive analysis of the ice layer's influence on the propagation characteristics of the guided waves within the CFRP laminate, thereby providing a robust foundation for the quantitative assessment of ice accumulation on the material's surface [46].

From Fig. 15.18, it can be found that the experimental signals for the case with ice (shown by the red dashed line) have a slight time delay compared to the experimental signal for the case without ice (shown by the black solid line), resulting in a lower slope in the frequency-ToF curves.

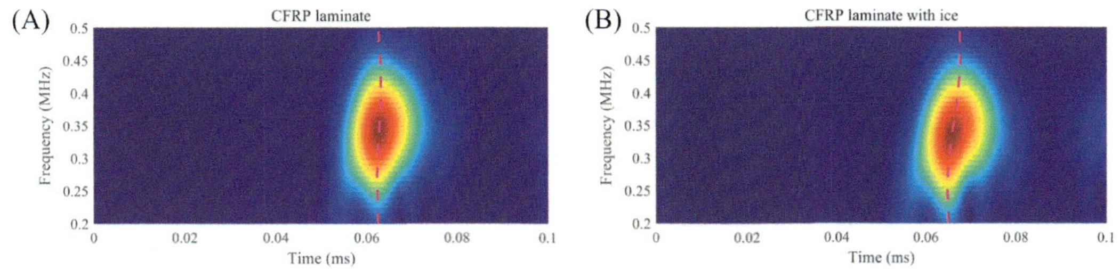

FIGURE 15.19 The distribution in the time-frequency domain of the sampling signal (A) without ice and (B) with ice.

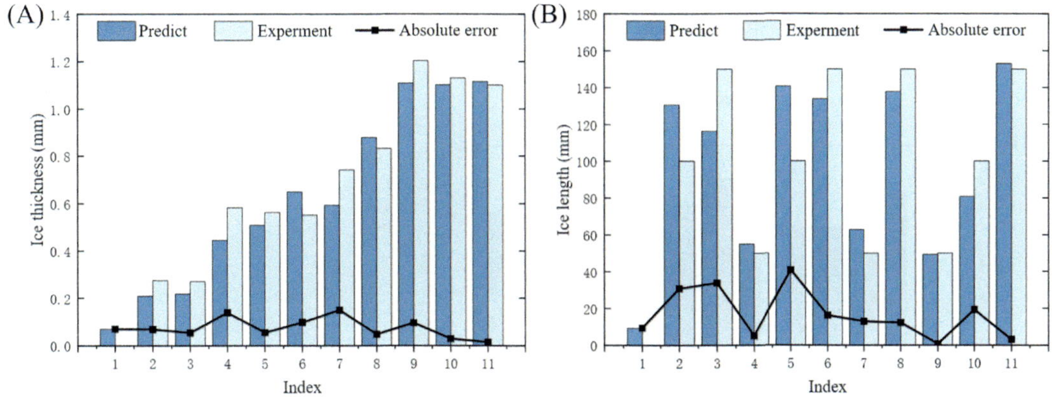

FIGURE 15.20 Predicted results of (A) ice thickness and (B) ice length.

The prediction of ice size was obtained by applying an artificial neural network (ANN) model, with the results presented in Fig. 15.20. The analysis demonstrates that increasing the size of the ice particles used in the training data set improves both the accuracy and stability of the predictions generated by the ANN model. These findings suggest that the performance of ANN models for this application is sensitive to the size distribution of ice particles represented in the training data, with larger sizes correlating with enhanced predictive capabilities. Further research is needed to characterize the optimal range of ice sizes for maximizing the effectiveness of ANN-based prediction in this domain.

15.4.3 Ice removal

deicing experiments were conducted in an icing wind tunnel to investigate the performance of wing surface structures under various temperature and wind speed conditions. The deicing performance results are presented in Fig. 15.21, while comparative analyses of

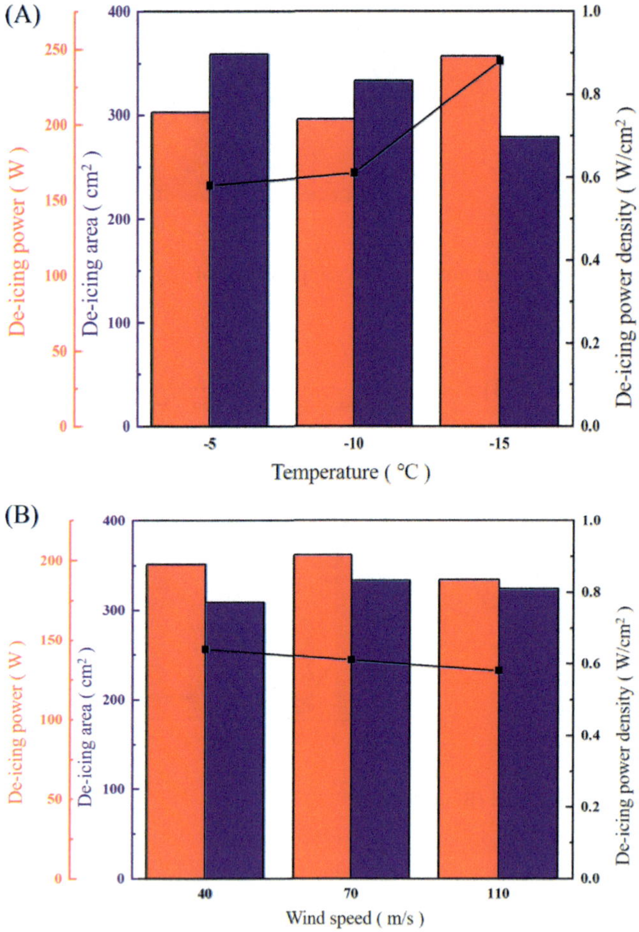

FIGURE 15.21 Deicing result comparisons of (A) temperature and (B) wind speeds.

the wing surface before and after deicing at different temperatures and wind speeds are illustrated in Figs. 15.22 and 15.23. The experimental results indicate that as the temperature decreases, the actual deiced area decreases, and the deicing power density increases. This phenomenon can be primarily attributed to the increased ice adhesion strength at lower temperatures, which consequently leads to higher deicing power consumption. The influence of wind speed on deicing power and the deiced area is relatively minor, with the deicing power density slightly decreasing (around 0.61 W/cm) as the wind speed increases. This suggests that wind speed does not play a decisive role in the ultrasonic deicing process on wing surfaces; however, higher wind speeds may have a certain mechanical promoting effect on the deicing process.

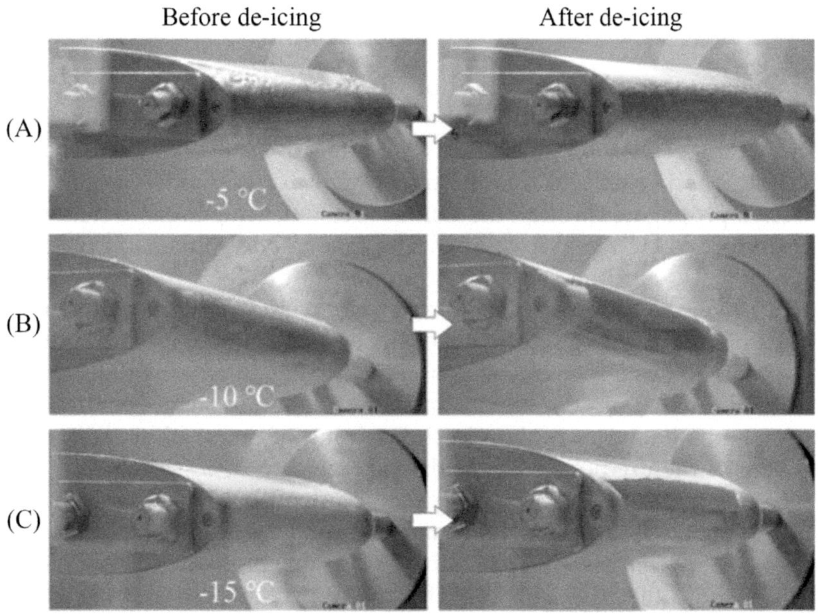

FIGURE 15.22 Comparison of wing structure before and after deicing at different temperatures: (A) $-5°C$; (B) $10°C$; (C) $-15°C$.

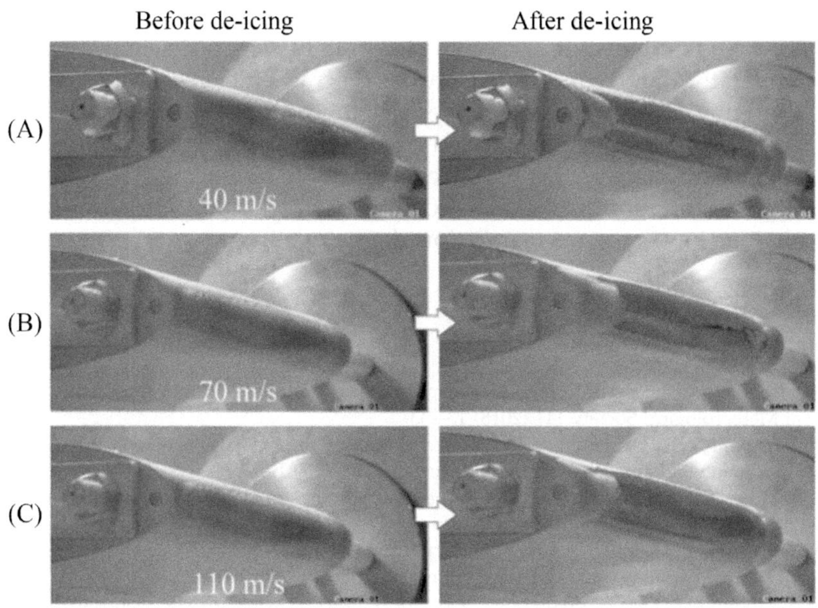

FIGURE 15.23 Comparison of wing structure before and after deicing at different wind speeds: (A) 40 m/s; (B) 70 m/s; (C) 110 m/s.

15.5 Concluding remarks

This chapter describes the application of UGWs for ice detection and deicing technology. In terms of detection, it focuses on the effect of ice accumulation on the propagation of UGWs and the method of guided-wave mode conversion analysis based on FDFE, based on which it is possible to quantitatively analyze the mode conversion that occurs when the UGWs propagate through the nonice area and icing area, to determine the propagation velocity of the UGWs in each area. The pure mode conversion process can reduce the signal complexity, and the selection of appropriate UGWs frequency mode pairs can effectively improve the detection accuracy. Moreover, in the experimental study, the frequency-ToF curve of the guided wave signal is obtained by introducing the time-frequency domain analysis method, which is able to retain as much effective information as possible while reducing the data dimension. The excellent prediction results are obtained from a small amount of data in the training of the machine learning model, and the relative error of the prediction size of the ice layer can be less than 5%.

In terms of deicing, The DEF was proposed to quantitatively characterize the deicing efficiency of the plate structure for the limited boundary and incomplete debonding, then the relationship between steady-state stress and deicing efficiency was studied. The research results show that the DEF can effectively evaluate the deicing efficiency of the system and realize the analysis and prediction of the deicing efficiency of different working conditions and deicing parameters. The ultrasonic deicing of a three-layer plate structure composed of ice-coating- substrate was studied, and then ISCC were introduced to quantitatively evaluate the deicing ability of Lamb wave and SH wave in different modes. Based on these, the effects of coating Young's modulus on ultrasonic deicing were revealed. The wind tunnel tests were carried out to verify deicing efficiency, and the effects of temperature and wind speed on the deicing efficiency were given. The results show that the wing structure has a higher deicing efficiency in clear ice, and the wind speed has a certain mechanical promotion effect on deicing. The uneven icing caused by the attack angle will increase the power consumption of deicing. Ultrasonic deicing can reduce power consumption above 56% compared to traditional electro-thermal deicing, which has obvious low energy consumption characteristics.

References

[1] E.M. Samuelsen, Ship-icing prediction methods applied in operational weather forecasting, Quarterly Journal of the Royal Meteorological Society 144 (2018) 13–33.

[2] A.R. Dehghani-Sanij, S.R. Dehghani, G.F. Naterer, et al., Marine icing phenomena on vessels and offshore structures: Prediction and analysis, Ocean Engineering 143 (2017) 1–23.

[3] Y. Cao, W. Tan, Z. Wu, Aircraft icing: An ongoing threat to aviation safety, Aerospace Science and Technology 75 (2018) 353–385.

[4] S. Nadeau, F. Morency, De-icing of aircraft: incorporating business risks and occupational health and safety, International Journal of Safety and Security Engineering 7 (2017) 247–266.

[5] O. Yirtici, S. Ozgen, I.H. Tuncer, Predictions of ice formations on wind turbine blades and power production losses due to icing, Wind Energy 22 (2019) 945–958.

[6] J.Y. Jin, M.S. Virk, Study of ice accretion and icing effects on aerodynamic characteristics of DU96 wind turbine blade profile, Cold Regions Science and Technology 160 (2019) 119–127.

[7] K. Wei, Y. Yang, H. Zuo, et al., A review on ice detection technology and ice elimination technology for wind turbine, Wind Energy 23 (2020) 433–457.
[8] O. Fakorede, Z. Feger, H. Ibrahim, et al., Ice protection systems for wind turbines in cold climate: characteristics, comparisons and analysis, Renewable and Sustainable Energy Reviews 65 (2016) 662–675.
[9] Z. Goraj, An overview of the deicing and anti-icing technologies with prospects for the future. In: 24th international congress of the aeronautical sciences. 2004.
[10] Y. Cao, Z. Wu, Y. Su, et al., Aircraft flight characteristics in icing conditions, Progress in Aerospace Sciences 74 (2015) 62–80.
[11] E. Esmaeilifar, L. Prince Raj, R.S. Myong, Computational simulation of aircraft electrothermal de-icing using an unsteady formulation of phase change and runback water in a unified framework, Aerospace Science and Technology 130 (2022) 107936.
[12] R.-Z. Szasz, M. Ronnfors, J. Revstedt, Influence of ice accretion on the noise generated by an airfoil section, International Journal of Heat and Fluid Flow 62 (2016) 83–92.
[13] S.A. Bagherzadeh, D. Asadi, Detection of the ice assertion on aircraft using empirical mode decomposition enhanced by multi-objective optimization, Mechanical Systems and Signal Processing 88 (2017) 9–24.
[14] W. Stocksreiter, H. Zangl, Compact multifunctional wireless capacitive sensor system and its application in icing detection, Sensors 19 (2019) 5562.
[15] D. Zheng, Z. Li, Z. Du, et al., Design of capacitance and impedance dual-parameters planar electrode sensor for thin ice detection of aircraft wings, IEEE Sensors Journal 22 (2022) 11006–11015.
[16] I. Aicardi, A. Lingua, L. Mazzara, et al., Ice detection on airlpane wings using a photogrammetric point cloud: a simulation, in: Int Arch Photogramm Remote Sens Spatial Inf Sci 2020; XLIII-B2-2020: 183–189.
[17] C.Q. Gómez Muñoz, F.P. Garcia Marquez, J.M. Sanchez Tomas, Ice detection using thermal infrared radiometry on wind turbine blades, Measurement 93 (2016) 157–163.
[18] S. Gantasala, J.-C. Luneno, J.-O. Aidanpaa, Detection of ice mass based on the natural frequencies of wind turbine blade, Wind Energy Science Discussions (2016) 1–17.
[19] R.U. Tariq, M. Ye, X.-L. Zhao, et al., Microwave sensor for detection of ice accretion on base station antenna radome, IEEE Sensors Journal 21 (2021) 18733–18741.
[20] Y. Liu, L.J. Bond, H. Hu, Ultrasonic-attenuation-based technique for ice characterization pertinent to aircraft icing phenomena, AIAA Journal 55 (2017) 1602–1609.
[21] J. Rao, L. Zeng, M. Liu, et al., Ultrasonic defect detection of high-density polyethylene pipe materials using FIR filtering and block-wise singular value decomposition, Ultrasonics 134 (2023) 107088.
[22] H. Liu, M. Liu, L. Zhang, et al., Directed acoustic shearography for crack detection around fastener holes in aluminum plates, NDT & E International 100 (2018) 124–131.
[23] Y. Shen, V. Giurgiutiu, WaveFormRevealer: An analytical framework and predictive tool for the simulation of multi-modal guided wave propagation and interaction with damage, Structural Health Monitoring 13 (2014) 491–511.
[24] Y. Wang, Y. Xu, Q. Huang, Progress on ultrasonic guided waves de-icing techniques in improving aviation energy efficiency, Renewable and Sustainable Energy Reviews 79 (2017) 638–645.
[25] O. Parent, A. Ilinca, Anti-icing and de-icing techniques for wind turbines: Critical review, Cold Regions Science and Technology 65 (2011) 88–96.
[26] E. Möhle, M.C. Haupt, P. Horst, Coupled numerical simulation and experimental validation of the electroimpulse de-icing process, Journal of Aircraft 50 (2013) 96–102.
[27] M. Budinger, V. Pommier-Budinger, G. Napias, et al., Ultrasonic ice protection systems: analytical and numerical models for architecture tradeoff, Journal of Aircraft 53 (2016) 680–690.
[28] Z. Wang, Recent progress on ultrasonic de-icing technique used for wind power generation, high-voltage transmission line and aircraft, Energy and Buildings 140 (2017) 42–49.
[29] H. Tan, G. Xu, T. Tao, et al., Experimental investigation on the defrosting performance of a finned-tube evaporator using intermittent ultrasonic vibration, Applied Energy 158 (2015) 220–232.
[30] H. Gao, J.L. Rose, Ice detection and classification on an aircraft wing with ultrasonic shear horizontal guided waves, IEEE Trans Ultrason, Ferroelect, Freq Contr 56 (2009) 334–344.
[31] S. Gupta, S. Sutrave, Scattering analysis of glaze ice accretion on CFRP laminated composite plate structures using ultrasonic lamb waves: towards aviation safety, Journal of Nondestructive Evaluation 43 (2023) 13.

[32] V. Memmolo, J. Moll, Investigation on guided waves propagating across ice layers, in: Fromme P, Su Z (eds) Health Monitoring of Structural and Biological Systems IX. Online Only, United States: SPIE, p. 126.

[33] J. Palacios, E. Smith, J. Rose, et al., Instantaneous de-icing of freezer ice via ultrasonic actuation, AIAA Journal 49 (2011) 1158–1167.

[34] R.A. Seppings, Investigation of ice removal from cooled metal surfaces. Imperial College London (University of London), 2006.

[35] Y. Zhu, Structural tailoring and actuation studies for low power ultrasonic de-icing of aluminum and composite plates, in: The Pennsylvania State University, 2010.

[36] A. Mlyniec, Ł. Ambroziński, P. Packo, et al., Adaptive de-icing system – numerical simulations and laboratory experimental validation, International Journal of Applied Electromagnetics and Mechanics 46 (2014) 997–1008.

[37] J.L. Palacios, E.C. Smith, H. Gao, et al., Ultrasonic shear wave anti-icing system for helicopter rotor blades, Annual Forum Proceedings-American Helicopter Society, American Helicopter Society, Inc, 2006, p. 1492.

[38] E. Monaco, F. Ricci, V. Memmolo, et al., Development of a de-icing system for aerodynamic surfaces based on ultrasonic waves, in: Kundu T (ed) Health Monitoring of Structural and Biological Systems XII. Denver, United States: SPIE, p. 16.

[39] J.L. Rose, Ultrasonic Guided Waves in Solid Media, Cambridge University Press, New York, 2014.

[40] A. Vary, The acousto-ultrasonic approach, Acousto-Ultrasonics: Theory and Application, 1988, pp. 1–21.

[41] Y. Tian, Q. Wang, M. Liu, et al., Robust and quantitative characterization of aircraft icing with mode and frequency selective ultrasonic guided wave, Ultrasonics 127 (2023) 106846.

[42] Z. Shi, Z. Kang, Q. Xie, et al., Ultrasonic deicing efficiency prediction and validation for a flat deicing system, Applied Sciences 10 (2020) 6640.

[43] V. Giurgiutiu, Structural health monitoring (SHM) of aerospace composites, Polymer Composites in the Aerospace Industry, Elsevier, 2020, pp. 491–558.

[44] Z. Shi, Y. Zhao, C. Ma, et al., Parametric study of ultrasonic de-icing method on a plate with coating, Coatings 10 (2020) 631.

[45] J. Palacios, Y. Zhu, E. Smith, et al., Ultrasonic shear and lamb wave interface stress for helicopter rotor de-icing purposes, in: 47th AIAA/ASME/ASCE/AHS/ASC Structures, Structural Dynamics, and Materials Conference 14th AIAA/ASME/AHS Adaptive Structures Conference 7th. 2006, p. 2282.

[46] Y. Tian, A. Duongthipthewa, Q. Chen, et al., Quantitative monitoring of icing on CFRP laminate with guided wave combining forward modeling and inverse characterization, Structural Health Monitoring (2024). 14759217231222544.

CHAPTER

16

Experimental investigation of thermal deicing characteristics and energy efficiency

Jiankai Dong, Yiqiang Jiang, Jiqin Li and Teng Xie

School of Architecture and Design, Harbin Institute of Technology, Harbin, Heilongjiang, P.R. China

16.1 Introduction

Ice accumulation and adhesion on exposed equipment, such as aircraft, trains, wire netting, and wind turbines, can cause reduced reliability and personal injury [1,2]. To address these issues and ensure the safety of the equipment operation, it is necessary to conduct more studies on effective deicing methods. There are four commonly used deicing methods: artificial deicing, mechanical deicing, chemical deicing, and thermal deicing. Artificial deicing is the original technology but is rarely adopted due to low efficiency. Mechanical deicing can be widely applied only when the issue of equipment damage and incomplete deicing is resolved. Chemical deicing is currently the most widely used method due to its high efficiency. However, the chemicals used in the process have a negative impact on the environment. Additionally, the high cost of chemical fluid recovery facilities has made it uneconomical for small and moderate deicing projects. Thermal deicing technologies are highly effective in removing ice and are environmentally friendly [3–7]. The efficiency of deicing is an important factor that determines the effectiveness of the deicing method used. However, there are limited studies on the energy consumption of deicing in comparison to defrosting. Therefore, it is essential to conduct experimental investigations on deicing characteristics that can help reduce the deicing time and improve efficiency. This chapter introduces an experimental setup to demonstrate the impact factors and energy efficiencies of two deicing methods—hot air and infrared ray (IR). The following section provides detailed explanations of these methods.

Frosting and Icing for Efficient Energy Use in Engineering Applications
DOI: https://doi.org/10.1016/B978-0-443-15495-9.00003-4

16.2 Experimental investigation of hot air deicing

Hot air deicing is an environment-friendly method that can effectively prevent equipment from icing up. However, its high energy loss and low deicing efficiency obstruct its widespread adoption. To address these issues, an experimental study was conducted on the ice-melting process using hot air as the heat source. In this section, the energy efficiency of the hot air deicing was calculated. Moreover, a waste heat recovery unit and an insulation layer on the air duct were applied to increase energy efficiency. Additionally, the impact of the distance between the hot air outlet and the ice was also revealed. The results have provided valuable insights into the hot air deicing characteristics and its energy consumption, which can help reduce the deicing energy consumption and improve deicing efficiency.

16.2.1 Analytical approach

Deicing using hot air was a complex heat transfer process. To design and analyze experimental results, the ice melt process was divided into three stages.

The first was a preheating stage, from the start of heating ice to the beginning of the ice melt. This stage was very short, which was about 30 seconds. In this stage, the temperature difference between the ice surface and hot air reduced with the rapid increase of ice surface temperature, which increased from −5.0°C to about 0°C in this experiment. Since the temperature difference was reduced, the heat flux at the ice surface decreased in this stage. The heat transfer between ice and hot air was expressed by Eq. (16.1):

$$\left(\rho_s c_s\right)\frac{\partial T}{\partial t} + Q_l = \frac{\partial}{\partial x}\left(k\frac{\partial T}{\partial x}\right) + \frac{\partial}{\partial y}\left(k\frac{\partial T}{\partial y}\right) + \frac{\partial}{\partial z}\left(k\frac{\partial T}{\partial z}\right) \tag{16.1}$$

where ρ_s is the density of the ice; c_s the specific heat capacity of ice; Q_l is the latent heat of ice melting; k is the heat transfer coefficient between ice and hot air, which is related to hot air temperature and air velocity. Therefore, the temperature and velocity of the hot air were two key factors affecting ice melting.

The second stage was from the start of melting to the stabilization thickness of the melted water layer. In this stage, the heat flux is further reduced with the temperature increase of the melted water layer. One part of the absorbed heat was used for melting ice, while the other was conducted to the inside ice. Due to the movement of the ice-water interface, the heat transfer of the ice-water interface became more complex. In this process, the rate of ice melting depended on heat flux from hot air and ice's internal heat conduction. The heat transfer of the ice-water interface was expressed by Eq. (16.2):

$$\begin{cases} T_l = T_s = T_m \\ \rho_s H_m \dfrac{ds}{dx} = \lambda_s \frac{\partial T_s}{\partial x}\Big|_{x=s} - \lambda_l \frac{\partial T_l}{\partial x}\Big|_{x=s} \\ x = S(t) \end{cases} \tag{16.2}$$

where T_l is the temperature of melted water; T_s is the ice temperature; T_m is the melting temperature of ice; H_m is the latent heat of ice melting; λ_s is the thermal conductivity of ice; λ_l is the thermal conductivity of water; $S(t)$ is the displacement of the solid–liquid interface, which is related to the ice thickness. Therefore, the ice thickness was another factor affecting ice melting.

Heat transfer between hot air and melted water layer could be expressed by Eq. (16.3):

$$\lambda_l \left.\frac{\partial T_l}{\partial x}\right|_{x=s} + \rho_l \frac{ds}{dt} c_l (T_b - T_l) = h_l (T_a - T_b) \tag{16.3}$$

where ρ_l is the water density; c_l is the specific heat capacity of water; t is the time; T_a is the temperature of hot air; T_b is the temperature of water close to hot air; h_l is the heat transfer coefficient between hot air and melted water.

The third stage was from the stabilization thickness of the melted water layer to the end of melting. In this stage, the heat flux at the ice surface was constant. Therefore, the rate of ice melting was also constant. Heat transfer between ice and water was expressed by Eq. (16.4):

$$h_s (T_c - T_s) = \lambda_s \left.\frac{\partial T_s}{\partial x}\right|_{x=s} + \rho_l \frac{ds}{dt} c_l (T_c - T_s) + \rho_s H_m \frac{ds}{dt} \tag{16.4}$$

where h_s is the heat transfer coefficient between ice and melted water; T_c is the temperature of water close to the ice.

Obviously, melting ice was a complex heat transfer process; it was accompanied by latent heat, heat, mass transfer, and boundary moving. Based on the analysis mentioned above, the factors affecting ice melting included air temperature, air velocity, and ice thickness. Afterward, the experimental study was about the property of ice melting at different air temperatures, air velocity and ice thickness.

16.2.2 Experimental setup and procedures

16.2.2.1 Experimental setup

An experimental setup was specifically established for carrying out the experimental work reported in this section. It included a hot air system, a sample of experimental ice, a cold air system, and a data acquisition system. Fig. 16.1 shows the schematic of the experimental setup.

The hot air system was composed of a centrifugal fan (air quantity 5000 m^3/h, rated power 1500 W), electric heaters (five electric heating wires, each rated power 3000 W), and an air duct with 1.5 m length and 0.15 m diameter. Moreover, a hot air waste heat recovery unit was installed to improve the system's energy efficiency. The heat recovery unit was composed of a centrifugal fan, an air duct, and a return air cover.

For the ice used in the experiment, it was frozen on an aluminum plate under the nature outdoor environment (the outdoor air temperature was between −10.0°C and −30.0°C during an experiment in Harbin, China) and placed vertically with 40 cm to the hot air tuyere during the experiment. In addition, the thicknesses of the used ice were set at 2.0 cm and 5.0 cm, respectively.

The cold air system was used to preclude the influence of ambient air, whose temperature was much higher than the ice freezing point, and maintain a relatively low-temperature environment of ice surrounding during the experiment. A cold air mouth, which was connected with the outdoor environment by using an insulated duct, was

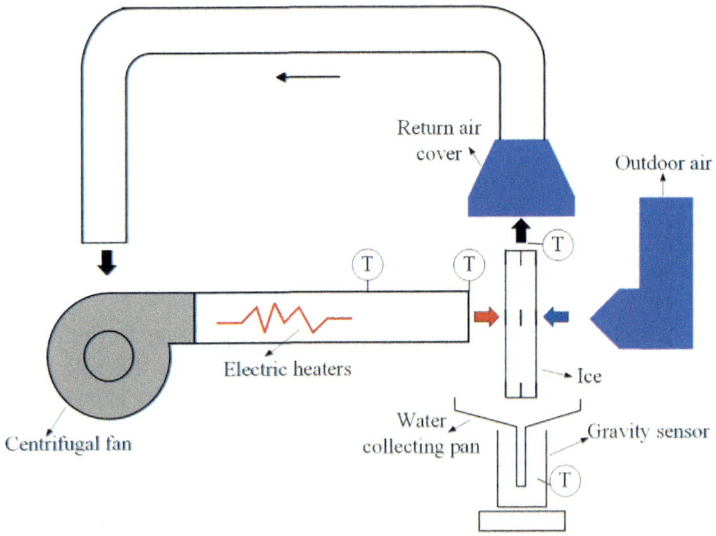

FIGURE 16.1 The schematic of the experimental setup.

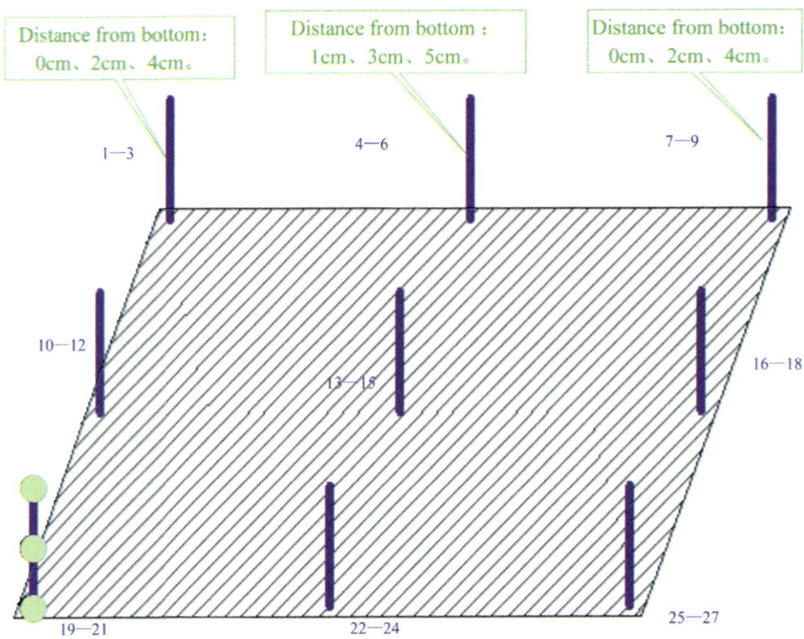

FIGURE 16.2 Position arrangement of thermocouples.

placed near the substrate to blow cold air under the effect of the air pressure difference between outdoor and indoor air. During the experiment, the surrounding temperature of ice was at about −5.0°C.

The data acquisition system was composed of a temperature sensor, wind speed sensor, and gravity sensor. During the experiment, nine T-type thermocouples were placed at different locations of ice inner to observe the temperature variation, which is shown in Fig. 16.2. At each location, the thermocouples were placed at three different heights. The bottom thermocouples at each location were used to measure the temperature of the aluminum plate, and the other two thermocouples were used to measure the temperature of the ice. The wind speed sensor was used to measure the velocity of hot air. The gravity sensor was used to measure the mass of ice-melted water.

During experiments, the temperatures of ice, melted water, and ambient air were measured using thermocouples, the accuracy of which was $\pm 0.2°C$. The velocity of hot air was measured using hot-wire anemometer, the accuracy of which was $\pm$ 0.1 m/s. The temperature of the supplied hot air was regulated by the temperature controller. Sensors and measuring devices used in this experiment were able to output voltage signals; the output voltage signal of the gravity sensor was 2.0–10.0 V, and the output voltage signal of thermocouples was 0–50 mV. The sensors and measuring devices were transferred to a data acquisition unit for logging and recording. The measured data throughout an experimental process were collected and recorded at an interval of 5 s.

16.2.2.2 Experimental procedures and conditions

The rates of melting ice were measured under four different hot air temperatures ranging from 30.0°C to 60.0°C and four different velocities ranging from 5.0 m/s to 14.0 m/s.

First, the eligible ice needed to be produced. The length and width of ice were both 0.4 m, and the thickness was 2.0 and 5.0 cm, respectively. When deicing, firstly switched on electric heaters to heat air, this was the beginning of a deicing operation. The temperature of hot air was regulated by the temperature controller. while the velocity of hot air was regulated by the opening degree of air inlet. In addition, a waste heat recovery unit was installed to recover the discharged air energy and further improve the energy efficiency of the whole system. The inlet of the heat recovery unit was 30 cm directly above the ice, and the outlet was connected with a centrifugal fan. Deicing operation was manually terminated, when the measured lowest temperature value of thermocouples inside of ice reached 0°C.

During experiments, the temperature behind the ice was maintained at $-5.0 \pm 0.5°C$, in equilibrium with the initial temperature of the ice. A water container was laid out under the ice to obtain the melted water. The mass of melted water was measured by a gravity sensor. Then the melt rate of ice could be calculated by the change of melted water mass.

16.2.2.3 Data reduction

The ice melting rate could be calculated by the mass change of the melted water. The deicing energy was supplied by electric heaters and centrifugal fans. At the same time, to get the efficiency of the hot air deicing system, the energy absorbed by the melted water and the sum energy loss during the deicing process needed to be analyzed.

During a deicing operation, the energy supplied from the centrifugal fan was evaluated by Eq. (16.5):

$$E_f = P_f t \tag{16.5}$$

where E_f is the total energy consumption of a centrifugal fan; P_f is the rated power of the centrifugal fan; t is the time of ice melting.

The input power to the electric heaters Q_h was directly measured by the power logger.

The energy absorbed by melted ice was evaluated by Eq. (16.6):

$$E = c_s m\Delta t + m\Delta H \tag{16.6}$$

where E is the total heat absorption of ice melting; ΔH is the melting enthalpy of ice.

Deicing efficiency was expressed by Eq. (16.7):

$$\eta = \frac{E}{E_f + Q_h} \tag{16.7}$$

Deicing efficiency was an important parameter to evaluate the performance of the hot air deicing system. It was defined as a ratio of the actual amount of required energy consumption to melt the ice to the total amount of energy supplied by electric heaters and centrifugal fans.

16.2.3 Results and analysis

16.2.3.1 Characteristics of ice melting

Fig. 16.3 shows the variation of melted ice water mass and melting rate. In this experiment, the ice thickness was measured at 5.0 cm. At the same time, the heated air arriving at the ice surface was controlled at 60.0°C and 14.0 m/s, respectively. As shown in Fig. 16.3, in the first 48.0 min, the rate of melted water was kept at a stable value. During

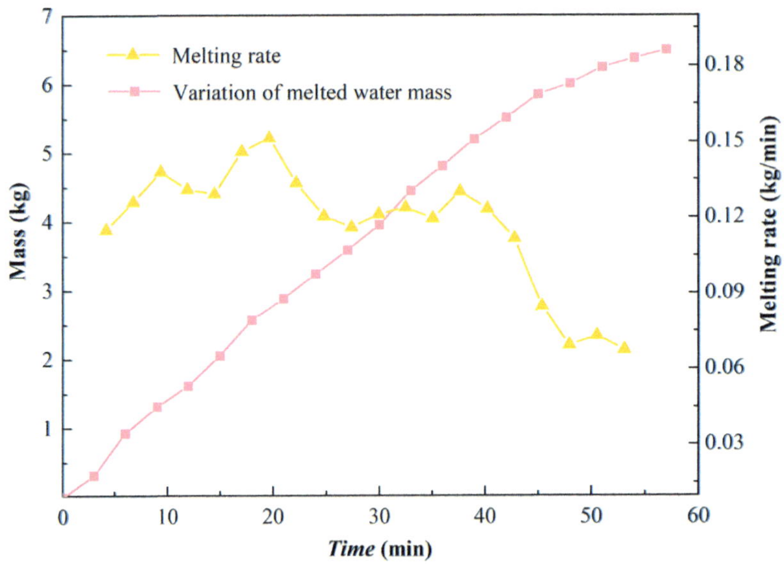

FIGURE 16.3 Variation of melted water mass and melting rate with a thickness of 5 cm, air temperature of 60.0°C, and air velocity of 14.0 m/s.

this period, the mass of melted water reached 6.0 kg, and the average deicing rate was 0.13 kg/min. In the last 10 minutes (the deicing operation was terminated at 58 minutes), the melting ice rate decreased from 0.1 kg/min to 0.06 kg/min. During this period, the total mass of meltwater was 0.5 kg, and the corresponding average deicing rate was 0.05 kg/min, which was much less than that in the prior period.

In the first few minutes, the melting rate gradually increased. After 7 minutes, the melting rate started to reduce and stabilize after a while because the convective heat transfer between the ice surface and hot air reduced with the decreased of temperature difference led by the rapid increase of ice surface temperature.

Finally, the decrease of ice surface heat flux resulted in the decline of melting rate. The increase of melted water mass was steady in the first 45 minutes. However, it had an obvious decrease after about 48 minutes due to the finish of part of the ice melted completely. The deicing time of 5.0 cm thickness of ice was approximately twice as much as 2.0 cm thickness. So, under fixed temperature and velocity experimental conditions, the ice melt rate was quite consistent and varied little with ice thickness.

Fig. 16.4 shows the deicing time under different air temperatures. In this experiment, the air velocity was set at 14.0 m/s, and the air temperatures were set at 30.0°C. 40.0°C, and 50.0°C, respectively. When the air temperature was 60.0°C, the deicing time reached its smallest value, which was 58.2 minutes. In contrast, it reached its biggest value of 138.6 minutes when the air temperature was 30.0°C. Therefore, with the increase in air temperature, the deicing time decreased due to more energy provided by high-temperature air.

When the air temperature increased from 30.0°C to 40.0°C, the deicing time decreased from 138.6 to 86.2 minutes, which decreased by about 37.8%. Nevertheless, when the air temperature increased from 50.0°C to 60.0°C, the deicing time only decreased from 64.0 to 58.2 minutes, which decreased by about 9.0%, which was much less than 37.8%. Therefore, when the temperature was higher than 50°C, the deicing time reduced slowly with the

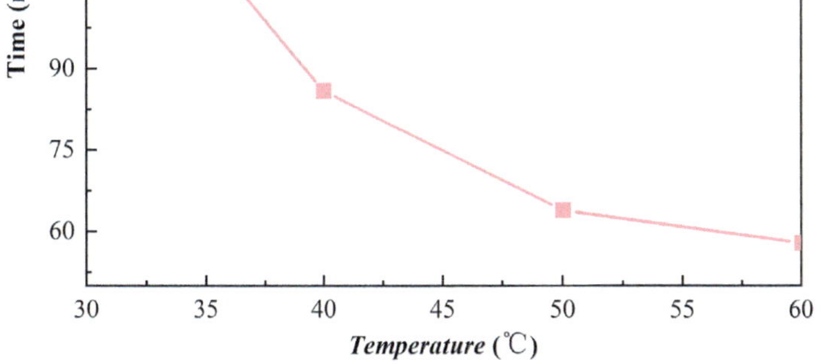

FIGURE 16.4 Deicing time under different air temperatures with a thickness of 5.0 cm and air velocity of 14.0 m/s.

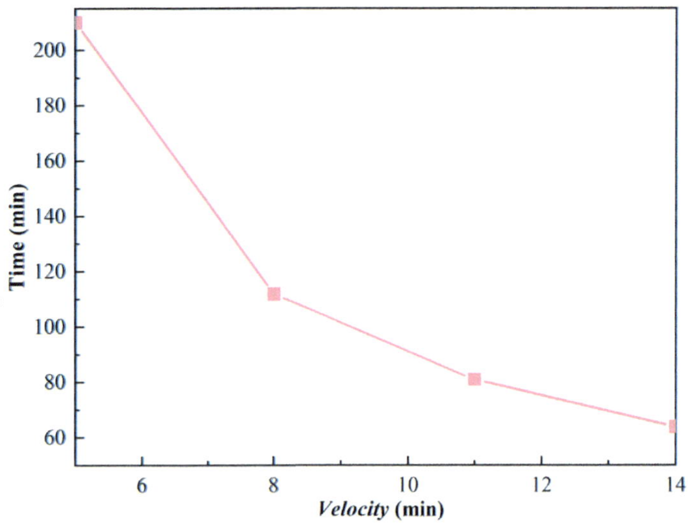

FIGURE 16.5 Deicing time under different velocities with a thickness of 5.0 cm and air temperature of 50.0°C.

increase in hot air temperature. This was because more heat transferred to indoor air with the increase of the hot air temperature. Therefore, the optimized air temperature was between 40.0°C and 50.0°C, considering both the energy consumption and deicing time.

Fig. 16.5 shows the variation of deicing time under different velocities. In this experiment, the air temperature was set at 50.0°C and the air velocity was set at 5.0, 8.0, 11.0, and 14.0 m/s, respectively. When the air velocity was 14.0 m/s, the deicing time reached its smallest value, which was 64.0 minutes. In contrast, it reached its biggest value, 210.3 minutes, when the air temperature was 30.0°C. Therefore, with the increase in air velocity, the deicing time decreased due to a higher convective heat transfer coefficient between ice and hot air.

However, when the air velocity increased from 5.0 to 8.0 m/s, the deicing time decreased from 210.3 to 112.5 minutes, which decreased by about 46.4%. Nevertheless, when the air velocity increased from 11.0 to 14.0 m/s, the deicing time only decreased from 81.7 to 64.0 minutes, which decreased by about 21.7%, which was much less than 46.4%. Therefore, when the velocity was higher than 11.0 m/s, the deicing time reduced slowly with the increase of hot air velocity. Because when the hot air velocity was nearby 5.0 m/s, the heat transfer form between hot air and ice was natural convection. Once the velocity was more than 8.0 m/s, the heat transfer form was forced convection. The forced convective heat transfer coefficient was much greater than that of natural convective. As a result, the optimized air velocity was between 11.0 and 14.0 m/s in this experiment.

As clearly shown in Figs. 16.6 and 16.7, ice melting time did decrease, corresponding to heater temperature increased and hot air rate increased. When the rate of hot gas is less than or equal to 5.0 m/s, it needs too long time to melt the ice. As we can see, the curve of 60.0°C was closely next to the curve of 50.0°C. So, at a fixed wind rate and temperature higher than 50.0°C, the efficiency of the melting ice will rarely rise with increasing temperature. So, the favorable deicing temperature was between 40.0°C and 50.0°C, and the appropriate air velocity was between 8.0 and 14.0 m/s.

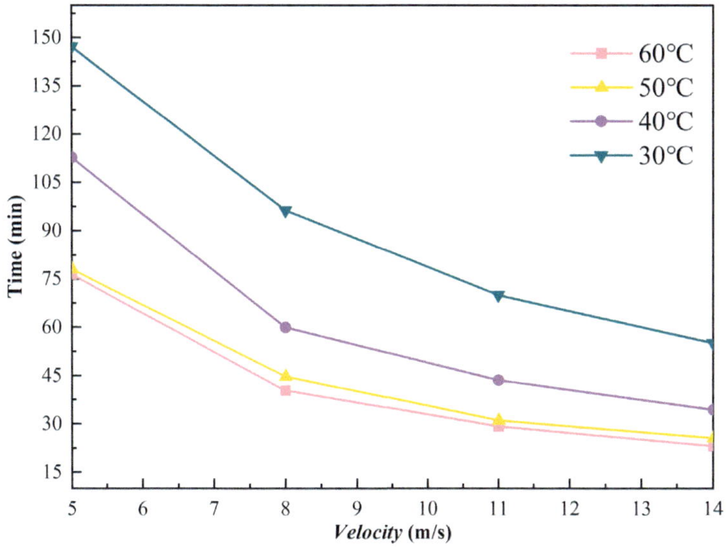

FIGURE 16.6 Deicing time under different experimental conditions for 2.0 cm ice thickness.

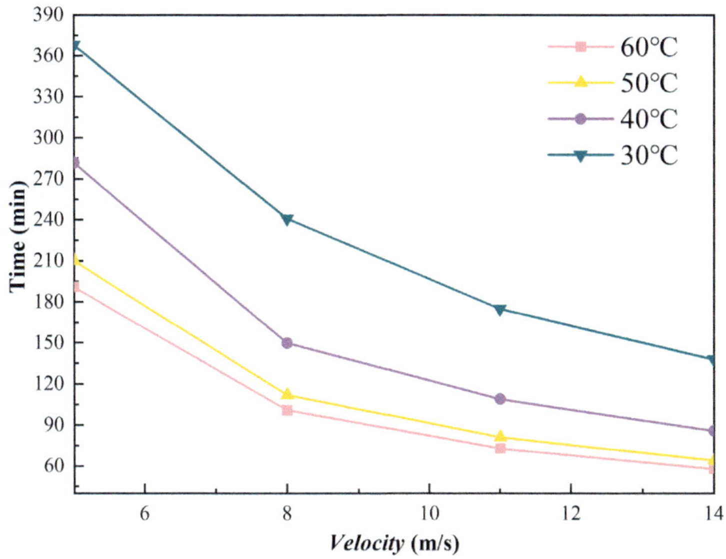

FIGURE 16.7 Deicing time under different experimental conditions for 5.0 cm ice thickness.

Fig. 16.8 shows the temperature variation of the inner ice. During the experiment, the selected ice thickness was 5.0 cm, and the hot air temperature and velocity were 50.0°C and 14.0 m/s, respectively. The temperature of external ice increased at the beginning of the deicing process. Initially, the temperature of the middle and interface ice both slowly rose until the temperature was approximately 0°C. The melting of the ice and subsequent increase of the thermocouples temperature was associated with the absorption of hot air convection

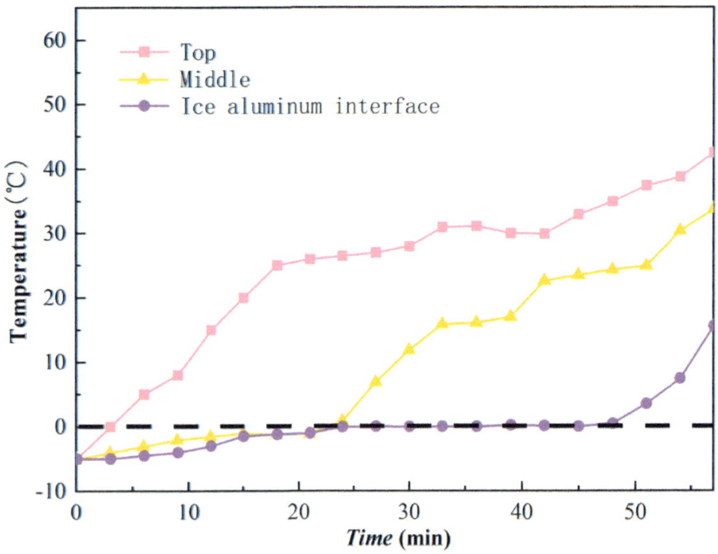

FIGURE 16.8 Temperature variation of the inner ice.

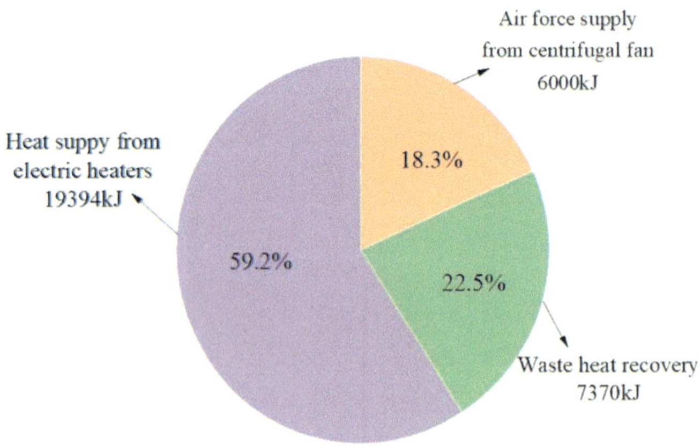

FIGURE 16.9 The measured heat supplies during deicing under air temperature of 50.0°C and air velocity of 14.0 m/s.

thermal energy. Conduction did not play a significant role in melting the ice. However, it did play a role in heating the ice when the ice temperature was below freezing.

In the ice melting process, part of the absorbed heat was used for melting ice, and the other part was conducted to the inside ice. In summary, when the ice temperature was below freezing, the increase in the inner ice temperature was caused mainly by conduction. This illustrated that latent heat holds a greater proportion than sensible heat in ice melting.

16.2.3.2 *Evaluation of deicing efficiency*

Fig. 16.9 summarizes the measured heat supplies during deicing. The total heat supply from these three sources was 26,764 kJ. As seen, the heat supply from electric heaters was

TABLE 16.1 Deicing efficiency with different air temperatures.

Temperature(°C)	30.0	40.0	50.0	60.0
Energy efficiency (%)	9.7	10.2	12.1	10.9

19,394 kJ, accounting for 59.2% of the total. The waste heat recovery was 7370 kJ, accounting for 22.5% of the total. The waste heat recovery reference temperature was indoor temperature. The air force supplied from the centrifugal fan was 6000 kJ, accounting for 18.3% of the total. Therefore, electric heaters provided the foremost share of the total energy supply.

The total effective energy absorbed by ice was 2320 kJ, which was far less than the total energy supply. Among the total effective energy, the energy consumption for melting ice was 2170 kJ, and the energy consumption for vaporizing melted water was 150 kJ. The deicing energy efficiencies were as low as 12.1% and 8.8% when with and without recovering the waste energy.

Table 16.1 shows the deicing efficiency varies with different air temperatures. The minimum energy efficiency was 9.7% with an air temperature of 30.0°C, and the maximum energy efficiency was 12.1% with an air temperature of 50.0°C, which was in the range of optimized air temperatures.

16.2.4 Discussion

Experimental and analytical results indicated that hot air could effectively melt ice. Deicing by hot air was an environmentally promising approach. However, the energy efficiency was as low as 12.1%, because lots of heat dissipated to indoor air. To elevated energy efficiency of hot air deicing, a waste heat recovery unit was necessary. On the one hand, waste heat recovery could increase hot air deicing energy efficiency. On the other hand, it could prevent the temperature of the operating environment from increasing. Adding air duct insulation and shortening the distance between the outlet and ice could also elevate deicing energy efficiency. Practically, the main energy loss channel was from the outlet to the ice surface. So, during the deicing operation, the distance between the outlet and the ice should be approaching as much as possible.

In this experimental study, the deicing time for 5.0 cm thickness ice was approximately twice as much as 2.0 cm thickness. So, under the experimental conditions of fixed temperature and velocity, the ice melt rate was quite consistent and varied little with ice thickness. That was because the melting process only exists at the surface of the ice. When the melted water layer thickness was stable, the ice surface heat flux would be constant. However, in both the analytical analysis and the analysis of experimental data, the flowing of melted water on the ice surface was not considered.

16.3 Experimental investigation of infrared ray deicing

IR deicing has advantages in energy consumption and environmental protection. Compared to hot air/water deicing, it is more efficient for aircraft deicing as the rays do not penetrate the aircraft's surface or warm up its internal space [8]. However, IR deicing is only suitable for specific climates [9]. The characteristics of the ice melt process and the energy consumption of IR deicing have not been fully understood. Therefore, more analysis and experiments are needed to theoretically and experimentally reveal the heat transfer mechanism and increase the application of IR deicing technology. In this section, an experiment setup was built to investigate the melt rate, temperature distribution, mass variation, and heat transfer mechanism during IR deicing. The energy consumption and energy efficiency of IR deicing were also calculated. The results of this study can help in designing a new IR deicing system for aircraft or other equipment that requires deicing.

16.3.1 Experimental approach

16.3.1.1 Experimental setup

An experiment setup was built to study the characteristics of IR deicing under different experiment conditions. The experiment setup mainly consists of the IR emission device, sample of experimental ice, measure and monitor system, and cold air distribution system. Fig. 16.10 shows the schematic of the experiment setup.

Three same-size customized Wolfram filaments were chosen as the IR emission device to radiate thermal energy, whose setting temperature was 1073K, 1773K, and 2273K, respectively. Furthermore, to reduce IR energy waste, especially the energy irradiating to ambient environment, a square poly mask was designed to gather thermal radiation energy. The Wolfram filaments were fastened at the mouth of the poly mask, and each filament was equidistant.

For the ice used in the experiments, it was frozen on an aluminum plate under the nature outdoor environment (the outdoor air temperature was between $-10.0°C$ and

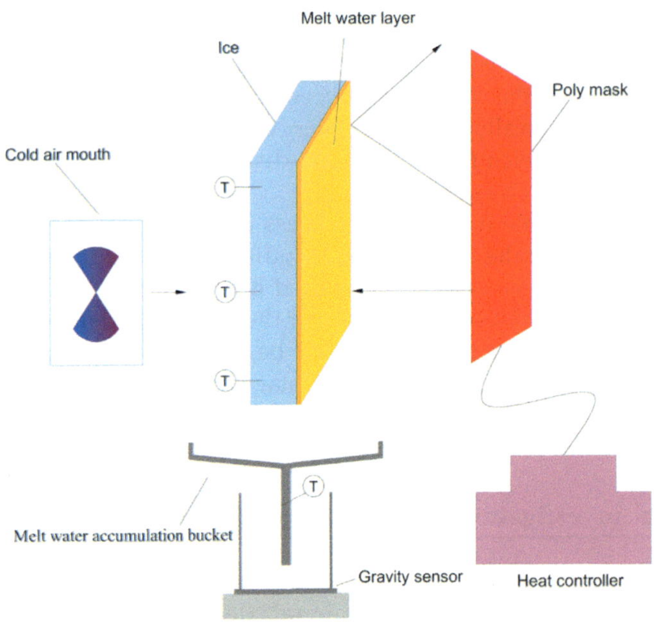

FIGURE 16.10 Schematic of the experiment setup.

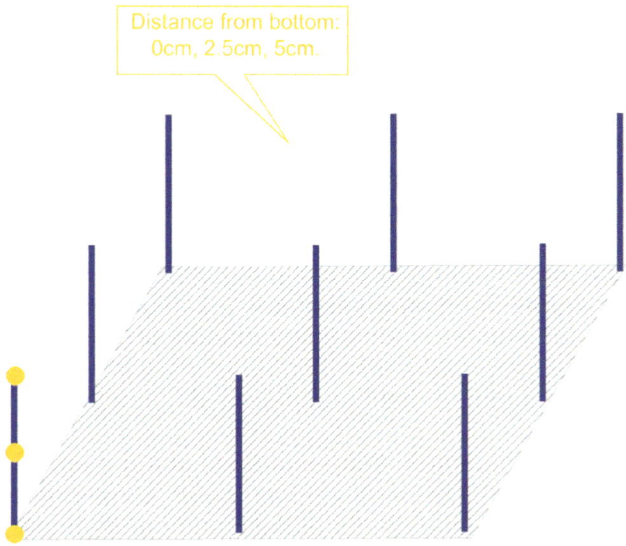

FIGURE 16.11 Placement of temperature measuring points.

−30.0°C during an experiment in Harbin, China) and placed vertically on the front of an infrared emission device during experiments. In addition, the thicknesses of the used ice were set at 2 cm (3.0 kg) and 5 cm (6.5 kg), respectively.

During experiments, nine T-type (of ±0.2°C accuracy) thermocouples were placed at different locations of ice inner to observe the temperature variation, which is shown in Fig. 16.11. At each location, the thermocouples, of which the temperature measurement range was from 223K to 573K, were placed at three different heights. The bottom thermocouples at each location were used to measure the temperature of the aluminum plate, while the other two thermocouples were used to measure the ice temperature. To measure the temperature of meltwater, a bucket was installed under the aluminum plate to accumulate the meltwater. Meanwhile, the PT1000 (of ±0.1°C accuracy) was chosen as the temperature sensor and inserted into the bucket to measure the meltwater temperature, which could help calculate the energy absorbed by the meltwater and evaluate the efficiency of IR deicing technology.

The cold air distribution system was used to preclude the influence of ambient air, whose temperature was much higher than the ice freezing point, and to maintain the relatively low temperature of the ice surrounding environment during experiments. A cold air mouth, which was connected with the outdoor environment by using an insulated duct, was placed nearly the substrate to blow cold air under the effect of the density difference between outdoor and indoor air. During experiments, the surrounding temperature of ice was at about 268K (−5°C).

16.3.1.2 Experimental procedure and conditions

In this experiment, all the experiments were done under dry weather [10]. For deicing, especially airport deicing, the wet weather meant that rain and snow had an impact on deicing. However, the dry weather meant that there were few cold droplets in the environment, and the droplets did not influence deicing. The total radiative flux increased proportionally according to Stefan–Boltzmann's Law. Therefore, to get a shorter wavelength, a higher temperature of IR emission was chosen to generate peak radiative fluxes in the wavelength close to 1.4 μm or even shorter than 1.4 μm [11].

Pure ice or impure ice was, respectively, used in each experiment. The pure ice was directly frozen in the outdoor environment where the temperature was −10.0°C to −30.0°C. Processed water was infused into the customized mold. After 4.0 hours being put in an outdoor environment, pure ice was obtained. The pure ice, which was made from processed water, was considered to have no bubbles in it and be very clean. It meant that the transmission of IR was not be degraded by bubbles or dust in it. The impure ice was made by a similar process as pure ice. However, some snow directly acquired from the natural environment was put into the processed water. Because the impurities contained in snow could not melt, the mass of impure content was measured after the IR deicing operation.

During the deicing process, the ambient environment was maintained at a temperature of 18.0 ± 0.5°C. At the same time, to preclude the ice melt, cold air mouth blew cold air from the outdoor environment. The surrounding temperature of the ice was maintained at −5.0 ± 0.1°C during the whole experiment.

When the IR emission device was put through power, it started to radiate thermal energy. Immediately, ice began to melt, and meltwater dropped alongside ice into the bucket. T-type and PT1000 temperature sensors, respectively, measured the temperature variation of ice and meltwater. The IR emission device was terminated when the mass changing curve of meltwater did not change. At that time, the ice was considered to melt completely.

16.3.1.3 *Optimization of infrared ray emission device*

To find the favorable IR emission, one of the solutions is to optimize the radiation temperature of the IR emission device.

Initially, based on Warren's values, the reflectance coefficient and absorption coefficient of ice at 0°C were obtained from Eqs. (16.8) and (16.9).

$$R = \frac{(n-1)^2 + k^2}{(n+1)^2 + k^2} \tag{16.8}$$

$$\alpha = \frac{4\pi k}{\lambda} \tag{16.9}$$

where R is the reflectance coefficient of ice, k is the Boltzmann constant, n is the real part of the complex index of refraction, α is the absorption coefficient of ice. Koenig's research [11] showed that the absorption coefficient could reach $10^7\ m^{-1}$ for ice when the wave lengthen was higher than 3 μm. In addition, the ice had a higher absorption rate when the wavelength was during 3–12 μm. The deicing effect would be better when the wavelength of the IR emission device according to the ice optimization absorption coefficient. Therefore, the optimization wavelength range of the IR emission device was from 3 to 12 μm.

The setting temperature of the IR emission device, whose material was wolfram, was from 1073K to 2273K in this experiment. The emissivity of the wolfram increased as the temperature increased at wavelength 3–12 μm. The temperature value 2273K was a better choice for the IR emission device, considering the ice melt rate and deicing energy efficiency.

16.3.2 Experimental results

First, the different factors, including the setting temperature, the distance, and the impure content that influenced the ice melt process were investigated. Second, the variation of meltwater temperature under different setting temperatures was measured and analyzed. At last, based on the variation of the ice's temperature, the heat transfer mechanism of IR deicing was studied.

16.3.2.1 Rate of ice melt

Fig. 16.12 shows the variation of ice melt rate. For the ice under the setting temperature of Wolfram filament 1073K, the whole experiment was divided into three stages. In the first stage, from initiation to 1300 seconds, the ice melt rate was approximately fixed at 0.39 kg/min. In the second stage, from 1300 to 1400 seconds, the gradient of the curve changed rapidly, and the ice melt rate became to 1.2 kg/min because the ice broke off from the substrate and confounded into meltwater. It proved that the IR could reach the substrate of ice and cut the bond between the ice and substrate. In the third stage, from 1400 seconds to the end, the ice melt rate returned to 0.39 kg/min.

Fig. 16.13 shows the variation of melt time for the impure ice and pure ice, respectively. First, the impure content was evenly mixed into ice. The mass of impure content of ice was quantified, which was 0.4 kg. Therefore, the purity of the impure ice was 93.8%. For pure ice, the time for the ice to completely melt was 1746 seconds (1073K), 1488 seconds (1773K), and 1002 seconds (2273K), while those were 1572 seconds (1073K), 1386 seconds (1773K), 984 seconds (2273K) for impure ice, respectively. Under a low setting temperature (1073K), the

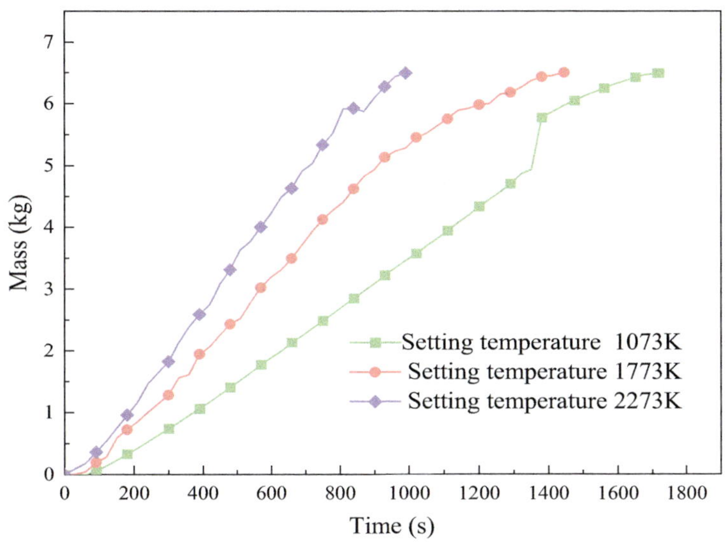

FIGURE 16.12 Variation of ice melt rate with a thickness of 5 cm.

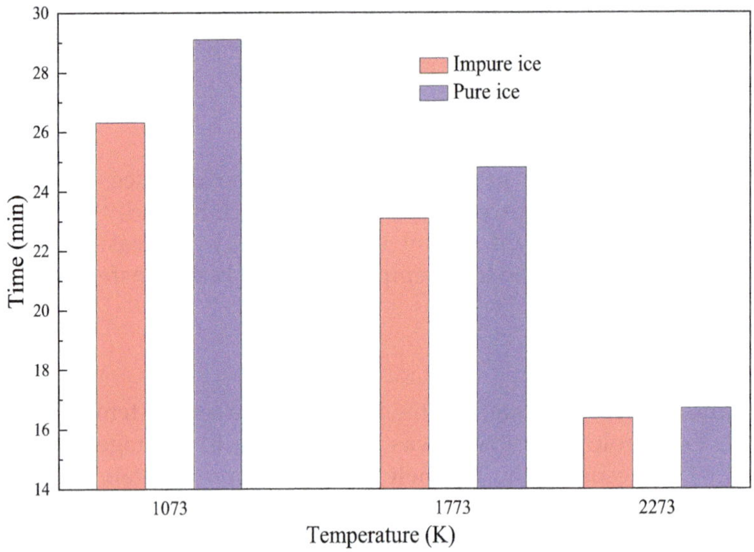

FIGURE 16.13 Variation of melt time for the impure ice and pure ice with a thickness of 5 cm.

TABLE 16.2 Variation of melt rate for the impure ice and pure ice (thickness: 5 cm).

		Melt rate	
		Pure ice (g/min)	**Impure ice (g/min)**
Setting temperature	1073K	223.4	226.0
	1773K	262.1	259.9
	2273K	389.2	382.3

melt time of the impure ice was approximately 90% of that of pure ice. Under the middle setting temperature (1773K), the proportion was approximately 95%. However, under a high setting temperature (2273K), the melt time for these two ices was almost the same.

However, it seems plausible that the impure content would help the ice melt faster because, at each setting temperature, the melt time of impure ice was shorter than that of pure ice. Actually, the conclusion that impure content in ice would help the ice melt was unconvincing. Table 16.2 shows the variation in the melt rate of impure ice and pure ice. Because the impure content in ice did not melt, only the mass of ice in impure ice was used to calculate the melt rate. The melt rates of both pure and impure ice were approximately the same under the same setting temperature. It indicated that the impure content had little influence on the process of ice melt.

During the above experiments, the distance between the ice surface and the IR emission device was fixed at 40 cm. Fig. 16.14 shows the variation of melt time under different distances between these two surfaces. During experiments, two different distances were set, which were 40 and 100 cm, respectively.

The melt time for distances 40 cm were 1746 seconds (1073K), 1488 seconds (1773K), and 1002 seconds (2273K), respectively. While those for distance 100 cm were 3222 seconds (1073K), 3006 seconds (1773K), and 2046 seconds (2273K), respectively. It was obvious that

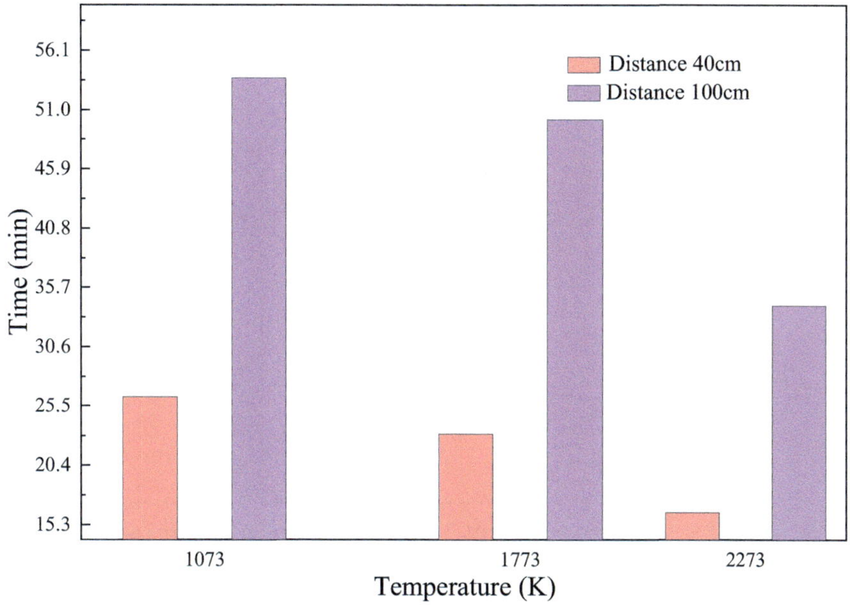

FIGURE 16.14 Variation of melt time under different distances with a thickness of 5 cm.

the melt time for a distance of 100 cm were about two times that of those for a distance of 40 cm. However, the common air environment is the nonparticipating media, which will not influence the thermal radiation heat transfer. Meanwhile, the factors, including temperature, zenith, circumferential angular, wavelength, and area, which influenced incident radiation energy from the IR emission device surface to the ice surface for both two experiments, were the same. However, the solid angles for both two experiments were not the same; as the distance became further the solid angle became smaller. The solid angle was a three-dimensional curvature. When the setting temperatures of the IR emission device were the same, the solid angle was the only factor that influencing incident thermal radiation. When the solid angle became small, the incident thermal radiation also became small. It meant that the incident thermal radiation energy to the ice surface diminished.

A comparison of the experimental results about the relationship between the distance and the ice melt rate obtained from Koenig's experiment [11] and this study was presented. In both two experiments, the setting temperature of the IR emission device and the relative distance between the device and the ice surface were similar. The experimental conditions for both experiments were the same, and the surrounding temperatures were both under freezing point. However, there were some slight differences, such as the IR emission device and the ice was horizontally placed in Koenig's experiment, while the device and the ice were vertically placed in this study. In George's experiment, the heater surface was placed nominal 0.79 m above the substrate surface, and a series of temperatures were set to the heater, from 774K to 1114K. Fig. 16.15 shows that variation of melt rate with the distance of 79 and 100 cm. The setting temperature of our experiment was 1073K, which was lower than George's experiment. While the distance was 100 cm, it was further than that used in

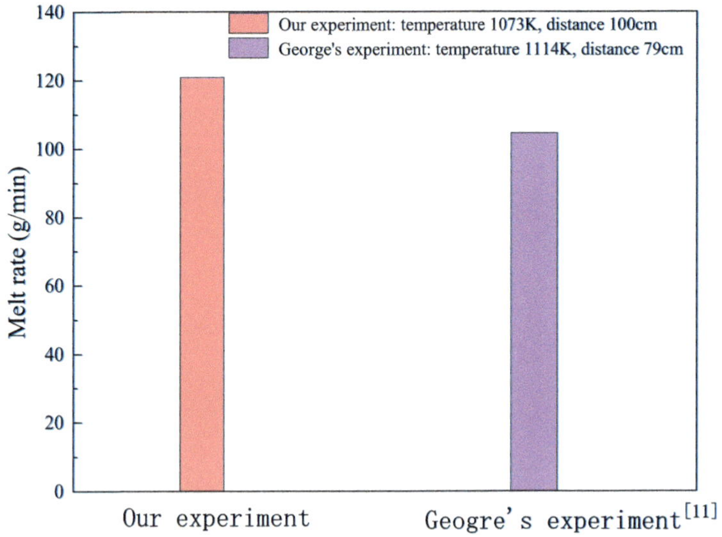

FIGURE 16.15 Variation of melt rate with the distance of 79 and 100 cm.

George's experiment, according to the conclusion that a far distance between the ice and IR emission device surface would attenuate IR energy and reduce the ice melt rate. It seems that because of the lower setting temperature and the far distance, the ice melt rate in our experiment was supposed to be lower than George's. However, the melt rate in our experiment reached 120.7 g/min and was greater than that in George's experiment, which was 104.5 g/min due to the use of a ploy mask, and the poly mask gathered the IR energy, which caused more IR energy reached the ice surface. Further explanations have been added. The ploy mask was made of smooth aluminum, which had a high reflectance for IR; as we all know, for the wolfram IR emission device, the light radiation of the IR device was diffused. So when a ploy mask was equipped, it reflected part of the radiant energy, which should have been diffused, to gather on the ice surface.

16.3.2.2 Variation of meltwater temperature

Fig. 16.16 shows the variation in meltwater temperature. The explanation has been added. The initial water temperatures at different setting temperatures were different, which were 0.1°C, 0.7°C and 1.5°C. And there was no water in the bucket before the experiment started. The initial temperature in the bucket was equal to the ambient air, which was about 291K (18.0°C). During the ice melt process, the variation of meltwater temperature underwent three stages. Taking the setting temperature at 1073K as a sample, in the first 90 seconds into the melt, the measured meltwater temperature dropped down rapidly, which was from 291K (18.0°C) to 273.1K (0.1°C). One thing to point, the value of 291K (18.0°C) was the measured ambient environment temperature because the thermocouple was exposed to the air before the ice melt. From 90 to 1440 seconds, the temperature of meltwater was maintained at a relatively stable value, which was about 273.1K (0.1°C). In the last 360 seconds, it rapidly increased from 274.1K (1.1°C) to 277.5K (4.5°C). Because the ice broke off from the

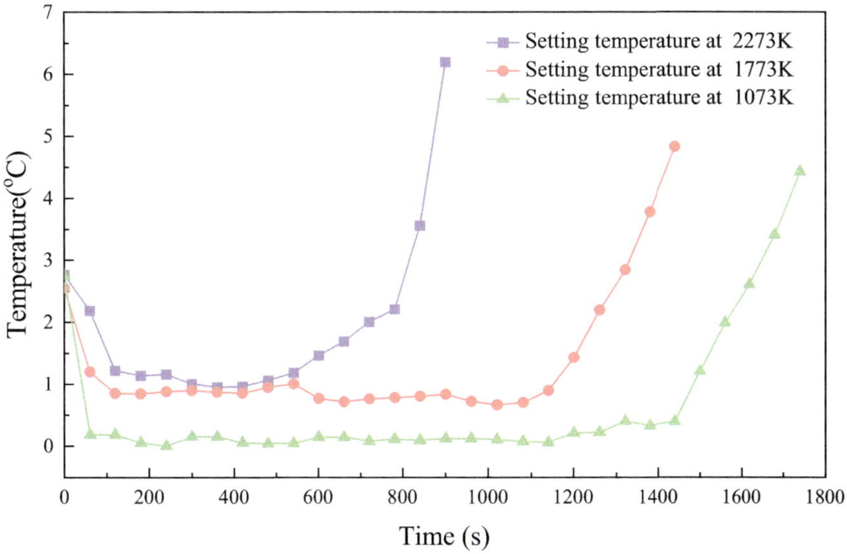

FIGURE 16.16 Variation of meltwater temperature with a thickness of 5 cm.

substrate at the end of the IR deicing experiment, some part of the substrate was directly exposed to the irradiation of IR and was heated. As the substrate was heated, the temperature of the substrate rose. Finally, the conduction between meltwater and the substrate became intense. The temperature of the meltwater increased rapidly.

When the setting temperatures were at both 1773K and 2273K, the meltwater temperature had the same variation trend with that at 1073K. But they had less melt time, which were 960 and 1500 seconds, respectively. Meanwhile, the average temperatures of meltwater in the second stage were 273.7K (0.7°C) and 274.5K (1.5°C), which were both higher than that at 1073K. This could be explained by that higher setting temperature caused a higher heat influx irradiate to the water layer surface, and then water layer could absorb more thermal energy.

The conclusion was obtained that the setting temperature was the main factor influencing the meltwater temperature. When a higher setting temperature was set, the emissive power of the heater device was more powerful. Thermal radiation heat transfer between the ice surface and the heater devices became more intense, and thus, the temperature of meltwater rose.

16.3.2.3 Heat transfer mechanism between ice and infrared ray emission device

Thermal radiation, conduction and convection are three forms of heat transfer affecting the energy exchange in IR deicing. However, during the IR deicing process, thermal radiation heat transfer was the main form. Whether conduction or convection plays an important role in the IR deicing process is still a question [11]. Measuring the temperature distribution of ice inner could help understand the heat transfer mechanism about IR deicing.

Fig. 16.17 shows the variation of ice temperature at the ice top, middle, and bottom, respectively. In the first 60 seconds, the temperature of the top ice grew rapidly from 268K (−5.0°C) to 273K (0°C) because it was heated by the IR rapidly. After 60 seconds, as the ice

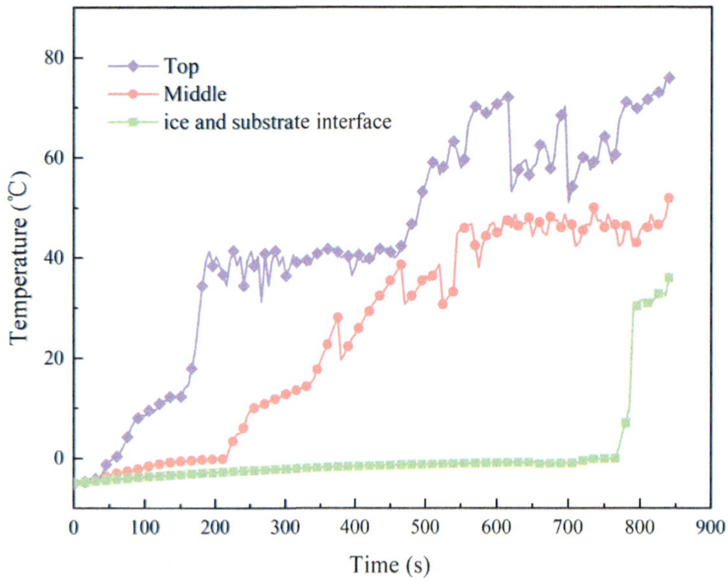

FIGURE 16.17 Variation of ice temperature with a thickness of 4 cm and setting temperature of 2273K.

was melt, the thermocouples were exposed to IR and heated, and the measured temperature grew rapidly, which reached 40.0°C at 222 seconds.

For the middle ice, its temperature was maintained at about −2.5°C in the first 240 seconds. With the conduct of melt, the middle ice started to melt at about 240 seconds. The temperature curve of ice in the middle had a fluctuation. Because when the ice was melted, both the convective and thermal radiation had contradictory effects on ice melt.

For the bottom ice, Because of the cold air mouth, the aluminum plate's temperature was constantly kept at about 268K (−5.0°C). So before 780 seconds, the ice near the ice aluminum plate interface did not melt. Mainly because the thermal energy, which transmitted through the ice and was absorbed by the substrate, did not make the temperature of the aluminum plate rise. Only after 780 seconds, some part of the substrate was directly exposed to IR and the temperature of the substrate grew rapidly.

16.3.3 The energy efficiency of infrared ray deicing

16.3.3.1 Physical description of heat transfer between ice and infrared ray emission device surface

When IR impinged on the interface of ice and meltwater, three-phase conditions (droplets, unfrozen liquid layer, and ice) occurred at the ice surface. When IR reached the ice surface, some were absorbed by the ice, some were reflected to the ambient, while others were transmitted through the ice and reached the substrate, which heated the aluminum plate. As the substrate temperature rose, heat conduction between ice and substrate also influenced the ice melt. It was good for deicing transmitting some energy to the substrate, because it can cut the bond between the ice and solid substrate.

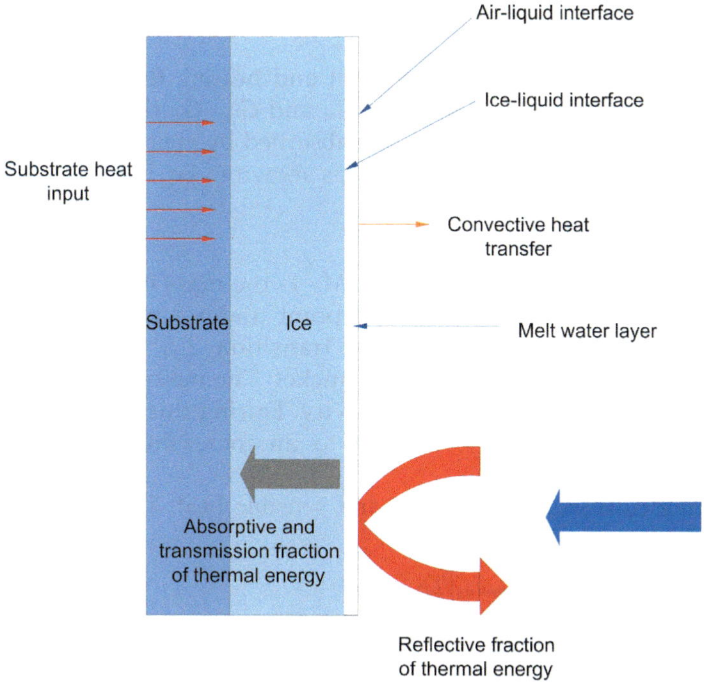

FIGURE 16.18 Heat transfer during ice melting.

The optical property of ice strongly influenced the heat transfer between the ice surface and the IR emission surface. However, during the IR deicing process, a water layer formed at the ice surface, further leaded to the change of the ice's optical property and influenced the heat transfer between both surfaces. Simultaneously, even though thermal radiation was the dominant form of heat transfer, the heat convection happened at the three-phase interface, and the conduction also had an impact on the heat transfer. Because the experiments were done in dry weather, the influence of cool droplets had minimal impact on the heat transfer happening at the ice surface. Also, because of the complexity of calculating heat transfer, to simplify this question, some assumptions were given:

1. The viscous heat and kinetic energy of super-cooled droplets were ignored.
2. The medium between the ice surface and the emission device surface was considered a vacuum.

Fig. 16.18 shows the schemcatic of the heat transfer process during ice melt.

16.3.3.2 Calculation of the measured energy consumption and heat supply

To calculate the energy efficiency of the whole IR deicing process, based on assumptions and experiment conditions, the measured energy consumption and heat supply were supposed to be calculated at first. So the energy efficiency can be defined as Eq. (16.10):

$$\zeta = \frac{G_{mec}}{G_{mhs}} \tag{16.10}$$

where ζ denotes the energy efficiency, G_{mec} is the measured energy consumption, G_{mhs} is the measured heat supply.

The measured energy consumption was used for ice melt and heating the meltwater. So, the measured energy consumption included two parts, G_m and G_{hw}. G_m denotes sensible heat and latent heat used for ice melt, G_{hw} the IR energy absorbed by meltwater.

The calculation of the IR energy absorbed by meltwater was shown in Eq. (16.11),

$$G_{hw} = M_w c_{p,w}(T_{w2} - T_{w1}) \tag{16.11}$$

where G_{hw} denotes the IR energy absorbed by meltwater, M_w is the mass of meltwater, $C_{p,w}$ is the specific heat of meltwater, T_{w1} is the ice melt point under a normal atmosphere. Because temperature changed a little under phase transition, T_{w1} was equal to 0°C. T_{w2} is the measured temperature of meltwater in the bucket. The meltwater flowed down along the ice surface under the effect of natural gravity. During this process, the meltwater was heated by the infrared energy, leading to an increasing meltwater temperature.

The heat absorbed by ice melt consists of latent heat and sensible heat. The total heat energy for ice melt could be calculated by Eq. (16.12):

$$G_m = M_w H_{fus} + M_w c_{p,i}(T_{i2} - T_{i1}) \tag{16.12}$$

where G_m denotes the heat energy for ice melt, M_w is the mass of meltwater, H_{fus} is the heat of fusion of ice, $c_{p,i}$ is the specific heat of ice, T_{i1} is the initial measured temperature of ice, T_{i2} is the temperature of the ice melt point.

Taken a sample of ice with a mass of 6.5 kg as an example to calculate the G_m and G_{conv}. Because the ambient temperature was 268K (−5°C) and the setting temperature was 2273K. According to Fig. 16.16, the final meltwater temperature was 274.5K (1.5°C). Based on Eq. (16.11), the convective heat was obtained:

$$G_{hw} = M_w c_{p,w} \Delta T_w = 4.2 \times 6.5 \times (274 - 273) = 40.95\ kJ \tag{16.13}$$

Furthermore, using Eq. (16.12), the sensible and latent heat needed for ice melt were calculated by Eq. (16.14):

$$G_m = M_i H_{fus} + M_i c_p \Delta T_i = 0.335 \times 6.5 + 6.5 \times 2.1 \times 4 \times 10^{-3} = 2.23\ MJ \tag{16.14}$$

Compared to melt heat, G_{hw} was very small, G_{hw} was approximately 2.0% of the G_m and hence could be ignored.

To calculate the measured heat supply, initially, the conduction heat G_{cond} was supposed to be considered. The conduction heat G_{cond} was the heat transfer between the ice and the aluminum plate. It was caused by the temperature difference between the substrate and the ice. In fact, because of the cold air distribution system, the temperature of the substrate was maintained beneath the freezing point. Fig. 16.17 shows that the temperature of the interface between ice and substrate was also beneath the freezing point, and the conduction heat only heated ice for a little series of times. So, the conduction heat could be ignored.

Because the conduction heat was ignored, the IR emission device transferred 95% of the electric power to radiate the thermal radiation energy. The electric power was considered

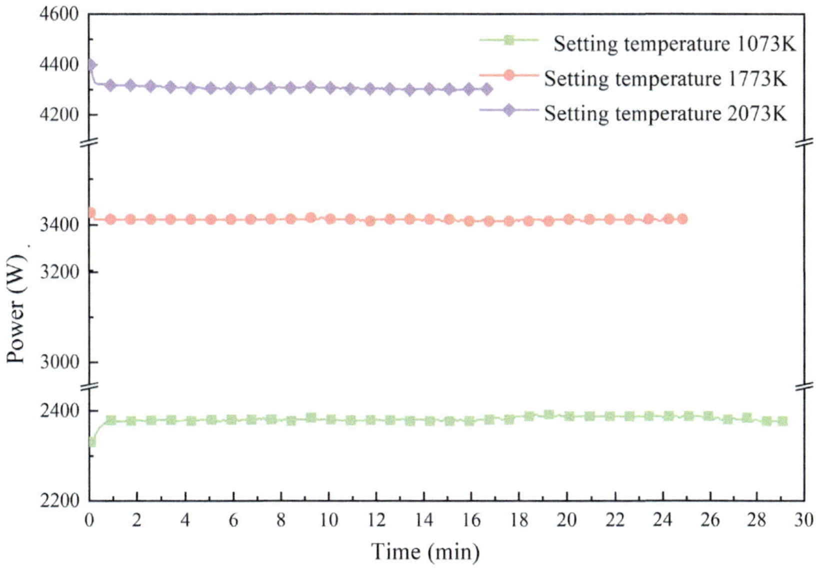

FIGURE 16.19 Real-time electric power of IR emission device. *IR*, Infrared ray.

as the only heat supply. Fig. 16.19 shows the variation of input electric power to the IR emission device.

The energy G_i inputted into the whole IR deicing system can be calculated by Eq. (16.15):

$$G_i = 0.95 \times q_i \times t \tag{16.15}$$

where 0.95 denotes the electric-heat transfer efficiency of the IR emission device, q_i is the real-time electric power, t is the time duration of measuring the real-time electric power.

Also, the convective heat transfer between the meltwater and the ice is considered to calculate the measured heat supply. Because the temperature of meltwater was higher than the freezing point of ice. As the fluency of meltwater, the heat was transferred from meltwater to ice as a form of convection. Based on the Nussle correlation, the calculation of the convective heat transfer between the meltwater layer and ambient environment was shown in Eq. (16.16):

$$\phi_{conv} = A_i h_w (T_w - T_i) \tag{16.16}$$

where A_i denotes the area of ice, h_w is the heat transfer coefficient of meltwater, T_w is the temperature of meltwater, T_i is the temperature of ice. Under the setting temperature 2273K, the heat transfer between the ice and the meltwater was shown in Eq. (16.17):

$$\phi_{conv} = A_i h_w (T_w - T_i) = 0.16 \times 200.5 \times (274 - 273) = 32.08\ W \tag{16.17}$$

Compared to the electric power q_i, the convective heat transfer between ice and meltwater could be ignored.

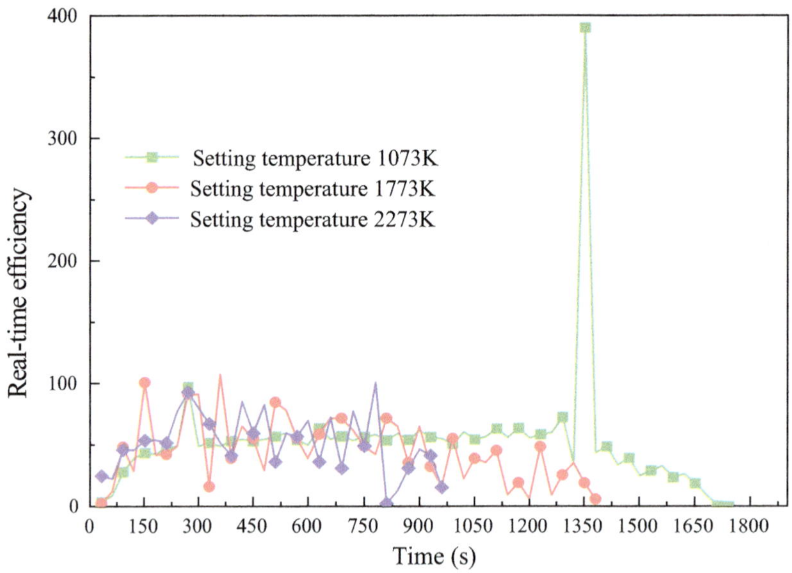

FIGURE 16.20 Variation of real-time energy efficiency.

TABLE 16.3 The energy efficiency under different setting temperatures (ice thickness: 5 cm).

Temperature (K)	1073	1773	2273
Energy efficiency	0.553	0.509	0.559

The real-time energy efficiency was explained that the measured energy used for ice melt to the measured heat supply. The real-time energy efficiency of IR deicing was obtained by Eq. (16.18):

$$\xi_{real-time} = \frac{G_m}{G_i} \tag{16.18}$$

where $\xi_{real-time}$ denotes the real-time energy efficiency. According to the statistics accumulated about ice mass changing and electric power, G_m was derived from Eq. (16.12), G_i was derived from Eq. (16.15).

Fig. 16.20 shows the variation of real-time efficiency. For ice melt under setting temperature 1073K, from the beginning of the experiment to 1350 seconds, the real-time efficiency was maintained at approximately 50%. At time 1350 seconds, the last period of deicing, the real-time efficiency reached 404.1%, just because the ice broke off from the substrate and dropped into the bucket, which made the ice melt rate increased rapidly. After 1350 seconds, the real-time efficiency decreased.

For ice melts under setting temperatures 1773K and 2273K, the real-time efficiency had a sharp fluctuation because the ice melted and became alveolate at higher setting temperatures, which meant that a small block of ice fell down and caused the sharp fluctuation of real-time efficiency.

The overall energy efficiency could be derived from Eq. (16.19):

$$\xi_a = \frac{\sum_{f=1}^{n} G_{e,f}}{\sum_{j=1}^{n} G_{i,j}} \tag{16.19}$$

where $\sum_{f=1}^{n} G_{e,f}$ denotes the sum of the energy for ice melt, $\sum_{j=1}^{n} G_{i,j}$ is the sum of the real-time electric energy. Table 16.3 shows the overall efficiency of IR deicing under different setting temperatures with a total ice mass of 6.5 kg. The energy efficiencies for both setting temperatures, 1073K and 2273K, are almost the same, they are 0.553 and 0.559, respectively. Actually, for setting temperatures at 1073K and 2273K, the energy used for ice melt was approximately the same. While calculating the electric energy consumption, the melt time was taken into consideration. The measured electric power was smaller at a lower setting temperature, but its melt time was longer. Therefore, the measured electric energy consumption for setting temperatures 1073K and 2273K was similar, and the total energy efficiency was also similar.

16.3.4 Discussion

The ice melt rate was investigated under different experiment conditions. The temperature of meltwater and ice were also measured and analyzed. The experiment results indicated that the ice melt rate became fast with the temperature increase of the infrared emission device. And the impure content did not influence ice melt. However, the closer distance between the ice and the IR emission surface would increase the melt rate of ice. Thermal radiation energy was mainly absorbed by the ice. The thermal radiation energy absorbed by the meltwater was approximately 2.0% of the thermal radiation energy absorbed by the ice. The dominant heat transfer form was thermal radiation. However, the convective cooling effect also interfered with thermal radiation heat transfer.

16.4 Conclusion

The energy efficiency of deicing is crucial for the advancement of thermal deicing technology. However, there has been a lack of research regarding the characteristics of the ice-melting process and the energy consumption involved. This chapter presents experimental investigations into the deicing characteristics and energy efficiency using hot air and IR as heat sources.

The experimental results of hot air deicing suggested that the optimal deicing temperature and velocity were between 40.0°C and 50.0°C and 8.0 m/s and 14.0 m/s, respectively. During the deicing operation, the waste heat recovery system contributed to 22.5% of the total energy required for deicing. The maximum deicing energy efficiency achieved in this study was 12.1%, which was higher than the energy efficiency of 8.8% observed without the waste heat recovery system.

The experiment results of IR deicing indicated that the minimum time required to melt 5 cm thickness ice was 1002 seconds with a temperature of 2273K and at a distance of 40 cm. The maximum time required to melt the same amount of ice was 3222 seconds with a temperature of 1073K and at a distance of 100 cm. During the experiment, the maximum overall energy efficiency of the deicing process was 55.9%, with a temperature of 2273K and a distance of 40 cm in the experiment. It was observed that almost half of the thermal energy was radiated to the ambient environment and was not used for deicing.

References

[1] V.F. Petrenko, C.R. Sullivan, V. Kozlyuk, Variable-resistance conductors (VRC) for power-line de-icing, Cold Regions Science and Technology 65 (1) (2011) 23–28.

[2] S.H. Fan, X.L. Jiang, C.X. Sun, Z.J. Zhang, L.C. Shu, Temperature characteristic of DC ice melting conductor, Cold Regions Science and Technology 65 (1) (2011) 29–38.

[3] R. Menini, Z. Ghalmi, M. Farzaneh, Highly resistant ice phobic coatings on aluminum alloys, Cold Regions Science and Technology 65 (1) (2011) 65–69.

[4] V.F. Petrenko, C.R. Sullivan, V. Kozlyuk, V. Veerasamy, Pulse electro-thermal deicer (PETD), Cold Regions Science and Technology 65 (1) (2011) 70–78.

[5] G.G. Koenig, C.C. Ryerson, An investigation of infrared deicing through experimentation, Cold Regions Science and Technology 65 (1) (2011) 79–87.

[6] O. Parent, A. Ilinca, Anti-icing and de-icing techniques for wind turbines: Critical review, Cold Regions Science and Technology 65 (1) (2011) 88–96.

[7] C.C. Ryerson, Ice protection of offshore platforms, Cold Regions Science and Technology 65 (1) (2011) 97–110.

[8] D.A. Cancilla, J. Martinez, G.C. van Aggelen, Detection of aircraft deicing/antiicing fluid additives in a Perched water monitoring well at an international airport, Environmental Science and Technology 32 (1998) 3834–3835.

[9] Advisory Circular FAA, Ground deicing using infrared energy. Washington, DC: AC No120-89 Federal Aviation Administration, 2012.

[10] ACRP, Technical support document: Deicing planning guidelines and practices for stormwater management systems (fact sheet 6), 1999.

[11] G.G. Koenig, C. Ryerson, An investigation of infrared deicing through experimentation, Cold Regions Science and Technology 65 (2011) 79–88.

CHAPTER

17

Dynamic ice-melting characteristics of carbon dioxide laser irradiation under natural convection

Zekang Zhen[1], *Mengjie Song*[1,2], *Long Zhang*[1], *Xuan Zhang*[1], *Keke Shao*[1,2], *Sirui Yu*[1], *Han Shi*[1] *and Runmiao Gao*[1]

[1]Department of Energy and Power Engineering, School of Mechanical Engineering, Beijing Institute of Technology, Beijing, P.R. China [2]School of Mechanical Engineering, Hanyang University, Seoul, South Korea

Nomenclatures

Symbol	Description	Unit
A_f	melting angle in the front view	°
A_s	melting angle in the side view	°
A_t	melting angle in the top view	°
D	melting depth of ice	mm
E_i	energy absorbed by ice melting	J
E_l	energy supplied by the laser	J
H	melting height	mm
n	grid numbers	/
P	laser power	W
Q	melting latent heat	J/kg
R	melting rate	mm/s

Frosting and Icing for Efficient Energy Use in Engineering Applications
DOI: https://doi.org/10.1016/B978-0-443-15495-9.00002-2

RH	relative humidity of air	/
S_1	area of melting hole in the top view	mm^2
S_2	area of melting hole in the front view	mm^2
T	temperature	°C
t	laser irradiation time	s
V	volume of melting ice	mm^3
W	melting width	mm
$X_1, X_2, \ldots, X_n$	independent variable	/
Greek symbol	**Description**	**Unit**
δ	deviation	/
ξ	uncertainty	/
ρ	density	kg/m^3
η	energy efficiency	%
ϕ	gas content of bubble ice	%
Subscript	**Description**	**Unit**
a	air	/
ax	axial	/
A	average value	/
b	bubble	/
d	deicing equipment	/
i	ice	/
l	laser	/
s	size	/

17.1 Introduction

Icing is a common and unavoidable natural phenomenon that occurs widely in the fields of power communication [1,2], rail transportation [3,4], and aerospace [5,6]. The accumulation of ice layers on the surface of production equipment can lead to serious accidents and disasters. For example, the transmission line can be loaded with ice up to 25 kg/m under extreme weather conditions [7]. Under extensive snow and ice coverage, the power lines are potentially endangered of disconnection and power pole collapse, which may even lead to widespread communication interruption [8]. In cold areas, when the body or bottom of the train freezes, it is prone to the risk of derailment out of control during its driving and even causing major traffic accidents [9]. The icing on wings and engine intakes affects aircraft maneuverability and may lead to irreparable losses and serious flight accidents [10,11]. In addition, the accumulation of frost and ice layers on the surface of heat pumps [12–14] and wind turbines [15,16] can reduce the operating efficiency of the unit and even result in system downtime. In recent years, due to the frequent occurrence of extremely cold weather, the impact of icing on industrial equipment is more serious. Therefore, it is necessary to explore more safe, convenient, and efficient deicing technology.

To address the negative effects of equipment icing, a variety of deicing methods have been proposed by researchers, which could be roughly divided into three types: mechanical deicing, chemical deicing, and thermal deicing. In mechanical deicing methods, electrical pulses [17,18] and ultrasonic waves [19,20] are commonly used to break the ice on the surface of an aircraft and wind turbine blade. This method consumes less energy but there are problems of difficult maintenance and incomplete deicing. In chemical deicing, salt or ethylene glycol can effectively reduce the thermodynamic freezing point of water [21,22], so it is commonly used to melt ice on roads and airports. However, the widespread use of deicing agents could cause pollution to the surrounding water. Thermal deicing methods are widely used in industry due to their efficiency and environmental friendliness [23,24]. As one of the most widely used methods of deicing, electric heating is often used to melt ice on power wires and aircraft surfaces because of its easy operation and good reliability [25]. But there are also high energy consumption and uneven temperature distribution problems [26]. Hot water [27] and hot gas [28] are usually used for ice drilling. However, due to the need for heavy heating and power equipment to assist, resulting in the practicality of deicing equipment is greatly reduced. In summary, the above deicing methods have been widely studied and applied on different occasions, but there are still application limitations due to low durability, high energy consumption, and pollution problems.

In recent years, the laser has been gradually used in power grid deicing and polar ice and snow drilling because of its noncontact, clean, high energy, and strong directional characteristics. As an electromagnetic wave, the energy is converted into heat when the laser acts on the target. The absorption effect of ice on different wavelengths of laser light is different, resulting in great differences in ice melting characteristics. In the study of Stone et al. [29], a 5000 W laser with a wavelength of 1070 nm was mounted to a smart ice penetrator for the detection of planetary and polar ice sheets. In the study of Luo et al. [30], a 75 W laser with a wavelength of 1940 nm was developed. It was found that when the laser power density was 25 W/cm^2, the icicle with a diameter of 8 mm was cut off within 28 seconds. In the deicing experiment of Yang et al. [31], an Nd: YAG laser with a wavelength of 1064 nm was developed and used. The experimental results show that 16 ice cracks are generated when the 120 J $/cm^2$ laser is applied to a thickness of 5 cm ice due to large thermal stress. The absorption coefficient of ice increases significantly in the near-infrared to mid-infrared region [32]. For a CO_2 laser with a wavelength of 10,600 nm, the absorption coefficient of ice is about 1476.5 cm^{-1}, which means that it can absorb almost 100% of the light intensity at a distance of 0.01 cm. Compared with the above wavelengths of lasers, the CO_2 laser can not only heat the ice layers effectively but also has the advantage of low energy consumption [33].

The feasibility of CO_2 lasers for deicing has been explored and validated in various experiments. To solve the problem of ice breakage of sea ice, the CO_2 laser deicing technology was used by Clark et al. [34] for ice-cutting experiments. As a result, when the power density is 500 W/cm^2, the cutting rate of the laser is around 1 cm/s, and the average cutting width is 0.3 cm. To eliminate icing on telecommunication networks, a continuous wave CO_2 laser deicing experiment was conducted by Qi et al. [33]. It was found that when the laser power increases from 500 to 2000 W, the melting rate of ice volume increases from 486 to 1639 mm^3/s. In the experiment of Sakurai et al. [35], a bubble-free ice block with the size $25 \times 13 \times 6\ cm^3$ was irradiated by a horizontal CO_2 laser with 106 W/m^2, and the linear melting rate was 1.2 mm/s. However, the ice melting rates were 0.8 and 1.7 mm/s when the laser incidence angle was changed to -5.7 and 5.7 degrees, respectively. This experiment shows that the melting rate of the CO_2 laser decreases by almost 50% due to the accumulation of meltwater in the melting hole, which should be considered in applications. The feasibility of CO_2 laser deicing has been demonstrated, but the energy consumption of this technology has been calculated and evaluated less. Therefore, to better evaluate the practicability of CO_2 laser, it is necessary to study the morphology, volume, and deicing efficiency of CO_2 laser-induced melting holes. In addition, icing on some high-altitude equipment is usually more difficult to remove and requires vertical laser irradiation. To explore a better deicing angle and compare it with horizontal irradiation, it is necessary to study the ice melting characteristics under vertical irradiation.

The formation mechanisms of ice are complicated, and the ice grown on industrial equipment is not always a compact solid [36] but tends to be a porous medium containing needle-shaped or egg-shaped bubbles [37,38]. A theoretical model for predicting the bubble formation of water droplets was proposed by Chu et al. [39]. It is found that the bubble formation rate is closely related to gas solubility, subcooling, and freezing time. At low ambient temperatures, supercooled liquid droplets and ultra-thin water films tend to form opaque frost ice and mixed rime on the surfaces of aircraft and power grids [40,41]. These ice layers with different porosity and bubble content have different thermal conductivity under different microstructures [42]. The numerical investigation study of Lei et al. [43] showed that the ice melting rate and melting volume all rise with the increase of porosity. In the melting simulation study of Sleptsov et al. [44], it was also concluded that the larger the bubble volume, the faster the ice melted. There are many studies on deicing, but the melting characteristics of bubble ice have not been adequately investigated. Therefore, to further study the melting characteristics of ice, it is essential to conduct experiments on bubble ice by vertical irradiation of CO_2 laser.

To solve the icing problem of power grids, aircraft, and railways in cold weather, lots of deicing methods have been developed and applied. However, as a safe and efficient deicing technology, the CO_2 laser deicing method has not been fully studied. The melting characteristics of clear ice under horizontal and vertical laser irradiation are first compared in this study. Then, the melting characteristics of clear ice and bubble ice under vertical laser irradiation are analyzed and compared. As shown in Fig. 17.1, the melting characteristics of ice are studied, such as melting depth, melting height, melting width, melting angle, and melting volume. Meanwhile, the energy efficiency of CO_2 laser deicing under different powers and irradiation times is emphatically analyzed, and the deicing efficiency is compared with other deicing methods. The experimental results can provide a reference for the optimization and application of CO_2 laser deicing technology in the engineering field.

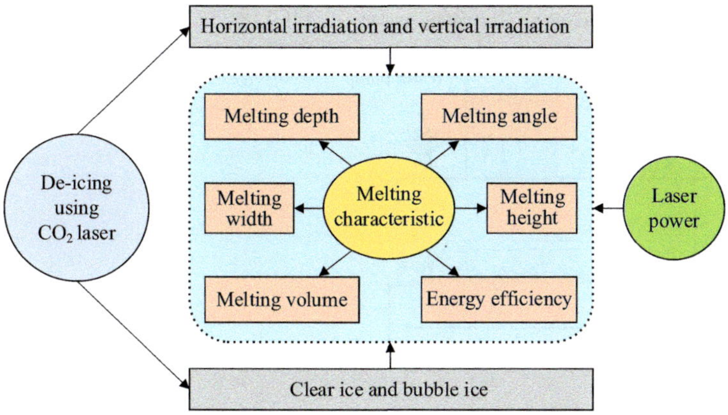

FIGURE 17.1 Framework of the present experimental study.

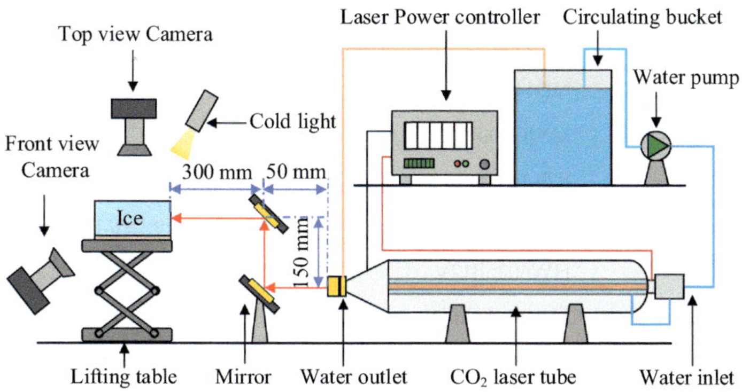

FIGURE 17.2 Schematic diagram of horizontal CO_2 laser ice-melting experimental system.

17.2 Experimental methodology

17.2.1 Experimental setup

To investigate the melting characteristics of clear ice and bubble ice under horizontal and vertical CO_2 laser irradiation, two experimental systems were designed and constructed at different output powers of CO_2 laser, as shown in Figs. 17.2 and 17.3. As shown in Fig. 17.2, the horizontal CO_2 laser ice-melting experiment platform consists of a CO_2 laser emission system, cooling system, reflection system, and image acquisition system. In the CO_2 laser emission system, it is equipped with a laser excitation power supply with a rated power of 100 W. The alternating current power supply could be converted into direct current high-voltage electricity by the laser power supply, which ionizes the gas in the laser tube and excites the CO_2 molecular transition to produce a laser light. In the cooling system, the cooling water is pressed into the water-cooled channel of the laser tube through a circulating pump. The continuous flow of cryogenic water could take away the heat generated by the gas and maintain the stability of the laser output power. In the reflection system, the CO_2 laser is reflected by two silicon mirrors with a diameter of 25 mm, which in turn radiate horizontally onto the clear ice surface. During each ice

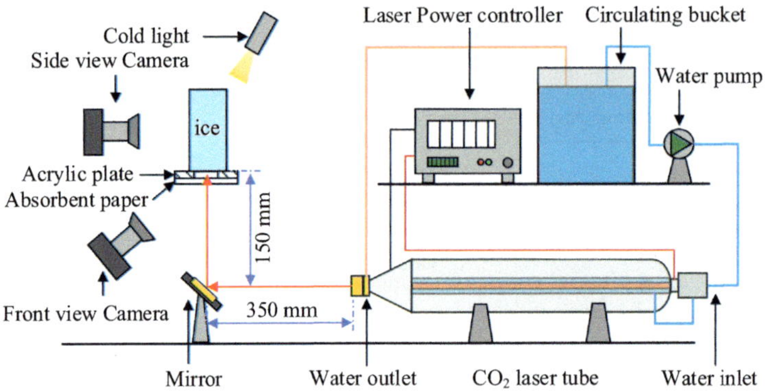

FIGURE 17.3 Schematic diagram of vertical CO_2 laser ice-melting experimental system.

TABLE 17.1 Descriptions of the experimental equipment used in this study.

Item	Equipment	Unit type	Parameter
1	CO_2 laser tube	ND1450	Diameter of 80 mm
2	Laser power supply	MYJG100W-X	100 W
3	Water pump	P25-DB380A	0.05 L/s
4	Laser power meter	HLP-200B	Accuracy of ± 0.1 W
5	Cold light source	HW6025B12V	65 W
6	Optical mirror	Made of silicon	Diameter of 25 mm
7	Top/Side view camera	Canon EOS 60D	Pixel resolution 1920×1080
8	Front view camera	Canon EOS 90D	Pixel resolution 1920×1080

melting process, the total distance between the laser outlet and the ice incident surface is set to 500 mm. To prevent the bottom of the clear ice from melting, an insulating layer is set under the clear ice. In the image acquisition system, two cameras and a cold light source are used to capture the top view and the front view at the same time.

The experimental system of clear ice and bubble ice melting under vertical CO_2 laser irradiation is shown in Fig. 17.3. As seen, the same CO_2 emission system and cooling system are used in the experiment platform for vertical laser ice melting. In the reflection system, the laser emitted from the laser tube is vertically reflected onto the lower surface of the bubble ice by the silicon mirror. The irradiation path between the laser tube outlet and the bottom of the bubble ice is also fixed at 500 mm in each experiment. Moreover, the ice is placed vertically in the center of an acrylic plate with an aperture of 22 mm to ensure that the laser would irradiate the bottom of the ice. To prevent water drops from falling on the mirror, affecting the results of the experiment, the absorbent paper is placed below the plate to absorb the melted water. In the image acquisition system, two cameras are used simultaneously to capture the front and side views, respectively. All the equipment used in the experiment and their parameters are summarized in Table 17.1.

17.2.2 Experimental procedures and conditions

To compare and analyze the melting effect of different types of ice, both clear ice and bubble ice used in this experiment need to be frozen, cut, and sampled in advance. Deionized water was placed in a well-insulated foam box around and at the bottom, with the top open, and frozen in a freezer at − 18 °C. After about 48 hours of freezing, both clear ice and bubble ice will be formed. As shown in Fig. 17.4A, dissolved gases escape from the top to the bottom due to the slow freezing rate, resulting in the formation of clear ice in the upper layer. As shown in Fig. 17.4B, with the increase in freezing time, the insoluble gases become trapped in the ice and form bubble ice in the lower layer. To make the laser melting time long enough, the size of clear ice used in the experiment is fixed at $25 \times 25 \times 110\ mm^3$. Bubble ice is not transparent, so the size is fixed at $25 \times 25 \times 65\ mm^3$ to observe the melting process better. The frozen clear ice and bubble ice were cut into the size of the experimental ice sample and put into the refrigerator for cooling. To maintain the consistency of ice samples, the bubble ice was taken from the same freezing device and the same location. At the same time, the gas content of ice mass was measured by the method of melting [37]. With a maximum deviation of 4.1%, the average gas content of bubble ice is 2.6%.

Before both horizontal and vertical laser ice-melting experiments begin, some preparation work is required. First, the circulating water pump should be opened in advance to ensure that the water-cooled jacket is full of circulating cooling water. Then, to ensure the accuracy of the direction and power of the laser beam in the experiment, the laser is calibrated horizontally and vertically, respectively, and the power meter is used to calibrate different laser powers. Next, the ice sample is taken out from the freezer and its incident surface is aligned with the path of the laser beam. Finally, the dual cameras, cold light, and laser power supply are turned on to start the melting experiment within 5 minutes to prevent the rapid melting of ice. To investigate the melting characteristics of clear ice and bubble ice at different CO_2 laser powers under horizontal and vertical irradiation, nine experimental cases are designed. Their detailed conditions are listed in Table 17.2. The laser power changed at a range of 20–60 W with a gap of 20 W. The experiment was conducted under natural convection conditions, and the ambient temperature and relative humidity were controlled at 15 °C and 30%, respectively. To ensure the reliability of the experimental results, the experiment at each power is repeated at least three times.

17.2.3 Data reduction

During the ice-melting experiment, the feature dimensions in the top, side, and front views of ice are collected in real-time. As shown in Fig. 17.5(a1), the horizontal laser ice melting is taken as an example. The horizontal distance D from the laser incident point to

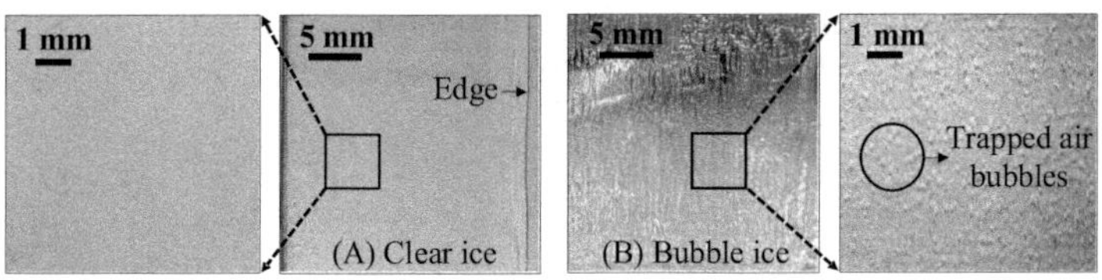

FIGURE 17.4 Comparison of frozen (A) clear ice and (B) bubble ice in this study.

TABLE 17.2 Experimental cases and conditions designed in this study.

Case No.	P (W)	T_a (°C)	RH (%)	I_S (mm³)
CH1	20	15 ± 0.3	30 ± 2	25 × 25 × 110
CH2	40			
CH3	60			
CV1	20			
CV2	40			
CV3	60			
BV1	20			25 × 25 × 65
BV2	40			
BV3	60			

B, Bubble ice; *C*, clear ice; *H*, horizontal irradiation; *V*, vertical irradiation.

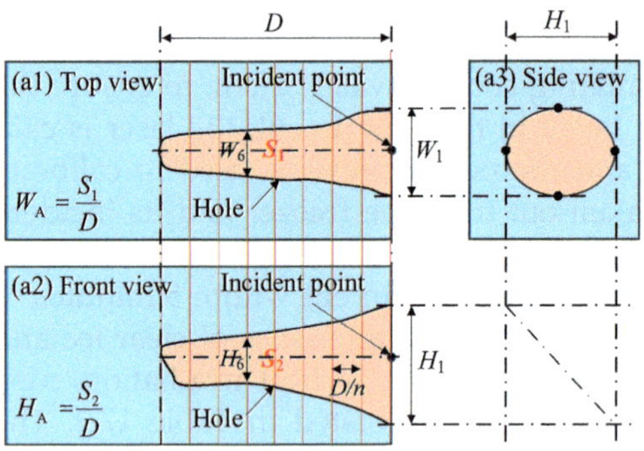

FIGURE 17.5 Schematic diagram of the feature dimensions and melting volume extraction.

the melting top point is defined as the melting depth. The average melting width, W_A, is defined as the ratio of the melting hole area S_1 to the melting depth. As shown in Fig. 17.5 (a2), the average melting height, H_A, is defined as the ratio of the melting hole area S_2 to the melting depth in the front view. Similarly, the average melting width is defined in the side view, and the average melting height is defined in the front view of vertical melting ice. Moreover, the laser beam is deflected due to the accumulation of meltwater in the melt holes. The detailed definition of the angle is presented in reference [3] to describe the degree of deflection of the laser. As shown in Fig. 17.5, to better calculate the volume of melting ice, three views of melting holes are drawn. The melting hole is approximated as an ellipse, with the melting height and width used as the major and minor axes, respectively. For volume calculations, the melting hole is approximately divided into several elliptical cylinders. Therefore, the volume of melted ice can be calculated according to Eq. (17.1).

$$V = \sum_{i=1}^{n} \left[\frac{\pi D (H_i + H_{i+1})(W_i + W_{i+1})}{16n} \right] \tag{17.1}$$

The grid numbers directly affect the uncertainty of melting volume. As the grid numbers increase, the extracted melting profile is closer to the actual melting profile, and the calculated melting volume is more accurate while the data processing effort increases exponentially. To better capture the change in the volume and reduce the data processing workload, it is necessary to check the independence of the grid size. The uncertainty of melting volume is calculated by the following equation [45]

$$\delta_V = \left[\left(\frac{\partial V}{\partial X_1}\delta_{X_1}\right)^2 + \left(\frac{\partial V}{\partial X_2}\delta_{X_2}\right)^2 + \cdots + \left(\frac{\partial V}{\partial X_n}\delta_{X_n}\right)^2\right]^{\frac{1}{2}} \tag{17.2}$$

where a dependent variable, V, is a function of several independent variables, X_1, X_2, ..., X_n. Each variable has a deviation of δ_{X1}, δ_{X2}, ..., δ_{Xn}. According to Eqs. (17.1) and (17.2), the uncertainty of the melting volume could be calculated as follows:

$$\xi_{V,i}(t) = \frac{\delta_V}{V} = \frac{\pi D\sqrt{\sum_{i=1}^{n}\left[(H_i+H_{i+1})^2 + (W_i+W_{i+1})^2\right]\delta_{W,i}^2}}{8nV} \tag{17.3}$$

The dynamic ice-melting process leads to the variation of the melting volume over time and its uncertainty. According to the experimental results of the melting process, it is found that the more complex the profile, the greater the uncertainty. When the laser power was 60 W and the time was 42 seconds, the experimental results showed a curved profile and were used to calculate the uncertainty of the melting volume. The melting volume and the uncertainties of different grid numbers are calculated by Eq. (17.3) and presented in Fig. 17.6. As seen, when the number of grids reaches 35, the uncertainty is 0.98%, and the calculated volume of melting ice can be considered acceptable.

During the process of laser irradiation, energy supplied by the laser, E_l, energy absorbed by melted ice, E_i, and energy efficiency, η is evaluated as follows [37]:

$$E_l = P_l t \tag{17.4}$$

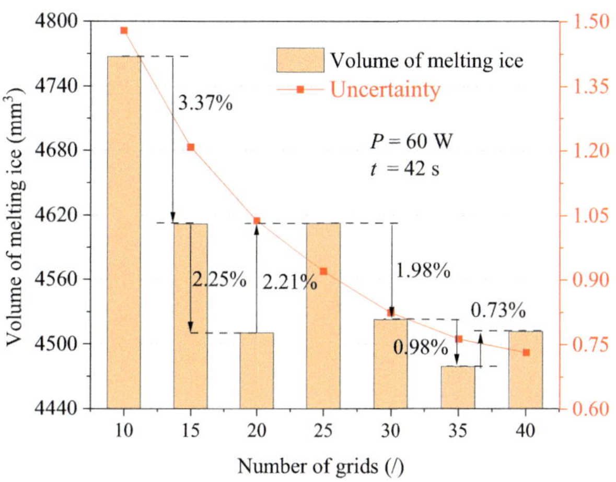

FIGURE 17.6 Decision process of selected grids on the melting volume of ice.

$$E_i = \rho_i V(1 - \phi)Q_i \tag{17.5}$$

$$\eta = \frac{E_i}{E_l} \times 100\% \tag{17.6}$$

where P_l is the laser power, t is the laser irradiation time, ρ_i is the density of clear ice, ϕ is the gas content in the bubble ice, and Q_i is the melting latent heat of ice.

17.3 Melting characteristics of horizontal and vertical laser irradiation

17.3.1 Melting depth and rate at different laser powers

The variation of clear ice in melting depth and rate with laser irradiation time is shown in Fig. 17.7. As shown in Fig. 17.7A, under horizontal laser irradiation, when the laser power is 20, 40, and 60 W, the melting time to reach the melting depth of around 100 mm is 78, 54, and 42 seconds, respectively. During 0–30 seconds, the difference between the melting depth of each power is small. After 30 seconds, the melting depth rises sharply, and the higher the laser power, the earlier the sudden rising point occurs. From 20 to 60 W, the time points are 51, 36, and 30 seconds, respectively. Moreover, the higher the laser power, the steeper the melting depth curve, and the faster the ice melts. During the entire melting process, the average melting rates are 1.64, 3.30, and 4.75 mm/s, respectively. In addition, the gap between the two adjacent melting depth curves is getting smaller, indicating the influence of the laser power in the melting depth is reduced. Under vertical laser irradiation, in the first 4 seconds, the melting depth at each power increase by around 5 mm. From 20 to 60 W, the melting depth began to rise sharply at 7, 7, and 4 seconds, respectively. The laser irradiation times to reach around 100 mm in three cases are 32, 17, and 17 seconds, respectively. During the entire melting process, the average melting rates in three cases are 3.11, 5.85, and 5.84 mm/s, respectively. By comparison, it could be found that the average melting rates from 20 to 60 W under vertical irradiation are 115.63%, 77.27%, and 22.95% higher than that under horizontal irradiation, respectively.

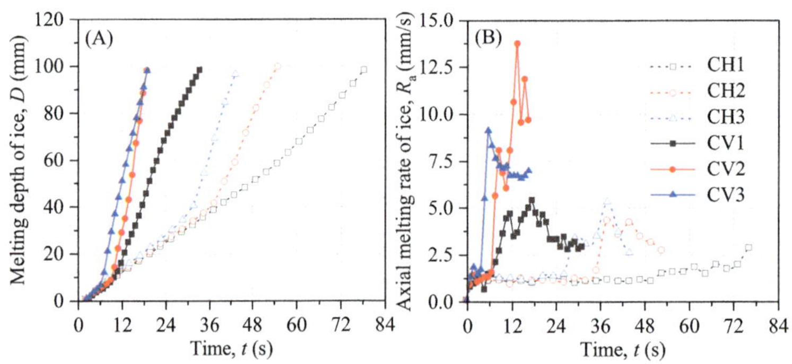

FIGURE 17.7 (A) Melting depth and (B) axial melting rate under horizontal and vertical irradiation.

The variation of ice axial melting rate with laser irradiation time is shown in Fig. 17.7B Under horizontal laser irradiation, during 0–30 seconds, the axial melting rate of different laser powers maintains at around 1.2 mm/s. After 30 seconds, the curve of melting rate showed a sharp increase, and the higher the laser power, the more obvious the increase in ice melting rate. From 20 to 60 W, the melting rate increases rapidly at 51, 36, and 30 seconds, respectively. The maximum axial melting rates are 2.84, 4.35, and 5.34 mm/s at the three laser powers, respectively. Under vertical laser irradiation, the axial melting rate increases first and then decreases gradually at each laser power. During 0–4 seconds, the axial melting rate fluctuates above and below 1.25 mm/s at different laser powers. After 4 seconds, from 20 to 60 W, the axial melting rate curve increases sharply at 7, 7, and 4 seconds, respectively. It is worth noting that after 10 seconds, the axial ice melt rate becomes smaller as the laser power increases from 40 to 60 W. The reason is that the higher the laser power, the more heat is conducted, but the more meltwater is produced, which could prevent further melting of the ice. From 20 to 60 W, the maximum axial melting rates are 5.41, 13.95, and 9.22 mm/s, respectively. By comparison, it could be found that the maximum axial melting rates from 20 to 60 W under vertical irradiation are 90.41%, 220.46%, and 72.66% higher than those under horizontal irradiation, respectively.

17.3.2 Melting width and height at different laser powers

The aperture of the melt hole changes with the increase of melting depth during laser irradiation. Therefore, to more accurately evaluate the influence of different laser powers on the melting aperture, the average melting width and average melting height are calculated and shown in Fig. 17.8. As shown in Fig. 17.8A, under horizontal laser irradiation, the average melting width curve first increases, then decreases, and finally becomes stable. The higher the laser power, the larger the average melting width extreme value and the earlier it appears. From 20 to 60 W, the maximum average melting widths are 6.03, 7.17, and 7.58 mm, respectively. In the first 30 seconds, the higher the laser power, the larger the average melting width. After 30 seconds, the melting depth increases rapidly, and the laser heat does not diffuse to the surroundings in time, so the average melting width decreases significantly. The final average melting width values in the three cases are 5.64,

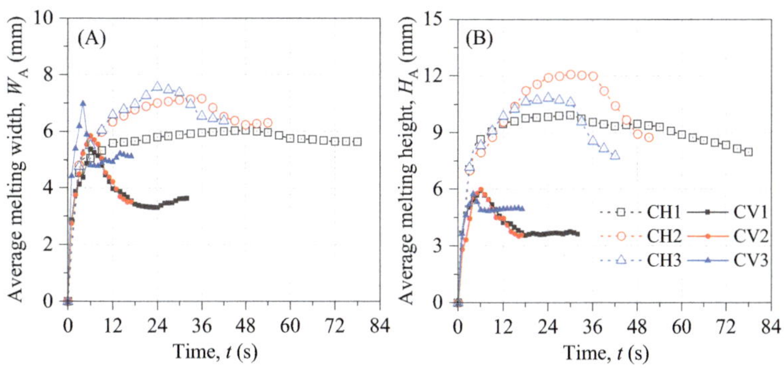

FIGURE 17.8 (A) Average melting width and (B) average melting height under horizontal and vertical irradiation.

6.32, and 6.39 mm, respectively. As shown in Fig. 17.8A, under vertical laser irradiation, the average melting width first increases then decreases, and finally becomes stable. When the laser power is 60 W, the average melting width reaches the maximum at 4 seconds, while at all other powers, the peak occurs at 6 seconds. The maximum average melting width values from 20 to 60 W are 5.37, 5.86, and 6.99 mm, respectively. The final average melting width values in three cases are 3.63, 3.51, and 5.14 mm, respectively. By comparison, it could be found that the final values of average melting width from 20 to 60 W under horizontal irradiation are 55.37%, 80.07%, and 24.32% higher than that under vertical irradiation, respectively.

As shown in Fig. 17.8B, under horizontal laser irradiation, the average melting height first increases then decreases, and finally becomes stable, which is similar to the changing process of the average melting width. In the early stage of melting, the meltwater and steam cannot be discharged in time, causing the average melting height to continue to expand. The maximum average melting height values from 20 to 60 W are 9.92, 12.08, and 10.85 mm, respectively. However, in the late melting period, the melt hole forms a slope that gradually decreases from inside to outside, causing the meltwater to flow out along the hole instead of melting down. Therefore, the aperture of the melting hole began to decrease, and the average melting height decreased significantly. The final average melting height values in the three cases are 7.98, 8.76, and 7.79 mm, respectively. As shown in Fig. 17.8B, under vertical laser irradiation, the average melting height is similar to the changing process of the average melting width. From 20 to 60 W, the maximum average melting heights are 5.93, 5.99, and 5.76 mm, respectively. The final average melting height values in three cases are 3.63, 3.58, and 4.92 mm, respectively. By comparison, it could be found that the final values of average melting height from 20 to 60 W under horizontal irradiation are 119.83%, 144.69%, and 58.33% higher than that under vertical irradiation, respectively.

17.3.3 Melting angles at different laser powers

To quantitatively describe the deflection angle of the laser in the horizontal and vertical directions, the melting angles in the top, side, and front view are measured and shown in Fig. 17.9. As shown in Fig. 17.9A, under horizontal laser irradiation, the melting angle in the top view gradually fluctuates in the range of −2 to 2 degrees with the increase of

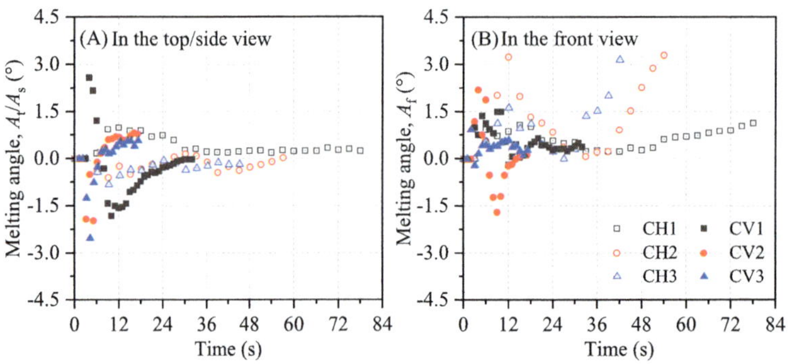

FIGURE 17.9 (A) Melting angle in the top/side view and (B) melting angle in the front view under horizontal and vertical irradiation.

irradiation time; the overall trend is to increase first and then decrease, and it goes to 0 degree in all three cases. When the laser power is 20 W, the melting angle in the top view fluctuates above 0 degree, while 40 and 60 W fluctuate below 0 degree, which has a certain randomness. The maximum melting angles in the top view from 20 to 60 W are 0.99, 0.61, and 0.8 degrees, respectively. Under vertical laser irradiation, the melting angle in the side view fluctuates in the range of − 3 to 3 degrees with the increase of irradiation time. In all three cases, the melting angle in the side view increases rapidly, then decreases gradually, and finally becomes stable. After 10 seconds, when the laser power is 20 W, the melting angle in the side view fluctuates below 0 degrees and finally tends to 0 degrees, while when the laser power is 40 and 60 W, it basically fluctuates above 0 degrees and finally tends to 0.7 degrees. The maximum melting angle in the side view values in three cases are 2.58, 1.97, and 2.53 degrees, respectively. By comparison, it could be found that the maximum melting angles in the side view under vertical irradiation from 20 to 60 W are 160.61%, 222.95%, and 216.25% higher than that in the top view under horizontal irradiation, respectively.

As shown in Fig. 17.9B, under horizontal laser irradiation, the melting angle in the front view of three cases increases rapidly at first and then gradually decreases to about 0 degrees in 0–30 seconds. After 30 seconds, the melting angle in the front view gradually increased. From the above analysis, it can be seen that in the later stage of melting, the meltwater continues to flow out, which will cause the laser irradiation path to be deflected. The higher the laser power is, the earlier the melting angle in the front view appears to increase significantly, which are 51, 39, and 30 seconds, respectively. The reason is that the higher the power of the laser, the more energy it transmits, which will lead to more meltwater production, and the greater the probability and angle of laser deflection in the front view. The final melting angle values in the front view of the three cases are 1.14, 3.31, and 3.16 degrees, respectively. Under vertical laser irradiation, due to the influence of gravity, meltwater could flow out of the hole smoothly, and the melting angle in the front view fluctuates in the range of − 3 to 3 degrees. In all three cases, the melting angle in the front view tends to increase first and then decrease, and it goes to 0 degrees. The maximum angle values in the front view from 20 to 60 W are 1.49, 1.89, and 0.93 degrees, respectively. By comparison, it could be found that the maximum melting angles in the front view under horizontal irradiation are 75.13% and 239.78% higher than that under vertical irradiation at 40 and 60 W, respectively. However, at 20 W, the maximum melting angle in the front view of vertical irradiation is 30.70% higher than that of horizontal irradiation.

17.4 Melting characteristics of clear ice and bubble ice

17.4.1 Melting depth and rate at different laser powers

The variation of ice melting depth and rate with laser irradiation time is shown in Fig. 17.10. As shown in Fig. 17.10A, when the clear ice is vertically illuminated by the laser, the time to reach the melting depth of around 60 mm is 20, 14, and 12 seconds from 20 to 60 W. The melting depth increases rapidly at 7, 7, and 6 seconds, respectively. From the turning point to the end of the melting, the average melting rate is 4.06, 7.41, and

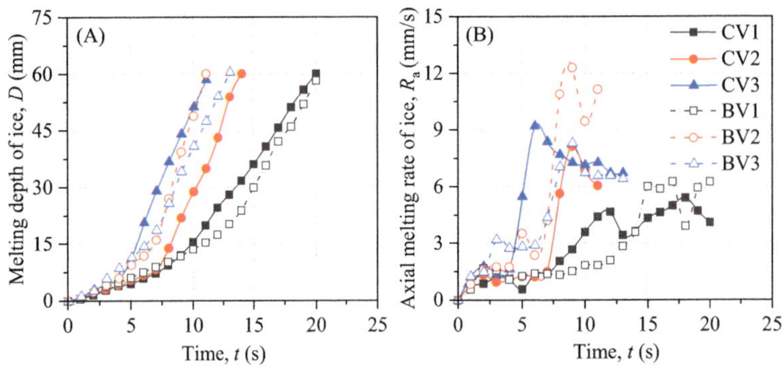

FIGURE 17.10 (A) Melting depth and (B) axial melting rate of clear ice and bubble ice.

7.36 mm/s. During the entire melting process, the average melting rates in three cases are 3.00, 4.29, and 5.33 mm/s, respectively. When the bubble ice is illuminated by the laser, from 20 to 60 W, the times to reach the melting depth of around 60 mm are 20, 11, and 13 seconds, respectively. During the first 2 seconds, the melting depth increases only about 2 mm at each power. After 2 seconds, from 20 to 60 W, the melting depth began to rise sharply at 12, 6, and 6 seconds, respectively. From the turning point to the end of the melting, the average melting rate is 5.09, 9.63, and 6.59 mm/s. During the entire melting process, the average melting rates are 2.91, 5.46, and 4.66 mm/s, respectively. It could be seen from the comparison that the turning point of the melting depth of bubble ice is more delayed than that of clear ice. At 20 and 60 W, the melt depth of clear ice is higher than that of bubble ice at the end of melting, but the slope of the melt depth curve of bubble ice is larger in the later melting period. By comparison, it could be found that the average melting rate calculated at the late turning point for vertically irradiated bubble ice is 25.37% and 29.96% higher at 20 and 40 W than for clear ice.

The variation of the axial melting rate at different laser powers is shown in Fig. 17.10B. The axial melting rate fluctuated at different laser powers but showed an overall trend of first increasing and then decreasing. When the clear ice is illuminated by the laser, from 20 to 60 W, the maximum axial melting rates are 5.41, 8.15, and 9.22 mm/s, respectively. When the bubble ice is illuminated by the laser, from 20 to 60 W, the maximum axial melting rates at each power are 6.26, 12.32, and 8.32 mm/s, respectively. From 20 to 40 W, the axial melting rate of ice increases significantly with increasing power. It is worth noting that after 7 seconds of laser irradiation, with increasing laser power from 40 to 60 W, the axial melting rate of the ice decreased. The reason is that when the laser power is greater than 40 W, the higher energy laser produces more steam. It is difficult for the internal steam to flow out due to the melting hole being deep, which seriously hinders the conduction of laser heat to the ice. By comparison, it could be found that at 20 W, the melting rate of bubble ice is slow in the early melting stage, but after 14 seconds, the melting rate is higher than that of clear ice. At 40 W, the melting rate of bubble ice is higher than that of clear ice. At 60 W, the melting rate of the two ice samples is basically the same at the end of the melting. Moreover, it could be found that the maximum axial melting rate of bubble ice at 20 and 40 W are 15.71% and 51.16% higher than that of clear ice, respectively.

17.4.2 Melting width and height at different laser powers

The average melting width and average melting height are calculated and shown in Fig. 17.11. As shown in Fig. 17.11A, the average melting width curve first increases, then decreases, and finally becomes stable. When the clear ice is illuminated, the maximum average melting width values from 20 to 60 W are 5.37, 5.86, and 6.99 mm, respectively. The final average melting width values in the three cases are 3.35, 4.38, and 4.93 mm, respectively. When the bubble ice is illuminated, from 20 to 60 W, the average melting width reaches the maximum value at 10, 4, and 6 seconds, respectively. The maximum average melting width values are 6.15, 6.54, and 6.85 mm, respectively. The final average melting width values in the three cases are 4.02, 4.19, and 5.78 mm, respectively. By comparison, it could be found that the final values of the average melting width of bubble ice at 20 and 60 W are 20.00% and 17.24% higher than that of clear ice, respectively. However, at 40 W, the final average melting width value of clear ice is 4.53% higher than that of bubble ice. As shown in Fig. 17.11B, the average melting height first increases then decreases, and finally becomes stable, which is similar to the changing process of the average melting width. When the clear ice is illuminated, from 20 to 60 W, the maximum average melting heights are 5.93, 5.99, and 5.76 mm, respectively. The final average melting height values from 20 to 60 W are 3.66, 4.45, and 4.97 mm, respectively. When the bubble ice is illuminated, from 20 to 60 W, the average melting height reaches the maximum value at 10, 4, and 4 seconds, respectively. The maximum average melting height are 6.27, 6.11, and 6.41 mm, respectively. The final average melting height values from 20 to 60 W are 3.69, 3.60, and 5.59 mm, respectively. By comparison, it could be found that the final values of the average melting height of vertically irradiated bubble ice at 20 and 60 W are 0.82% and 12.47% higher than that of clear ice, respectively. However, at 40 W, the final value of vertically irradiated clear ice is 23.61% higher than that of bubble ice.

17.4.3 Melting angles at different laser powers

To accurately describe the deflection of the laser in the vertical directions, the melting angles in the side and front view are extracted and shown in Fig. 17.12. As shown in Fig. 17.12A,

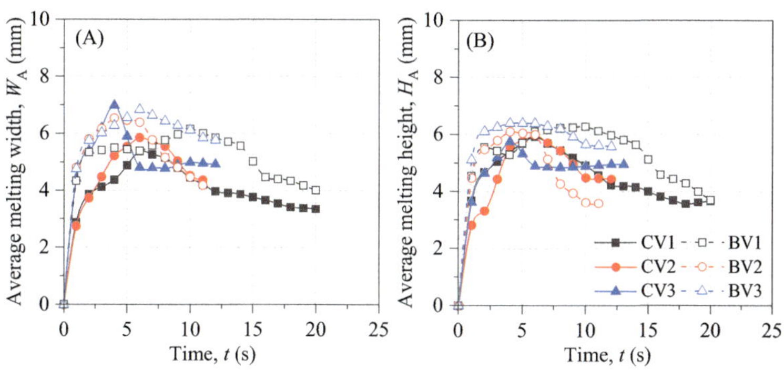

FIGURE 17.11 (A) Average melting width and (B) average melting height of clear ice and bubble ice.

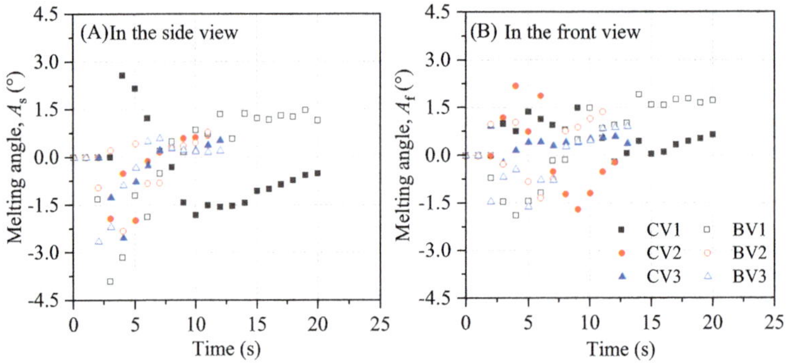

FIGURE 17.12 (A) Melting angle in the side view and (B) melting angle in the front view of clear ice and bubble ice.

when the clear ice is illuminated, the melting angle in the side view fluctuates in the range of −3 to 3 degrees during the melting process. From 20 to 60 W, the maximum melting angle values in the side views are 2.58, 1.97, and 2.53 degrees, respectively. When the bubble ice is illuminated, the melting angle in the side view fluctuates in the range of −5 to 2 degrees. In the first 5 seconds, the melting angle in the side view increases rapidly, then decreases gradually after 5 seconds, and finally becomes stable. When the laser power is 20 W, the melting angle in the side view may be affected by the bubble due to the low energy and finally stabilizes at about 1.4 degrees. The maximum melting angle in the side view values are 3.91, 2.31, and 2.64 degrees, respectively. By comparison, it could be found that the maximum melting angles in the side view of vertically irradiated bubble ice from 20 to 60 W are 51.55%, 17.26%, and 4.35% higher than that of clear ice, respectively. Moreover, the melting angle in the side view between clear ice and bubble ice is different in size and direction, but they both tend to 0 degrees in the end. As shown in Fig. 17.12B, the melting angle in the front view fluctuates in the range of −3 to 3 degrees. When the clear ice is illuminated, the maximum melting angles in the front view from 20 to 60 W are 1.49, 1.89, and 0.93 degrees, respectively. When the bubble ice is illuminated, at 20 W, the melting angle in the front view may be affected by the bubble due to the low energy and finally stabilizes at about 1.5 degrees. From 20 to 60 W, the maximum melting angle in the front view values are 1.91, 1.37, and 1.60 degrees, respectively. By comparison, it could be found that the maximum melting angle in the front view of vertically irradiated bubble ice at 20 and 60 W are 28.19% and 72.04% higher than that of clear ice, respectively. However, at 40 W, the maximum melting angle value of clear ice is 37.96% higher than that of bubble ice.

17.5 Melting volume and energy efficiency

The variation of the melting ice volume under the nine cases is shown in Fig. 17.13. It can be seen that the melting volume of ice increases uniformly in nine cases with time. During the whole laser irradiation time, the volume melting rate at each laser power is calculated. Fig. 17.13A shows the variation in the melting volume of clear ice under horizontal laser irradiation. As seen, during the entire laser irradiation time, the volume melting

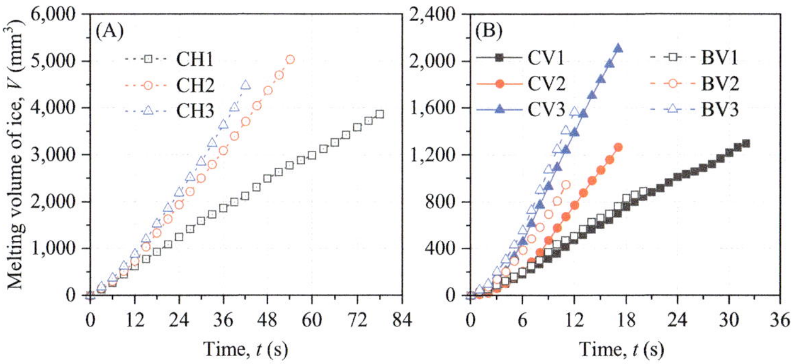

FIGURE 17.13 (A) Melting volume of clear ice under horizontal irradiation and (B) melting volume of clear ice and bubble ice under vertical irradiation.

rates from 20 to 60 W are 49.51, 93.36, and 106.65 mm^3/s, respectively. The slope of melting ice volume also increases as the increases of laser power and the gap between the two adjacent melting volume curves gets smaller. Fig. 17.13B shows the variation in the melting volume of clear ice and bubble ice under vertical laser irradiation. As seen, during the entire laser irradiation time, the volume melting rates of clear ice from 20 to 60 W are 40.45 mm^3/s, 74.65 mm^3/s, and 124.18 mm^3/s, respectively. By comparison, it could be found that the average volume melting rates within 100 mm of clear ice at 20 and 40 W under horizontal irradiation are 22.4% and 25.06% higher than those under vertical irradiation, respectively. However, at 60 W, the average volume melting rate under vertical irradiation was 16.43% higher than that under horizontal irradiation. When the bubble ice is illuminated vertically, from 20 to 60 W, the volume melting rates are 44.46, 86.28, and 130.74 mm^3/s, respectively.

In deicing equipment applications, it is important not only to ensure the high rate of ice melting of the deicing technology but also to consider the low energy consumption and losses. To evaluate the performance of the CO_2 laser ice melting system in this study, the variation of the energy efficiency under the nine Cases is shown in Fig. 17.14. It can be seen that with the increase of laser irradiation time, the energy efficiency under all cases gradually increases, and the growth rate gradually slows down. The reason is that the laser energy is not easily reflected to the outside due to the deepening of the thaw holes, and more energy is transferred to the inside of the ice. Fig. 17.14A shows the variation in the energy efficiency of clear ice under horizontal laser irradiation. As seen, when the laser power is 40 and 60 W, the energy efficiency reaches the maximum value at the last melting ice time points, which are 71.72% and 54.62%, respectively. However, the maximum energy efficiency is 81.5% at 20, which appears in 30 seconds. At the end of ice melting, the final energy efficiency at 20 W is 76.07%.

Fig. 17.14B shows the variation in the energy efficiency of clear ice and bubble ice under vertical laser irradiation. As seen, when clear ice is illuminated vertically, the energy efficiency reaches the maximum values at 40 and 60 W at the last melting ice time point, which are 57.34% and 63.59%, respectively. At 20 W, the maximum energy efficiency is 64.8%, which appears in 20 seconds. At the end of ice melting, the final energy efficiency at 20 W is 62.15%. By comparison, it could be found that the final energy efficiency of clear

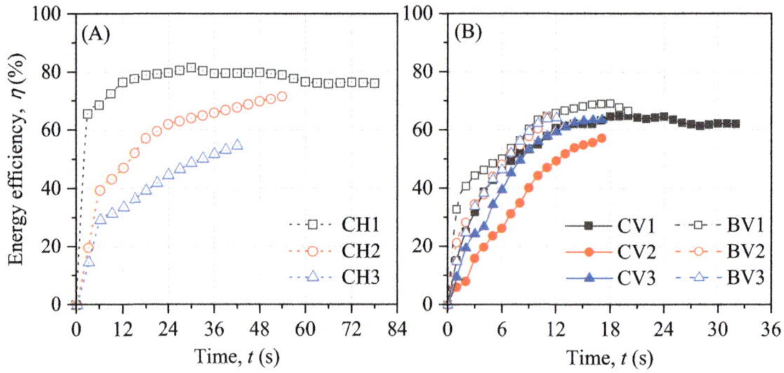

FIGURE 17.14 (A) Energy efficiency of clear ice under horizontal irradiation and (B) energy efficiency of clear ice and bubble ice under vertical irradiation.

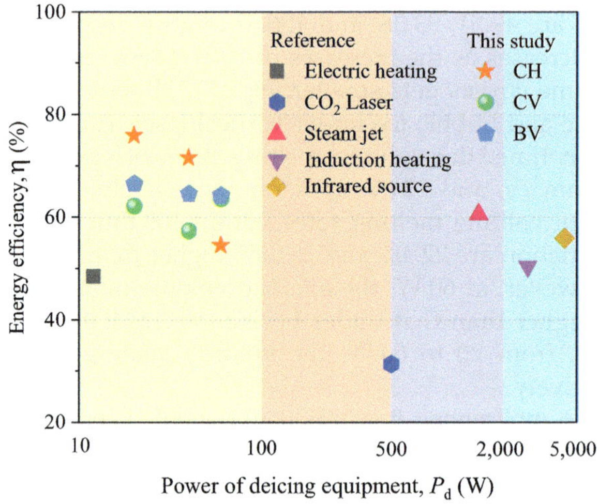

FIGURE 17.15 Comparison of energy efficiency on different deicing methods.

ice at 20 and 40 W under horizontal irradiation is 22.40% and 25.06% higher than that under vertical irradiation, respectively. However, at 60 W, the energy efficiency under vertical irradiation was 16.43% higher than that under horizontal irradiation. As shown in Fig. 17.14B, when bubble ice is illuminated vertically, the energy efficiency of 18 seconds is the highest at 20 W, which is 69.1%. At the end of ice melting, the final energy efficiency at 20 W is 66.54%. When the laser powers are 40 and 60 W, the energy efficiency reaches the maximum at the last melting time point, which is 64.56% and 64.22%, respectively.

The comparison of the energy efficiency for various deicing methods in existing experimental studies is shown in Fig. 17.15. It could be seen that the energy efficiency of all deicing methods is below 70% at different input powers. For electric heating deicing, a low-voltage heating film was proposed by Guadagno et al. [46] for deicing aircraft and motor vehicles, with an energy efficiency of 48.5% at 12 W in the experiment. For laser deicing, a deicing experiment of a 500 W high-power CO_2 laser was conducted by Qi et al. [33], and the energy efficiency calculated is 31.4%. In the deicing experiment of Ji et al. [47], the maximum energy efficiency of the steam jet was 60.5% when the input power of the electric boiler was 1472 W. In the experiment of Liu et al. [48], a new conductive layer induction heating system is used to remove the ice layer of the asphalt pavement, and the energy efficiency is 50.4% when the input power is 2700 W. In the experiments of Xie et al. [49], the maximum energy efficiency was only 12.1% when 50 °C and 14 m/s of hot air were used for deicing. In another experimental study by Xie et al. [50], when an infrared heat source at 2000 °C was used for deicing, the maximum energy efficiency reached 55.9%. In this study, the ice melting efficiency of seven cases reached more than 60% when the CO_2 laser was used. Compared with other deicing methods, the CO_2 laser in this study has less energy loss and greater energy efficiency. In addition, the CO_2 laser system is low-cost and easy to operate, so it is expected to be widely used in the field of deicing.

17.6 Concluding remarks

In this chapter, the effect of CO_2 laser on the melting characteristics of cube ice under natural convection is experimentally studied. The influencing factors mainly include laser power, irradiation direction, and ice type. First, the melting characteristics of clear ice under horizontal and vertical laser irradiation are compared. Then, the melting characteristics of clear ice and bubble ice under vertical laser irradiation are analyzed and compared.

When clear ice is irradiated horizontally and vertically, the maximum melting rates at 40 W are 4.35 and 13.95 mm/s, respectively. The average melting rate within 100 mm at 40 W under vertical irradiation is 77.27% higher than that under horizontal irradiation. When the clear ice is irradiated horizontally, the meltwater accumulates in the melt holes, resulting in a large difference in the melting height from vertical irradiation. Under horizontal and vertical laser irradiation, the final average melting heights of clear ice at 40 W are 8.76 and 3.58 mm, respectively.

When the clear ice and bubble ice are irradiated vertically, the average melting rates within 60 mm at 40 W are 4.29 and 5.46 mm/s, respectively. At 60 W, the average melting height and average melting width of the bubble ice are 12.47% and 17.24% higher than that of the clear ice, respectively. The energy efficiency gradually increases with the increase of melting hole depth. When the depth of the melting hole reaches about 100 mm, the energy efficiency of clear ice under horizontal and vertical irradiation is 76.07% and 62.15% at 20 W, respectively.

References

[1] S.R. Yu, M.J. Song, R.M. Gao, J. Bae, X. Zhang, S.Q. Zhou, A review of icing prediction techniques for four typical surfaces in low-temperature natural environments, Applied Thermal Engineering 241 (2024) 122418.

[2] R.M. Gao, M.J. Song, C.Y.H. Chao, S.L. Lin, L. Zhang, X. Zhang, Review on condensation frosting and defrosting experiments for superhydrophobic surfaces, Applied Thermal Engineering 236 (2024) 121691.

[3] Z.K. Zhen, R.M. Gao, M.J. Song, X. Zhang, Y.J. Xu, T.Z. Zhan, An experimental study on the effect of CO_2 laser powers on melting characteristics of clear ice - Part II: Vertical irradiation, Applied Thermal Engineering 246 (2024) 122946.
[4] X. Zhang, Y.B. Gao, X. Liu, J.C. Min, K.L. Li, M.J. Song, Modelling of the icing processes of sessile supercooled water droplets on cold spheres, Applied Thermal Engineering 243 (2024) 122596.
[5] S.N. Chang, W.D. Yu, M.J. Song, M.Y. Leng, Q.Y. Shi, Investigation on wavy characteristics of shear-driven water film using the planar laser induced fluorescence method, International Journal of Multiphase Flow 118 (2019) 242–253.
[6] D.Y. Ma, S.N. Chang, K. Wu, M.J. Song, Experimental study on the water film thickness under spray impingement based on planar LIF, International Journal of Multiphase Flow 130 (2020) 103329.
[7] S. Paul, D. Lee, K. Kim, J. Chang, Nonlinear modeling and performance testing of high-power electromagnetic energy harvesting system for self-powering transmission line vibration deicing robot, Mechanical Systems and Signal Processing 151 (2021) 107369.
[8] X. Zhang, K.L. Li, Z.B. Zhu, W.Z. Fang, F.Q. Zhu, C. Yang, Droplet impact and freezing dynamics on ultracold surfaces: a scaling analysis of central-concave pattern, Applied Thermal Engineering 239 (2024) 122135.
[9] J.B. Wang, J. Zhang, F. Xie, Y. Zhang, G. Gao, A study of snow accumulating on the bogie and the effects of deflectors on the de-icing performance in the bogie region of a high-speed train, Cold Regions Science and Technology 148 (2018) 121–130.
[10] W. Sun, Y.T. Wei, Y.H. Feng, F.Q. Chu, Anti-icing and deicing characteristics of photothermal superhydrophobic surfaces based on metal nanoparticles and carbon nanotube materials, Energy 286 (2024) 129656.
[11] Y.Y. Tian, Y.L. Liu, H. Sayyed, M.J. Song, J. Shen, L. Zhang, An experimental study on edge-affected frosting characteristics on a vertical cold plate at different air velocities, Applied Thermal Engineering 236 (2024) 121882.
[12] J.Y. Hao, Y.J. Xu, T.B. He, N. Mao, M.J. Song, J.F. Tang, Numerical study on heat transfer of oily wastewater spray falling film over a horizontal tube in a sewage source heat pump, International Journal of Heat and Mass Transfer 142 (2019) 118423.
[13] L.Z. Huang, M.J. Song, J. Shen, D.R. Kim, L. Zhang, L. Pekar, Experimental study on the effect of inverted aperture plate temperature on frosting characteristics under rising flow considering edge effect, International Journal of Heat and Mass Transfer 224 (2024) 125343.
[14] M.A. Moradkhani, R.M. Gao, S.H. Hosseini, M.J. Song, Estimation of frost layer characteristics during natural convection on vertical and inverted cold surfaces based on smart approaches, Applied Thermal Engineering 239 (2024) 122090.
[15] R. Patrice, L. Jean, R. Jean, B.D. André, The effect of icing severity on wind turbine ice protection systems efficiency, Cold Regions Science and Technology 207 (2023) 103774.
[16] L.Z. Huang, Y.Y. Tian, M.J. Song, J. Shen, X. Zhang, L. Zhang, An experimental study on edge-affected frosting characteristics on a vertical cold plate at different surface temperatures, Applied Thermal Engineering 236 (2024) 121538.
[17] Q. Dang, M.J. Song, L. Zhang, X. Zhang, K.K. Shao, J. Shen, Experimental study about solidification processes of sessile deformed water droplets on the biaxial inclined cold plate surface under natural convection, Applied Thermal Engineering 236 (2024) 121883.
[18] Y.J. Zhang, L. Zhang, G.S. Luo, Study on de-icing criterion of anti-icing coating and simulation analysis method of mechanical de-icing process for polar ship superstructure, Ocean Engineering 288 (2023) 115811.
[19] X.L. Liu, L.Z. Huang, X. Zhang, L. Zhang, M.J. Song, J. Shen, A modeling study on freezing characteristics of sessile and pendant water droplets1on cold plate surfaces, Applied Thermal Engineering 236 (2024) 121693.
[20] Y. Li, H. Shen, W.F. Guo, Simulation and experimental study on the ultrasonic micro-vibration de-icing method for wind turbine blades, Energies 14 (24) (2021) 8246.
[21] R.G. Michael, H. Bernhard, H. Markus, S. David, M.S. Teresa, G. Hinrich, Deicing performance of common deicing agents for winter maintenance with and without corrosion-inhibiting substances, Cold Regions Science and Technology 208 (2023) 103795.
[22] H.X. Ma, H.Y. Ma, H.F. Yu, B. Da, W. Gong, N. Zhang, The physical properties of organic aircraft deicers under low temperature, Cold Regions Science and Technology 196 (2022) 103471.
[23] Z.K. Zhen, K.K. Shao, M.J. Song, L. Zhang, L.M. Shen, L. PEKAŘ, An experimental study on the effect of CO_2 laser powers on melting characteristics of clear ice–part I: Horizontal irradiation, Applied Thermal Engineering 248 (2024) 123208.

[24] Y.D. Wang, G. Zhang, Z.B. Tian, R.C. Qiu, Z.G. Liu, An online thermal deicing method for urban rail transit catenary, IEEE Transactions on Transportation Electrification 7 (2) (2020) 870–882.
[25] X. Jiang, S. Fan, Z. Zhang, Simulation and experimental investigation of DC ice-melting process on an iced conductor, IEEE Transactions on Power Delivery. 25 (2) (2010) 919–929.
[26] Y. Zhang, J. Wei, C. Liu, Y.B. H, F.Y. S, Reduced graphene oxide modified Ti/CFRP structure-function integrated laminates for surface Joule heating and deicing, Composites Part A: Applied Science and Manufacturing 166 (2023) 107377.
[27] P.G. Talalay, Perspectives for development of ice-core drilling technology: a discussion, Annals of Glaciology 55 (68) (2014) 339–350.
[28] R.S. Wang, L. An, P.L. Cao, B.Y. Chen, M. Sysoev, D.Y. Fan, et al., Rapid ice drilling with continual air transport of cuttings and cores: general concept, Polar Science 14 (2017) 21–29.
[29] W.C. Stone, B. Hogan, V. Siegel, S. Lelievre, C. Flesher, Progress towards an optically powered cryobot, Annals of Glaciology 55 (65) (2014) 1–13.
[30] J. Luo, B. Li, X.D. Zhou, Z.P. Qin, G.Q. Xie, Laser deicing for high-voltage composite insulator by high-power mid-infrared fiber laser, Optical Engineering 61 (2) (2022) 21005.
[31] S. Yang, Y.Y. Yang, J.Y. Zhang, Z.Y. Zhang, L. Zhang, X.C. Lin, Laser-induced cracks in ice due to temperature gradient and thermal stress, Optics & Laser Technology 102 (2018) 115–123.
[32] G.W. Stephen, Optical constants of ice from the ultraviolet to the microwave, Applied Optics 23 (8) (1984) 1206–1255.
[33] L.J. Qi, X. Zhu, C.H. Zhu, F. Guo, G.Z. Zhu, S.Q. Gu, Deicing with Nd: YAG and CO_2 lasers, Optical Engineering 49 (11) (2010) 792–796.
[34] A.F. Clark, J.C. Moulder, R.P. Reed, Ability of a CO_2 laser to assist ice breakers, Applied Optics 12 (1973) 1103–1104.
[35] T. Sakurai, H. Chosrowjan, T. Somekawa, M. Fujita, H. Motoyama, O. Watanabe, et al., Studies of melting ice using CO_2 laser for ice drilling, Cold Regions Science and Technology (2016) 11–15.
[36] K.K. Shao, M.J. Song, X. Zhang, L. Zhang, Growth and distribution characteristics of trapped air bubbles in ice slices, Physics of Fluids 35 (11) (2023) 113319.
[37] Z.K. Zhen, M.J. Song, J. Shen, L. Zhang, X. Zhang, An experimental study on the effect of CO_2 laser powers on melting characteristics of ice with trapped air bubbles under vertical irradiation, Applied Thermal Engineering 236 (2024) 121533.
[38] K.K. Shao, M.J. Song, J. Shen, X. Zhang, P. Libor, Experimental study on the distribution and growth characteristics of trapped air bubbles in ice slices at different freezing temperatures, Applied Thermal Engineering 244 (2024) 122600.
[39] F.Q. Chu, X. Zhang, S.K. Li, H.C. Jin, J. Zhang, Bubble formation in freezing droplets, Physical Review Fluids 4 (7) (2019) 071601.
[40] S.N. Chang, H.K. Zheng, M.J. Song, M.Y. Leng, Heat transfer characteristics of micron ultrathin shear-driven water film flowing on a horizontal metal surface, International Journal of Heat and Mass Transfer 148 (2020) 119065.
[41] Z. Zhang, H. Zhang, S. Yue, W.H. Zeng, A review of icing and anti-icing technology for transmission lines, Energies 16 (2) (2023) 601.
[42] J. Hruba, G. Kletetschka, Environmental record of layers of bubbles in natural pond ice, Journal of Glaciology 64 (248) (2018) 866–876.
[43] G.L. Lei, W. Dong, M. Zheng, Z.Q. Guo, Y.Z. Liu, Numerical investigation on heat transfer and melting process of ice with different porosities, International Journal of Heat and Mass Transfer 107 (2017) 934–944.
[44] S.D. Sleptsov, N.A. Savvinova, M.A. Grishin, Numerical study of the thermal state of an ice layer containing air bubbles, Journal of Applied Mechanics and Technical Physics 62 (2021) 451–457.
[45] M.J. Song, S.W. Lei, S.H. Hosseini, X.Y. Luo, Z.H. Wang, An experimental study on the effect of horizontal cold plate surface temperature on frosting characteristics under natural convection, Applied Thermal Engineering 211 (2022) 118416.
[46] L. Guadagno, F. Foglia, R. Pantani, M.D. Romero-Sanchez, B. Calderón, L. Vertuccio, Low-voltage icing protection film for automotive and aeronautical industries, Nanomaterials 10 (7) (2020) 1343.
[47] W.T. Ji, X.D. Lu, L. Chen, Y.W. Zhang, W.Q. Tao, Experimental investigation on the ice melting heat transfer with a steam jet impingement method, International Journal of Heat and Mass Transfer 118 (6) (2020) 104901.

[48] K. Liu, P.X. Xu, F. Wang, C. Jin, M.Y. Huang, D.L. Dai, et al., Deicing efficiency analysis and economic-environment assessment of a novel induction heating asphalt pavement, Journal of Cleaner Production 273 (2020) 123123.

[49] T. Xie, J.K. Dong, H.W. Chen, Y.Q. Jiang, Y. Yao, Experimental investigation of deicing characteristics using hot air as heat source, Applied Thermal Engineering 107 (2016) 681–688.

[50] T. Xie, J.K. Dong, H.W. Chen, Y.Q. Jiang, Y. Yao, Experiment investigation on deicing characteristics and energy efficiency using infrared ray as heat source, Energy 116 (2016) 998–1005.

CHAPTER

18

Conclusions and future work

Mengjie Song, Xuan Zhang and Long Zhang

Department of Energy and Power Engineering, School of Mechanical Engineering, Beijing Institute of Technology, Beijing, P.R. China

18.1 Conclusions of the present work

The present research work is devoted to the research, analysis, and methods of frosting and icing in different industrial fields, with an emphasis on the whole frosting/icing process. This study covers six space scales, from trapped air bubbles during freezing, freezing of sessile and impact water droplets, frosting/icing on cold plates, frosting on heat exchangers of air source heat pump units, icing on the wing surface of airplanes, and icing in Polar Regions. The following five achievements are obtained.

The first is the qualitative and quantitative evaluation of the freezing process of sessile pure water and multicomponent droplets under natural convection at the early frosting/icing stage. Droplet solidification is the start of a frosting/icing process on the cold surface, which is the fundamental topic for frost and ice. Many freezing behaviors were studied with numerical and experimental methods, including the dynamical interface capturing, the evolution of freezing front radius and mixture volume, the initial and final profiles of droplets, etc. Lattice Boltzmann simulation for water droplet freezing was developed and finally validated by experimental data, with all mentioned freezing characteristics discussed. The conclusions are reported in Chapters 2–4.

The second is the realization and detection of the whole icing process on cold, flat surfaces and spheres. In this study, the impact process of water droplets is experimentally and numerically investigated, and the influence of Weber number and contact angle on the impact characteristics of water droplets is analyzed. This is followed by the impacting and icing process of the water droplet on a spherical surface. A volume of fluid method coupled with a dynamic contact angle model and solidification/melting model was established and validated. The effects of sphere curvature, sphere shape, contact angle, and surface temperature on the droplet impacting morphologies and the maximum spreading characteristics were investigated. Based on the freezing process of a very thin water film, characteristics of trapped air bubbles were investigated. Specially, the frosting and icing process of trace water in a closed environment was investigated, with different highly accurate methods reported. Conclusions are reported in Chapters 5–8.

Frosting and Icing for Efficient Energy Use in Engineering Applications
DOI: https://doi.org/10.1016/B978-0-443-15495-9.00019-8

The third is to fully understand the frost formation process and the influencing parameters for the further development of antifrosting techniques. Experimental investigations on frost formation and growth on flat, cylinders, and heat exchangers cold surfaces under natural and forced convection were surveyed, reporting the frosting influence evaluation results of cold surface temperature, air relative humidity, surface structures, etc. A sustainable defrosting air source heat pump system, which adds a phase change energy storage component around the compressor to store the heat dissipated by the compressor, is proposed and experimentally investigated in detail. Compared to the traditional reverse cycle defrosting mode, predefrosting, and postdefrosting stages become unnecessary. A novel frost-free air-source heat pump system, coupled with coupled with an extra heat exchanger coated with a solid desiccant and an energy storage device was proposed, with the experimental and simulation results of the system's performance given. The conclusions are reported in Chapters 9–11.

2The fourth is case studies in some industrial fields, including the development and optimization of atomization equipment for simulating icing environments and the thermodynamic evolution of landfast ice in Polar Regions. In the icing cloud simulation system, the effect of the nozzle arrangement of atomization equipment was numerically analyzed, with the secondary development model of FLUENT. The inhomogeneous sizes, gravity and evaporation effects of water droplets were considered, and ice shape and maximum ice thickness on the test rod as two validation parameters. Using field measurements from the wintering team, the regional and temporal evolution of the landfast ice, including FYI and SYI near the Antarctic ZS from December 2011 to December 2012, were examined. Regular site measurements were made during the whole year, and winter surveys were carried out at various locations throughout the Nella Fjord. Finally, ice prediction technologies for two types of stationary surfaces: roads and power transmission lines, and two types of moving surfaces: wind turbine blades and aircraft, were reviewed. The conclusions are reported in Chapters 12–14.

Based on the aforementioned four achievements, the last one was reached, which is the experimental and numerical studies around deicing technologies with ultrasonic guided waves method, hot air, infrared ray, and carbon dioxide laser irradiation, etc. Based on the frequency-domain finite element method, the propagation process of certain ultrasonic-guided wave mode-frequency pairs in the icing plate was simulated, and the mode conversion of ultrasonic-guided waves was quantitatively analyzed. To reduce deicing time and improve efficiency, experiments on hot air and infrared deicing were carried out, and the ideal air temperature and velocity for hot air deicing were given. The effect of carbon dioxide laser on the melting characteristics of clear and bubble cube ice under natural convection was experimentally studied. The influencing factors mainly include laser power, irradiation direction, and ice type. The conclusions are reported in Chapters 15–17.

18.2 Proposal of the future work

Based on the results presented in this book, the following five potential areas for future research work related to frosting and icing are recommended.

1. *Frosting and icing mechanism study*. Among the reported frost characteristics, the thickness and density of the frost layer seem most fundamental. However, many methods can

measure the frost layer thickness from a single or multipoint, single or multiline, while none of them can reach real-time high accuracy owing to the frost layer's unevenly distribution. The mass of the frost layer is always calculated using the melted frost after the defrosting process. To solve this long-standing problem by accurately measuring frosting characteristics, effective frost layer thickness and accumulation methods should be explored by developing dynamic three-dimensional frosting models. To finally solve the frosting/icing problem, the formation process should be firstly clear, from condensation and solidification to the freezing or frosting process.

2. *New frosting/icing model development*. Although certain parameters can be physically measured, most of them could be calculated using empirical or semiempirical correlations. Several correlations of frost/ice thermal conductivity and convective heat and mass transfer coefficients have been used in the development of frosting/icing models, which should be validated with real-time experimental data. During frosting/icing model development, some assumptions could be deleted by understanding mechanisms, such as the freezing front in a droplet that is not flat but curves and the distribution of micro-trapped air bubbles.
3. *Prediction technology of frost and ice*. Most studies are about anti/defrosting/icing, while the real-time prediction of frost and ice is much more important for industrial applications. If the frosting and icing could be accurately predicted, such as the location, time, duration, growing rate, type, and mass of frost or ice, the cold surface condition could be changed, for example, the mode of the air source heat pump, the fly condition of the airplane, the mode of the wind turbine, the road of the ship, etc. The development of frost and ice prediction technology relies on the development of frost and ice models, and emerging computing technologies such as artificial intelligence will become important directions for development.
4. *Detection technology of frost and ice*. If the frost and ice could be real-time accurately detected, the data could be used for further prediction of frosting and icing. The cost of detection and prediction of frost/ice is much lower than any defrosting or deicing technologies. Various methods have previously been implemented for ice sensing in aerospace, power energy, and transportation scenarios, including capacitive sensing, ultrasonic-based methods, and guided wave approaches. No one is the best, but some of them could be combined to detect more features of frosting and icing. For example, infrared or visible light imaging methods can detect distribution, resonance methods can detect mass, and microwaves can detect the thickness of frost and ice, etc.
5. *Control strategy optimization*. Although there are many articles on the frosting and freezing of hydrophilic and hydrophobic surfaces, they have not yet been effectively applied in any industrial field. It is easy to understand that after frosting or icing, the surface is changed, and the original hydrophilic and hydrophobic surfaces become invalid. But these surfaces could be effective when coupled with a heating surface, which is the control strategy optimization. Another example is the defrosting strategy of the air source heat pump. As the start of defrosting, the frosting control strategy of refrigerant and air distribution adjustment should be studied; then, four typical mal-defrosting phenomena could be effectively avoided. Furthermore, the operation control strategy could be coupled with big data on climates. Highly accurate, convenient, automatic, and low-cost physical measurements for system anti/defrosting/icing are always expected.

Index

Note: Page numbers followed by "*f*" and "*t*" refer to figures and tables, respectively.

G

I

J

K

L

Y

Z

9780443154959